— Handbook of —
Materials for Nanomedicine
Lipid-Based and Inorganic Nanomaterials

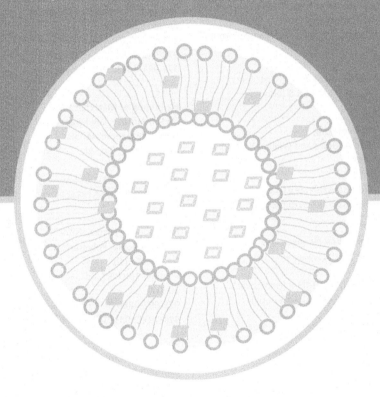

Jenny Stanford Series on Biomedical Nanotechnology

Series Editors

Vladimir Torchilin and Mansoor Amiji

Titles in the Series

Published

Vol. 1
Handbook of Materials for Nanomedicine
Vladimir Torchilin and Mansoor Amiji, eds.
2010
978-981-4267-55-7 (Hardcover)
978-981-4267-58-8 (eBook)

Vol. 2
Nanoimaging
Beth A. Goins and William T. Phillips, eds.
2011
978-981-4267-09-0 (Hardcover)
978-981-4267-91-5 (eBook)

Vol. 3
Biomedical Nanosensors
Joseph Irudayaraj, ed.
2013
978-981-4303-03-3 (Hardcover)
978-981-4303-04-0 (eBook)

Vol. 4
Nanotechnology for Delivery of Therapeutic Nucleic Acids
Dan Peer, ed.
2013
978-981-4411-04-2 (Hardcover)
978-981-4411-05-9 (eBook)

Vol. 5
Handbook of Safety Assessment of Nanomaterials: From Toxicological Testing to Personalized Medicine
Bengt Fadeel, ed.
2014
978-981-4463-36-2 (Hardcover)
978-981-4463-37-9 (eBook)

Vol. 6
Handbook of Materials for Nanomedicine: Lipid-Based and Inorganic Nanomaterials
Vladimir Torchilin, ed.
2020
978-981-4800-91-4 (Hardcover)
978-1-003-04507-6 (eBook)

Vol. 7
Handbook of Materials for Nanomedicine: Polymeric Nanomaterials
Vladimir Torchilin, ed.
2020
978-981-4800-92-1 (Hardcover)
978-1-003-04511-3 (eBook)

Vol. 8
Handbook of Materials for Nanomedicine: Metal-Based and Other Nanomaterials
Vladimir Torchilin, ed.
2020
978-981-4800-93-8 (Hardcover)
978-1-003-04515-1 (eBook)

Jenny Stanford Series on Biomedical Nanotechnology Volume 6

—— Handbook of ——
Materials for
Nanomedicine
Lipid-Based and Inorganic Nanomaterials

edited by
Vladimir Torchilin

JENNY STANFORD
PUBLISHING

Published by

Jenny Stanford Publishing Pte. Ltd.
Level 34, Centennial Tower
3 Temasek Avenue
Singapore 039190

Email: editorial@jennystanford.com
Web: www.jennystanford.com

British Library Cataloguing-in-Publication Data
A catalogue record for this book is available from the British Library.

Handbook of Materials for Nanomedicine: Lipid-Based and Inorganic Nanomaterials

ISBN 978-981-4800-91-4 (Hardcover)
ISBN 978-1-003-04507-6 (eBook)

Contents

5. Solid Lipid Nanoparticles   **173**

Karsten Mäder

Chapter 1

Liposome-Scaffold Systems for Drug Delivery

Sandro Matosevic

Department of Industrial and Physical Pharmacy,
Purdue University, West Lafayette, Indiana 47907, USA
sandro@purdue.edu

The combination of liposomes and scaffolds can afford unprecedented functional and safety enhancements to the delivery of drugs, protein therapeutics, or genes. Combining liposomes with scaffolds creates three-dimensional, bi- or multi-modal systems that can achieve spatiotemporal control of the delivery process, support sustained delivery, and enhance stability and targetability, ultimately significantly improving the pharmaceutical potential of drug and gene delivery mediated by liposomes for applications in regenerative medicine and cell therapy.

1.1 Introduction

While liposomes have been successfully used as delivery systems for therapeutically active compounds for a number of decades,

Handbook of Materials for Nanomedicine: Lipid-Based and Inorganic Nanomaterials
Edited by Vladimir Torchilin
Copyright © 2020 Jenny Stanford Publishing Pte. Ltd.
ISBN 978-981-4800-91-4 (Hardcover), 978-1-003-04507-6 (eBook)
www.jennystanford.com

challenges with their use in vivo still exist. Poor stability, potential toxicity, and rapid reticuloendothelial system (RES) clearance are some of the more prominent obstacles to their therapeutic use. For these reasons, providing liposomes with a structural and functional support can enhance their persistence in vivo, but also improve their targetability and therapeutic effect.

Bioengineered scaffolds have been used as bioactive delivery agents in tissue engineering and regenerative medicine [1]. In these contexts, scaffolds serve two main purposes: they enable the integration of cells or tissues with the in vivo vasculature, and they support the growth of cells and tissues with the overall aim of enhancing their regeneration, reconstruction, or replacement in vitro or in vivo. In order for scaffolds to be able to support these functions, they must possess well-defined physicochemical features including appropriate biocompatibility, mechanical properties, porosity, biodegradability, safety, ease of fabrication and suitable surface chemistry for cell attachment, proliferation, and differentiation [2].

As support materials, scaffolds can support the three-dimensional integration of cells into tissues to ultimately achieve regeneration of complex organs [3]. These functions can be augmented by the ability to incorporate biological cargo—drugs, genes or therapeutic proteins, such as growth factors, enzymes or antibodies—within scaffolds. On one hand, liposomes serve a conceptually similar role to scaffolds: in simplest terms, they are vessels for the delivery of biological cargo [4]. At the same time, scaffolds and liposomes can synergistically be employed to achieve improved therapeutic delivery of bioactive compounds by combining the benefits associated with each delivery system. For instance, local delivery of liposome-encapsulated biomolecules can maintain appropriate drug levels at a healing site, thereby reducing adverse side effects associated with continued bolus administration. Scaffolds can provide additional biocompatibility and stability to liposomes thereby increasing their in vivo retention and reducing or eliminating their clearance. Studies have shown that combining liposomes with scaffolds can indeed enhance therapeutic delivery [5, 6].

1.1.1 Benefits of Combining Liposomes and Scaffolds for Therapeutic Delivery

Although liposomes have been extensively demonstrated to be attractive carriers for therapeutically active agents, they suffer from drawbacks, which have significantly limited their widespread use.

These include low cargo entrapment efficiency, leakage and loss of hydrophilic cargo when delivered in vivo, poor stability, elevated toxicity particularly for cationic lipid-based liposomes, and poor reproducibility and instability of formulations. Among these, poor biological stability and rapid clearance in vivo have been two of the biggest hindrances to their more widespread use. The lack of mechanical support has resulted in adverse pathology related to delayed healing due to poor delivery of bioactive molecules to healing sites [7]. In response to these drawbacks, modification of liposome structure by attachment of polyethylene glycol (PEG) to the surface of lipids has been a common strategy aimed at extending liposomes' blood circulation time [8]. However, reducing RES clearance and stability, especially over time frames practical for therapy, remains a major issue.

More effective is the endowment of scaffold support to liposomes. Scaffolds can be manufactured with a range of materials imparting significant mechanical rigidity that can be modulated based on the required end use. Hybrid nanoscale composites combining liposome encapsulation with scaffold-supported sustained release can also offer a three-dimensional substrate suitable for biologically active support of cell and tissue growth. When used by themselves, the release kinetics of liposomal cargo are affected by the nature of the bioactive agent, the liposome composition, the method of encapsulation, and the interaction between cargo and liposome [9, 10]. These properties change when liposomes are combined with scaffolds, the latter inducing a significant change in release kinetics, biological stability and functional performance.

The most attractive and, so far, therapeutically promising, feature of combining tissue-engineered scaffolds with liposomes for enhanced delivery of bioactive molecules is the ability to control delivery in a sustained and targeted manner. Strategies to achieve targeted delivery have relied on the use of receptor-targeted biomaterials [11], alongside the design of new ligand-supported polyvalent scaffolds [12]. Despite these benefits, little knowledge exists on the exact release kinetics of entrapped molecules as a function of composite structure and the effect that scaffold– liposome interactions exert on the activation of post-release signaling pathways. Rheological studies into scaffold–liposome composites [13, 14] have investigated the physicochemical features of these systems, including liposome-hydrogels' non-Newtonian, pseudoplastic and thixotropic behavior, which is important for topical applications.

1.1.2 Clinical Considerations

Though liposomes have proven to be clinically effective—they were the first "nanoscale" drug delivery system to achieve regulatory approval in the USA—hybrid systems, such as scaffold–liposome combinations, bear a different regulatory burden. Because incorporation of a scaffold component can alter the function and purpose of the composite structure, the regulatory framework requires different types of testing to be performed to prove efficacy, safety, and biocompatibility. To achieve clinical translation of liposome-scaffold constructs from the laboratory to mainstream clinical practice, a multidisciplinary approach that successfully integrates material science, engineering, and drug delivery. Though scaffolds have traditionally been developed as systems embedded within tissue engineered applications in regenerative medicine, when combined with liposomes they can take on different functional purposes. This can alter their regulatory designation. The FDA outlined the designation of regenerative medicine advanced therapies (RMATs) within the 21st Century Cures Act. As more pre-clinical reports combining these hybrid delivery systems appear, the regulatory framework for combination products that are aimed at drug delivery while supporting in vivo homing is expected to mature. This, it is hoped, will ultimately result in a clearer path to translation for these next-generation products.

1.2   Cargo Delivery from Liposome-Scaffold Composites

It makes sense to combine two distinct functional materials, such as liposomes and scaffolds, when the ultimate goal is to achieve a functionality which none of the individual components is capable of by itself. Incorporating liposomes inside polymeric matrices generates functionally diverse two-component systems that enhance therapeutic cargo delivery. The scaffold component preserves the structural integrity of the nanoparticle-stabilized liposomes, while allowing for controllable visco-eleasticity and tunable, sustained liposome release rate.

Release of liposome and encapsulated cargo can occur via passive diffusion or with the aid of external triggers such as temperature or pH. The nature and type of liposome dictates the type of stimulus that can be used to release the cargo. Alongside sustained delivery, liposome/scaffold complexes allow for enhancement of stability of entrapped liposomes which can, in turn, enhance their bioavailability. At the same time, the nature of the scaffold and the physicochemical properties of the biological cargo both affect the release profile from liposomes [15].

Advances in biomaterial science have enabled the development of a wide array of materials that can modulate cargo delivery kinetics from liposome/scaffold composites. For all of these, however, release kinetics follow strict rules. The two main types of release mechanisms from a liposome or scaffold-based vehicle are diffusion-based release—described by Fick's law of diffusion—and degradation-dependent release [16]. Most release profiles are described by zero order kinetics [17] and include a burst release period, which is often thought as undesirable as it may decrease half-life and contributes to local or systemic toxicity. Generally, cargo release is thought to be slower from liposomes composed of rigid bilayer membranes (e.g., phosphatidylcholine (PC) compared to 1,2-distearoyl-sn-glycero-3-phosphocholine (DSPC)/cholesterol) [4], while the liposomal gel phase is considered more conducive to slower permeation [18]. Release kinetics were also found to be dependent on size of the entrapped liposome when charge is also in play, with smaller ($\sim$100 µm) liposomes with negative zeta potential exhibiting sustained release compared to larger (~300 μm) liposomes [5]. Similarly, chitosan-coated phospholipid-based liposomes were shown to follow square root time-dependent Higuchi release—a Fick's law-based diffusion model—of cargo from the liposome interior [1].

Among scaffolds, hydrogels—three-dimensional networks of polymers—is the collective catch-all name for the most widely studied group of materials [19, 20]. Hydrogels have favorable rheological properties, including a high viscosity, which acts as a protective mechanism to stabilize liposomes. For that reason, hydrogels have been particularly attractive for their use as three-dimensional scaffolds.

Liposomal hydrogels have been characterized for use in topical, ocular [21] or vaginal delivery [22]. In topical delivery, for instance, the liposome enables encapsulation of the drug to retain high local concentration, while hydrogels afford good stability, moisture, oxygen permeability and act as a barrier against contaminants such as bacteria. The application defines the mode of interaction of liposome/hydrogel composites with the biological system. In topical delivery, for instance, interaction of biological carriers with the *stratum corneum* is material-specific and defines the mode of entry and efficacy of transdermal delivery [3].

Generally, it is thought that addition of liposomes increases the viscous modulus and improves overall hydrogel strength [15]. This was shown to be true for a variety of hydrogel materials, including poly(2-vinyl pyridine)-b-poly(acrylic acid)-b-poly(n-butyl methacrylate) (P2VPPAA-PnBMA) hydrogels, thiolated chitosan-coated liposomes [2], as well as carrageenan/carboxymethylcellulose gels. The latter, when mixed with chitosan-coated oleic acid liposomes, showed that the presence of liposomes increased the elasticity of the gel and displayed both a higher complex viscosity and higher spreadability [23]. These, taken together, improve the release of liposomes from the gel and enhance the stability of the dispersed liposomes during storage.

Interactions between the liposome and the hydrogel are key to stabilizing the composite structure. Within poly(N-isopropylacrylamide) (PNIPAm) hydrogels, hydrophobic interactions between the liposome and the polymer were shown to be critical in maintaining structural integrity and delivery profile [21]. Carbopol 940 hydrogels encapsulating PC/cholesterol liposomes showed significant charge-dependent entrapment efficiency [24]. Positively charged liposomes, obtained with the addition of stearylamine, showed improved entrapment efficiency over negatively charged liposomes containing dicetyl phosphate when used as transcorneal delivery vehicles [25].

The release rate of lipophilic drugs from the interior of hydrogel-entrapped liposomes was found to be more dependent on the polymeric hydrogel dissolution rate, or erosion, than the diffusion rate by up to 30%. Gel dissolution was shown to follow zero-order kinetics, while increasing liposome concentration could be used to tune the release kinetics, thereby eliminating the need for elevating polymer concentrations [26]. The development of therapeutic applications for various liposome/scaffold composites over the last few years has been remarkable.

1.3   Hydrogels as Delivery Scaffolds

Hydrogels, crosslinked networks of hydrophilic polymers with distinct three-dimensional structures, are highly attractive as materials for biomedical applications. Their favorable properties, including biodegradability, biocompatibility, porosity and swelling behavior makes them suited as scaffolds for use as delivery support systems for a variety of purposes. Hydrogels can be manufactured from both natural and synthetic polymers with different structural features and physicochemical properties. They can be tailored to different mechanical properties, and are usually prepared by chemical or physical crosslinking [27], all of which has been extensively reviewed elsewhere [28, 29]. Hydrogels encompass a variety of materials which can be thermally and pH- responsive, and includes poloxamers as well as natural and synthetic polymers such as poly(lactic acid) (PLA) and poly(glycolic acid) (PGA) [30]. These materials can show remarkable stability under harsh conditions, making them particularly suitable for in vivo use. And though hydrogels are good candidates as tissue replacement materials, their versatility makes them attractive in drug delivery as well. For pharmaceutical delivery, the ability of incorporating liposomes within hydrogels enables the entrapment and delivery of hydrophobic drugs [31]. This, effectively, overcomes the rapid and ineffective release that as typically observed with hydrophilic cargo-containing hydrogels [32]. Cargo release from hydrogel-based substrates can be tuned by pH, temperature, light, and electric or magnetic field [33] through either crosslinked or either crosslinked hydrogels or those formed from semi-interpenetrating polymer networks. Recently, the development of "nanogels"—nanometer-scale hydrogel-based systems for drug delivery—has allowed the avoidance of macrophage uptake while mediating rapid release of cargo intracellularly [34]. In anti-cancer therapy, in situ gelling hydrogels have enabled the avoidance of pharmacokinetic restrictions typically encountered by intravenous injection. This was demonstrated through hyperthermia-induced release of doxorubicin from dipalmitoyl phosphatidylcholine (DPPC), monostearoyl phosphatidylcholine (MSPC) and distearoyl phosphatidylethanolamine-poly(ethylene)glycol 2000 (DSPE-PEG2000) liposomes incorporated within hydrogels [35]. In the following sections, we discuss the most commonly used hydrogels as scaffolds for the delivery of liposome-encapsulated drugs.

1.3.1   Chitosan-Based Scaffold–Liposome Composites

Chitosan is a naturally derived, linear, water-soluble polysaccharide which has been extensively used in the medical field. It is obtained by the deacetylation of chitin. Among hydrogels, chitosan-based gels have been one of the most studied systems, in large part thanks to their bio- and mucoadhesiveness. Chitosan's favorable biodegradability, thermal-responsiveness, and in situ gelation have been extensively exploited in biomedical applications, particularly in ocular delivery, tissue regeneration and wound healing applications [36]. Chitosan-liposome composites have also been used in transdermal drug delivery [37], topical vaginal therapy [38] and topical cancer therapy [39]. Degradation of chitosan increases with degree of acetylation, and occurs by chemically mediated catalysis and, in vivo, enzymatic catalysis [40]. In vivo, chitosan is predominantly degraded by lysozyme and other bacterial enzymes. While chitosan is considered safe and highly biocompatible, its toxicity increases with increasing charge density.

Liposomes have frequently been incorporated within chitosan-based supports. Coating liposomes with chitosan has been one strategy used to improve the performance of liposome-based drug delivery systems. Many reports have studied the release parameters from entrapped liposomes within chitosan hydrogels. Chitosan-DSPC/cholesterol liposome composites, the release rate was shown to depend on the liposome size and composition [41]. The release rate was slower for larger liposomes (280 nm) than smaller (100 nm) liposomes, and slowest for the largest (580 nm) liposomes, owing to a reduction in diffusion coefficient. The addition of cholesterol resulted in compaction of the membrane, which reduced its permeability, an effect that was compounded by replacement of egg PC with DSPC. While negatively charged lipids were shown to affect the gelation of the hydrogel, they exhibited similar release kinetics to liposomes prepared with lipid having a high phase transition temperature. Other reports have shown that the plastic viscosity of DPPC liposomes increased after coating with chitosan to support sustained release over two weeks [42]; however, a decrease in the stability of the liposomes was observed in the presence of excess chitosan [42]. Specifically, a concentration of 0.3% chitosan showed slowest release kinetics (approximately 20% slower than 0.6% chitosan-containing constructs) and a higher retention time measured by MRT (mean residence time) = 62.10 ± 2.15 min and $t_{1/2}$ = 50.22 ± 1.04 min, while an increase in chitosan to 0.6% reduced these values to MRT = 47.74 ± 0.96 min and $t_{1/2}$ = 46.73 ± 2.84 min. Release kinetics from covalently crosslinked composites were described by a multi-scale mechanism which starts with a burst phase followed by swelling of the gel, and, finally, an equilibrium phase described by Peppas-Sahlin and Weibull equations [43]. Crosslinking was described to be critical to directing the rate of liposomal cargo release, with faster release observed for a lower degree of crosslinking [44]. Chitosan-based liposome-hydrogels were also shown to exhibit 60% fluid uptake from exuding wound-modeling agar gels, contributing to the upkeep of a moist environment in the wound environment during healing, as compared to plain chitosan hydrogels [16].

The incorporation of a chitosan-based support to enhance liposome-mediated delivery was recently also for the delivery of Substance P [45], a neuropeptide that plays an important role in wound healing, and for the delivery of curcumin, using thiolated chitosan-coated phosphocholine/cholesterol liposomes [46]. Release studies with chitosan-liposomal hydrogels have been carried out both in vitro and in vivo. In vivo studies have benefited from chitosan's mucoadhesiveness, which was shown to improve the gastrointestinal stability of the entrapped liposomes encapsulating hydrophilic drugs. Orally administered risedronate from chitosan-coated liposomes [47] resulted in the targeted detection of 10.2 ±3.0 µg of drug released from the liposomes, compared to only 5.5±1.9 µg from non-coated liposomes. This was attributed to the effect of chitosan's mucoadhesiveness which improved the gastrointestinal stability of the liposomal membrane, leading to enhanced membrane transport of the encapsulated drug. Elsewhere, chitosan-coated liposomes encapsulating Cyclosporin A enabled enhanced delivery of the drug to cornea, conjunctiva, and sclera of rabbits [48].

The ability to engineer composite scaffolds that can be triggered by external stimuli is an attractive feature that has successfully been demonstrated with chitosan-based gels. Among these stimuli, temperature and pH have been most widely explored. Temperature, in particular, is attractive for its ease of implementation, not requiring external gelling agents. Crosslinking chitosan to β-glycerophosphate generates a two-component blend that exists in liquid phase at ambient temperature, but turns into a semi-solid hydrogel at elevated temperatures. This feature has been exploited to generate responsive liposome/chitosan composites that have been tested both in vitro and in vivo. Pharmacokinetics of thermoresponsive chitosan-β-glycerophosphate hydrogels encapsulating liposomal doxorubicin showed enhanced antitumor activity in blood and in tumor, alongside lower toxicity [49]. Combined with DSPC/chol/1,2-Dioleoyl-sn-glycero-3-phosphoethanolamine (DOPE) liposomes, chitosan-β-glycerophosphate hydrogels showed, in in vivo mouse studies, an accumulation of 51.2% model cargo released from the liposomal hydrogel, while this was 2.5 and 3.8% for free label and plain hydrogel cargo, respectively. Similarly, accumulation of liposomes in RES during 24 h ranged from 1.1–35.1 (%ID/g) for liposomal hydrogel, significantly higher than non-encapsulated cargo [50].

Though chitosan has been extensively studied, it is not yet approved by the FDA for drug delivery. Despite that, preclinical work is increasing, and clinical effects of chitosan-liposome composites, with a focus on such parameters as biodegradability and long-term stability [51] are showing clinical relevance.

1.3.2   Collagen-Based Scaffold–Liposome Composites

Collagen is the major insoluble fibrous protein found in the extracellular matrix and connective tissues, where it plays a major role in maintaining structural integrity and providing support. As such, it has been recognized as an attractive tissue engineered-biomaterial owing to its favorable properties, which include good biocompatibility, porosity, permeability, good biodegradability, and low immunogenicity. Composite collagen-liposome systems for drug delivery have been studied since the 1980s [52]. From these studies, it emerged that collagen either forms a protein layer at the liposomal surface, or becomes incorporated within the lipid bilayer via its hydrophobic residues [53]. Collagen was indicated to provide a 2-fold greater enhancement in stability, as measured by a higher retention of encapsulated matter, in negatively charged liposomes compared to neutral liposomes [54].

Collagen has been combined with liposomes via coating interactions, as well as by incorporating liposomes within collagen matrices. Because of its abundance in connective tissue, it makes sense that the most common applications with collagen-liposome composites have been focused on wound healing applications. Nunes et al. [55] described collagen films incorporating usnic acid-encapsulating liposomes as a wound dressing for dermal burn healing. Wounds treated with this therapeutic composite system exhibited a dramatically increased formation of type-I and type-III collagen fibers to resemble the native dermis after 21 days of treatment. Collagen deposition grew to almost 50% in the same period, while myofibroblast content was reduced by over 15%—higher than what had been observed for non-liposome containing collagen dressings.

It was also used as a ternary drug delivery system when combined with chondroitin sulfate-encapsulating liposomes [56]. In this instance, high encapsulation efficiency was observed, attributed to ionic interactions between O-sulfate groups of chondroitin sulfate and the amino terminus of cationic stearylamine within the lipid bilayer. In the study, Craciunescu et al. also observed higher cytocompatibility in L929 cells, alongside improved chondroitin sulfate release from liposome-containing composites compared to chondroitin sulfate alone.

Similarly to other hydrogels, stimuli-responsive composites have been generated with collagen-complexed liposome systems. Lopéz-Noriega [57] et al. developed a temperature-sensitive composite by incorporating thermoresponsive liposomes within collagen-hydroxyapatite scaffolds. In this example, the scaffolds were functionalized with sulfhydryl groups, and subsequently complexed to DPPC-MSPC-DSPE-PEG2000-maleimide liposomes encapsulating the parathyroid hormone-related protein (PTHrP 107–111). In vitro, this system exhibited pro-osteogenic and anticatabolic effects on MC3T3-E1 bone cells. Hyperthermic pulses controlled cargo release, which otherwise occurs at a steady rate (50% release after 14 days). Alkaline phosphatase activity, alongside expression of OPN and OCN genes, demonstrated effectiveness of the applied pulses.

Most systems discussed so far have employed collagen in association with liposomes via coating-like interface. Such system enables loading of matrix-stabilized liposome cores which reduce internal kinetic cargo release rate and achieve improved spatiotemporal control of release rate. Collage matrix-embedded liposome composites have thus far been developed by encapsulating collagen inside laurdan liposomes, thereby generating liposome-entrapped hydrogel meshes to control drug delivery which could be modulated in a systematic manner by tuning collagen mesh properties [58].

In large part thanks to its biocompatibility, collagen continues to be an attractive material for regenerative medicine. The FDA has approved a number of collagen-based products, though for very specific medical applications. Nonetheless, complexation of collagen with liposomes remains a less-studied though potentially attractive avenue of research.

1.4   Gelatin-Based Scaffold–Liposome Composites

In its simplest form, gelatin is a form of denatured, partially hydrolyzed collagen. Its most well-known use is in the food industry, though its use as a wound dressing, in drug delivery, and as a scaffold in tissue engineering is growing. Gelatin has been used as a material for biomedical scaffold assembly primarily to overcome some of the drawbacks associated with collagen, namely its poor in vivo swelling properties, low mechanical strength, and low elasticity. Compounding this is Type I collagen's high price. Chemically, gelatin contains favorable cationic, anionic, and hydrophobic groups in 1:1:1 ratio.

Though gelatin may offer some advantages over collagen, it has not been as extensively used as a support scaffold for drug delivery as other more prominent hydrogels, such as chitosan. Gelatin is typically only derived from sources rich in Type I collagen and generally contains no Cys. Above a certain concentration in water, gelatin crosslinks and undergoes a conformational change from a random coil to a triple helix. The ability to support compressive forces is higher for more crosslinked gelatin, while its random-coil structure has lower mechanical strength. However, gelatin hydrogels are still associated with low shape stability, poor mechanical strength, and low elasticity, significantly limiting their potential use in biomedical applications. Because of that and due to its rapid enzymatic degradation in vivo, gelatin is typically blended or mixed with other polymers to enhance its performance and swelling behavior.

Liposomes have been complexed with gelatin in various configurations—by both embedding liposomes within gelatin matrices, and by encapsulating gelatin nanostructures within liposome cores. The rate and degree of liposome release from gelatin matrices was shown to depend on liposome concentration, size and composition, while interaction with gelatin hydrogels was shown to not compromise thermal stability nor interfere with the resiliency of the scaffolds to tensile force [59]. On the other hand, liposomes could also support the encapsulation of gelatin nanocarriers encapsulating stavudine [60], showing potential applicability in HIV-1 treatment. Liposome-gelatin scaffolds were also used to study drug release kinetics from collagen-associated liposomes as well as biodegradability of gelatin scaffolds. For this purpose [61], Peptu et al. used rhodamine-labeled liposomes encapsulating model cargo (calcein) that were crosslinked to gelatin/sodium carboxymethylcellulose scaffolds. The authors then monitored calcein for up to 25 days. Scaffold density, liposome size (larger liposomes slowed down release), degree of swelling, and degree of crosslinking were shown to affect the rate of release. By crosslinking cholesteryl group-modified tilapia gelatins to dioctadecyldimethylammonium bromide (DODAB) liposomes, Taguchi et al. [62] investigated the effect of the modification of cholesterol content on the degradation rate. Collagenase degradation was 2× slower for composites containing 69 mol% cholesterol, and growth of human hepatocellular carcinoma HepG2 cells was over 2-fold higher on liposome composites containing a 69 mol% cholesterol after 3 days in culture. Gelatin scaffolds were also rendered pH-responsive by blending them with pH-sensitive liposomes of sodium oleate [63]. These blends demonstrated pH-triggered release of encapsulated calcein which was most effective at a pH of 10, consistent with the transition of sodium oleate vesicles into micelles. The rather rapid release of cargo at an elevated pH could be retarded via the gradual control of pH transition from a pH of 8.3, where sodium oleate vesicles are intact, to 10.

Similarly to collagen, gelatin–liposome composites are a fairly emerging combination system. However, chemically modified gelatin can exhibit tunable mechanical properties, which can be modified with various crosslinkers, such as glutaraldehyde, genipin, and dextran dialdehyde. This broadens its potential appeal. Moreover, the ability to process gelatin hydrogels using additive manufacturing can be an attractive prospect for personalized medicine and the assembly of personalized medical devices and delivery systems.

1.5 Dextran-Based Scaffold–Liposome Composites

Dextran, a natural linear polymer of glucose, includes a range of polysaccharides encompassing three classes, and has been widely used as polymeric scaffolds in drug delivery. Its reactivity involves derivatization via hydroxyl groups to improve its performance or mechanical properties [64]. Hydroxyl group modifications can form various spherical, tubular, and 3D network structures, and its good biocompatibility has enabled dextran to be used in many biopharmaceutical applications. Some studies have suggested that the introduction of amine groups with dextran results in hydrogels with better biocompatibility and more favorable release properties for use in biomedical applications.

The combination of dextran and liposomes for drug delivery has been studied since the early 2000s. In recent years, its use as a matrix has increased. Many of these studies have investigated the effect of various dextran modifications on release parameters and functional performance. In one of the earliest examples of such combination delivery systems, Stenekes et al. [65] developed injectable liposome-dextran composites made of DPPC-DPPG-cholesterol liposomes entrapped inside methacrylated dextran microspheres to study delivery of liposome-encapsulated calcein. They found that lower water content liposomes exhibited a more sustained release pattern over 100 days. When methacrylated dextran microspheres contained lactate, they resulted in a higher degree of hydrolysis of the microspheres which, in turn, caused rapid release of encapsulated cargo.

Stimuli-responsive composites were also developed. Epstein-Barash et al. [66] generated ultrasound-sensitive dextran-liposome composites by encapsulating microbubbles—gas-filled lipid monolayers—within the structures and evaluated their drug release performance in vitro and in vivo. Elsewhere, aldehyde-modified dextran:adipic hydrazide-modified carboxymethylcellulose hydrogels containing 6% (w/v) polymer were constructed incorporating either DSPC:DSPG (distearoylphosphatidylglyce rol):cholesterol (negatively charged), DSPC:DODAB:cholesterol (positively charged) or DSPC:cholesterol (neutral) liposomes, alongside DSPC-PEG40S microbubbles. The composites were shown to successfully carry and release model cargo (fluorescent dyes) while the microbubbles enhanced the drug release kinetics by increased cavitation, which was dependent on liposome content both in vitro and in vivo in male CD-1 mice. With application of ultrasound, dye-containing liposomes showed on-demand (after six

10-sec bursts separated by 1-sec off at 6 W/cm^2 administered 48 h apart) release for two weeks. The addition of microbubbles resulted in a higher release of trypan blue in response to ultrasound—23 ± 3% for hydrogels with liposomes, compared to 56 ± 2.7% for hydrogels with liposomes and microbubbles. Dextran gels have also been polymerized in the interior of liposomes to create so-called nanogels. UV-based polymerization of dextran hydroxyethylmethacrylate inside 1-stearoyl-2-oleoyl-sn-glycero-3-phosphocholine liposomes was demonstrated to result in sustained release kinetics of BSA and lysozyme which could be tuned by controlling the crosslinking density, while avoiding serum aggregation [67]. More recently, self-assembling dextran nanohydrogels were developed, which could mediate the pH- and redox-controlled release of anti-cancer drug doxorubicin [68]. And though these systems were devoid of liposomes, construction of more sophisticated multi-modal systems could further enhance in vivo performance. Elsewhere, N-maleoyl-b-alanine-modified dextran was also used as a tunable hydrogel to control the size of giant unilamellar vesicles as potential biological cell models [69]. Michael addition crosslinking of PEG to dextran hydrogels generated a substrate which was shown to promote the generation of free-floating vesicles and their growth.

Though dextran is very attractive in biomedical applications, its ultimate success as a clinically useful scaffold material will depend largely on the degree, nature, and quality of chemical modifications, which would not only allow it to exhibit improved complexation with liposomes, but promote durable therapeutic effect.

1.6 Alginate-Based Scaffold–Liposome Composites

Alginate is a linear, naturally occurring, hydrophilic polysaccharide composed of repeating units of 4 β-D-mannuronic acid and α-L-guluronic acid, which has enjoyed extensive use across a number of application areas, from the food industry to biomedicine. It possesses favorable biocompatibility, biodegradability, chelatability, and non-antigenicity, making it a suitable material as a scaffold for biological delivery. In tissue engineering, it has most commonly been crosslinked with calcium chloride. Its mechanical rigidity and degradation rate can be controlled by varying the molecular weight distribution of alginates at different oxidation degrees [70]. Alginate's responsiveness to calcium-rich environments has made it attractive for bone tissue engineering applications. Owing to its affinity for divalent cations, ionically crosslinked alginate's mechanical strength increases as a function of these ions [71]. Moreover, the acid insolubility of alginate gels makes them good candidates for use in applications where the stability of liposomes can be compromised, such as in the stomach or intestine. Studies have investigated the effect of calcium ions on the release from alginate-embedded liposomes [72], as well as the effect of coating on pH and ionic strength stability on sodium alginate and chitosan-coated multilayered liposomes as a means to understand the effect of environment conditions on drug release during activities such as digestion [73]. The authors found that, while changes in pH and ionic strength (as high as 200 mM NaCl) affected the appearance and size of alginate/chitosan-coated liposomes, their cores and delivery rates remained unaltered, suggesting good protective stability of the polymers. In simulated gastric fluid, release of vitamin C from chitosan and alginate-coated liposomes was 14±1% after 2 h, lower than that observed with uncoated liposomes (25±4%). On the other hand, after digestion in simulated intestinal fluid with pancreatin and bile salts, the release of vitamin C from uncoated liposomes was 82±8%, compared to <20% for chitosan and alginate-coated liposomes.

Alongside pH, temperature was also investigated as a trigger for the release of liposome-encapsulated cargo from within alginate hydrogels. Holmium ion-crosslinked alginate hydrogels were prepared encapsulating DSPC-MSPC-DSPE-PEG2000 liposomes for the delivery of chemotherapeutic agents [74]. The use of holmium ions rendered these composites imageable by MRI, alongside liposome-encapsulated MRI contrast agent [Gd(HPDO3A)- (H_2O)]. Release of liposome-encapsulated doxorubicin, was demonstrated successfully following mild hyperthermia in ex vivo embolization experiments involving sheep kidney. Temperature-sensitive liposomes released 20% doxorubicin at 37°C in HEPES after 3 h. In 50% FBS, the release was faster (30% after 3 h at 37°C). For both systems, release was proportionally faster at 42°C, with the FBS system showing almost complete release of doxorubicin in 3 min. This was explained by the presence of proteins which desorb doxorubicin from the negatively charged alginate, while holmium-doxorubicin interaction was shown to minimally affect effectiveness of the liposomes. Cholesterol:DPPC liposomes embedded in a calcium alginate gel matrix were developed to investigate release kinetics from temperature-triggered liposomes [75]. The temperature and the cholesterol:DPPC ratio were shown to affect release kinetics of a carboxyfluorescein, while leakage during the crosslinking process was highest for liposomes composed of DPPC and devoid of cholesterol, corroborating earlier held notions that cholesterol decreases spontaneous leakage from liposomes through membrane rigidization.

Davis et al. [76] developed alginate-encapsulated liposomal bupivacaine as a local anesthetic in conjunction with mesenchymal stem cell administration. This formulation resulted in the prevention or preservation of MSC-mediated anti-inflammatory PGE2 secretion compared to either bolus or liposomal bupivacaine.

Alginate gels have also been developed by entrapping alginate within the core of liposomes [77]. Guo et al. [78] developed an alginate-encapsulating liposomal nanolipogel, which contained alginated in the core, and used it to study the effect of nanolipogel rigidity on tumor accumulation. In an orthotopic breast cancer model in vivo, soft alginate nanolipogels were shown to preferentially accumulate in tumors, while elastic nanolipogels accumulated mainly in the liver. Smith et al. [41] developed DPPC liposomes encapsulating sodium alginate and alkaline phosphatase to study the enzyme's release in an acidic environment. Alginate-liposome conjugates have also been investigated as matrices for delivery of drugs with low aqueous solubility. Wang et al. [79] developed cisplatin-conjugated sodium alginate entrapped in epidermal growth factor (EGF)-modified liposomes to specifically target epidermal growth factor receptor (EGFR)-expressing tumors. In vitro, the composite system was able to selectively recognize EGFR-positive SKOV3 cells and penetrate tumor spheroids, while in vivo the delivery of cisplatin into ovarian tumor tissues was achieved.

The growing body of work of liposome–alginate composites is fueled by the marriage of alginate's favorable biocompatibility and its ability to generate robust, tunable composites when complexed with liposomes.

1.7 Hyaluronic Acid-Based Scaffold–Liposome Composites

Hyaluronic acid, a naturally occurring, high molecular weight linear polysaccharide composed of repeating units of D-glucuronic acid and N-acetyl-D-glucosamine disaccharide, is biocompatible, non-toxic and non-inflammatory. It is typically chemically modified to alter its physical properties for specific applications. Alongside its use as a support scaffold for drug delivery, hyaluronic acid has been exploited as a polymer for protein and peptide conjugation. Hyaluronic acid is biogenic, distributed widely in the extracellular matrix. The presence of hyaluronic acid receptors in the liver has been particularly attractive for the development of targeted-delivery systems, supported by modifications of hyaluronic acid derivatives and hyaluronic acid-drug conjugates as a means to target specific intracellular sites [80]. Hyaluronic acid binds cations and strongly affects cell mobility and tissue regrowth. Physical properties of hyaluronic acid hydrogels depend on the polymer's molar mass distribution and concentration, ultimately affects hydrogel permeability and mechanical properties.

When preferential liver accumulation is not desired, PEGylation of hyaluronic acid nanoparticles has allowed tumor tissue targetability [81]. Paliwal et al. [82] developed targeted, pH-sensitive PC/DOPE (1,2-Dioleoyl-sn-glycero-3-phosphoethanolamine)/DSPE-PEG$_{2000}$ liposomes for doxorubicin delivery to tumor sites of mice in vivo via phospholipid functionalization of liposomes and polymer. Dual-targeting configurations have also been demonstrated. Jiang et al. developed cell penetrating peptide-targeted liposomes incorporating hyaluronic acid as a shield which protected the liposomes from plasma protein-induced degradation [83]. These liposomes, encapsulating paclitaxel and constructed with arginine and histidine-rich cell-penetrating peptides, demonstrated stronger cytotoxicity toward hepatic cancer HepG2 cells at pH 6.4 than pH 7.4, while coumarin 6-loaded hyaluronic acid-liposomes showed efficient intracellular trafficking including endosomal escape. Targeting was also achieved using hyaluronic acid/ceramide-coated DOPE/DOTAP (N-[1-(2,3-Dioleoyloxy)propyl]-N,N,N-trimethylammonium methyl-sulfate) liposomes to CD44$^+$ cancer cells [84].

Hyaluronic acid-liposome systems have been shown effective in anticancer therapy in combinations that included entrapping liposomes within hyaluronic acid scaffolds as well as complexing lipid bilayers and hyaluronic acid derivatives. As an example of the latter approach, Miyazaki et al. [85] modified liposome with hyaluronic acid-based polymers that were both pH sensitive and which could direct liposomes to target CD44$^+$ cancer cells. Under acidic conditions, these polymers mediated release of anti-cancer drugs in a target-specific manner.

Other therapeutic applications of hyaluronic acid liposomal gels have recently demonstrated efficient delivery of corticosteroids to the inner ear [86]. The drug dexamethasone was incorporated into PEGylated liposomes entrapped in a hyaluronic acid hydrogel and sustained release of the drug was observed in the perilymph of a guinea pig model for a remarkable 30 days, owing to the enhanced stability of the liposomal delivery vehicle within the gel.

Control of spatial distribution and temporal release of hydrophobic and hydrophilic therapeutic cargo from hyaluronic acid-liposomes could be achieved by incorporating liposomes onto solid supports. One such system involves the fabrication of polyelectrolyte multilayers of poly-L-Lysine and poly(sodium styrene sulfonate) onto which hyaluronic acid covalently functionalized PC-DPPE (1,2-dipalmitoyl-sn-glycero-3-phosphoethanolamine)-cholesterol liposomes are embedded [87]. Hydrophobic doxorubicin and hydrophilic cholesterol cargo were delivered locally and in a sustained manner to 21MT-1 breast cancer cells. Localized cellular uptake of the liposomal nanoparticles was observed 36 h and 60 h after cell seeding. Additionally, after 60 h in a cell-free system, 10% uptake of nanoparticles was observed via flow cytometry, which was attributed to diffusion/film swelling. With 21MT-1 cells present, however, the 60 h time point indicated uptake of 100% liposome nanoparticles, suggesting cell-dependent uptake to be a strong mechanism driving the release of the nanoparticles.

Similarly to the previously discussed example in anti-cancer therapy, hyaluronic acid-liposome complexes with temperature- and pH- responsiveness have been extensively studied for a variety of applications. In one such example, Ren et al. [88] encapsulated horseradish peroxidase inside liposomes embedded in a hyaluronic acid–tyramine conjugate and hydrogen peroxide-based gel. The system was responsive to temperature-induced phase changes: At room temperature, the solution was injectable, while exposure to body temperature caused crosslinking of the scaffold and release of the enzyme. Rheological investigations into the effect of liposomes on hyaluronic acid gel behavior [89] upon local injection revealed a pronounced increase in viscosity exerted by PEGylated liposomes, while higher lipid concentrations lead to lower viscosity and elasticity for lipids larger than 200 nm. No size-dependent rheological effect was observed at a lipid concentration of 10 mM. The viscoelasticity and shear-thinning behavior of hyaluronic acid gels were additionally found to favor fast immobilization and improved injectability, respectively. Liposome-hyaluronic acid composites were also investigated as a potential intranasal vaccine platforms. Ionic complexation between DOTAP and thiolated hyaluronic acid (HA-SH) was used to generate a composite that was further stabilized by reacting the HA-SH layer with thiolated PEG [90]. The liposome-hyaluronic acid nanoparticle encapsulating monophosphoryl lipid A (MPLA), a toll-like receptor 4 agonist, showed significant improvement in bone marrow-derived dendritic cell activation alongside a 20-fold reduction in cytotoxicity of the liposomes. Importantly, when F1-V, a candidate vaccine for *Y. pestis*, was incorporated in the liposomal nanoparticles, the authors were able to show balanced Th1/Th2 humoral immune responses using low doses of F1-V (1–5 µg), unlike an equivalent vaccine dose in soluble form. This resulted in an 11-fold increase in total IgG titers after 77 days for the liposome nanoparticle system compared to the soluble F1-V vaccine. The liposomal nanoparticle composites, moreover, showed prolonged release of protein antigen over at least 3 weeks at 37°C.

1.8 Other Materials for Scaffold–Liposome Composites

Though hydrogels have been most commonly used, other materials have also been studied for the controlled delivery of bioactives for specific therapeutic applications. The choice of material other than hydrogels is primarily dictated by the chemistry that can be achieved by their use. Examples of hydrogel alternatives include carrageenan, pullulan and xanthan gum. Carrageenans, particularly ι and κ variants, have proven attractive owing to their high flexibility, thermosensitivity and anticoagulant effect due to the presence of sulfonyl groups [91]. Carrageenan was studied as a support scaffold for PC/cholesterol liposomes carrying antifungal drug Ciclopirox olamine either alone or in combination with agar for the treatment of dermal infections [92]. More recently, Kulkarni et al. [93] developed κ-carrageenan-Pluronic® F127-stabilized lipid-based carrier films for therapeutic drug delivery. In these studies, the authors used oil/water lipid emulsions, rather than fully formed bilayer liposomes. A thin-film structure was achieved via dehydration of the composite, while FTIR spectroscopy was used to observe release kinetics, which were superior to those from non-lipid-containing hydrogels.

Pullulan, a neutral water-soluble polysaccharide, has also been complexed with liposomes. Hydrophobized pullulan can self-aggregate in water to form colloidally stable nanoscale structures. Pullulan can be modified to generate derivatives with different chemistries. Cholesterol-bearing pullulan nanogels formed via self-aggregation have the capacity to form stable complexes with various biological molecules and serve as robust supports for drug and gene delivery. To prove this, Sekine et al. demonstrated the Michael addition-mediated complexation of pullulan nanostructures to pentaerythritol tetra(mercaptoethyl) polyoxyethylene can create hybrid hydrogels in association with liposomes which can mediate effective drug delivery [94]. Elsewhere, pullulan was also studied for the delivery of terbinafine hydrochloride for the treatment of onychomycosis [95]. Topical administration of pullulan-liposome films resulted in therapeutic-level accumulation of active drug at the application site both in vitro and in ex vivo permeation studies. The in vitro release percentage of drug was 71.6 ± 3.28 from liposome-containing pullulan films, while therapeutic range of accumulated drug (31.16 ± 4.22) in nail plates was observed.

Xanthan gum is a natural high molecular weight, branched-chain polysaccharide which is highly stable with respect to temperature and pH. For these reasons, it is considered valuable as a drug delivery substrate. However, its dissolution and poor swelling behavior requires it to be chemically modified for optimal utility. Xanthan gum was shown to be a suitable coating scaffold for the support of biomolecule-encapsulating liposomes with applications that have included delivery of liposomal rifampicin [96] from chitosan-xanthan gum scaffolds as well as nasal curcumin [97]. In the latter application, xanthan gum-associated liposomal curcumin administered intranasally showed significantly higher drug distribution in the brain (1240 ng) compared to free curcumin (65 ng). A recent study [98] also looked at developing biodegradable and injectable aldehyde-modified xanthan gum-supported liposome delivery systems. The complex was formed by crosslinking aldehyde modified xanthan gum hydrogels to phosphatidylethanolamine liposomes to form an injectable hydrogel by Schiff base bond formation between aldehyde groups of aldehyde-modified xanthan gum and amino groups of liposomes. Other materials for liposome-hydrogel composites have also included hydroxyethyl-cellulose [99], poly-2-hydroxyethylmethacrylate [100], poly(N-isopropylacrylamide) [101] and poly(2-vinyl pyridine)-b-poly(acrylic acid)-b-poly(n-butyl methacrylate) (P2VP25-PAA576-PnBMA36) ABC terpolymer [102].

Mimicry of extracellular milieu environments was also achieved with self-assembling hydrogel nanofiber peptides [103] using short chain oligopeptides with extracellular matrix protein-mimicking sequence. Wickremasinghe et al. [104] developed a multi-domain peptide, $K(SL)_3RG(SL)_3KGRGDS$, entrapping DOPC (1,2-dioleoyl-sn-glycero-3-phosphocholine)-DPPC-cholesterol liposomes encapsulating three growth factors—EGF, monocyte chemotactic peptide-1 (MCP-1) and placental growth factor-1 (P1GF-1). Such a system supports bimodal release mechanisms, wherein the liposomes provide a controlled release system, while the supramolecularly assembled peptides present, in this case, the RGD (Arg-Gly-Asp) adhesion sequence, as well as enzyme-mediated degradation. The liposome-containing composite systems significantly reduced the release time for biological cargo—for EGF, 15% was released from liposomes after 7 days compared to 80% from control gels, for MCP-1, the release was 80% within 24 h from control gels, while it took 5 days for the same amount to be released from liposome-containing gels. Finally, P1GF-1-encapsulating liposomes maintained a 4–5 day delay in release consistently over 2 weeks.

Various crosslinking options to improve the complexation between liposome and scaffold have also been explored [105]. For instance, liposome-crosslinked hydrogels generated via Michael-type addition of maleimide-functionalized liposome crosslinkers and thiolated PEG were rendered degradable upon exposure to glutathione and shown suitable for the differential release of doxorubicin and cytochrome C cargo (Fig. 1.1) [106]. In this system, release rate constants for liposome-encapsulated doxorubicin (5.33 × 10^{-3} h^{-1}) and cytochrome c (2.64 × 10^{-2} h^{-1}) were found to be comparable whether released from the same or separate hydrogels. Moreover, the mechanism of doxorubicin release was shown to be dominated by a degradation-mediated release, while Fickian diffusion governed cytochrome c release from the hydrogel. Other studies (Fig. 1.2) have sought to improve fusion of liposomes with bacterial membranes as a potential topical strategy to treat bacterial infections, wherein egg-PC-DOTAP liposome-stabilized carboxyl-functionalized gold nanoparticles are entrapped inside thermo-gelling hydrogels to demonstrate their release into bacterial culture, followed by pH-dependent fusion with the bacterial membrane [107]. Release of the gold nanoparticle/liposomes, tested at different crosslinker concentrations (0.6%, 0.7% and 0.8%), was shown to occur at a pH = 7.4, while at a pH below the pKa of the carboxylic group (pKa ~ 5), the gold nanoparticles detached from the liposomes, with no cytotoxicity following in vitro application to a mouse skin model after 7 days.

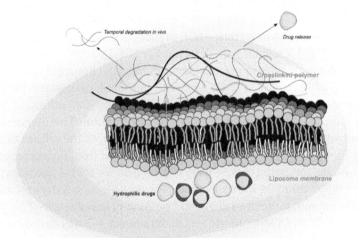

Figure 1.1 Diagram of liposome encapsulated in crosslinked chitosan nanoparticle. Hydrophilic drugs, encapsulated in the liposome interior, is released from the liposome core upon degradation of nanoparticle following delivery. Spatiotemporal release is a function of nanoparticle properties. Various crosslinkers can be used to stabilize chitosan nanoparticles, including genipin and tripolyphosphate.

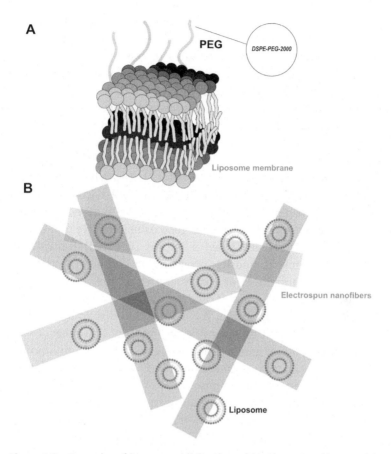

Figure 1.2 Examples of liposome stabilization with polymers or biomaterials. (A) Stealth liposomes. These liposomes are composed of lipid bilayers which contain polyethylene glycol (PEG)-conjugated lipids. A common such lipid is DSPE-PEG-2000. (B) Liposome entrapped in electrospun nanofibers. Liposomes can be immobilized on the surface of electrospun nanofibers to enhance their stability.

1.8.1 Ceramics

The development of bone substitute materials for maxillofacial, orthopedic, and reconstructive surgery [108] has been a substantial effort in regenerative medicine. Alongside bone grafts and growth factors, ceramics used for this purpose should be biocompatible and should not evoke an adverse inflammatory response. Such materials have included calcium phosphate derivatives, particularly hydroxyapatite, as well as calcium sulfate. These materials have emerged as suitable bone substitutes or bone-filling materials—bone itself contains 70% calcium phosphate mineral—owing to their favorable physicochemical properties. Calcium plays an important role in endocytosis, and has excellent biocompatibility and osteoconductivity. Bone graft substitutes based on these materials are characterized by significant structural porosity, which facilitates vascularization, a feature that has been exploited for the delivery of genetic material [109] and growth factors [110]. While structurally attractive, the potential for innate inflammatory reactions induced by calcium phosphate has shifted the attention to derivatives such as tricalcium phosphate, which has good osteoinductive properties.

Zhu et al. [111] developed a porous (0.5 mm) β-tricalcium phosphate scaffold composite incorporating liposomal gentamicin sulfate for the treatment of post-operative osteomyelitis. Release kinetics of the drug were shown to depend on the size of the liposomes, which varied from 0.1 – 5 μm, and displayed an initial burst phase followed by slow release from the liposomes. Doses of 2.5–800 μg/mL of gentamicin were tested for their anti-biofilm properties on in vitro *S. aureus* biofilms. The highest inhibiting effect was observed when liposomes were smallest (100 nm) at the lowest drug concentration (2.5 μg/mL), while at higher drug concentrations (800 μg/mL) the maximum antibacterial activity was achieved with the liposomes were approximately 800 nm in size. Elsewhere, anti-biofilm activity was also evaluated by loading cationic gentamicin-encapsulating liposomes into calcium sulfate scaffolds [112]. In vivo studies showed a 2-fold improvement in antibacterial activity compared to liposome-free scaffolds at low dosages. Zhou et al. [113] prepared ceramic scaffolds based on nano-hydroxyapatite/β-tricalcium phosphate (HA/β-TCP) encapsulating liposomal ceftazidime. The authors obtained a mean nanoparticle size of 161.5 ± 5.37 nm while encapsulation of the drug was 16.57 ± 0.13%. Antimicrobial activity on *S. aureus* biofilms was observed at concentrations as low as 6.00 μg/mL.

Wang et al. [114] developed a composite system made up of collagen and hyaluronic acid with bisphosphonate (2-(3-mercaptopropylsulfanyl)-wthyl-1, 1-bisphosphonic acid)-derivatized DSPE-PEG-cholesterol liposomes for bone regeneration.

The bisphophonate increased the liposomes' binding to bone tissue. The liposomes were engineered to encapsulate carboxyfluorescein, doxorubicin, and lysozyme. Owing to the higher affinity between the drugs and the scaffolds, bisphosphonate-derivatized composites displayed slower release kinetics of the drug from the liposomes than PEG-liposomes without bisphosphonate (while the bisphosphonate-liposomes induced up to approximately 85% release over 48 h, PEG-liposomes released >90% drug after 12 h).

1.8.2   Nanofiber-Liposome Structures

Nanofibers are a new class of materials with features that resemble the extracellular matrix. Composed of fibers that are typically less than 100 nm in diameters, nanofibers can provide a biomimetic environment suitable for a number of biomedical applications— particularly in tissue engineering—owing to their favorable physicochemical properties and innovative manufacturing [115]. Nanofibers are typically made by electrospinning synthetic, natural, or hybrid materials [116], with new material options being actively investigated [117, 118]. The morphology, high surface-to-volume ratio and fibrous structure of nanofibers, alongside the ability to manipulate and functionalize their surface properties with drugs and various biomolecules [119], has fueled the rapid ascent of these nanofabricated structures as tissue engineering supports [120] and as drug delivery systems [121] for a variety of biomedical applications [122, 123]. Bioactive molecules, such as drugs, are released from electrospun nanofibers under kinetics that are dependent on drug loading [124], surface deposition, drug aggregation and fiber diameter [125]. Because of their growing use in regenerative medicine as substrates for the regeneration of damaged tissues, nanofibers have been recognized for their ability to incorporate liposomes as multi-modal combination delivery systems.

Bioactive materials can be incorporated within nanofibers using either blend or coaxial electrospinning. In blend electrospinning, the bioactive molecule and polymer mix before the electrospinning process. In coaxial electrospinning, on the other hand, core-sheath structures are generated within individual fibers, with biomolecules within each layer designed to diffuse out sequentially [126]. Coaxial electrospinning was shown to result in the preservation of intact liposomes, unlike blend electrospinning [127]. Core-sheath features of electrospun nanofibers can avoid undesirable burst release of entrapped liposomal cargo by burying the liposomes in the core of core-sheath nanofibers [128]. Coaxially prepared electrospun nanofibers of sodium hyaluronate and cellulose acetate entrapping naproxen-loaded PC-cholesterol liposomes showed release kinetics marked by a burst release within 8 h followed by sustained release for up to 12 days in vitro measured by dialysis experiments [129]. In this case, sodium hyaluronate contributed to additional stabilization of liposomes by adsorption onto their surface. Empty nanofibers, on the other hand, displayed non-Fickian diffusion.

While most applications of liposome/nanofiber hybrid systems have by and large focused on biomedical uses, applications of these systems in the food industry have also been suggested. De Freitas Zômpero et al. [130] described the incorporation of liposomes encapsulating β-carotene within polyvinyl alcohol and polyethylene oxide electrospun fibers during the nanofiber assembly process, and revealed the contribution of liposomes to the rheological properties and size of the resultant fibers, with greater stability from UV degradation.

Drug-bearing liposomes have also been incorporated within nanofibers by immobilizing liposomes onto the surface of the fibers. Liposomes encapsulating bone morphogenic factor (BMP-2) have been covalently attached to electrospun poly(L-lactic acid) nanofibers, resulting in sustained release of BMP-2 from the nanofibers for over 21 days [131]. Elsewhere, Monteiro et al. [132] immobilized dexamethasone-encapsulated liposomes on the surface of polycaprolactone nanofibers and demonstrated sustained release over 3 weeks which resulted in successful osteogenic differentiation of human bone marrow-derived mesenchymal stem cells. Similarly, DPPC-cholesterol liposomes (2:1 mol) encapsulating gentamicin, a drug used to treat bacterial infections, were immobilized onto multi-functionalized fibers of thiolated chitosan [133]. Thiolated chitosan is attractive as a nanofiber material as it is versatile and mucoadhesive via disulfide bond formation with mucus glycoproteins allowing it to move across mucosal layers, and has favorable gelling characteristics. Gentamicin has also been encapsulated within liposomes incorporated inside thiolated chitosan nanofibers, and displayed burst-dependent initial release over 16 h, followed by sustained release up to 24 h. These hybrid composites displayed a log reduction in the amount of bacteria of 3.87 ± 0.33 for *S. aureus*, 4.87 ± 0.21 for *E. coli*, and 4.20 ± 0.24 for *P. aeruginosa*, translating to more than 99.9% kill rate.

The release kinetics of drugs encapsulated inside liposomes incorporated within nanofibers have distinct release characteristics which depend not only on the drug and liposome, but significantly on the nature, structure and physical characteristics of the nanofiber support. To study the biophysical localization of entrapped nanocarriers within electrospun nanofibers, Zha et al. [134] encapsulated liposomal cerasomes inside electrospun gelatin Type A nanofibers using a custom-built electrospinning device. Cerasomes, the size of which ranged from 70 ~ 100 nm, featured a triethoxysilyl head moiety, a hydrophobic double-chain tail, and a urea link. Fluorescent microscopy using nitrobenzoxadiazole-1,2-dipalmitoyl-sn-glycero-3-phosphoethanolamine (NBD-DPPE)-labeled lipid allowed the visualization of centrally localized cerasomes in the electrospun nanofibers. This supported the derivation of a controlled model for the centering of cerasomes within nanofibers based on cerasome geometry and composition.

The most common applications of liposome-entrapping electrospun nanofibers have related to biomedicine, particularly tissue engineering and regenerative medicine. Rampichová et al. [135] anchored fetal bovine serum-loaded liposomes to poly(2-hydroxyethyl methacrylate) (PHEMA) nanofibers, resulting in improved chondrocyte attachment and proliferation. Nanofibers have also been investigated within multi-component drug delivery systems in vivo. Filová et al. [136] developed electrospun nanofiber-based composites made of asolectin liposomes encapsulating basic fibroblast growth factor and insulin embedded in a fibrin/type I collagen/hyaluronic acid composite hydrogel. Insulin displays preferential distribution toward unsaturated phospholipid hydrocarbon chains, a fact that was confirmed in this study. In vitro observations suggested that the composites promoted superior viability and recruitment of bone marrow-derived mesenchymal stem cells onto the scaffolds over 7 days of culture, unlike control scaffolds, which displayed no such activity. When implanted into seven osteochondral defects of miniature pigs, the composite scaffold stimulated differentiation of the defect toward hyaline cartilage and fibrocartilage, which was confirmed to be type II collagen-positive.

Peptide nanofibers have been studied for their ability to form well-organized nanostructures by controllable self-assembly, without the use of electrospinning. These structures can also act as carriers for liposome-encapsulated growth factors. Wickremasinghe et al. [137] developed an angiogenic hybrid system composed of multi-domain nanofiber peptides entrapping DPPC/DPPG (1,2-Dipalmitoyl-sn-glycero-3-phosphoglycerol)/cholesterol liposomes carrying placental growth factor-1, a pro-angiogenic growth factor. These hybrid nanofiber peptide-based structures showed to be able to temporally control both in vitro and in vivo angiogenic receptor activation and promote local microvasculature. Blood vessel development was shown to begin after 2 days, with retention of mostly robust mature vessels after 10 days.

One of the major drawbacks to the electrospinning process is its relatively low productivity. However, the market of electrospinning equipment is expanding with new units now able to achieve high-throughput manufacturing, thus enabling large sheets of electrospun fibers to be made in relatively short times. Challenges associated with process control, including solvent selection and loading amounts as well as control of parameters including encapsulation efficiency, fiber diameter and uniformity, degradation rate and release kinetics, remain to be addressed. Nonetheless, the potential of incorporating liposomes into 3D nanofibers has been proven to be of great interest to both biomedical and food industries.

1.9 Liposome-Scaffold Systems for Gene Delivery

Gene therapy has shown clinical promise in treating serious disorders by delivering corrective genes to target tissues. Genetic material that has been delivered intracellularly includes plasmid DNA, oligonucleotide fragments, messenger RNA, short interfering RNA, short hairpin RNA, microRNA, peptide nucleic acids, RNA-based adjuvants, and clustered regularly interspaced short palindromic repeats/Cas gene editing systems. In order to be effective, endogenous genetic material must cross the cell membrane to reach the nucleus. This has most commonly been achieved using viral vectors. Viral vectors are considered the workhorses of gene therapy, mediating genomic incorporation of exogenous genetic material with high efficiency across a broad range of cell types. However, their use is marred by concerns over their safety, which includes potential genotoxicity and unwanted replication in the host.

Owing to rapid advances in materials synthesis and characterization, non-viral gene carriers are being developed as alternatives to viral vectors for the delivery of genetic cargo [138]. Liposomes have quickly grown to become clinically promising non-viral carriers of genetic material, as evidenced by over 100 clinical trials currently ongoing worldwide utilizing cationic liposomes as gene delivery vehicles. Cationic lipids are preferred to neutral or negatively charged lipids as they promote improved complexation with DNA as well as with mammalian cells, ultimately leading to more efficient uptake. The formation of liposome-nucleic acid complexes relies on charge-charge interactions between the positive charges on the liposome membrane and negative charges on the nucleic acid. These systems have mostly included the lipids 1,2-dioleoyl-3-trimethylammonium propane (DOTAP) and 1,2-di-O-octadecenyl3-trimethylammonium propane (DOTMA).

Similarly to liposome/scaffold complexes used in drug delivery, some of the earliest examples of combining liposomal gene delivery systems with scaffolds have been described for bone tissue engineering applications [139]. In early studies, hydroxyapatite, a common bone substitute, combined with cationic liposomes carrying bone morphogenic factor (BMP-2) cDNA showed therapeutic efficacy in vivo, upon injection of BMP-2 DNA-carrying liposome/scaffold complexes into rabbits bearing cranial defects. Significant bone formation was observed in defect sites 3 weeks following administration. Similarly, type II collagen-glycosaminoglycan scaffolds incorporating the commercial liposomal transfection reagent GenePORTER® and carrying insulin-like growth factor-1 plasmid DNA showed improved transfection of chondrocytes, compared to scaffolds carrying naked DNA [140]. This correlated to a prolonged IGF-1 overexpression during a 2-week culture period in vitro. Similar results were reported for collagen scaffolds delivering endostatin plasmid DNA to chondrocytes and mesenchymal stem cells from the same liposomal delivery reagent [141].

Strategies to improve the complexation between lipids and nucleic acids have included incorporating sticky overhangs [142], while enhancing the stability of nucleic acid/liposome carriers has been achieved by incorporating anionic biodegradable polymers such as hyaluronic acid (HA), chondroitin sulfate, or polyglutamic acid. HA is an unbranched glycosaminoglycan also present in the microenvironment of many solid tumors, and attractive as it lacks immunostimulatory CpG motifs. Choco et al. [143] developed DOTAP-cholesterol liposome-HA/siRNA composite nanoparticles. Condensation of siRNA and hyaluronic acid was achieved by the addition of protamine, while coating with PEGylated lipid containing anisamide rendered these nanoparticles amenable to direct targeting to B16F10 melanoma cells. In vivo luciferase gene silencing in a B16F10 lung metastasis model achieved suppression of 80% luciferase activity. Commercial lipofection reagent Lipofectamine 2000® was also successfully coated with an HA-based shell [4], also including alginic acid, pectin and polyglutamic acid to shield positive charges. This coating resulted in improved transfection efficiencies to a variety of cells types.

Chitosan has also been used to generate complexes with DNA-bearing liposomes. This is supported by electrostatic interactions between chitosan and DNA. Kang et al. [144] generated amine-PEG-PE (phosphatidylethanolamine)-POPC (1-palmitoyl-2-oleoyl-sn-glycero-3-phosphocholine)-POPG (1-palmitoyl-2-oleoyl-sn-glycero-3-phosphoglycerol) liposomes conjugated to targeting ligand folate encapsulating chitosan/oligodeoxynucleotide polyplexes. In vitro delivery of nucleic acid to melanoma B16F10 cells confirmed improved targeting of the complex while mediating enhanced gene transfer efficiency.

Apart from artificial scaffolds, naturally occurring protein-based materials have also been used as substrates for liposome/DNA complexes. Fibrin scaffolds—obtained by polymerization of human plasma-derived protein fibrinogen—has been successfully complexed to nucleic acid-bearing liposomes (Lipofectin®) for efficient transfection in vitro and in vivo of enhanced green fluorescent protein, luciferase and β-galactosidase plasmid DNA [145]. Other examples include the study of neurite outgrowth by transfection of NIH3T3 and HEK293T cells with nerve growth factor-encoding plasmid DNA using Lipofectamine 2000® reagent encapsulated inside PGL (polyglycerin) scaffolds.

Most examples discussed so far have included systems wherein liposome/DNA nanocarriers are embedded within the scaffold support. As an alternative complexation strategies, liposome-nucleic acid complexes can also be attached onto solid support substrates [146].

1.10   Conclusions

As next-generation drug and gene delivery systems, liposome-scaffold composites have emerged as promising platforms to achieve sustained therapeutic efficacy while avoiding the drawbacks that are associated with the use of either liposome or scaffold components alone in the delivery of cargo. Both synthetic and natural polymers have been studied for their use in such systems. Hydrogels, in particular chitosan and collagen, are among the most common materials that have so far been utilized in combination with liposomes. More recently, electrospun nanofibers—which benefit from a large surface area, high loading capacity, porosity, and encapsulation efficiency—have expanded the material repertoire. This has resulted in a wide range of options, though material choice should be informed by the application and the benefits that the use of a specific scaffold can bring to the system. This is particularly important as many polymers are therapeutically responsive. When combined with chemical modification options, the use of scaffolds as supports for the liposome-mediated delivery of bioactive materials is virtually limitless. Additionally contributing to the rapid expansion in materials options is the coming-of-age of 3D fabrication and printing technologies, which are showing remarkable advances in achieving physical architectures with previously unthinkable ease.

More clinical work is needed to fully ascertain the benefit of these combination delivery platform in vivo, as well as any long term effects. And though the goal of fully replacing living tissues and organs is still a ways away, advances in biofabrication and delivery that we are witnessing today were unthinkable only a decade ago.

References

1. Zhu J. Biomimetic hydrogels as scaffolds for tissue engineering. *J. Biochip. Tiss. Chip.*, 2, e119 (2012).

2. O'Brien F. J. Biomaterials & scaffolds for tissue engineering. *Mater. Today*, 14(3), 88–95 (2011).

3. Stratton S., Shelke N. B., Hoshino K., Rudraiah S., Kumbar S. G. Bioactive polymeric scaffolds for tissue engineering. *Bioactive Mater.*, 1(2), 93–108 (2016).

4. Wang S. S., Yang M. C., Chung T. W. Liposomes/chitosan scaffold/human fibrin gel composite systems for delivering hydrophilic drugs--release behaviors of tirofiban in vitro. *Drug Deliv.* 15(3), 149–157 (2008).

5. Hurler J., Berg O. A., Skar M., Conradi A. H., Johnsen P. J., Skalko-Basnet N. Improved burns therapy: Liposomes-in-hydrogel delivery system for mupirocin. *J. Pharm. Sci.*, 101(10), 3906–3915 (2012).

6. Jose A., Mandapalli P. K., Venuganti V. V. Liposomal hydrogel formulation for transdermal delivery of pirfenidone. *J. Liposome Res.*, 26(2), 139–147 (2015).

7. Santo V. E., Gomes M. E., Mano J. F., Reis R. L. From nano- to macroscale: Nanotechnology approaches for spatially controlled delivery of bioactive factors for bone and cartilage engineering. *Nanomedicine*, 7, 1045–1066 (2012).

8. Immordino M. L., Dosio F., Cattel L. Stealth liposomes: Review of the basic science, rationale, and clinical applications, existing and potential. *Int. J. Nanomed.*, 1(3), 297–315 (2006).

9. Matosevic S., Paegel B. M. Layer-by-layer cell membrane assembly. *Nat. Chem.*, 5, 958–963 (2013).

10. Matosevic S., Paegel B. M. Stepwise synthesis of giant unilamellar vesicles on a microfluidic assembly line. *J. Am. Chem. Soc.*, 133(9), 2798–2800 (2011).

11. Garg T., Rath G., Goyal A. K. Biomaterials-based nanofiber scaffold: Targeted and controlled carrier for cell and drug delivery. *J. Drug Target.*, 23(3), 202–221 (2015).

12. Vance D., Martin J., Patke S., Kane R. S. The design of polyvalent scaffolds for targeted delivery. *Adv. Drug. Deliv. Rev.*, 61(11), 931–939 (2009).

13. Ortan A., Parvu C. D., Ghica M. V., Popescu L. M., Ionita L. Rheological study of a liposomal hydrogel based on carbopol. *Rom. Biotech. Lett.*, 16(1), 47–54 (2011).

14. Grassi G., Farra R., Noro E., Voinovich D., Lapasin R., Dapas B., Alpar O., Zennaro C., Carraro M., Giansante C., Guarnieri G., Pascotto A., Rehimers B., Grassi M. Characterization of nucleic acid molecule/ liposome complexes and rheological effects on pluronic/alginate matrices. *J. Drug Deliv. Sci. Technol.*, 17(5), 325–331 (2007).

15. Hurler J., Žakelj S., Mravljak J., Pajk S., Kristl A., Schubert R., Škalko-Basnet N.. The effect of lipid composition and liposome size on the release properties of liposomes-in-hydrogel. *Int. J. Pharm.*, 456(1), 49–57 (2012).

16. Peppas N. A., Hilt J. Z., Khademhosseini A., Langer R. Hydrogels in biology and medicine: From molecular principles to bionanotechnology. *Adv. Mater.*, 18, 1345–1360 (2006).

17. Buhus G., Popa M., Desbrieres J. Hydrogels based on carboxymethylcellulose and gelatin for inclusion and release of chloramphenicol. *J. Bioact. Compat. Pol.*, 24(6), 525–545 (2009).

18. Zasadzinski J. A., Wong B., Forbes N., Braun G., Wu G. Novel methods of enhanced retention in and rapid, targeted release from liposomes. *Curr. Opin. Colloid In.*, 16(3), 203–214 (2011).

19. Cohen R., Kanaan H., Grant G. J., Barenholz Y. Prolonged analgesia from Bupisome and Bupigel formulations: From design and fabrication to improved stability. *J. Control. Release*, 160(2), 346–352 (2012).

20. Mourtas S., Fotopoulou S., Duraj S., Sfika V., Tsakiroglou C., Antimisiaris S. G. Liposomal drugs dispersed in hydrogels. Effect of liposome, drug and gel properties on drug release kinetics. *Colloids Surface. B*, 55, 212–221 (2007).

21. Widjaja L. K., Bora M., Chan P. N., Lipik V., Wong T. T., Venkatraman S. S. Hyaluronic acid-based nanocomposite hydrogels for ocular drug delivery applications. *J. Biomed. Mater. Res. A*, 102(9), 3056–3065 (2014).

22. Vanić Ž., Hurler J., Ferderber K., Golja Gašparović P., Škalko-Basnet N., Filipović-Grčić J. Novel vaginal drug delivery system: Deformable propylene glycol liposomes-in-hydrogel. *J. Liposome Res.*, 24(1), 27–36 (2014).

23. Tan H. W., Misni M. Effect of chitosan-modified fatty acid liposomes on the rheological properties of the carbohydrate-based gel. *Appl. Rheol.*, 24(3), 34839 (2014).

24. Hosny K. M. Optimization of gatifloxacin liposomal hydrogel for enhanced transcorneal permeation. *J. Liposome Res.*, 20(1), 31–37 (2010).

25. Hosny KM. Ciprofloxacin as ocular liposomal hydrogel. *AAPS PharmSciTech*, 11(1), 241–246 (2010).

26. Nie S., Hsiao W. L. W., Pan W., Yang Z. Thermoreversible Pluronic® F127-based hydrogel containing liposomes for the controlled delivery of paclitaxel: In vitro drug release, cell cytotoxicity, and uptake studies. *Int. J. Nanomed.*, 6, 151–166 (2011).

27. Thirumaleshwar S., Kulkarni K. P., Gowda D. V. Liposomal hydrogels: A novel drug delivery system for wound dressing. *Curr. Drug Ther.*, 7, 212–218 (2012).

28. Vashist A., Ahma S. Hydrogels: Smart materials for drug delivery. *Orient. J. Chem.*, 29(3), 861–870 (2013).

29. Vashist A., Vashist A., Gupta Y. K., Ahmad S. Recent advances in hydrogel based drug delivery systems for the human body. *J. Mater. Chem. B*, 2, 147–166 (2014).

30. Li J., Mooney D. J. Designing hydrogels for controlled drug delivery. *Nat. Rev. Mater.*, 1, 16071 (2016).

31. McKenzie M., Betts D., Suh A., Bui K., Kim L. D., Cho H. Hydrogel-based drug delivery systems for poorly water-soluble drugs. *Molecules*, 20(11), 20397–20408 (2015).

32. Hoare T. R., Kohane D. S. Hydrogels in drug delivery: Progress and challenges. *Polymer*, 49(8), 1993–2007 (2008).

33. Ashley G. W., Henise J., Reid R., Santi D. V. Hydrogel drug delivery system with predictable and tunable drug release and degradation rates. *Proc. Nat. Acad. Sci. USA*, 110(6), 2318–2323 (2013).

34. Gao D., Xu H., Philbert M. A., Kopelman R. Bio-eliminable Nano-hydrogels for drug delivery. *Nano Lett.*, 8(10), 3320–3324 (2008).

35. López-Noriega A., Hastings C. L., Ozbakir B., O'Donnell K. E., O'Brien F. J., Storm G., Hennink W. E., Duffy G. P., Ruiz-Hernández E. Hyperthermia-induced drug delivery from thermosensitive liposomes encapsulated in an injectable hydrogel for local chemotherapy. *Adv. Healthcare Mater.*, 3(6), 854–859 (2014).

36. Wang J. J., Zeng Z. W., Xiao R. Z., Xie T., Zhou G. L., Zhan X. R., Wang S. L. Recent advances of chitosan nanoparticles as drug carriers. *Int. J. Nanomed.*, 6, 765–774 (2011).

37. Li L., Zhang Y., Han S., Qu Z., Zhao J., Chen Y., Chen Z., Duan J., Pan Y., Tang X. Penetration enhancement of lidocaine hydrochloride by a novel chitosan coated elastic liposome for transdermal drug delivery. *J. Biomed. Nanotechnol.*, 7(5), 704–713 (2011).

38. Jøraholmen M. W., Vanić Z., Tho I., Skalko-Basnet N. Chitosan-coated liposomes for topical vaginal therapy: Assuring localized drug effect. *Int. J. Pharm.*, 472(1–2), 94–101 (2014).

39. Wang W., Zhang P., Shan W., Gao J., Liang W. A novel chitosan-based thermosensitive hydrogel containing doxorubicin liposomes for topical cancer therapy. *J. Biomater. Sci. Polym. Ed.*, 24(14), 1649–1659 (2013).

40. Kean T., Thanou M. Biodegradation, biodistribution and toxicity of chitosan. *Adv. Drug. Deliv. Rev.*, 62(1), 3–11 (2010).

41. Ruel-Gariépy E., Leclair G., Hildgen P., Gupta A., Leroux J. C. Thermosensitive chitosan-based hydrogel containing liposomes for the delivery of hydrophilic molecules. *J. Control. Release*, 82(2–3), 373–383 (2002).

42. Zhuang J., Ping Q., Song Y., Qi J., Cui Z. Effects of chitosan coating on physical properties and pharmacokinetic behavior of mitoxantrone liposomes. *Int. J. Nanomed.*, 5, 407–416 (2010).

43. Desbrieres J., Popa M., Peptu C., Bacaita S. Liposome loaded chitosan hydrogels. A promising delayed release kinetics mechanism. *10th World Biomaterials Congress*, Montréal, Canada, 17 May - 22 May (2016).

44. Rao K. P., Alamelu S. Effect of crosslinking agent on the release of an aqueous marker from liposomes sequestered in collagen and chitosan gels. *J. Membrane Sci.*, 71(1–2), 161–167 (1992).

45. Mengoni T., Adrian M., Pereira S., Santos-Carballal B., Kaiser M., Goycoolea F. M. A Chitosan—based liposome formulation enhances the in vitro wound healing efficacy of substance P neuropeptide. *Pharmaceutics,* 9(4), 56 (2017).

46. Li R., Deng L., Cai Z., Zhang S., Wang K., Li L., Ding S., Zhou C. Liposomes coated with thiolated chitosan as drug carriers of curcumin. *Mater. Sci. Eng. C*, 80, 156–164 (2017).

47. Jung I.-W., Han H.-K. Effective mucoadhesive liposomal delivery system for risedronate: Preparation and in vitro/in vivo characterization. *Int. J. Nanomed.*, 9, 2299–2306 (2014).

48. Li N., Zhuang C. Y., Wang M, Sui C. G., Pan W. S. Low molecular weight chitosan-coated liposomes for ocular drug delivery: In vitro and in vivo studies. *Drug Deliv.*, 19(1), 28–35 (2012).

49. Ren S., Dai Y., Li C., Qiu Z., Wang X., Tian F., Zhou S., Liu Q., Xing H., Lu Y., Chen X., Li N. Pharmacokinetics and pharmacodynamics evaluation

of a thermosensitive chitosan based hydrogel containing liposomal doxorubicin. *Eur. J. Pharm. Sci.*, 92, 137–145 (2016).

50. Alinaghi A., Rouini M. R., Johari Daha F., Moghimi H. R. Hydrogel-embedded vesicles, as a novel approach for prolonged release and delivery of liposome, in vitro and in vivo. *J. Liposome Res.*, 23(3), 235–243, (2013).

51. Kozhikhova K., Ivantsova M., Tokareva M., Shulepov I., Tretiyakov A., Shaidarov L., Rusinov V., Mironov M. Preparation of chitosan-coated liposomes as a novel carrier system for the antiviral drug triazavirin. *Pharm. Dev. Technol.*, 1–28 (2016).

52. Weiner A. L., Carpenter-Green S. S., Soehngen E. C., Lenk R. P., Popescu M. C. Liposome-collagen gel matrix: A novel sustained drug delivery system. *J. Pharm. Sci.*, 74(9), 922–925 (1985).

53. Mady M. M. Biophysical studies on collagen-lipid interaction. *J. Biosci. Bioeng.*, 104(2) 144–148 (2007).

54. Pajean M., Herbage D. Effect of collagen on liposome permeability. *Int. J. Pharm.*, 91(2–3), 209–216 (1993).

55. Nunes P. S., Albuquerque-Júnior R. L. C., Cavalcante D. R. R., Dantas M. D., Cardoso J. C., Bezerra M. S., Souza J. C., Serafini M. R., Quitans L. J. Jr, Bonjardim L. R., Araújo A. A. Collagen-based films containing liposome-loaded usnic acid as dressing for dermal burn healing. *J. Biomed. Biotechnol.*, 2011, 761593 (2011).

56. Craciunescu O., Gaspar A., Trif M., Moisei M., Oancea A., Moldovan L., Zarnescu O. Preparation and characterization of a collagen-liposome-chondroitin sulfate matrix with potential application for inflammatory disorders treatment. *J. Nanomater.*, Article ID 903691, 9 p. (2014).

57. López-Noriega A., Ruiz-Hernández E., Quinlan E., Storm G., Hennink W. E., O'Brien F. J. Thermally triggered release of a pro-osteogenic peptide from a functionalized collagen-based scaffold using thermosensitive liposomes. *J. Control. Rel.*, 187, 158–166 (2014).

58. Papi M., Palmieri V., Maulucci G., Arcovito G., Greco E., Quintiliani G. Controlled self assembly of collagen nanoparticle. *J. Nanopart. Res.*, 13(11), 6141–6147 (2013).

59. DiTizio V., Karlgard C., Lilge L., Khoury A. E., Mittelman M. W., DiCosmo F. Localized drug delivery using crosslinked gelatin gels containing liposomes: Factors influencing liposome stability and drug release. *J. Biomed. Mater. Res.*, 51(1), 96–106 (2000).

60. Nayak D., Boxi A., Ashe S., Thathapudi N. C., Nayak B. Stavudine loaded gelatin liposomes for HIV therapy: Preparation, characterization and

in vitro cytotoxic evaluation. *Mater. Sci. Eng. C Mater. Biol. Appl.*, 73, 406–416, (2017).

61. Peptu C., Popa M., Antimisiaris S. G. Release of liposome-encapsulated calcein from liposome entrapping gelatin-carboxymethylcellulose films: A presentation of different possibilities. *J. Nanosci. Nanotechnol.*, 8(5), 2249–2245 (2008).

62. Taguchi T., Endo Y. Crosslinking liposomes/cells using cholesteryl group-modified tilapia gelatin. *Int. J. Mol. Sci.*, 15, 13123–13134 (2014).

63. Dowling M. B., Lee J. H., Raghavan S. R. pH-responsive jello: Gelatin gels containing fatty acid vesicles. *Langmuir*, 25(15), 8519–8525 (2008).

64. Sun G., Mao J. J. Engineering dextran-based scaffolds for drug delivery and tissue repair. *Nanomedicine*, 7(11), 1771–1784 (2012).

65. Stenekes R. J., Loebis A. E., Fernandes C. M., Crommelin D. J., Hennink W. E. Degradable dextran microspheres for the controlled release of liposomes. *Int. J. Pharm.*, 214(1–2), 17–20 (2001).

66. Epstein-Barash H., Orbey G., Polat B. E., Ewoldt R. H., Feshitan J., Langer R., Borden M. A., Kohane D. S. A microcomposite hydrogel for repeated on-demand ultrasound-triggered drug delivery. *Biomaterials*, 31(19), 5208–5217 (2010); this article describes an innovative scaffold/ liposome composite system for drug delivery triggered via ultrasound pulses with minimal in vivo cytotoxicity.

67. Van Thienen T. G., Raemdonck K., Demeester J., De Smedt S. C. Protein release from biodegradable dextran nanogels. *Langmuir*, 23(19), 9794–9801 (2007).

68. Wang H., Dai T., Zhou S., Huang X., Li S., Sun K., Zhou G., Dou H. Self-assembly assisted fabrication of dextran-based nanohydrogels with reduction-cleavable junctions for applications as efficient drug delivery systems. *Sci. Rep.*, 7, 40011 (2017).

69. López Mora N., Hansen J. S., Gao Y., Ronald A. A., Kieltyka R., Malmstadt N., Kros A. Preparation of size tunable giant vesicles from cross-linked dextran(ethylene glycol) hydrogels. *Chem. Commun.*, 50(16), 1953–1955 (2014).

70. Kong H. J., Kaigler D., Kim K., Mooney D. J. Controlling rigidity and degradation of alginate hydrogels via molecular weight distribution. *Biomacromolecules*, 5(5), 1720–1727 (2008).

71. Torres A. L., Gaspar V. M., Serra I. R., Diogo S. G., Fradique R., Silva A. P., Correia I. J. Bioactive polymeric–ceramic hybrid 3D scaffold for

application in bone tissue regeneration. *Mater. Sci. Eng. C*, 33(7), 4460–4469 (2013).

72. Hong Y. J., Kim J. C. PH- and calcium ion-dependent release from egg phosphatidylcholine liposomes incorporating hydrophobically modified alginate. *J. Nanosci. Nanotechnol.*, 10(12), 8380–8386 (2010).

73. Liu W., Liu W., Ye A., Peng S., Wei F., Liu C., Han J. Environmental stress stability of microencapsules based on liposomes decorated with chitosan and sodium alginate. *Food Chem.*, 196, 396–404 (2016).

74. Van Elk M., Ozbakir B., Barten-Rijbroek A. D., Storm G., Nijsen F., Hennink W. E., Vermonden T., Deckers R. Alginate microspheres containing temperature sensitive liposomes (TSL) for MR-guided embolization and triggered release of doxorubicin. *PLoS ONE*, 10(11), e0141626 (2015).

75. Ullrich M., Hanuš J., Dohnal J., Štěpánek F. Encapsulation stability and temperature-dependent release kinetics from hydrogel-immobilised liposomes. *J. Colloid Interf. Sci.*, 394, 380–385 (2013).

76. Davis M. S., Marrero-Berrios I., Perez X. I., Maguire T., Radhakrishnan P., Manchikalapati D., Schianodi Cola J., Kamath H., Schloss R. S., Yarmush J. Alginate-liposomal construct for bupivacaine delivery and MSC function regulation. *Drug Deliv Trans Res.*, 8(1), 226–238 (2018).

77. Smith A. M., Jaime-Fonseca M. R., Grover L. M., Bakalis S. Alginate-loaded liposomes can protect encapsulated alkaline phosphatase functionality when exposed to gastric pH. *J. Agric. Food Chem.*, 58, 4719–4724 (2010).

78. Guo P., Liu D., Subramanyam K., Wang B., Yang J., Huang J., Auguste D. T., Moses M. A. Nanoparticle elasticity directs tumor uptake. *Nat. Commun.*, 9(1), 130 (2018).

79. Wang Y., Zhou J., Qiu L., Wang X., Chen L., Liu T., Di W. Cisplatin-alginate conjugate liposomes for targeted delivery to EGFR-positive ovarian cancer cells. *Biomaterials*, 35(14), 4297–4309 (2014).

80. Oh E. J., Park K., Kim K. S., Kim J., Yang J. A., Kong J. H., Lee M. Y., Hoffman A. S., Hahn S. K. Target specific and long-acting delivery of protein, peptide, and nucleotide therapeutics using hyaluronic acid derivatives. *J. Control. Release*, 141(1), 2–12 (2009); a comprehensive, critical review describing the use of hyaluronic acid and its derivatives as carriers for the therapeutic delivery of protein, peptide and nucleic acid pharmaceuticals.

81. Choi K. Y., Min K. H., Yoon H. Y., Kim K., Park J. H., Kwon I. C., Choi K., Jeong S. Y. PEGylation of hyaluronic acid nanoparticles improves tumor targetability in vivo. *Biomaterials*, 32(7), 1880–1889 (2011).

82. Paliwal S. R., Paliwal R., Agrawal G. P., Vyas S. P. Hyaluronic acid modified pH-sensitive liposomes for targeted intracellular delivery of doxorubicin. *J. Liposome Res.*, 26(4), 276–287 (2016).

83. Jiang T., Zhang Z., Zhang Y., Lv H., Zhou J., Li C., Hou L., Zhang Q. Dual-functional liposomes based on pH-responsive cell-penetrating peptide and hyaluronic acid for tumor-targeted anticancer drug delivery. *Biomaterials*, 33(36), 9246–9258 (2012); relevant paper describing the development of liposome/hyaluronic acid scaffold composites complexed to pH-sensitive cell penetrating peptides for enhanced targeting to tumor sites.

84. Mallick S., Park J. H., Cho H. J., Kim D. D., Choi J. S. Hyaluronic acid–ceramide-based liposomes for targeted gene delivery to CD44-positive cancer cells. *Bull. Korean Chem. Soc.*, 36, 874–881 (2015).

85. Miyazaki M., Yuba E., Hayashi H., Harada A., Kono K. Hyaluronic acid-based pH-sensitive polymer-modified liposomes for cell-specific intracellular drug delivery systems. *Bioconj. Chem.*, 29(1), 44–55 (2017).

86. El Kechai N., Mamelle E., Nguyen Y., Huang N., Nicolas V., Chaminade P., Yen-Nicolaÿ S., Gueutin C., Granger B., Ferrary E., Agnely F., Bochot A. Hyaluronic acid liposomal gel sustains delivery of a corticoid to the inner ear. *J. Control. Release*, 226, 248–257 (2016).

87. Hayward S. L., Francis D. M., Sis M. J., Kidambi S. Ionic driven embedment of hyaluronic acid coated liposomes in polyelectrolyte multilayer films for local therapeutic delivery. *Sci. Rep.*, 5, 14683 (2015).

88. Ren C. D., Kurisawa M., Chung J. E., Ying J. Y. Liposomal delivery of horseradish peroxidase for thermally triggered injectable hyaluronic acid–tyramine hydrogel scaffolds. *J. Mater. Chem. B*, 3, 4663–4670 (2015).

89. Kechai N. E., Bochot A., Huang N., Nguyen Y., Ferrary E., Agnely F. Effect of liposomes on rheological and syringeability properties of hyaluronic acid hydrogels intended for local injection of drugs. *Int. J. Pharm.*, 487(1–2), 187–196 (2015).

90. Fan Y., Sahdev P., Ochyl L. J., Akerberg J., Moon J. J. Cationic liposome-hyaluronic acid hybrid nanoparticles for intranasal vaccination with subunit antigens. *J. Control. Release*, 208, 121–129 (2015).

91. Sharma A., Bhat S., Vishnoi T., Nayak V., Kumar A. Three-dimensional supermacroporous carrageenan-gelatin cryogel matrix for tissue engineering applications. *Biomed Res. Int.*, 2013, 478279 (2013).

92. Verm A. M. L., Palani S. Development and in-vitro evaluation of liposomal gel of Ciclopirox olamine. *Int. J. Pharma. Bio. Sci.*, V1(2), 1–6 (2010).

93. Kulkarni C. V., Moinuddin Z., Patil-Sen Y., Littlefield R., Hood M. Lipid-hydrogel films for sustained drug release. *Int. J. Pharm.*, 479(2), 416–421 (2015).

94. Sekine Y., Moritani Y., Ikeda-Fukazawa T., Sasaki Y., Akiyoshi K. A hybrid hydrogel biomaterial by nanogel engineering: Bottom-up design with nanogel and liposome building blocks to develop a multidrug delivery system. *Adv. Healthcare Mater.*, 1(6), 722–728 (2012).

95. Tanrıverdi S. T., Hilmioğlu Polat S., Yeşim Metin D., Kandiloğlu G, Özer Ö. Terbinafine hydrochloride loaded liposome film formulation for treatment of onychomycosis: In vitro and in vivo evaluation. *J. Liposome Res.*, 26(2), 163–173 (2016).

96. Manca M. L., Manconi M., Valenti D., Lai F., Loy G., Matricardi P., Fadda A. M. Liposomes coated with chitosan-xanthan gum (chitosomes) as potential carriers for pulmonary delivery of rifampicin. *J. Pharm. Sci.*, 101(2), 566–575 (2012).

97. Samudre S., Tekade A., Thorve K., Jamodkar A., Parashar G., Chaudhari N. Xanthan gum coated mucoadhesive liposomes for efficient nose to brain delivery of curcumin. *Drug Deliv. Lett.*, 5(3), 201–207 (2015).

98. Ma Y. H., Yang J., Li B., Jiang Y. W., Lu X., Chen Z. Biodegradable and injectable polymer–liposome hydrogel: A promising cell carrier. *Polym. Chem.*, 7, 2037–2044 (2016).

99. Mourtas S., Fotopoulou S., Duraj S., Sfika V., Tsakiroglou C., Antimisiaris S. G. Liposomal drugs dispersed in hydrogels: Effect of liposome, drug and gel properties on drug release kinetics. *Colloid Surf. B*, 55(2), 212–221 (2007).

100. Gulsen D., Li C. C., Chauhan A. Dispersion of DMPC liposomes in contact lenses for ophthalmic drug delivery. *Curr. Eye Res.*, 30(12), 1071–1080 (2005).

101. Liu Y., Li Z., Liang D. Behaviors of liposomes in a thermo-responsive poly(N-isopropylacrylamide) hydrogel. *Soft Matter*, 8, 4517–4523 (2012).

102. Popescu M. T., Mourtas S., Pampalakis G., Antimisiaris S. G., Tsitsilianis C. H-Responsive hydrogel/liposome soft nanocomposites for tuning drug release. *Biomacromolecules*, 12, 3023–3030 (2011).

103. Aulisa L., Dong H., Hartgerink J. D. Self-assembly of multidomain peptides: Sequence variation allows control over cross-linking and viscoelasticity. *Biomacromolecules*, 10(9), 2694–2698 (2009).

104. Wickremasinghe N. C., Kumar V. A., Hartgerink J. D. Two-step self-assembly of liposome-multidomain peptide nanofiber hydrogel for time-controlled release. *Biomacromolecules*, 15(10), 3587–3595 (2014).

105. Mufamadi M. S., Pillay V., Choonara Y. E., et al., A review on composite liposomal technologies for specialized drug delivery. *J. Drug Deliv.*, 2011, 939851 (2011); comprehensive review focusing on the physicochemical aspects of the development of scaffold carriers for liposomal delivery of drugs.

106. Liang Y., Kiick K. L. Liposome-cross-linked hybrid hydrogels for glutathione-triggered delivery of multiple Cargo molecules. *Biomacromolecules*, 17(2), 601–614 (2016).

107. Gao W., Vecchio D., Li J., Zhu J., Zhang Q., Fu V., Li J., Thamphiwatana S., Lu D., Zhang L. Hydrogel containing nanoparticle-stabilized liposomes for topical antimicrobial delivery. *ACS Nano*, 8(3), 2900–2907 (2014).

108. Yuan H., Fernandes H., Habibovic H. P., de Boer J., Barradas A. M., de Ruiter A., Walsh W. R., van Blitterswijk C. A., de Bruijn J. D. Osteoinductive ceramics as a synthetic alternative to autologous bone grafting. *Proc. Nat. Acad. Sci. U. S. A.*, 107, 13614–13619 (2010).

109. Kingston R. E., Chen C. A., Okayama H. Calcium phosphate transfection. *Curr. Protoc. Immunol.*, Greene Publishing and Wiley-Interscience, New York, 10.13.1–10.13.9 doi: 10.1002/0471142735.im1013s31 (2001).

110. Bose S., Tarafder S. Calcium phosphate ceramic systems in growth factor and drug delivery for bone tissue engineering: A review. *Acta Biomater.*, 8, 1401–1421 (2012).

111. Zhu C. T., Xu Y. Q., Shi J., Li J., Ding J. Liposome combined porous beta-TCP scaffold: Preparation, characterization, and anti-biofilm activity. *Drug Deliv.*, 2010. 17(6), 391–398.

112. Hui T., Yongqing X., Tiane Z., Gang L., Yonggang Y., Muyao J., Jun L., Jing D. Treatment of osteomyelitis by liposomal gentamicin-impregnated calcium sulfate. *Arch. Orthop. Trauma Surg.*, 129(10), 1301–1308 (2009).

113. Zhou T. H., Su M., Shang B. C., Ma T., Xu G. L., Li H. L., Chen Q. H., Sun W., Xu Y. Q. Nano-hydroxyapatite/β-tricalcium phosphate ceramics scaffolds loaded with cationic liposomal ceftazidime: Preparation, release characteristics in vitro and inhibition to Staphylococcus aureus biofilms. *Drug Dev. Ind. Pharm.*, 38(11), 1298–1304 (2012).

114. Wang G., Babadağli M., Uludağ H. Bisphosphonate-derivatized liposomes to control drug release from collagen/hydroxyapatite scaffolds. *Mol Pharm.* 8, 1025 (2011).

115. Monteiro N., Ribeiro D., Martins A., Faria S., Fonseca N. A., Moreira J. N., Reis R. L., Neves N. M. Instructive nanofibrous scaffold comprising runt-related transcription factor 2 gene delivery for bone tissue engineering. *ACS Nano*, 8(8), 8082–8094 (2014).

116. Hu X., Liu S., Zhou G., Huang Y., Xie Z., Jing X. Electrospinning of polymeric nanofibers for drug delivery applications. *J. Control. Release*, 185, 12–21 (2014).

117. Suyitno S., Purwanto A., Hidayat R. L. L. G., Sholahudin I., Yusuf M., Huda S., Arifin Z. Fabrication and characterization of zinc oxide-based electrospun nanofibers for mechanical energy harvesting. *J. Nanotechnol. Eng. Med.*, 5(1), 011002 (2014).

118. Agarwal S., Greiner A., Wendorff J. H. Functional materials by electrospinning of polymers. *Prog. Polym. Sci.*, 38(6), 963–991 (2013).

119. Ramakrishna S., Fujihara K., Teo W.-E., Yong T., Ma Z., Ramaseshan R. Electrospun nanofibers: Solving global issues. *Materials Today*, 9(3), 40–50 (2006).

120. Vasita R., Katti D. S. Nanofibers and their applications in tissue engineering. *Int. J. Nanomed.*, 1(1), 15–30 (2006).

121. Son Y. J., Kim W. J., Yoo H. S. Therapeutic applications of electrospun nanofibers for drug delivery systems. *Arch. Pharm. Res.*, 37(1), 69–78 (2014).

122. Sridhar R., Venugopal J. R., Sundarrajan S., Ravichandran R., Ramalingam B., Ramakrishna S. Electrospun nanofibers for pharmaceutical and medical applications. *J. Drug. Deliv. Sci. Tech.*, 21(6), 451–468 (2011).

123. Yan E., Fan Y., Sun Z., Gao J., Hao X., Pei S., Wang C., Sun L., Zhang D. Biocompatible core-shell electrospun nanofibers as potential application for chemotherapy against ovary cancer. *Mater. Sci. Eng. C Mater. Biol. Appl.*, 41, 217–223 (2014).

124. Tunngprapa S., Jungchud I., Supapol P. Release characteristics of four model drugs from drug loaded electrospun cellulose acetate fiber mats. *Polymer*, 48, 5030 (2007).

125. Zong X., Kim K., Fang D., Ran S., Hsiao B. S., Chu B. Structure and process relationship of electrospun bioabsorbable nanofiber membranes. *Polymer*, 43, 4403 (2002).

126. Xie Q., Jia L., Xu H., Hu X., Wang W., Jia J. Fabrication of core-shell PEI/pBMP2-PLGA electrospun scaffold for gene delivery to periodontal ligament stem cells. *Stem Cells Int.*, 2016, 5385137 (2016).

127. Mickova A., Buzgo M., Benada O., Rampichova M., Fisar Z., Filova E., Tesarova M., Lukas D., Amler E. Core/shell nanofibers with embedded

liposomes as a drug delivery system. *Biomacromolecules*, 13(4), 952–962 (2012).

128. Li Z., Kang H., Li Q., Che N., Liu Z., Li P., Zhang C., Liu R., Huang Y. Ultrathin core-sheath fibers for liposome stabilization. *Colloid. Surface. B*, 122, 630–637 (2014).

129. Li Z., Kang H., Che N., Liu Z., Li P., Li W., Zhang C., Cao C., Liu R., Huang Y. Controlled release of liposome-encapsulated Naproxen from core-sheath electrospun nanofibers. *Carbohydr. Polym.*, 111, 18–24 (2014).

130. de Freitas Zômpero R. H., López-Rubio A., de Pinho S. C., Lagaron J. M., de la Torre L. G. Hybrid encapsulation structures based on β-carotene-loaded nanoliposomes within electrospun fibers. *Colloids Surf. B Biointerfaces*, 134, 475–482 (2015).

131. Mohammadi M., Alibolandi M., Abnous K., Salmasi Z., Jaafari M. R., Ramezani M. Fabrication of hybrid scaffold based on hydroxyapatite-biodegradable nanofibers incorporated with liposomal formulation of BMP-2 peptide for bone tissue engineering. *Nanomedicine*, doi: 10.1016/j.nano.2018.06.001 (2018).

132. Monteiro N., Martins A., Pires R., Faria S., Fonseca N. A., Moreira J. N., Reis R. L., Neves N. M. Immobilization of bioactive factor-loaded liposomes on the surface of electrospun nanofibers targeting tissue engineering. *Biomater. Sci.*, 2, 1195–1209 (2014).

133. Monteiro N., Martins M., Martins A., Fonseca N. A., Moreira J. N., Reis R. L., Neves N. M. Antibacterial activity of chitosan nanofiber meshes with liposomes immobilized releasing gentamicin. *Acta Biomater.*, 18, 196–205 (2015).

134. Zha Z., Leung S. L., Zhifei Dai Z., Wu X. Centering of organic-inorganic hybrid liposomal cerasomes in electrospun gelatin nanofibers. *Appl. Phys. Lett.*, 100, 033702 (2012).

135. Rampichová M., Martinová L., Koštáková E., Filová E., Míčková A., Buzgo M., Michálek J., Přádný M., Nečas A., Lukáš D., Amler E. A simple drug anchoring microfiber scaffold for chondrocyte seeding and proliferation. *J. Mater. Sci. Mater. Med.*, 23(2), 555–563 (2012).

136. Filová E., Rampichová M., Litvinec A., Držík M., Míčková A., Buzgo M., Koštáková E., Martinová L., Usvald D., Prosecká E., Uhlík J., Motlík J., Vajner L., Amler E. A cell-free nanofiber composite scaffold regenerated osteochondral defects in miniature pigs. *Int. J. Pharm.*, 447(1–2),139–149 (2013).

137. Wickremasinghe N. C., Kumar V. A., Shi S., Hartgerink J. D. Controlled angiogenesis in peptide nanofiber composite hydrogels. *ACS Biomater. Sci. Eng.*, 1(9), 845–854 (2015).

138. Nayerossadat N., Maedeh T., Ali P. A. Viral and nonviral delivery systems for gene delivery. *Adv. Biomed. Res.*, 1, 27 (2012).

139. Ono I., Yamashita T., Jin H. Y., Ito Y., Hamada H., Akasaka Y., Nakasu M., Ogawa T., Jimbow K. Combination of porous hydroxyapatite and cationic liposomes as a vector for BMP-2 gene therapy. *Biomaterials*, 25(19), 4709–4718 (2004).

140. Shah R. N., Spector M. Collagen scaffolds for nonviral IGF-1 gene delivery in articular cartilage tissue engineering. *Gene Ther.*, 14(9), 721–732 (2007).

141. Jeng L., Olsen B. R., Spector M. Engineering endostatin-producing cartilaginous constructs for cartilage repair using nonviral transfection of chondrocyte-seeded and mesenchymal-stem-cell-seeded collagen scaffolds. *Tissue Eng. Pt. A*, 16(10), 3011–3021 (2010).

142. Bolcato-Bellemin A. L., Bonnet M. E., Creusat G., Erbacher P., Behr J. P. Sticky overhangs enhance siRNA-mediated gene silencing. *Proc. Nat. Acad. Sci. U. S. A.*, 104(41), 16050–16065 (2007).

143. Chono S., Li S.-D., Conwell C. C., Huang L. An efficient and low immunostimulatory nanoparticle formulation for systemic siRNA delivery to the tumor. *J. Control. Release*, 131(1), 64–69 (2008); key paper developing one of the early systemic delivery systems composed of a mixture of liposome-protamine-siRNA-hyaluronic acid nanoparticles for tumor targeting.

144. Kang J. H., Battogtokh G., Ko Y. T. Folate-targeted liposome encapsulating chitosan/oligonucleotide polyplexes for tumor targeting. *AAPS PharmSciTech*, 15(5), 1087–1092 (2014).

145. Kulkarni M., Breen A., Greiser U., O'Brien T., Pandit A. Fibrin–lipoplex system for controlled topical delivery of multiple genes. *Biomacromolecules*, 10(6), 1650–1654 (2009).

146. Bengali Z., Pannier A. K., Segura T., Anderson B. C., Jang J. H., Mustoe T. A., Shea L. Gene delivery through cell culture substrate adsorbed DNA complexes. *Biotech. Bioeng.*, 90(3), 290–302 (2005).

Chapter 2

Elastic Liposomes for Drug Delivery

Nicole J. Bassous,[a] Amit K. Roy,[a] and Thomas J. Webster[a,b]
[a]*Department of Chemical Engineering, Northeastern University,
Boston, Massachusetts, USA*
[b]*Wenzhou Institute of Technology, Wenzhou, China*
th.webster@neu.edu

The integration of pharmaceutics, materials science, and nanotechnology has prompted the maturation of sophisticated drug delivery devices, among which liposomes have been extensively investigated. Liposomes have been acknowledged as multifaceted entities with tunable characteristics that could be adapted to help ameliorate medical conditions or to aid with clinical diagnostics. Although liposomal systems have been preeminently engineered with the object of transporting specialized therapeutic compounds, particles containing one or more lipid bilayers have been utilized extensively for modeling biological cells or vesicular systems. Following their manufacture, liposomes adopt an architectural configuration in which a hydrophobic membrane core is insulated from internal and external aqueous media by hydrophilic phospholipid headgroups. Stemming from this characteristic

Handbook of Materials for Nanomedicine: Lipid-Based and Inorganic Nanomaterials
Edited by Vladimir Torchilin
Copyright © 2020 Jenny Stanford Publishing Pte. Ltd.
ISBN 978-981-4800-91-4 (Hardcover), 978-1-003-04507-6 (eBook)
www.jennystanford.com

membrane arrangement, which permits the regulated encapsulation of hydrophilic, lipophilic, and/or hydrophobic molecules, liposomes have been recognized as versatile medical devices with broad applications. Several auxiliary initiatives are associated with the clinical initiative of applying liposomes as drug carriers. Namely, optimally engineered structures are capable of (1) regulating localized or prolonged therapeutic release, and preventing regional peaks in drug concentration intensity, (2) inhibiting dilution or degradation of active ingredients within the vasculature, and (3) restricting the cytotoxicity of pernicious therapeutics to diseased sites only. The intent of this chapter is to provide an overview of the technologies and the technical challenges associated with the production of liposomes, as well as to provide a report on the potential applicability of liposomes as clinical agents or analytical tools. Due to the protracted research history of liposomes, with their discovery dating back to 1965, key topics will be broadly addressed here, especially as they pertain to synthesis materials, routes, and the therapeutic uses of discrete particle systems.

2.1 Introduction

Cellular entities have enabled life through the compartmentalization of motile or otherwise functional biological materials. Cell-specific assemblies facilitate a cascade of complex biochemical actions and potentials that maintain life. In particular, characteristics that are fundamentally attributable to variable cell types include (1) the ability to self-replicate and grow, (2) the capacity to self-feed or otherwise acquire vital nutrients, (3) inherent chemical signaling pathways and response competencies, (4) differentiability, and (5) the potential to evolve and adjust to ambient environments or conditions. The fundamental anchor for any cell type is the semi-permeable, selcctive, and non-polar cellular membrane [1]. Amphiphilic phospholipids naturally assemble into the bilayer cell membrane that sequesters the intracellular environment of the cell [2]. This sequestration helps to localize small molecules or sub-compartments, including organelles, that interact to direct critical cellular responses. Hydrophobic interactions that transpire as a result of membrane amphiphilicity provide a driving force for the selective transfer of proteins or other molecules into or out of the cell.

The organizational framework specific to eukaryotic cells, and associated systematic functional modes, have attracted attention for over half a decade from researchers seeking to emulate mammalian cellularity for clinical applications. The fabrication of so-called "artificial cells," constructed of purely synthetic materials or hybrid biologics, is anticipated to offset a new generation of therapeutics that process or perform biological functions as the result of engineered complexes [1]. The historical basis for artificial cell engineering dates back to 1957, when medical researcher Chang proposed a generalized emulsion drop method for preparing red blood cell (RBC) enzyme and hemoglobin composites surrounded by ultrathin polymeric membranes [3]. The principles of this initial discovery have been extended to implicate variable preparation materials and methods, and the potential for bioencapsulating unique entities, including genetically engineered or otherwise standard healthy cells, has since attracted researchers to the field of artificial cellularity. Figure 2.1 provides a visual representation of the discrete sizes, membrane materials, and functionalization species that delivery vehicles can adopt.

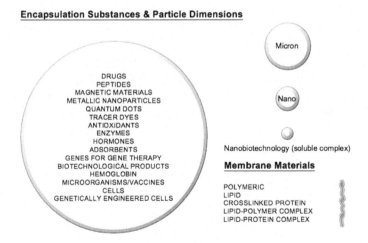

Figure 2.1 Particle size, constitution, and therapeutic load can be adapted to meet desired standards for performance and end operation.

In his monograph titled *Artificial Cells,* Chang further capitalized on the intrinsic nature of these complexes: "[The] Artificial Cell is not a specific physical entity. It is an idea involving the preparation of artificial structures of cellular dimensions for possible replacement or supplement of deficient cell functions. It is clear that different approaches can be used to demonstrate this idea [3]". Ideally, synthetic constructs will have the capacity to perform native functions that are deficient as the result of disease or to enable remedial operations that are inherently impractical for natural cellular systems functioning independently. An engineered prototype may see substantial clinical application as a targeted drug delivery device, a medical imagining diagnostic, a pseudo RBC or immune cell, a gene transplantation accessory, or as an enzyme replacement therapy for regulating the symptoms of inherited metabolic disorders [1].

For several years of research activities in diverse fields, including drug delivery, biotechnology, pharmacology, cell therapy, nanotechnology, and nanomedicine, researchers have speculated on the conceptualization of small composite structures that perform physiological functions on demand. As such, the term "artificial cell" has often been lost in the literature, supplanted instead by terminologies reflecting the vernacular of disparate scientific disciplines [1]. For example, frequently investigated delivery systems include capsules, micelles, dendrimers, silica particles, and protein carriers. Variable membrane materials lend versatility to configuration, permeability, and composition of artificial cells, and affect terminal function and functionalizability. The primary focus of this review is on a class of nanocapsules referred to as liposomes. Liposomes comprising one or more lipid bilayers have been utilized extensively for modeling biological cells or vesicular systems. In fact, following their discovery in 1965 by Bangham, Standish, and Watkins, and the conceptualization of the "liposome" moniker in 1968, liposomes have been acknowledged as multifaceted entities possessing adaptable characteristics that can be suited for the amelioration of medical complications [4, 5]. In addition to simulating the outer membrane features of living cells using liposomes organized from variable phospholipids, researchers have successfully studied membrane responsiveness to extracellular signals as well as intercellular transport of biological materials [6]. Moreover, natural cells have demonstrated the capacity to adsorb, fuse, and endocytose liposomes and/or liposomal constituents, in addition to exchanging lipids or other small molecules [6]. Refer to Fig. 2.2 for a representation of common nanoparticle types, liposome membrane configurations and morphologies, and mechanisms through which liposomes deliver their cargoes into cells.

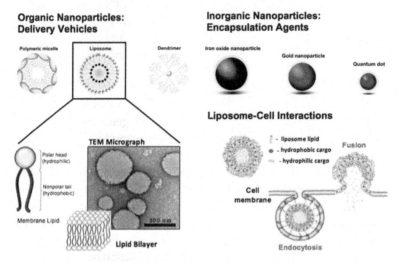

Figure 2.2 Common types of organic and inorganic nanoparticles, which can be used as drug delivery vehicles or loadable therapeutic agents, respectively, are depicted (adapted from [8], published by the Royal Society of Chemistry). The membrane aggregation of lipids to form liposomes is shown in the accompanying expansion, and a transmission electron micrograph [9], with a scale bar corresponding to 200 nm in length, demonstrates the spherical shape and average size distribution of liposome capsules. Two possible modes of action (adapted from [10]) through which cells can accept the therapeutic load of liposome carriers, namely, fusion and endocytosis, are also represented under the heading, liposome–cell interaction

Liposome constructs adhere to a basic architectural scheme, in which a hydrophobic membrane core is insulated from an internal and external aqueous environment by hydrophilic phospholipid headgroups [10]. Available aqueous compartments within the liposomal interior enable the efficient encapsulation of hydrophilic drugs or the unperturbed progression of vital enzymatic reactions [11]. The hydrophobic membrane complex can alternatively accommodate lipophilic and hydrophobic molecules, including therapeutic agents [12], organic drugs [13], or sensing probes [14]. Liposomes that are engineered to maintain polar edge functional groups have the potential for conjugation with targeting ligands or other active molecules [15]. These features essentially reinforce the claim that liposomes are versatile agents with broad medical applications. Since the 1970s, for example, the role of liposomes as analytical tools has been well publicized [16]. Liposomes have been utilized in cosmetic products in order to enhance the uptake of active ingredients related to skin care [17]. Recently, the systematic delivery of drugs and other disease therapeutics using liposomes has been investigated and brought to the clinic. Several incentives are associated with this medical initiative. Namely, tunable delivery vehicles are capable of (1) regulating localized or prolonged therapeutic release, and preventing regional peaks in drug concentration intensity, (2) inhibiting dilution or degradation of active ingredients within the vasculature, and (3) restricting cytotoxicity of relatively pernicious therapeutics to diseased sites only. The objectives of this review are to provide introductory details on the constitution of liposomes and to describe areas in which liposomes have demonstrated analytical and clinical efficacy. Due to the profoundly protracted history and the widely researched nature of liposome particles, key topics will be broadly addressed here, especially as they pertain to synthesis routes and applications.

2.2 Overview

It is widely accepted that optimum treatment of a disease requires the availability of suitable drugs which can be delivered to desired targets without any toxic side effects. One of the important carriers of drugs are liposomes. Liposomes have been extensively studied since 1970 as carriers for drugs, toxins, enzymes, proteins, peptides and other bioactive materials [18]. These are basically colloidal vesicles with lipid bilaycr membranes and aqueous interiors.

There are several advantages of liposomes as drug carriers, and these include reduced allergic or immunological responses, lower drug dosages, increased cellular permeability, and delayed drug elimination. It has been observed that liposomes vary greatly in size, lamellarity, charge, membrane fluidity, and encapsulation efficiency, which influences their functions during delivery [19]. As a result, numerous improvements have been made, thus making this technology potentially useful for the treatment of certain diseases in the clinics. Numerous liposome-based formulations have become commercially available, and several more are actively being studied in the clinic, thus substantiating the success of these particles as drug delivery vehicles.

2.2.1 Liposome Conditioning

Within the past two decades, a few major breakthroughs in liposome technology have prompted the emergence of new pharmaceutical liposomal applications. In the effort to produce delivery vehicles that are optimized to dispense their contents for maximum efficacy, strategies have been designed to temporarily augment drug permeation rates and to facilitate the targeted delivery of therapeutic agents. Novel methods to assemble next-generation liposomes for drug delivery have addressed, in particular, the regulation of biophysical parameters such as particle charge, stability, size, composition, and lamellar density [20].

Various types of liposomes have been used as carriers for drugs, enzymes, proteins, peptides, toxins, and other bioactive materials. The success of a liposome as a drug carrier is contingent on the drug encapsulation and retention efficiencies. A high ratio of drug-to-lipid per liposome is desired in order to maximize delivery of the target compound, to reduce the possibility of lipid-induced toxicity in patients, and also to lessen the cost of the treatment. Commonly, liposome encapsulation optimization research is based on trial-and-error, rather than on a complete analysis of influencing parameters. However, since the encapsulation efficiency of a system dictates whether the liposome–drug combination is suitable for administration in the clinic, factors affecting both the liposome and the drug need to be assessed early in the research process. Overall, factors that affect the encapsulation of drugs in liposomes include liposome size and type, charge on the liposome surface, bilayer rigidity, method of preparation, remote loading, ion pairing, the use of complexing agents, and characteristics of the drug to be encapsulated [21]. A few of these are described in this review.

2.2.2   The Translation of Nanotechnology to Liposome Delivery Systems

The use of nano-sized liposomes for drug delivery has shown significant promise for the intracellular delivery of antigenic and adjuvant materials. Due to the physicochemical properties of phospholipids (particularly the hydrophobicity of the hydrocarbon tail and hydrophilicity of the polar head group), liposomes are formed through the self-assembly of phospholipids into large unilamellar vesicles (LUVs) in the presence of aqueous solutions. Hence, drug candidates (antigenic/adjuvant materials in this case) with hydrophobic structures similar to those of the hydrocarbon tails may be embedded within the phospholipid bilayer of the liposome during self-assembly. In doing so, the drug candidate has the potential for increased cellular-uptake and greater overall efficacy; this is most likely due to the compatibility between liposomes and the phospholipid bilayer of mammalian cells.

An effective drug delivery system is a critical issue in health care and medicine. Recent advances in cell-molecular biology and nanotechnology have allowed the introduction of nanomaterial-based drug delivery systems. A major field of academic-industrial research is nanomedicine. The impact of nanomedicine on human health is enormous. While the outset for liposome product commercialization and use has been quite strong, contemporary materials science and pharmaceutical advancements are paving the way toward more sophisticated vehicle assemblies. In the biomedical field, nano-liposomes are additionally being explored for their potential to be integrated within crosslinked polymeric network hydrogels for implantation purposes. Due to the excellent properties associated with modern liposomes, including their tunable size, uniformity, high drug encapsulation, stability in serum suspensions, minimal toxicity, and stimuli responsiveness, coupled with facileness in their production, liposomes have been described as next generation drug delivery systems by many [22]. The diverse application range of liposomes supports this claim, as they have been explored not only for the delivery of bioactive compounds, but also as chemotherapy agents, diagnostic tools, and organ targeting devices. Therefore, modern research is directed toward the improvement of technological or medical shortcomings associated with liposome systems, as well as the optimization of drug encapsulation, targeted delivery, and cost parameters.

The pharmaceutical and biotechnology industries have been focusing over the past decades on the design and manufacturing of advanced drug delivery systems. Indeed, delivery technologies and nanomedicine have enabled a new era of disease treatment, governed by a rigorous regulation of therapeutic dosages, delivery rates and kinetics, and distribution sites, in addition to improved bioavailability of compounds that are hydrophobic or permeable [23].

2.3 Liposome Membrane Materials

2.3.1 Phospholipids as Basic Building Blocks

The assemblage of the lipid bilayer that embodies the principal internal framework of the liposome nanoparticle is accomplished by the incorporation of naturally occurring or synthetic phospholipid molecules into synthesis solutions, often together with supplementary molecules that promote particle stability or functionality. Phospholipids are amphiphilic in nature, and their constitution is such that two hydrophobic fatty acid chains are held to a phosphate and hydrophilic head group by ester linkages. This dual nature accorded to phospholipid molecules facilitates the potential for particle self-assembly. The non-polar hydrophobic tails pack so as to impede contiguity with an aqueous medium, and often, the polar hydrophilic head groups arrange in a manner that supports this low-energy state. In particular, liposomes conform to an organizational scheme in which an aqueous core and external environment are partitioned by a bilayer membrane. The hydrophilic headgroups form an insulation barrier by spontaneous alignment that protects the hydrophobic tails from contact with the water-rich solution both inside and outside of the particles.

2.3.2 Amphiphilicity Effects

Nanoparticles can be designed to accommodate a diverse group of therapeutic molecules, depending on their membrane properties.

In the case of nanocapsules, it is generally possible to incorporate hydrophilic and/or hydrophobic species. One conceivable advantage of a particle configuration retaining a hydrophobic core is the potential for solubilizing hydrophobic drugs within the core and, as a result, increasing the concentration and enhancing the effectiveness of this hydrophobic drug in an aqueous environment. Generally, excipients such as Kolliphor EL, which cause autoimmunity or hypersensitivity, have been used to deliver poorly water-soluble drugs [24–28]. The solubilization and delivery of hydrophobic drugs utilizing a hydrophobic particle core would help circumvent comparable undesirable reactions or side effects.

2.3.3   Integral Membrane Fabrics

Several materials can successfully self-assemble, and form the fundamental framework for guided biological, chemical, or sensory delivery [29–34]. Liposomes are very commonly prepared using phosphatidylcholine (PC) and its derivatives. PCs are zwitterionic phospholipids that are found naturally in mammalian and plant tissues, and their utility inside liposome membranes has been historically linked to organic solvent extraction from powdered egg yolks [35]. Egg PCs typically contain saturated and unsaturated hydrocarbon tails with approximately 16–18 carbons each [36]. Despite the potential for saturated fatty acyl chains to stabilize liposomal structures, the heterogeneity and susceptibility to peroxidative damage of unsaturated egg PC chains prompted investigators to identify alternative phospholipids that would be more suitable for the construction of liposome membranes [37]. Contemporary research predispositions favor the use of synthetic phospholipids that can be engineered to meet certain design specifications, such as terminal particle stability or uniformity. Moreover, after the modeling and testing of unique phospholipids inside the laboratory, strategies could be developed to produce large, highly pure yields of favorable complexes on-demand, that are not subjected to the heterogeneity conditions imposed due to the utilization of naturally occurring molecules. One group of phospholipids that have been synthetically produced and investigated are the symmetric 12, 14, 16, and 18 carbon-containing phosphocholinedipalmitoyls, which are respectively   1,2-dilauryl-sn-glycero-3-phosphocholinedipalmitoyl

(DLPC), 1,2-dimyristoyl-sn-glycero-3-phosphocholinedipalmitoyl (DMPC), 1,2-dipalmitoyl-sn-glycero-3-phosphocholinedipalmitoyl (DPPC), and 1,2-distearoyl-sn-glycero-3-phosphocholinedipalmitoyl (DSPC) [38]. More recent trends in liposome self-assembly have scrutinized the positive influence on particle performance of incorporating species having a net positive charge into the lipid bilayer, instead of zwitterionic molecules [39, 40]. An uneven charge distribution has been credited with deterring particle aggregation and favoring advantageous electrostatic interactions. In fact, this latter trait has been correlated with the improved encapsulation potential of multilamellar liposomes, as repulsive bilayer-to-bilayer forces expand the volume effective for species allocation [41]. Charged lipids that have been used to generate competent vesicles include the cationic 1,2-dioleoyl-3-trimethylammonium-propane (DOTAP) and the anionic dipalmitoylphosphatidyl-serine (DPPS) or dipalmitoylphophatidylglycerol (DPPG) [42, 43]. Refer to Fig. 2.3 for a structural representation of common lipids used to fabricate liposomes.

2.3.4 Cholesterol as a Membrane-Enhancing Additive

Often, cholesterol is incorporated into the lipid membrane as a spacer molecule in order to reduce particle permeability and to enhance the in vivo stability of the liposomes [11]. A modification like this is in line with the popularized concept of modulating lipid properties in order to generate second-generation particles having improved features and characteristics. In the situation directly correlated with cholesterol-membrane integration, the hydrophobicity of the organic sterol contributes to favorable interactions with the phospholipid core that promotes membrane stabilization. It is widely accepted within the scientific literature that the 3β-hydroxyl group found in the cholesterol molecules orients them toward the aqueous media inside and outside the bilayer membrane, whereas the steroid moiety fills defects contiguous to the hydrophobic tails [44]. It must be noted, however, that the hydroxyl function generally does not form a junction with the polar headgroups that comprise the lipid bilayer. Instead, it is theorized that the cholesterol hydroxyl groups occupy regions just short of the ester carbonyls constitutive of the hydrophilic phospholipid head [45]. As a result, the incorporation of cholesterols into lipid membranes confers some amphipathy to their hydrophobic segment.

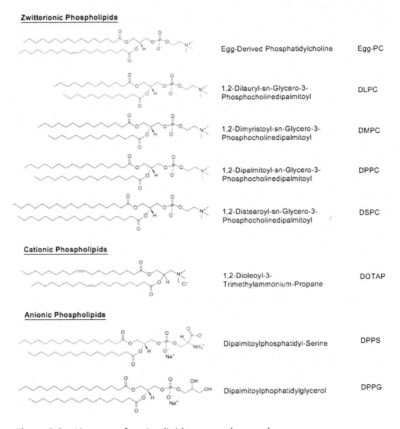

Zwitterionic Phospholipids

Egg-Derived Phosphatidylcholine   Egg-PC

1,2-Dilauryl-sn-Glycero-3-Phosphocholinedipalmitoyl   DLPC

1,2-Dimyristoyl-sn-Glycero-3-Phosphocholinedipalmitoyl   DMPC

1,2-Dipalmitoyl-sn-Glycero-3-Phosphocholinedipalmitoyl   DPPC

1,2-Distearoyl-sn-Glycero-3-Phosphocholinedipalmitoyl   DSPC

Cationic Phospholipids

1,2-Dioleoyl-3-Trimethylammonium-Propane   DOTAP

Anionic Phospholipids

Dipalmitoylphosphatidyl-Serine   DPPS

Dipalmitoylphophatidylglycerol   DPPG

Figure 2.3 Liposome-forming lipid types and examples.

Cholesterol integration contributes to more densely packed lamina, which are less likely to be taken up high- or low-density lipoproteins that naturally transport lipid molecules through the blood plasma. Moreover, cholesterol can aid in the anchoring of instrumental small molecules or polymers to liposomes [46]. These subsidiary species include deoxyribonucleic acids that enhance liposomal biosensing properties, or polymers, such as polyethylene glycol (PEG) or polyethylene oxide (PEO), that have been recognized to impede the occurrence of undesirable autoimmune responses in vivo and to increase the circulation half-life of vesicles. This ease of encapsulation supported by cholesterol supplementation inside the bilayer membrane could be attributed to covalent modifications associated with the hydroxyl function, or to structural hydrophobicity. For instance, DNA sequences that are cholesteryl-modified can be facilely incorporated into the hydrophobic cores of liposome bilayers [38]. Figure 2.4a–b depicts the possible orientations for cholesterol molecules either within liposome membrane or along their surfaces.

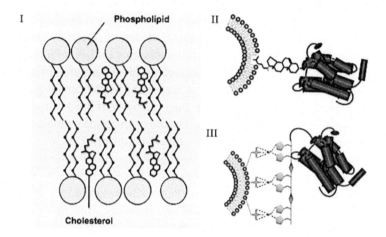

Figure 2.4 The integration of molecules or functional groups into or along liposome membranes is demonstrated, where I shows the alignment of liposome bilayers saturated with cholesterol, and II-III demonstrate the attachment of active species, such as peptides, to liposome surfaces by cholesterol anchoring or chemical conjugation, respectively [47].

2.3.5 Structural Saturation by Reactive Functional Groups

Alternatively, the covalent modification of liposomes can be directly achieved through the application of lipids having specialized functional groups. Refer to Fig. 2.4c for a visual depiction of active species chemical attachment. Covalent modification is often advocated in order to promote the attachment of active biomolecules, homing agents, or immobilization surfaces to the liposome periphery. Chemical crosslinkers that facilitate the attachment of reactive functional groups on lipids to sulfhydryls or amines include sulfosuccinimidyl-4-(N-maleimidomethyl)cyclohexane-1-carboxylate (sulfo-SMCC) and 1-ethyl-3-(3-dimethylaminopropyl) carbodiimide (EDC), respectively. Alternatively, molecules or surfaces that are modified by maleimides can readily bind to headgroups that possess the sulfhydryl functionality. Examples of lipids that retain covalently reactive species include the carboxylic acid-containing 1,2-distearoyl-*sn*-glycero-3-phosphoethanolamine-N-[(carboxy(polyethylene glycol)-2000)] (DSPE-PEG-COOH) [48], the amine-functionalized 1,2-dipalmitoyl-*sn*-glycero-3-phosphoethanolamine (DPPE) [49], and the 3-(2-pyridyldithio) propionate-modified phosphatidylethanolamine (SPDP-m PE) [15].

Covalent modifications to liposomes could be instituted prior or subsequent to particle synthesis, as prescribed by the pertinent end operation of the terminal vesicles, the sensitivity of the additives to fabrication conditions, and the effect of the conjugation on molecular functions. In the case of lipid functionalization prior to liposome production, the covalently attaching agents have the potential for incorporation within the lipid bilayer of the liposomes, or along the hydrophilic headgroups that circumscribe internal or external aqueous mixtures. Almost exclusively, however, this method is used to conjugate small molecules to the headgroup surfaces, at a displacement that is proportional with the length of any hydrophilic spacers that are used [38]. Since vesicular syntheses commonly involve the use of aggressive organic solvents or detergents, however, covalent modification after liposome assembly and purification is often favored, especially for the attachment of delicate biomolecules like antibodies or peptides. In the post-formation context, conjugating molecules are exposed to water-based liposome suspensions that are not intended for harsh conditions. However, coupling conditions must be maintained that avoid liposome aggregation, particle-to-particle crosslinking, and the loss of bilayer membrane integrity [38].

2.3.6 Environmentally Responsive Phospholipids

Other sets of lipids have been categorized according to their responsiveness to specialized signals or environmental changes. For instance, N-(4-carboxybenzyl)-N,N-dimethyl-2,3-bis-(oleoyloxy) propan-1-aminium (DOBAQ) is an ionizable lipid that responds to changes in pH [63]. Diacetylene-based liposomes may be UV-polymerized through a 1,4 addition reaction to yield colored suspensions containing robust vesicles with unique chromatic properties [51]. The intravascular extrication of vesicular loads due to temperature changes can be achieved using a DPPG-DPPC-DSPC lipid formulation [52]. Alternatively, homing devices can be appended to liposomes constitutive of unsaturated lipids, such as dioleoylphosphatidylethanolamine (DOPE), that undergo structural collapse following association with their intended target [53]. Lysolipids that lack a hydrocarbon chain have been investigated for their capacity to decrease phase transition temperatures and to contribute to an increase in the membrane permeability; in many cases, the lower solid-to-liquid phase transformation temperature facilitates the more facile production of liposomes, in addition to enhanced long-term storage stability at standard conditions [54]. Indeed, liposomes boast a wide applicability range, as prescribed by the combination of unique lipids and other molecules that are employed in their constitution.

2.3.7 Inferences on Liposome Circulation Times and Permissible Influencing Conditions

Despite the types of materials used to construct the bilayer membrane, the in vivo fate of liposome particles is subjected to dimensionality constraints. Particle size must be carefully tuned and is contingent on administration route and discharge routine. The intrinsic dimensionality of moderately sized nanoparticles brings about diminished renal clearance [55]. Cell bodies injected into or otherwise taken up by the vasculature should be designed to cap at 1–2 μm in their outer diameter in order to bypass possible agglomeration in small blood vessels and capillaries [56]. As a reference, a 10 nm effective pore size cutoff and a 30–50 kDa molecular weight cutoff are typically associated with glomerular filtration [57]. Therefore, prior to biodegradation, particles <200 nm in outer diameter will avoid filtration by the kidneys. Literature sources suggest that particles having a diameter of 200 nm or less experience an extended stay in circulation as compared to larger particles, due to a reduced rate of clearance from the body [56]. Extended circulation times may be attributed to a decline in the efficiency of opsonin binding. The high radius of curvature characteristic of smaller particles has been directly correlated to this phenomenon. This general coupling, of a sufficiently small particle diameter and extended circulation times, has been demonstrated to enhance therapeutic reservoir discharge within tissue that is presumptively diseased. Presumptively diseased tissue constitutes regions of abnormal lymphatic drainage, as in tumors or inflammatory sites, or otherwise excessively permeable vascular structures. This drug targeting strategy passively assists in the identification of disease and the concurrent evasion of healthy tissue, and is conventionally termed the enhanced permeability and retention (EPR) effect [58, 59]. In particular, as the result of a ubiquitous understanding that optimal quantum effects and enhanced surface area to volume ratios are associated with dimensions on the order of 100 nm or less, it is generally assumed that liposomes or other delivery vehicles should preferentially adopt nanodimensions [60, 61].

Particle circulation time is further augmented by the incorporation of hydrophilic PEG, or PEO, molecules, which have been until recently described as "stealth" molecules, or immune system evaders [62]. Scientific evidence does demonstrates that water inside the body forms a dense barrier around PEGylated surfaces that impedes opsonin adhesion [62]. PEG has been acknowledged as a satisfactory hydrophilic surface coating for attenuating accelerated renal clearance and evading the mononuclear phagocyte system (MPS) [62]. The ability of PEG to attract and solubilize in water molecules creates a hydration zone that is fundamentally not vulnerable to attack by certain enzymes and immune system opsonins. Many PEGylated products have demonstrated augmented therapeutic efficacy, such as extended circulation half-lives and reduced cellular toxicity, and limited initiation of a foreign body immunogenic response. However, recent literature by Lila et al. suggests that anti-PEG antibodies are evolving in synchrony with the heightened utility of PEG in medical or other products. This anomaly, and the resulting so-called accelerated blood clearance (ABC) phenomenon, suggest that PEG is not completely inert, and that the conventional stealth characteristic attributed to PEG is only transitory [62, 63]. In other words, PEG is not non-immunogenic, as has been previously claimed. In fact, anti-PEG antibodies are produced by the adaptive immune system in response to PEGylated medical products, even in individuals who have never before been exposed to PEG. Furthermore, patients who have received medical intervention involving PEGylated drugs, molecules, or devices are anticipated to experience the ABC phenomenon, in which therapeutics administered after the first dose can be rapidly cleared by the body. This is especially disappointing given that several PEG-containing products are actively being produced for clinical applications. Nevertheless, PEG substitutes, including poly(vinyl pyrrolidone) (PVP), poly(acrylic acid) (PAA), and poly(2-ethyl-2-oxazoline) (PEtOx) are being investigated [64].

2.4 Liposome Fabrication Methods

2.4.1 Synthesis Mode Selection Criteria

The manufacture of liposomes is managed using a variety of techniques conformant with the principles of vesicular self-assembly. The precepts that guide the adoption of a synthesis mechanism could depend on one or several influencing parameters at varying stages of production or clinical use. Critically, the chemical architecture of the lipids selected for liposomal assembly provide some indication about the nature of the organic medium or detergent molecules that would produce an optimal yield of appropriately dimensioned particles. Synthesis conditions are further governed by the intended end operation of the vesicular structures, since properties pertaining to particle stability, dimensionality, polydispersity, and shelf-life factor significantly into their applicability to physiological systems. From an industrial standpoint, a production method that is adaptable to the large-scale manufacture of the output species, with batch-to-batch reproducibility, could factor critically into the initial phases of liposome synthesis. Further, the physiochemical attributes of additive drugs, therapies, or molecules affects particle synthesis conditions since solvents and processes should be tuned to maximize the degree and efficiency of encapsulation. With reference to these and auxiliary clinical distribution considerations, a few common liposome fabrication routes will be outlined in this section.

A prominent consideration in the synthesis of vesicular structures is the degree of active species encapsulation. Manufacturing designs could be adapted to accommodate entrapped agents into the liposome structure through active or passive mechanisms [65]. Passive loading refers to the encapsulation of molecules or other functional materials into liposomes during the synthesis process, or the attachment of these species to lipids prior to liposome production. For instance, the Bangham method, which will be discussed in more detail, has very successfully resulted in the production of liposomes having a broad selection of encapsulated molecules, with loading efficiencies dependent on solubility of drugs inside the lipid bilayer and on the effective trapping volume of the aqueous core. In the case of water- and lipid-soluble compounds, active or remote loading refers to the transmission of these molecules into liposomes that have been pre-fabricated. For instance, hydrophilic drugs with protonizable amine groups that are dispersed inside liposome suspensions are prone to intraparticle entrapment following the induction of pH gradients [66].

2.4.2 The Bangham Method (Thin-Film Hydration)

Spherulites having a lamellar structure were first observed under an electron microscope by Alec Bangham and research colleagues in 1964 [67]. In a subsequent manuscript prepared by Bangham et al., it was reported that the mechanical agitation of dispersed phosphatidylcholine, containing or lacking cholesterols, inside an aqueous medium produced vesicles that resembled biological membranes [68]. Univalent ion exchange across the lamellae, especially of lipid particles having uniform concentrations of K^+ and Na^+ ions and in the presence of ionophores, supported the closed membrane theory. This propagated a ubiquitous interest in the field of liposomology, and several unique strategies have since been adapted by researchers to produce vesicles having distinct properties of size, membrane material, or constitution.

A prevalent synthesis approach is the Bangham method, otherwise denominated as the thin-film hydration mechanism for liposome recovery [11, 69]. Shortly, synthetic or natural lipids are dissolved in a nonpolar organic solvent, and a lipid thin film is deposited along the surface of the containment device upon evaporation of the solvent. Chloroform in its absolute form, or in mixture with methanol, is a common organic solvent used to solubilize lipids. Alternatively, pure cyclohexane and tertiary butanol have been adopted for homogenous lipid preparations. Generally, clear lipids-in-organics solutions are obtainable at concentrations within the range of 10–20 mg/ml of lipid in an organic solvent. Dry nitrogen or argon gas infiltration and rotary evaporation are conventional methods for eliminating bulk organic solvents following lipid dissolution, and residual solvents can be displaced via drying under vacuum. Upon complete lipid mixing and subsequent solvent evaporation, the system is infiltrated by a rehydrating solution that promotes thin-film dispersion and liposome self-assembly. The constitution of the hydrating agent is influenced by the desired end operation, and suitable prototypes for biological delivery include distilled water, saline or buffer solutions, and non-electrolytic media. Optimal particle formation is achieved when the temperature of the hydrating solution is maintained above the phase transition temperature of the lipid being investigated [70]. Hydration times, of typically 1 h, under continuous mechanical agitation are recommended for most lipids. Extended aging should be avoided for lipids having high gel-liquid crystal transition temperatures due to the increased possibility for hydrolysis to occur [69]. The thin-film hydration method has been credited with a partiality for promoting the construction of large multilamellar vesicles, and further processing is generally recommended in order to improve particle homogeneity and to produce smaller unilamellar structures. Mechanical techniques for particle sizing and dispersion optimization are examined in the section on sonication and extrusion. Figure 2.5(I) provides a generalized representation of the thin-film hydration process.

2.4.3 Solvent Dispersion Mechanisms

In reverse phase evaporation, phospholipids are interspersed by sonication throughout a two-phase system of admixed aqueous and organic media, and the emergent water-in-oil emulsion is disrupted by the evaporation of the organic solvent under reduced pressure [11, 69]. Prolonged rotary evaporation at low-pressure conditions is performed to ensure the removal of excess residual solvent, a process that generally transforms the lipids mixtures from a viscous gel into a liposome suspension. Refer to Fig. 2.5(II) for an illustration of the systemic changes that are typical of a reverse phase evaporation process Macromolecules with varying size distributions have been demonstrated to encapsulate into liposomes using reverse phase evaporation, with loading efficiencies of up to 65%.

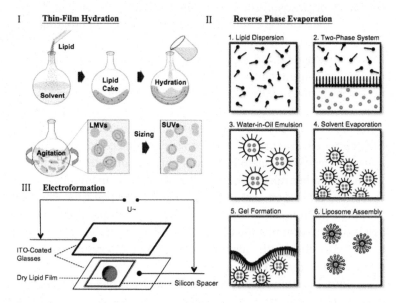

Figure 2.5 Standard liposome synthesis strategies, including: (I) thin-film hydration, (II) reverse phase evaporation, and (III) electroformation.

While several organic solvents or solvent mixtures, including isopropyl ether, isopropyl ether/chloroform, and diethyl ether, are suitable for use with the rotary evaporation method, synthesis routes that perform optimally via lipid dispersions inside specialized solvents have been adapted. In particular, strategies that implicate the controlled mixing of ether- or ethanol-and-lipid solutions into aqueous media have presented adequate liposome production [63, 65]. Summarily, the ether injection approach follows the structure of slow lipid-in-ether immersion into a hydrating solution at high temperature (55–65°C) or reduced pressure conditions, superseded by vacuum extraction of the ether. Alternatively, multilamellar vesicles instantaneously self-assemble following the rapid infusion of lipid-in-ethanol dispersions to bulk buffer solutions. Complex extractions must be performed to remove the ethanol from the terminal suspensions due to the azeotropic nature of alcohol and water mixtures, and it is generally impracticable to completely discharge ethanol from the particle system.

2.4.4   Electroformation

Electroformation has been investigated and applied as a viable formation method for generating giant vesicles, having diameters greater than 1 µm [72, 73]. In this method, electrodes are attached to a conductive coating along the bottom layer of a closed chamber, and a lipid/organic solvent solution is spread onto the conductive face of the apparatus. The chamber is next saturated with water and subjected to an alternating electrical field that quickly generates giant liposomes. Figure 2.5(III) provides an illustration of a prototypical electroformation setup. Although these vesicles are generally too large for most drug delivery or imaging applications, their dimensions are often favorable for membrane characterization and mechanical testing. Characteristic moduli that can be defined by studying larger vesicles, by either micropipette aspiration or similar techniques, include area expansion and membrane bending capacity. Giant vesicles produced using electroformation mechanisms are regularly evaluated by researchers studying the physical chemistries or physics of novel lipid bilayers.

2.4.5   Liposome Synthesis in the Absence of an Organic Solvent

Despite the reported success of these outlined methods at generating vesicular particles, it must be noted that processes that entail the direct mixing of the organic and aqueous phases are devalued due to the possibility for active ingredients to be degraded as the result of chemical exposure. This is especially true with regard to the ethanol injection strategy since biologic molecular functions are suspended under exposure to nominal ethanol concentrations. In the context of protein encapsulation, for example, organic solvents have displayed the potential to breach the hydrophobic interactions associating the nonpolar side chains of constituent amino acids. The consequential remodeling of the native protein structure augments the protein denaturation rate [74]. The sonication applied under reverse phase evaporation could also have a damaging effect on encapsulation materials. Moreover, the liposomal populations produced by ether or ethanol injection into an aqueous medium are generally heterogeneous and demonstrate high polydispersities. These deficiencies are aggravated by the safety concerns involved with the handling of organic solvents.

Detergents have been used to supplant conventional organic solvents as lipid solubilizing agents [69]. When the critical micelle concentration of a detergent has been approached or exceeded, removal of the detergent by dialysis, gel chromatography, or microsphere adsorption prompts the organization of large unilamellar vesicles. This method is credited with promoting good particle reproducibility and homogeneity, although trace volumes of the cleansing agent are undesirably maintained within the aqueous internal environment of the liposomes. Studies devised to evaluate the performance of liposomes in the presence of detergents, even in trace quantities, have concluded that vesicular functionality is effectively diminished as the result of changes in membrane permeability or conductance, or increased aggregation of encapsulated substances [74]. In consequence, alternative liposome synthesis strategies have been formulated that circumvent the requirement for organic solvents as dispersive media.

In a modified synthesis approach developed by Zheng et al. that does not involve the use of an organic solvent, liposome vesicles are assembled following a series of dispersion, homogenization, dehydration, and rehydration steps [74]. Specifically, as in the system investigated by Zheng and colleagues, the novel synthesis route is initiated with the addition of a surfactant, egg phosphatidylcholine, to deionized water at 3.0–10.0 wt%, and heating of the suspension to 60°C. A separate mixture containing cholesterol, dicetyl phosphate, and α-tocopherol is vortexed and then injected into the pre-heated egg phosphatidylcholine suspension under vigorous stirring. Following 30 min of rotary agitation with a magnetic bar, a milk-like suspension is obtained. Repeated passage of the mixture through a microfluidization apparatus with internal pressure differences of up to 14,000 psi yields a homogenous suspension, which can be manipulated further to facilitate the incorporation of active species into the liposomes. The lipids are dehydrated using an evaporator operating at 60°C, and the resulting thin film is rehydrated at room temperature using an aqueous solution containing the encapsulation species of interest. Post-processing measures include vigorous liposomal agitation for 30 min with a vortex mixer, followed by additional homogenization and possible particle washing and concentration. This novel synthesis method is amenable toward the incorporation of intact peptides or protein drugs into liposomal carriers, as substantiated by the absence of any organic solvent. In addition to maximizing the quality of encapsulation substances, the omission of an organic solvent was determined to improve liposome membrane stability, to contribute to the optimization of the active species encapsulation, and to minimize the risk and cost associated with process scale-up.

Table 2.1 Tabulation of the surface tension values, in units of dyne/cm, for different ratios of egg-PC mixed in water

Bulk solution	Temperature (°C)	Contact phase	Surface tension (dyne/cm)
Deionized Water	25	Wet air	71.0
Deionized Water/1.5 wt.% egg-PC	25	Wet air	25.0
Deionized Water/3.0 wt.% egg-PC	25	Wet air	25.0

2.4.6 Particle Sizing: Sonication and Extrusion

Regardless of the synthesis approach that is selected, generic post-formation strategies can be applied to improve the homogeneity of a liposome suspension, or to optimize constituent particle dimensions. These sizing strategies conventionally incorporate some form of particle sonication, passage through an extruder, or pressurized homogenization [70].

The application of sonic energy to liposome suspensions generates membrane disruptions that promote structural reconstruction. The high-energy input to large multilamellar vesicles generates an environment in which small unilamellar vesicles, having average size distributions within the range 15–50 nm, are favored [69]. Bath or probe tip sonicators are suitable for use on lipid suspensions. However, in the latter sonication approach, contamination of the liposome suspension by titanium particulates detached from the probe tip, or degradation of the lipid constituents due to process overheating, are common issues. More frequently, nanoparticles are loaded into containment devices, i.e., test tubes or vials, which are then fixed inside pre-heated bath sonicators. In the preparation of small unilamellar vesicles, sonic energy is normally supplied to liposome suspensions for 5–10 min at temperatures exceeding the liquid crystal transition temperatures of the membrane lipids. Perceivable visual cues, especially, variations in the opacity of the suspension during the sonication process, indicate the size redistribution of the nanoparticles. For instance, suspensions containing larger particles are more clouded, relative to those containing smaller particles, since they tend to scatter light to a greater extent. Following the sonication procedure, centrifugation is performed to eliminate any residual large multilamellar vesicles. The nature of the terminal particles is influenced by sonication tuning parameters, including exposure time and temperature, applied power, and bath volume. The particle composition and concentration prior to sonication play significantly into the final mean size and distribution [69. Due to the improbability of precisely replicating fabrication conditions from process-to-process, sonication strategies often yield product batches with marginally disparate size distributions of the constituent particles.

Alternatively, in the production of particles having reduced diameters, extrusion of liposomes through apparatuses having fixed orifices has been investigated. Specifically, in the extrusion method, large particle suspensions are passed through fixed-diameter valves at high pressures, using hydraulic pumps, as in the case of the French pressure cell press, or mechanical force [69]. Extrusion temperatures are often regulated at around 4°C, although the temperature can be adapted to approach the ideal synthesis conditions or to satisfy the constraints of the available equipment. Operations that implicate particle extrusion are generally rapid and reproducible, although particle processing times could be substantial since usable apparatuses typically can handle only small working volumes per run. Similar to extrusion, pressurized homogenization employs a specialized valve that partitions the issuing particles. These methods generate liposomes that are more uniform diametrically, as compared to sonication, and with a detectably larger radius. The larger radius implies a reduced curvature, which minimizes the possibility for particle breakage during storage at temperatures that are below the phase transition temperature of the constituent lipids.

2.5 Applications

Liposomal drug delivery systems have come of age since their humble beginnings over five decades ago when the British hematologist Dr. Alec D Bangham in 1964 first described them. The bilayer structures are called liposomes and the monolayer structures are called micelles. Liposomes are used for drug delivery due to their unique properties. In fact, they can contain a wide variety of hydrophilic and hydrophobic diagnostic or therapeutic agents, providing a larger drug payload per particle and protecting the encapsulated agents from metabolic processes. Over a dozen liposome-based drug delivery systems are currently approved by the FDA, with many more in various stages of development.

As described earlier, therapeutically, liposomes are used as carriers for biologically active molecules and are nontoxic to humans. General applications of liposomes include their role as drug delivery vehicles for medicine, adjuvants in vaccinations, signal enhancers/carriers in medical diagnostics and analytical biochemistry, solubilizers as well as support matrices for various ingredients, and penetration enhancers in cosmetics [75].

2.5.1 Applications of Liposomes in Medicine

Liposome encapsulation can alter the spatial and temporal distribution of the encapsulated drug molecules in the body, which may significantly reduce unwanted toxic side effects and increase the efficacy of the treatment. The utility of liposomes as therapeutic or diagnostic agents has been extensively researched, and the wide selection of elements and drugs available for loading into specially designed membrane structures has enabled the acquisition of substantial data about the interactions of disparate materials. Moreover, liposomes have been adapted as tools or models for studies at cellular interfaces and as recognition platforms [76]. Fundamentally, the in vivo fate of particles and their interactions with the biological circuitry define their usefulness and potential to enter the clinic. Nevertheless, a commonly observed phenomena surrounding cell-liposome relations is the transferal of therapeutic agents by either adsorption or endocytosis [77].

2.5.2   Liposomes in Anticancer Therapy

The non-specific distribution of cytotoxic anti-cancer drugs using conventional treatment modes can produce several side effects, as both malignant and normal cells are exposed. It has been demonstrated in the literature that the entrapment of these anti-cancer agents inside liposomes increases drug flow life span, improves deposition onto infected tissues, and serves as a defense from metabolic degradation of the drug. Moreover, targeting and controlled release properties of appropriately functionalized liposomes promote the improved uptake of the drug by organs that are rich in mononuclear phagocytes, such as the liver, spleen, and bone marrow, as well as decreased consumption in the kidney, brain, and myocardium [78].

Cytotoxic drugs can distribute non-specifically throughout the body and cause the death of normal as well as malignant cells, thereby giving rise to a variety of side effects. The entrapment of these drugs into liposomes results in increased flow life span, improved deposition in the infected tissues, and defense from the drug metabolic deprivation. Moreover, altered tissue release of the drug could aid in the improved uptake of vital constituents by organs rich in mononuclear phagocytic cells (liver, spleen and bone marrow) and decreased uptake in the kidney, myocardium, and brain [78]. To target tumors, liposomes must competently leave the blood and access the acidic microenvironment of tumor tissue. Liposomal entrapment of these drugs, including doxorubicin, showed reduced cardiotoxicity and dermal toxicity, in addition to better survival of experimental animals compared to the controls receiving free drugs [79]. Furthermore, a significant increase of the liposomal drug in tumor tissue was observed when TNF-α was co-administered [80]. The antitumor effects of liposomes against solid tumors were superior to those of only the TNF solution. The selective targeting of liposomal anticancer drugs to malignant cell antigens through the chemical attachment of specialized ligands to liposome membranes additionally contributes to improved efficacy of the particles, in addition to reduced toxicity toward healthy cells which is extensively observed among chemotherapy patients [78].

However, modifications are being explored in order to ensure the desired action of species that are enclosed by liposomes. Very early studies mostly showed decreased toxicity of the liposome-encapsulated drug, but in most of the cases, the drug molecules were not bioavailable [81], resulting not only in reduced toxicity but also in severely compromised efficacy. Unfortunately, this was also found to be true for primary and secondary liver tumors. In many cases it was shown that liposome formulations of various anticancer agents were less toxic than the free drug [82].

2.5.3   Immune Liposomes

An increased number of studies have indicated the possible use of liposomes in immunopotentiation [83]. The biodegradable, non-toxic, and immunostimulating properties of liposomes render them highly valuable as potential vaccine carriers for human use. The association of a protein antigen with liposomes internally (entrapment) or externally (surface-association) has been shown to be effective in promoting antibody formation [84].

Another possible function of liposomes in medicine is the potentiation of the immune response by acting as an immunological adjuvant. The reconstitution of antigens into liposomal membranes or their incorporation into the interior water core of the liposome would cause the enhancement of the immune response such as macrophage activation and antibody production [85]. Due to their special properties, liposomes are being studied for the treatment of other diseases, genetic therapies, vaccinations, and for diagnostic applications. Examples are liposomal antibiotics, antivirals, prostaglandins, steroidal and non-steroidal anti-inflammatory drugs, and insulin, among others.

2.5.4   Antimicrobial Therapy

A wide variety of bacteria are responsible for producing disease in humans by compromising organs or organ systems. Critical ailments such as tuberculosis, meningitis, and pneumonia are aggravated by several bacterial types, including *Mycobacterium tuberculosis, Neisseria meningitides,* and *Streptococcus pneumonia,* among others. Cell membranes are relatively impermeable to many antibiotics, and this circumstance, compounded by the toxicity of antibiotics, disallows the use of many antibiotics for intracellular bacterial infections. Liposomal encapsulation of toxic antibiotic drugs may be used in order to allow intracellular antibiotic delivery and thus increase the toxic antibiotic's therapeutic activity against intracellular pathogens while reducing unwanted side effects.

Researchers investigating liposome technologies have designed novel strategies for inhibiting bacterial proliferation and infectivity potential. For instance, in a recent study performed by Bartomeau Garcia et al., composite lipids having the constitution 1,2-dioleoyl-*sn*-glycero-3- phosphoethanolamine (DOPE): dipalmitoylphosphatidylcholine (DPPC): cholesteryl hemisuccinate (CHS): polyethylene glycol (PEG) in a 40%:15%:35%:10% molar ratio were used to fabricate anti-meningococcal liposomes through a thin-film rehydration mechanism [86]. Specifically, it was demonstrated by the authors that Tat peptide-functionalized liposomes enclosing carefully tuned concentrations of vancomycin or methicillin efficiently inhibited the growth of multi-drug resistant bacteria without impairing the proliferation of healthy astrocytes and endothelial cells.

The treatment of mycobacterial infections differs from that of other bacterial diseases because of several properties possessed by the mycobacteria and the host [87]. In a simple in vitro culture, liposomal neomycin and penicillin were found to be active beside bacteria, while the liposome trap clearly restricted the antimicrobial motion of chloramphenicol [78]. Liposome encapsulation changes the tissue distribution of gentamicin when given by the intravenous route to rabbits [78]. Compared to free rifabutin, the incorporation of rifabutin into liposomes produced a significant enhancement in the activity against *Mycobacterium avium* infection [78]. Furthermore, rifampin encapsulation inside egg-PC-based liposomes resulted in a significant increase in antitubercular efficacy [78].

2.5.5 Antiviral Therapy

The intracellular delivery of novel macromolecular drugs against viruses, including antisense oligodeoxynucleotides, ribozymes and therapeutic genes, may be achieved by encapsulation in or association with certain types of liposomes [88]. Liposomes may also protect these drugs against nucleases. Low-molecular-weight, charged antiviral drugs may also be delivered more efficiently via liposomes. Liposomes have been targeted to HIV-1-infected cells via covalently coupled soluble CD4 [88]. Human monocyte-derived macrophages produced less HIV-1 upon exposure to negatively charged multilamellar liposomes that were loaded with an HIV-1 protease inhibitor; in fact, the effect of the liposome-loaded drug was over ten times more potent than the free drug, and a lower EC90 was additionally observed. Alternatively, sterically stabilized liposomes that were loaded with the drug behaved just as well as the free drug. When delivered in pH-sensitive liposomes to HIV-2-infected macrophages, the reverse transcriptase inhibitor 9-(2-phosphonylmethoxyethyl)adenine (PMEA) had an EC50 value that was one order of magnitude lower. HIV-1 replication in macrophages via the delivery of a 15-mer antisense oligodeoxynucleotide inside pH-sensitive liposomes was diminished, where generally this oligonucleotide displays little to no activity at restricting viral reproduction in its free form. Oligodeoxynucleotide-encapsulated liposomes with sterically stabilized and pH-responsive membranes were additionally developed by the authors, and their prolonged in vivo performance was exceptional.

2.5.6 Ultrasonic Imaging

In more recent developments, researchers have investigated the utility of liposomes and other micro- or nano-bubbles in enabling ultrasound scattering via their ability to serve as echo contrast enhancers [89, 90]. Summarily, particle suspensions of gas micro- or nano-bubbles are injected into a vein that is subjected to an applied acoustical field. The particles' sound echo reaction aids in the acquisition of high-contrast and quality ultrasound images that enable enhanced blood flow characterization. Echogenic vesicles, including liposomes, can be chemically appended with antibodies that identify and bind atherosclerotic lesions, thrombi, plaques, and/or the unique features associated with these defects. Researchers have successfully demonstrated that atheroma fibers and thrombi are targeted by antifibrinogen-bound liposomes, and maturing atheromas are targeted by anti-ICAM-1-bound liposomes. The augmented ultrasonic imaging enabled by liposomal utility results in image panels that more coherently illustrate the landscape of targeted structures, as compared with direct backscatter images.

2.6 Conclusions and Future Directions

The progressiveness guiding contemporary research in drug delivery is by nature multidisciplinary. Advances in drug delivery will inevitably yield mature medical devices that greatly outperform contemporary analytical, treatment, and diagnostic methodologies. Specifically, natural and synthetic lipids, derivatized for their responsiveness to nano-assembly fabrications, are anticipated to expand the scope of competent tools available for various clinical applications. In this review, phospholipid and liposome particle types, as well as their manipulations to achieve therapeutic, biorecognition, and/or sensory benefits, were examined. Methodologies were outlined, that address the lipid modifications conducive toward structural improvement or functionalization, especially, by the incorporation of entrapment agents or stabilizers. It is clear from the status of contemporary liposome and nanoparticle research that the future of medicine may be concentrated within the sphere of influence encompassing nanotechnology and targeted delivery. Indeed, the conjugation of species like analytes, including amino acids, vitamins, and inorganic ions; disease inhibitors, from specialized antimicrobial peptides to anti-cancer drugs; and imaging agents, such as quantum dots or superparamagnetic metal nanoparticles, to biomimetic, biodegradable, and biofunctional liposomes implies the availability of a series of constructs that could be used to remedy medical dilemmas. The therapeutic competency of liposome particles has been augmented in recent years by the fabrication of compartmentalized particle systems, using layer-by-layer formation techniques, which can accommodate multiple conjugates per particle. Engineered compartmentalization has been credited with facilitating the generation of a new class of particles capable of performing complex functions. Complementary discoveries in synthetic organic chemistry are expected to improve the efficacy and stability of carrier/entrapment species interactions.

For instance, small molecule reactions utilizing recently elucidated concepts from "click" chemistry have been applied to liposomes, particularly in the conjugation of such species as sugars, peptides, and aptamers. Novel microfluidic-based techniques for synthesizing liposomes are regarded as suitable solutions for the optimization of product quality. The adaptability afforded to liposomes due to complex species encapsulations is amplified by research studies that examine novel controlled release mechanisms. Examples of liposomes that discharge their contents on demand, through unique process applications, are pervasive in the literature. Contingent on the constituent lipid properties and on any structural adaptations, triggered release has been accomplished through the application of magnetic fields or the incorporation of coordination complexes, modifications in the bulk solution pH or temperature, and light reactions. Despite the reported drawbacks that have disabled the transition of liposome formulations to the clinic, current research is avidly directed toward enhancing particle specificity and facilitating the transition to large-scale production. Admissible oversight entails the facile production and enrichment of bulk quantity liposome suspensions, which are refined to express greater stabilities and reduced polydispersities. Moreover, features to minimize patient sensitivity to prolonged or repeated particle exposure, i.e., as caused by an autoimmune response or toxicity effects, are being adapted. An index of safe and multifaceted particles that functionally surpass current treatment or diagnostic regimens is continuously expanding, and impendent developments are expected to lay the foundations for an innovative clinical scene in which otherwise irremediable complications or conditions are medicated through the specialized tailoring of drug delivery systems.

References

1. T. Ming Swi Chang, *Artif. Cells, Blood Substitutes Biotechnol.*, 2007, **35**, 545–554.
2. C. Martino, L. Horsfall, Y. Chen and E. Al., *ChemBioChem*, 2012, **13**, 792–795.
3. T. Ming Swi Chang, *Artificial Cells*, Charles C. Thomas, Springfield, IL, 1972.

4. A. D. Bangham and R. W. Horne, *J. Mol. Biol.*, 1964, **8**, 660–668.

5. A. D. Bangham, M. W. Hill and N. G. A. Miller, in *Methods in Membrane Biology*, 1974, pp. 1–68.

6. D. LASIC, in *Liposomes: From Physics To Applications*, 1992, vol. 80, pp. 20–30.

7. D. A. Richards, A. Maruani and V. Chudasama, *Chem. Sci.*, 2017, **8**, 63–77.

8. A. F. A. Aisha, A. M. S. A. Majid and Z. Ismail, *BMC Biotechnol.*, DOI:10.1186/1472-6750-14-23.

9. O. A. Andreev, D. M. Engelman and Y. K. Reshetnyak, *Front. Physiol.*, 2014, 5 March.

10. D. Papahadjopoulos and H. K. Kimelberg, *Prog. Surf. Sci.*, 1974, **4**, 141–144.

11. G. Bozzuto and A. Molinari, *Int. J. Nanomed.*, 2015, 10, 975–999.

12. C. M. Colley and B. E. Ryman, *Trends Biochem. Sci.*, 1976, 1, 203–205.

13. M. Alavi, N. Karimi and M. Safaei, *Adv. Pharm. Bull.*, 2017, 7, 3–9.

14. M. Petaccia, C. Bombelli, F. Paroni Sterbini, M. Papi, L. Giansanti, F. Bugli, M. Sanguinetti and G. Mancini, *Sensors Actuators B Chem.*, 2017, **248**, 247–256.

15. P. Marqués-Gallego and A. I. P. M. De Kroon, *Biomed Res. Int.*, 2014, article ID 129458, 12 pages, DOI:10.1155/2014/129458.

16. R. Wei, C. R. Alving, R. L. Richards and E. S. Copeland, *J. Immunol. Methods*, 1975, **9**, 165–170.

17. G. Betz, A. Aeppli, N. Menshutina and H. Leuenberger, *Int. J. Pharm.*, 2005, **296**, 44–54.

18. P. Kulkarni, J. Yadav and K. Vaidya, *Int. J. Curr. Pharm. Res.*, 2011, **3**, 10–18.

19. C. Zylberberg and S. Matosevic, *Drug Deliv.*, 2016, 23, 3319–3329.

20. L. Sercombe, T. Veerati, F. Moheimani, S. Y. Wu, A. K. Sood and S. Hua, *Front. Pharmacol.*, 2015, 6, 286.

21. S. B. Kulkarni, G. V. Betageri and M. Singh, *J. Microencapsul.*, 1995, **12**, 229–246.

22. H. Ragelle, F. Danhier, V. Préat, R. Langer and D. G. Anderson, *Expert Opin. Drug Deliv.*, 2017, 14, 851–864.

23. A. Gazzaniga, A. Maroni, A. Foppoli and L. Palugan, *Discov. Med.*, 2006, **6**, 223–228.

24. L. Simon-Gracia, H. Hunt, P. D. Scodeller, J. Gaitzsch, G. B. Braun, a.-M. a. Willmore, E. Ruoslahti, G. Battaglia and T. Teesalu, *Mol. Cancer Ther.,* 2016, **15**, 670–680.

25. X. Shuai, T. Merdan, A. K. Schaper, F. Xi and T. Kissel, *Bioconjug. Chem.,* 2004, **15**, 441–448.

26. J. Xie and C. H. Wang, *Pharm. Res.,* 2005, **22**, 2079–2090.

27. S. Y. Kim and Y. M. Lee, *Biomaterials,* 2001, **22**, 1697–1704.

28. X. Zhang, J. K. Jackson and H. M. Burt, *Int. J. Pharm.,* 1996, **132**, 195–206.

29. K. Fujioka, *Adv. Drug Deliv. Rev.,* 1998, **31**, 247–266.

30. Y. Tabata and Y. Ikada, *Adv. Drug Deliv. Rev.,* 1998, **31**, 287–301.

31. S. Dumitriu and E. Chornet, *Adv. Drug Deliv. Rev.,* 1998, **31**, 223–246.

32. D. Quintanar-Guerrero, E. Allémann, H. Fessi and E. Doelker, *Drug Dev. Ind. Pharm.,* 1998, **24**, 1113–28.

33. G. Lambert, E. Fattal and P. Couvreur, *Adv. Drug Deliv. Rev.,* 2001, 47, 99–112.

34. K. S. Soppimath, T. M. Aminabhavi, A. R. Kulkarni and W. E. Rudzinski, *J. Control. Release,* 2001, 70, 1–20.

35. P. van Hoogevest and A. Wendel, *Eur. J. Lipid Sci. Technol.,* 2014, 116, 1088–1107.

36. S. Tajima and R. Sato, *J. Biochem.,* 1980, **87**, 123–134.

37. Y. Y. Huang, T. W. Chung and C. I. Wu, *Int. J. Pharm.,* 1998, **172**, 161–167.

38. K. A. Edwards, *Liposomes in Analytical Methodologies*, Pan Stanford Publishing, Singapore, 2016.

39. S. Lesieur, C. Grabielle-Madelmont, M.-T. Paternostre and M. Ollivon, *Anal. Biochem.,* 1991, **192**, 334–343.

40. T. Schneider, A. Sachse, G. Rößling and M. Brandl, *Int. J. Pharm.,* 1995, **117**, 1–12.

41. S. Benita, P. A. Poly, F. Puisieux and J. Delattre, *J. Pharm. Sci.,* 1984, **73**, 1751–1755.

42. D. Simberg, S. Weisman, Y. Talmon and Y. Barenholz, *Crit. Rev. Ther. Drug Carrier Syst.,* 2004, **21**, 257–317.

43. M. J. Fonseca, E. C. A. Van Winden and D. J. A. Crommelin, *Eur. J. Pharm. Biopharm.,* 1997, **43**, 9–17.

44. P. L. Yeagle, *BBA - Rev. Biomembr.,* 1985, 822, 267–287.

45. D. L. Worcester and N. P. Franks, *J. Mol. Biol.,* 1976, **100**, 359–378.

46. L. Hosta-Rigau, Y. Zhang, B. M. Teo, A. Postma and B. Städler, *Nanoscale*, 2013, **5**, 89–109.

47. D. S. Watson, A. N. Endsley and L. Huang, *Vaccine*, 2012, 30, 2256–2272.

48. R. U. Palekar, J. W. Myerson, P. H. Schlesinger, J. E. Sadler, H. Pan and S. A. Wickline, *Mol. Pharm.*, 2013, **10**, 4168–4175.

49. B. Korchowiec, M. Paluch, Y. Corvis and E. Rogalska, *Chem. Phys. Lipids*, 2006, **144**, 127–136.

50. C. L. Walsh, J. Nguyen and F. C. Szoka, *Chem. Commun.*, 2012, **48**, 5575.

51. Y. R. Kim, S. Jung, H. Ryu, Y. E. Yoo, S. M. Kim and T. J. Jeon, *Sensors (Switzerland)*, 2012, 12, 9530–9550.

52. B. Kneidl, M. Peller, G. Winter, L. H. Lindner and M. Hossann, *Int. J. Nanomed.*, 2014, **9**, 4387–4398.

53. R. J. Y. Ho, L. Huang and B. T. Rouse, *Biochemistry*, 1986, **25**, 5500–5506.

54. J. K. Mills and D. Needham, *Biochim. Biophys. Acta Biomembr.*, 2005, **1716**, 77–96.

55. M. C. Jones and J. C. Leroux, *Eur. J. Pharm. Biopharm.*, 1999, 48, 101–111.

56. O. Ishida, K. Maruyama, K. Sasaki and M. Iwatsuru, *Int. J. Pharm.*, 1999, **190**, 49–56.

57. E. Gagliardini, S. Conti, A. Benigni, G. Remuzzi and A. Remuzzi, *J. Am. Soc. Nephrol.*, 2010, **21**, 2081–9.

58. M. Yokoyama, G. S. Kwon, T. Okano, Y. Sakurai and K. Kataoka, in *Polymeric Drugs and Drug Administration*, 1994, vol. 545, pp. 126–134.

59. Y.-L. Hao, Y.-J. Deng, Y. Chen, K.-Z. Wang, A.-J. Hao and Y. Zhang, *J. Pharm. Pharmacol.*, 2005, **57**, 1279–1287.

60. M. Bardosova and T. Wagner, in *Nanomaterials and Nanoarchitectures: A Complex Review of Current Hot Topics and their Applications*, 2015, pp. 1–343.

61. D. Williams, *Biomaterials*, 2008, **29**, 1737–1738.

62. R. Gref, Y. Minamitake, M. T. Peracchia, V. Trubetskoy, V. Torchilin and R. Langer, *Science (80-.)*, 1994, **263**, 1600–1603.

63. T. Ishida and H. Kiwada, *Int. J. Pharm.*, 2008, **354**, 56–62.

64. R. P. Brinkhuis, F. P. J. T. Rutjes and J. C. M. van Hest, *Polym. Chem.*, 2011, **2**, 1449.

65. A. Akbarzadeh, R. Rezaei-Sadabady, S. Davaran, S. W. Joo, N. Zarghami, Y. Hanifehpour, M. Samiei, M. Kouhi and K. Nejati-Koshki, *Nanoscale Res. Lett.*, 2013, **8**, 1–8.

66. I. Mayer, T. Madden, M. Bally and P. Cullis, in *Liposome Technology*, vol. 11, 1993, pp. 27–44.

67. N. Düzgüneş and G. Gregoriadis, *Methods Enzymol.*, 2005, 391, 1–3.

68. A. D. Bangham, M. M. Standish and J. C. Watkins, *J. Mol. Biol.*, 1965, **13**, IN26–IN27.

69. J. Dua, A. Rana and A. Bhandari, *Int. J. Pharm. Stud. Res.*, 2012, **III**, 7.

70. S. A. Roberts, N. Parikh, R. J. Blower and N. Agrawal, *J. Liposome Res.*, 2017, **28**, 331–340.

71. C. Jaafar-Maalej, R. Diab, V. Andrieu, A. Elaissari and H. Fessi, *J. Liposome Res.*, 2010, **20**, 228–243.

72. B. M. Discher, Y. Y. Won, D. S. Ege, J. C. M. Lee, F. S. Bates, D. E. Discher and D. A. Hammer, *Science*, 1999, **284**, 1143–1146.

73. M. Angelova, S. Soleau and P. Méléard, *Trends Colloid ...*, 1992, **89**, 127–131.

74. S. Zheng, H. Alkan-Onyuksel, R. L. Beissinger and D. T. Wasan, *J. Dispers. Sci. Technol.*, 1999, **20**, 1189–1203.

75. D. Umalkar, K. Rajesh, G. Bangale, B. Ratinaraj, G. Shinde and P. Panicker, *Pharma Sci. Monit.*, 2011, **2**, 24–39.

76. Y. Zhang, B. Ceh and D. Lasic, in *Polymeric Biomaterials*, ed. S. Dumitriu, Marcel Dekker, Second, 2002, p. 809.

77. D. D. Lasic, *Handb. Biol. Phys.*, 1995, **1**, 491–519.

78. P. Goyal, K. Goyal, S. Vijaya Kumar, A. Singh, O. Katare and D. Mishra, *Acta Pharm.*, 2005, **55**, 1–25.

79. J. W. Park, *Breast Cancer Res.*, 2002, **4**, 95–99.

80. T. L. M. ten Hagen, A. L. B. Seynhaeve, S. T. van Tiel, D. J. Ruiter and A. M. M. Eggermont, *Int. J. Cancer*, 2002, **97**, 115–20.

81. T. Feng, Y. Wei, R. J. Lee and L. Zhao, *Int. J. Nanomed.*, 2017, 12, 6027–6044.

82. K. Egbaria and N. Weiner, *Adv. Drug Deliv. Rev.*, 1990, 5, 287–300.

83. N. van Rooijen, in *Vaccines. NATO Science Series (Life Sciences)*, eds. G. Gregoriadis, B. McCormack and A. Allison, Springer, Boston, MA, 1995, vol. 282, pp. 15–24.

84. C. M. E. Lutsiak, D. L. Sosnowski, D. S. Wishart, G. S. Kwon and J. Samuel, *J. Pharm. Sci.*, 1998, **87**, pp. 1428–1432.

85. C. R. Alving, *J. Immunol. Methods*, 1991, 140, 1–13.

86. C. B. Garcia, D. Shi and T. J. Webster, *Int. J. Nanomed.*, 2017, **12**, 3009–3021.

87. W. W. Barrow, *Rev. Sci. Tech.*, 2001, **20**, 55–70.

88. N. Düzgüneş, E. Pretzer, S. Simões, V. Slepushkin, K. Konopka, D. Flasher and M. C. de Lima, *Mol. Membr. Biol.*, 1999, **16**, 111–118.

89. S. Hilgenfeldt, D. Lohse and M. Zomack, *Eur. Phys. J. B*, 1998, **4**, 247–255.

90. S. Hilgenfeldt and D. Lohse, *Ultrasonics*, 2000, **38**, 99–104.

Chapter 3

Recent Advances with Targeted Liposomes for Drug Delivery

Josimar O. Eloy,[a] Raquel Petrilli,[b] Fabíola Silva Garcia Praça,[b] and Marlus Chorilli[c]

[a]School of Pharmacy, Odontology and Nursing – Federal University of Ceara, Fortaleza, Rua Capitão Francsico Pedro, 1210, 60430-372 Fortaleza, Ceará, Brazil
[b]School of Pharmaceutical Sciences of Ribeirao Preto, University of Sao Paulo, Av. Cafe s/n, 14040-903 Ribeirao Preto, São Paulo, Brazil
[c]School of Pharmaceutical Sciences, UNESP- Sao Paulo State University, Araraquara, Rodovia Araraquara-Jau, km. 1, 14801-902 Araraquara, São Paulo, Brazil
josimar.eloy@ufc.br

Liposomes are lipid vesicles that can be formed by different compositions of lipids, resulting in a variety of liposomal sizes and lipid bilayer characteristics. These nanosystems are promising candidates to deliver encapsulated drugs because they can protect it against chemical degradation, which improves the stability. Liposomal surface can also be decorated with hydrophilic polymers like polyethylene glycol resulting in prolonged circulation time due to the escape from the reticular endothelial system. Thus, after the approval of Doxil®, the first liposomal product, a variety

Handbook of Materials for Nanomedicine: Lipid-Based and Inorganic Nanomaterials
Edited by Vladimir Torchilin
Copyright © 2020 Jenny Stanford Publishing Pte. Ltd.
ISBN 978-981-4800-91-4 (Hardcover), 978-1-003-04507-6 (eBook)
www.jennystanford.com

of commercial liposomes are currently available in the market. Although liposomes can potentially accumulate in cancerous sites due to the Enhanced Permeability and Retention (EPR) effect, several strategies can be used for targeting these nanoparticles to specific disease sites, resulting in improved drug cytotoxicity and cellular internalization with minimal off-target toxicity. In this chapter, different strategies such as targeting using monoclonal antibodies or antibody fragments, aptamers, folate, and transferrin are presented and discussed. Herein we will present concepts and focus on recent in vitro and in vivo applications (2015–2018). Furthermore, some recent applications in clinical trials will be highlighted.

3.1   Introduction

Liposomes are small and spherical vesicles constituted by phospholipid bilayers containing an aqueous core. This nanocarrier is useful to encapsulate both hydrophilic and lipophilic drugs and can be classified according to its dimensions, such as <100 nm (small), 100–250 nm (medium), and >250 nm (large). Furthermore, based on the liposomal lipid bilayer characteristics, they can be uni- or multilamellar vesicles (Noble et al., 2014; Zylberberg and Matosevic, 2016; Accardo and Morelli, 2015).

Liposomes are the most versatile and well-investigated nanocarriers for targeted drug delivery, and their use in medicine is increasing rapidly (Eloy et al., 2017; Noble et al., 2014; Sercombe et al., 2015). There are many examples of liposomes that are available in the clinic. The first nanoscale drug formulation approved by the Food and Drug Administration (FDA) was Doxil® in 1995 and currently there are several liposome-based products available on the market, such as DaunoXome®, Ambisome®, Myocet®, Marqibo®, Lipodox®, and Visudyne® (Barenholz, 2012; Dou et al., 2017; Zylberberg and Matosevic, 2016).

There are several advantages of encapsulating drugs into liposomes. For instance, the liposome is able to stabilize the encapsulated drug, preventing it from chemical degradation, improving the stability. Moreover, liposomes can have their lipid bilayer modified with amphiphiles or their surface can be coated with hydrophilic polymers, such as polyethylene glycol to increase the circulation time, in order to reduce the capture by the reticular endothelial system (Zylberberg and Malosevic, 2016). Consequently, the pharmacokinetics and biodistribution is improved, leading to better pharmacological effects (Eloy et al., 2017; Sercombe et al., 2015).

The literature reports different targeting strategies with these nanosystems to optimize the selective delivery. Liposomes with an average diameter between 100 and 200 nm can extravasate and passively accumulate in leaky pathological tissues, such as tumors and inflammation sites, due to the Enhanced Permeability and Retention (EPR) (Noble et al., 2014; Sercombe et al., 2015). Moreover, many pathological sites, such as cancer cells, exhibit cell receptors that can be recognized by a wide sort of ligands. Thus, liposomes can be actively targeted to these sites by using several moieties such as peptides (Accardo and Morelli, 2015; Liu et al., 2014; Apte et al., 2014; Nahar et al., 2014), monoclonal antibodies or antibody fragments (Moles et al., 2015; Moles et al., 2017; Loureiro et al., 2015), aptamers (Plourde et al., 2017; Ara et al., 2014; Alshaer et al., 2015), folate (Guo et al., 2017; Riaz et al., 2018; Barbosa et al., 2015) and transferrin (Lopalco et al., 2018; Johnsen et al., 2017; Deshpande et al., 2018). Targeting liposomes using these ligands result in improved selectivity of drug delivery, with several outcomes, including improved drug cytotoxicity, increased cell uptake and minimal effect in off-target regions (Eloy et al., 2017; Sercombe et al., 2015).

In this chapter, we will address recent advances on targeted liposomes for drug delivery published in the last 4 years. We will outline the targeting ligands monoclonal antibodies or antibody fragments, aptamers, peptides, folate and transferrin, giving emphasis on in vitro and in vivo evaluation of targeted liposomes and we will mainly discuss their application in medicine.

3.2 Liposomes Functionalized with Monoclonal Antibodies or Antibody Fragments

Monoclonal antibodies bind to a specific epitope and are usually generated through the hybridoma technology. In order to reduce immunogenicity, antibody fragments, particularly the Fab and scFv fragments, are also available. The functionalization of liposomes with monoclonal antibodies or antibodies fragments for disease targets has been widely used as a strategy for enhanced drug delivery, with improved cell uptake and decreased side effects. These functionalized liposomes, also known as immunoliposomes, have been reported for a variety of targets, including cancer, cardiovascular, infectious, autoimmune, and degenerative diseases. Furthermore, immunoliposomes have been also employed for vaccine delivery. Recently, we reviewed the functionalization methods, which usually include a thioether covalent bond formation between a thiolated antibody and the maleimide group (Fig. 3.1) (Eloy et al., 2017).

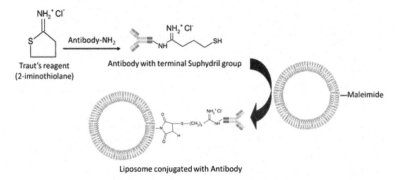

Figure 3.1 Schematic representation of immunoliposomes preparation. Reproduced with permission from Eloy et al. (2017).

Immunoliposomes have been reported as delivery systems for cancer diagnosis and therapy for a variety of targets, commonly including the HER2 and EGFR receptors (Eloy, Petrilli, Chesca, et al., 2017; Petrilli et al., 2016). The HER2 receptor, a transmembrane glycoprotein, is involved in the regulation of epithelial cell proliferation and survival, via the PI3K/AKT and RAS/MAPK pathways. HER2 is a target for breast cancer treatment, found in approximately 20% to 25% of cases and is usually associated with more aggressive tumors. The first anti-HER2 drug was the humanized monoclonal antibody trastuzumab (Herceptin®), which induces apoptosis in breast cancer cells, improving survival in metastatic patients (Tai et al., 2010; Labidi et al., 2016). We proved that trastuzumab-conjugated liposomes improved the cytotoxicity of co-delivered paclitaxel and rapamycin against HER2-positive cells (Eloy et al., 2016). Furthermore, the immunoliposomes were further demonstrated to have better cell uptake in HER2-positive SKBR3 cell line than in HER2-negative 4T1 cells and, in vivo, slowed down tumor growth in SKBR3-bearing xenograft mice compared to liposomes and free drugs (Eloy, Petrilli, Chesca, et al., 2017). A phase 2 clinical trial, HERMIONE, was planned for evaluation of the efficacy and safety of MM-302 formulation, a doxorubicin-loaded anti-HER2 scFv immunoliposome, plus trastuzumab, in locally advanced/ metastatic breast cancer; however, the trial was stopped in 2016 before conclusion, following a recent independent Data and Safety Monitoring Board (DSMB) recommendation and subsequent futility analysis (Miller et al., 2016).

Another potential target of immunoliposomes in cancer therapy is the EGFR, a member of tyrosine kinase receptors, involved in the pathogenesis of a variety of tumors, such as head and neck and colon. Clinically, the anti-EGFR human-murine chimeric IgG cetuximab (Erbitux®), has been used in combination with other drugs, such as cisplatin or 5- fluorouracil (FU), for the treatment of head and neck squamous cell carcinoma (SCC) and colorectal cancers (Normanno et al., 2006; Martinelli et al., 2009). 5-FU was encapsulated in cetuximab-functionalized liposomes in a previous paper of our research group. The immunoliposome had cytotoxic synergistic effect of cetuximab and 5-FU against A431 SCC cells (Petrilli et al., 2016). For colorectal cancer, oxaliplatin was loaded into liposomes functionalized with whole cetuximab or its Fab' fragment. Interestingly, liposomes functionalized with the fragment showed better drug accumulation in tumor tissue than liposomes conjugated with the whole antibody. Moreover, the same behavior was confirmed in colorectal cancer bearing mice, with better response for the Fab'-functionalized liposome (Zalba et al., 2015). Noteworthy, a large-scale process has been developed for the production process of EGFR-targeted immunoliposomes with several aspects covered considering the good manufacturing practice (GMP), meaning that these immunoliposomes are industrially viable (Wicki et al., 2015).

Owing to the biochemistry diversity and complexity of cancer, other targets have been reported for immunoliposomes and are presented in Table 3.1. Some papers reported that immunoliposomes had better cell uptake and enhanced cytotoxicity, while others showed significant antitumor activity following in vivo studies.

Table 3.1 Examples of targets and main results of immunoliposomes for cancer treatment

Drug	Target	Main results	Reference
Temozolomide	Blood-brain barrier and glioblastoma stem cells	In vivo study evidenced a significant reduction in tumor size after intravenous administration of immunoliposome to the orthotopically implanted brain tumor mice.	(Kim et al., 2018)
Indocyanine green (fluorescent probe) and doxorubicin	MUC-1 receptors	Rapid accumulation was observed for immunoliposomes at early time points mainly in the periphery of the tumor.	(Lozano et al., 2015)
Antisense oligonucleotide	CD-20	Selective reduction of the expression of BCL2 in target cells. Mice-engrafted tumors expressing the specific marker showed high efficiency of the liposome formulations against tumor development.	(Meissner et al., 2015)

Drug	Target	Main results	Reference
None	CD-105 and Fibroblast Activation Protein (FAP)	Bispecific scFv FAP/CD105-IL interacted stronger with cells expressing FAP and endoglin (both targets simultaneously) compared to the monospecific immunoliposomes.	(Rabenhold et al., 2015)
Paclitaxel	Vascular endothelial growth factor (VEGF)	Animals treated with mAb-liposomal paclitaxel showed better anticancer activity than the commercial formulation, Taxol, and non-targeted liposomal formulations.	(Shi et al., 2015)
Gemcitabine and bevacizumab	CD-133	The antitumor efficacy of the two drugs combined in the immunoliposome was better than monotherapy.	(Shin et al., 2015)
Doxorubicin	CD-147	Better antitumor effects of anti-CD147 doxorubicin-loaded immunoliposomes in both cancer cells and patient-derived xenograft models.	(Wang et al., 2017)

Vascular targeting is an interesting approach for both diagnosing and treating a variety of diseases that involve the endothelium, including cancer and atherosclerosis. Targets are very diverse and include the family of tumor endothelial markers (TEMs, TEM8 especially), cell adhesion molecules (CAMs), vascular/epidermal growth factors and their receptors (V/EGF and V/EGFR), selectins, cell surface nucleolin, and fibrin/fibronectin complexes (Atukorale et al., 2017). For instance, recently Gholizadeh et al. reported the fabrication of immunoliposomes directed against E-selectin activated endothelial cells for delivery of rapamycin, as an approach to reduce inflammatory signaling cascades. The results evidenced that upon internalization, rapamycin was able to inhibit proliferation and migration of endothelial cells, as well as decreased inflammation (Gholizadeh et al., 2017). Another interesting example of vascular targeting was provided by Kim and collaborators, who synthesized anti-VCAM-1- immunoliposomes containing nitric oxide. The delivery system, upon ultrasound stimulation, was enabled visualization of atheroma, thus serving as both for diagnosis and treatment (Kim et al., 2015).

Drug delivery to microorganisms and also viruses is promising once their surface targets are identified, paving the way for developing immunoliposomes for treating infections. Within this context, Moles and collaborators published two recent papers on immunoliposomal delivery to plasmodium-infected cells, as an approach to treat malaria. Their most significant data showed that, in vivo, immunoliposomes cleared the pathogen below detectable levels at a chloroquine dose of 0.5 mg/kg, whereas free drug administered at 1.75 mg/kg was 40-fold less efficient (Moles et al., 2015). Subsequently, the same research group developed immunoliposomes against parasite proteins exported to the surface of *Plasmodium falciparum*-parasitized red blood cells (pRBCs). Their results showed that the drug lumefantrine delivered by immunoliposomes caused potent inhibition growth of the parasites (Moles et al., 2016). Regarding the development of immunoliposomes to treat viral infections, two different research groups, in 2015, published papers on immunoliposomes for HIV infection, using different targeting strategies, HIV-1 virus-like particles (VLPs) or CD4 receptor. Despite presenting promising in vitro results, these papers did not evaluate the immunoliposomes in vivo (Ramana et al., 2015; Petazzi et al., 2015).

Drug delivery to the brain mediated by immunoliposomes has been recently studied and the biggest challenge is to cross the blood-brain barrier. Loureiro and collaborators in 2015 developed liposomes with dual targeting, using both anti-transferrin receptor antibody, to help cross the blood-brain barrier, and anti-amyloid beta peptide antibody, for Alzheimer's disease targeting. Very importantly, it was demonstrated, in vivo, the ability of the immunoliposomes to cross the blood-brain barrier (Loureiro et al.,

2015). More recently, another group developed immunoliposomes targeted with the anti-amyloid β antibody, which, in vivo, were able to reduce the levels of circulating amyloid β in aged, but not in adult mice (Ordóñez-Gutiérrez et al., 2016). Finally, another interesting application of targeting against Alzheimer's disease was the use of immunoliposomes, which were successfully employed for imaging amyloid β in transgenic mice using the technique of time-of-flight secondary ion mass spectrometry (Carlred et al., 2016).

Stimuli-responsive nanocarriers are able to trigger drug release upon intrinsic stimuli, such as pH or redox potential, or extrinsic stimuli, such as temperature or ultrasound (Eloy et al., 2015). Lately, many papers have reported the combination of immunoliposomes with stimuli-responsive abilities, creating multifunctional liposomes, which hold great potential for clinical application. For instance, Haeri and collaborators combined the stimuli-responsive and the EGFR targeting strategies. For this purpose, thermosensitive liposomes were targeted with either an EGFR peptide or a Fab' fragment of cetuximab. Results showed that liposomes functionalized with Fab' were able to more efficiently bind EGFR. Furthermore, EGFR-mediated cell uptake, as well as doxorubicin cytotoxicity, was enhanced upon hyperthermia (Haeri et al., 2016). Another logical approach is the combination of immunoliposomes with pH-stimuli responsive lipids, considering the acidic pH of tumors. For this purpose, pH-sensitive 1,5-dihexadecyl N,N-diglutamyl-lysyl-L-glutamate (GGLG) liposomes were functionalized with a Fab' fragment of an ErbB2 antibody and further loaded with doxorubicin. The authors demonstrated enhanced cell association of stimuli-responsive immunoliposomes, as well as improved cytotoxicity in breast cancer cells (T. Li et al., 2017). Finally, in 2018 a paper was published reporting liposomes loaded with SPION, a type of superparamagnetic iron nanoparticle and further functionalized with rituximab, an anti-CD20 antibody, as a novel theranostic agent for central nervous system lymphoma, with promising in vitro results (Saesoo et al., 2018).

3.3   Liposomes Functionalized with Aptamers

Systematic evolution of ligands by exponential enrichment (SELEX) is a well-established and efficient technology, able to produce oligonucleotides (DNA and RNA molecules), called aptamers, with high target affinity. Aptamers, also known as "chemical antibodies," have many advantages, including stability, having high affinity and selectivity, besides presenting low immunogenicity (Darmostuk et al., 2014). Thus, aptamers are considered promising alternatives to antibodies and can be used for cancer detection and targeted drug delivery (Zhou et al., 2018).

Considering the high prevalence and mortality, breast cancer is subject to intense research in the field of targeted drug delivery, and reports using aptamer-functionalized liposomes are available. Some studies employed aptamer-directed liposomes for delivery of doxorubicin, employing mostly DNA aptamers. Very recently, Liu and collaborators reported novel DNA aptamers (T1) for dual targeting of polymorphonuclear myeloid-derived suppressor cells and tumors cells, evaluated in vivo in breast cancer models treated with liposomal doxorubicin, with superior cytotoxic and immunomodulatory effects (Liu et al., 2018). Another DNA aptamer, against nucleolin, present in multidrug resistant breast cancer cells, was conjugated to liposomal doxorubicin containing ammonium bicarbonate for bubble generation upon heating and thus facilitating drug release. In vivo studies in xenograft models showed that functionalized liposomes were able to better accumulate doxorubicin in the tumor tissues, inhibiting tumor growth and reducing systemic side effects, including cardiotoxicity (Liao et al., 2015). Another, DNA-aptamer, SRZ1, with specific binding affinity to 4T1 breast cancer cells, was linked to liposomes for doxorubicin delivery, resulting in tumor growth suppression and increased animal survival rate (Song et al., 2015). Finally, a DNA aptamer against the HER3 receptor was linked to doxorubicin-loaded liposomes, causing greater breast tumor suppression than other groups and alleviated side effects such as weight loss, low survival rate, and organ (heart and liver) injury (Dou et al., 2018).

Prostate cancer is also being researched for drug delivery using aptamers as targeting moieties. In this field, a very interesting paper was recently published on the delivery of clustered regularly interspaced short palindromic repeats (CRISPRs)-Cas9, for inhibition of gene expression. In that work, a RNA aptamer (anti PSMA, prostate-specific membrane antigen) was conjugated to liposomes loaded with CRISPR-Cas9 and the outcome was significant cell binding, causing in vitro gene silencing, and, in vivo, there was great regression of prostate cancer (Zhen et al., 2017). Furthermore, the PSMA RNA aptamer had been previously used for functionalization of doxorubicin-containing liposomes. Interestingly, the authors evaluated the drug delivery efficiency between drug intercalated in the aptamer and encapsulated in the aptamer, and the conclusion was that the encapsulation strategy was better for drug delivery (Park et al., 2015).

Other cancer targets have been also explored for aptamer-mediated liposomal delivery. EGFR aptamers have been evaluated for the delivery of erlotinib to lung cancer, with some promising results in reversing the hypoxia drug resistance both in vitro and in vivo (F. Li, Mei, Xie, et al. 2017; F. Li, Mei, Gao, et al., 2017). Another target with great potential for clinical application is the CD44 receptor, which is one of the most common cancer stem cell surface markers of a variety of tumors. Alshaer and collaborators, in 2015, synthesized anti-CD44 liposomes based on a thiol-maleimide reaction. Flow cytometry and confocal microscopy revealed high selectivity and sensitivity for binding CD44 positive cell lines (Alshaer et al., 2015). More recently, the same research group published a paper on aptamer-guided liposomes for selective delivery of siRNA to CD44-expressing breast cancer cells. It was demonstrated the ability of siRNA to silence luciferase gene both in vitro and in vivo (Alshaer et al., 2018).

3.4 Liposomes Functionalized with Peptides

Peptides have been largely employed for efficient intracellular targeted delivery of drugs, particularly using Cell Penetrating Peptides (CPPs) and Tumor Targeting Peptides (TTPs) (Dissanayake et al., 2017). CPPs are classified into three different categories. They can be short cationic peptides, containing the amino acids arginine, lysine and histidine and they can also be hydrophobic and/or amphipathic molecules, typically with 5–30 amino acids. CPPs have been reported to mediate cellular uptake of drugs with different physicochemical characteristics, including small molecules, plasmid DNA and siRNA, mostly delivered by nanocarriers. These peptides promote cellular uptake through nonendocytotic or energy-independent pathways and the endocytotic pathways (Farkhani et al., 2014). Although the entry mechanism is still not clearly understood, it appears to be dependent on the peptide type, the peptide concentration, the cargo, and the cell type (Ramsey and Flynn, 2015).

Liposomes have been functionalized with a variety of CPPs, mostly for drug delivery to cancer tissue, with successful in vitro and in vivo results, recently reviewed (Kuang et al., 2017). One of the most commonly CPPs employed to functionalize liposomes is TAT, a peptide derived from the TAT protein, an 86-residue transcriptional regulator of the HIV-1 virus, which has been reported to traverse cell membranes (Krämer and Wunderli-Allenspach, 2003). Recent advances in targeting liposomes with TAT include the development of a doxorubicin-loaded liposome functionalized with TAT for skin delivery, which was proved as a good alternative to enhance skin permeation (Boakye et al., 2015). For cancer treatment, an innovation is the combination of targeting and stimuli-responsive materials. For instance, TAT-targeted liposomes with redox-sensitive cleavable PEG enhanced paclitaxel delivery both in vitro, increasing cell uptake and thus inhibiting the proliferation melanoma B16F10 cells and, in vivo, inhibiting tumor growth (Fu et al., 2015). A recent application of TAT-directed liposome is for treating Alzheimer's disease. For this purpose, Gregori and collaborators synthesized liposomes functionalized with retro-inverso peptide RI-OR2-TAT and observed that this formulation inhibited the aggregation and toxicity of Aβ peptide, protecting transgenic mice against memory loss in a novel object recognition test (Gregori et al., 2017).

Penetratin, a cationic short peptide derived from the *Antennapedia* homeoprotein, has been also extensive employed for drug delivery. A very interesting study was recently published regarding multifunctional liposomes, able to respond to thermal and magnetic stimuli, for delivery of siRNA–penetratin conjugates to treat cancer. The results showed effective cellular uptake, endosomal escape, resulting in gene silencing in breast cancer MCF-7 cells. Furthermore, in vivo, under alternating current magnetic field, it was demonstrated antitumor efficacy and gene silencing effect in a breast cancer xenograft model (Yang et al., 2016). Likewise, octarginine (R8) cationic peptide has been demonstrated to enhance delivery of siRNA. In 2017, a paper was published on the development and evaluation of siRNA-loaded liposomes functionalized with R8 peptide for delivery to vascular smooth muscle cells. The formulation was proved to have significant gene silencing; however, in vivo studies are needed for translational development (Fisher et al., 2017).

Another class of CPPs is represented by the amphiphilic peptides. An important member of this group is GALA, a 30 amino acid synthetic peptide with a glutamic acid-alanine-leucine-alanine (EALA) repeat that also contains a histidine and tryptophan residue (Li, Nicol, and Szoka, 2004). Recently, the PEGylation of the GALA peptide enhanced the lung delivery of siRNA-loaded liposomes, promoting enhanced gene silencing (Santiwarangkool et al., 2017). The KALA peptide (WEAKLAKALAKALAKHLAKALAKALKA) is an amphiphilic peptide that forms an α-helical structure at physiological pH. One example of use of KALA for liposomal drug delivery was the application to facilitate MHC class I antigen presentation as an approach to enhance anti-tumor effect. Interestingly, the subcutaneous injection of KALA-modified ovalbumin (OVA)-loaded liposomes was able to elicit more potent specific cytotoxicity and anti-tumor effect than liposomes that were functionalized with octa-arginine (R8) peptides, which was attributed to the better ability of KALA to escape from endosomes/lysosomes for cytoplasmic delivery (Miura et al., 2017). Finally, another example of amphiphilic peptide for liposome targeting is Pep-1, a short amphipathic peptide consisted of a hydrophobic tryptophan-rich domain and a hydrophilic lysine-rich domain separated by a spacer. Jiao et al. (2017) reported that Pep-1-decorated liposomes allowed enhanced delivery of cilengitide, resulting in higher apoptosis of glioma cancer cells, with consequent better in vivo anticancer effect (Jiao et al., 2017).

Owing to the intensive research of peptide-mediated liposomal delivery, other peptides have been synthesized and are gradually becoming available for research. Some of these examples include the functionalization of raft-like liposomes with chimeric peptides directed to HIV envelope (Gómara et al., 2017), the peptide CARSKNKDC (CAR), a pulmonary-specific targeting sequence, for targeting liposomes to isolated perfused rat lungs (Gupta et al., 2015), the EGFR specific peptide (GE11) for targeted delivery of thermosensitive liposomes to EGFR positive cancer cell lines (Haeri et al., 2016), among others.

TTPs are an efficient alternative for selective delivery of drugs or diagnostic probes to tumor cells, including a variety of peptides, with advantages over other targeting molecules, such as monoclonal antibodies, due to rapid blood clearance, increased diffusion and tissue penetration, as well as higher chemical stability and the ease of large-scale production (Liu et al., 2017). Table 3.2 shows recent papers on targeting liposomes using peptides. Most studies employed cyclic or linear RGD peptide, for targeting integrin $\alpha v \beta 3$. Overall, the RGD-modified liposomes enhanced cell uptake and cytotoxicity, improving the in vivo anticancer effect. Furthermore, other peptides have shown utility for cancer targeting, including the asparagine-glycine-arginine peptide (NGR) which selectively binds the highly expressed CD13 on tumor cells, as well as somatostatin analogue for somatostatin receptor binding, and the ATR7 peptide, which recognizes the vascular endothelial growth factor receptor 2 (VEGFR2) and neuropilin-1 (NRP-1) that are overexpressed on glioma cells.

Table 3.2   Recent examples of peptide-mediated drug delivery

Peptide	Drug	Main results	Reference
Cyclic RGD	Doxorubicin	Liposome internalization by HUVEC cells via integrin-mediated endocytosis. It was shown targeting of both tumor vasculature and tumor cells. Effective treatment with no side effects.	(Amin et al., 2015)
		Functionalized liposomes had increased cytotoxicity against glioblastoma cells and deeply penetrated into tumor spheroids. In vivo, the formulation had higher tumor distribution with greatest antitumor effect.	(Belhadj et al., 2017)

Peptide	Drug	Main results	Reference
RGD	Docetaxel	Peptide-decorated liposomes had higher cytotoxicity and cellular uptake. In vivo study showed better tumor accumulation.	(Chang et al., 2015)
		RGD-modified pH-sensitive liposomes had longer blood circulation than commercial docetaxel formulation and presented stronger inhibition of tumor growth.	(Zuo et al., 2016)
	Arsenic trioxide	RGD conjugated liposome-hollow silica hybrid nanovesicles had enhanced cellular uptake and higher cytotoxicity. In hepatic carcinoma-xenograft mice, the formulation improved targeting efficiency and drug anticancer activity.	(Fei et al., 2017)
Cyclic A7R	Doxorubicin	Liposomes functionalized with the linear peptide and cyclic peptides were compared and the latter enabled stronger antiproliferative effect and higher accumulation in glioma cells, resulting in better tumor growth suppression.	(Ying et al., 2016)
NGR	siRNA	Liposomes modified with CPP and NGR showed effective cellular uptake, endosomal escape and gene silencing. In vivo, the formulation delayed tumor growth.	(Yang et al., 2015)

(*Continued*)

Table 3.2 (*Continued*)

Peptide	Drug	Main results	Reference
	Docetaxel	NRG-modified pH-sensitive liposomes caused specific targeting ability and enhanced antitumor activity to HT-1080 cells. Liposomes could significantly and safely accumulate in tumor tissue in xenografted nude mice.	(Gu et al., 2017)
Somatostatin analogue	Diacerein	In vitro, better anticancer activity due to increased apoptosis and, in vivo, significant tumor growth suppression.	(Bharti et al., 2017)

3.5 Liposomes Functionalized with Folate

Folate (FR) is a member of the vitamins B class which is known as folic acid or vitamin B9 and is essential for normal cellular reproduction and DNA replication (Belfiore et al., 2018). Folate is transported across the cellular membrane by (i) reduced folate carrier; (ii) proton-coupled folate transporter which utilizes the transmembrane protons gradient to mediate folate transport into the cell; and (iii) folate receptor carrying folate-mediated endocytosis such as glycopolypeptide members with molecular weight ranging from 38 to 45 KDa divided into different isoforms: folate receptor alpha (FRα); folate receptor beta (FRβ) and folate receptor gamma (FRγ). The α and β isoforms are attached to cell membranes via glycosylphosphatidylinositol, whereas the gamma isoform is found only in hematopoietic cells (Liu et al., 2016).

As addressed earlier in this chapter, many receptors have been identified as being overexpressed in cancer cells and active targeting strategies using folate receptor-targeted liposomes have been explored extensively in the preclinical setting, showing improved efficacy over non-functionalized liposomes both in in vitro and in vivo models (Belfiore et al., 2018). Notwithstanding some cancer cells express the FRβ isoform, the FRα isoform has the most potential for targeted cancer therapy. With this in mind, the FRα is mentioned in the literature to be overexpressed in more than 40% of the human cancer cells, particularly ovary, breast, pleura, lung, cervix, kidney, bladder, brain and others (Fernandez et al., 2018; Eloy et al., 2017; Ucar et al., 2017; Cheung et al., 2016).

Conversely, the distribution of the FRα in normal human cells and tissues is restricted to low-level expression. The FRα expression in carcinomas is 100 to 300 times higher than on healthy cells since there are 1 to 10 million receptor copies per cell (Fernandez et al., 2018). In fact, folate trafficking via FRα occurs by receptor-mediated endocytosis pathway with caveolae vesicles due to specific folate binding to FRα. Once internalized, early endosomes are formed and undergo acidification and subsequent fusion with lysosomes to release folate (Siwowska et al., 2017). Using proper linkers, folate receptor-targeted carriers are able to release the drug inside their target cells and thus it can perform the desired cytotoxic effect.

Several drugs that have been linked to folic acid including (i) chemotherapy drugs such as paclitaxel (PTX), docetaxel (DTX), doxorubicin (DOX), carboplatin (CP) and others; (ii) anti-inflammatory drugs; (iii) photosensitizers used in photodynamic therapy; (iv) oligonucleotides for antisense therapy; (v) radioimaging and radiotherapeutic agents; (vi) immunotherapeutic agents, and others. In many cases, these folate receptor targeted agents have been proved as superior when carried by folate receptor-targeted liposomes (F-TLP) compared to unmodified plain liposomes (Poh et al., 2018; Gazzano et al., 2018; Riaz et al., 2018).

DOX-F-TLP showed an enhanced cytotoxicity and high cell uptake for ovarian, breast, lung, prostatic cancer cells line than the nontargeted liposomes even when different liposomes compositions or commercial liposome, Caelyx®, were used (Gazzano et al., 2018). In addition, higher cell uptake and sites of inflammation in mouse models were observed (Poh et al., 2018). In a similar way, PTX, DTX and CP have been widely studied in liposomes functionalized with folate as a targeting moiety (Monteiro et al., 2018; Ucar et al., 2017).

Gazzano and colleagues, (2018) produced the first liposomal formulations of nitrooxydoxorubicin decorated with folic acid (LNDF), in order to improve their active targeting against Pgp expressing (Pgp) tumors. By analyzing human and murine breast cancer cells with different expression of the folate receptor and Pgp, the authors demonstrated that LNDF is internalized in a folate receptor-dependent manner and achieve maximal anti-tumor efficacy against folate receptor-positive/Pgp positive cells. Upon uptake of LNDF, nitrooxy-doxorubicin was delivered within the nucleus, where it induced cell cycle arrest, causing DNA damage, and triggered mitochondria-dependent apoptosis. The LNDF reduced the growth of folate receptor-positive/Pgp-positive tumors and prevented in vivo tumor formation in mice, whereas doxorubicin and Caelyx® failed to perform the same effect. Meanwhile, nanostructured lipid carrier system loaded with PTX (PTX-NLC) and modified with folate (PTX-F-NLC) were investigated in vitro on three cancer cell lines, MCF7 (breast cancer cells), HeLa (cervical cancer cells) and A549 (lung cancer cells). There was higher uptake in every cell line to PTX-F-NLC than PTX-NLC. The highest cells incorporation ratio was observed on HeLa cell line and the lowest was on A549 cell line. Incorporation ratios on cell lines were consistent with the relative levels of folate receptor expressions in the cells. Additionally, in vivo animal studies were conducted to evaluate the biodistribution, showing that the folate-nanostructured lipid carrier had considerably higher uptake in folate receptor positive organs and distributed rapidly after intravenous injection in both normal and receptor blocked group (Ucar et al., 2017).

Folate-coated long-circulating and pH-sensitive liposomes loaded with PTX have been evaluated in healthy and breast tumor-bearing mice. The pharmacokinetic properties of non-functionalized lipid nanoparticle and PTX-F-TLP decreased in a monophasic manner showing half-life of 400.1 and 541.8 min, respectively. Biodistribution studies showed a significant uptake in liver, spleen, and kidneys, demonstrating that these routes are involved in excretion. At 8 h post-injection, the liposomal tumor uptake was higher than free PTX (Monteiro et al., 2018).

Overcoming multidrug resistance cancer cells (MDR) to chemotherapeutic agents using folic-conjugated liposomal formulation was also investigated (Williams et al., 2017; J. M. Li et al., 2015). Recently, the cytosolic delivery of DOX via F-TLP enhanced the killing of MDR cells, reducing the viability to 60% compared to 97% for MDR cells treated with free DOX (Williams et al., 2017).

Moreover, bioimaging application methods using folate F-TLP loaded with quantum-dots (QDs) or radiolabeled agents (technetium-99 m (99 mTc) or far-red/near-infrared (squaraine dye, SQR23) were recently explored in vivo and served as a promising diagnostic imaging tool (Dong et al., 2018; Ucar et al., 2017). Quantum dots are nanoparticles composed by semiconductor materials in an inorganic transition metal to form a core/shell system while SQR23 includes minimal interfering absorption and fluorescence from biological samples, enploys inexpensive diode excitation with reduced scattering, and enhanced tissue penetration depth. 99 mTc is an ideal radionuclide used in diagnostic nuclear medicine due to its appropriate half life (6 h) and gamma energy (140 keV) (Dong et al., 2018; Ucar et al., 2017). QDs, 99 mTc and SQR23 are able to anchor in the lipid bilayer membrane of lipid nanoparticles. Combining diagnostic and therapy from multifunctional liposomes containing DTX and QDs or DTX and polydiacetylene as fluorescence agent proved to be advantageous as they are superior in killing cancer cells compared to non-functionalized liposomes (L.Li, An, and Yan, 2015). In 2018, F-TLP containing a fluorescent molecule and betamethasone was used in inflammatory diseases and it was demonstrated specific binding to folate receptor positive cells, resulting in accumulation at inflammation sites in mouse models of colitis and atherosclerosis (Poh et al., 2018).

Porphysomes are all-organic bilayer liposome-like nanoparticles that self-assemble from porphyrin (photosensitizer used in photodynamic therapy) and lipid conjugates (cholesterol and PEG). These nanostructures were functionalized with folic acid for enhanced tumor targeting and better photodynamic therapy efficacy, showing intrinsic multimodal capabilities for both imaging and therapy. Because of the extremely high photosensitizer packing density in the phospholipid bilayer, both fluorescence and singlet oxygen generation can be suppressed in the range of 99%, however they can be restored once the nanoparticles are internalized into the cancer cells (Kato et al., 2017).

Targeted delivery of nucleic acids mediated by folate-modified lipid nanoparticles is a promising tool for clinical gene therapy. Improvements in gene therapy were reached by F-TLP to deliver small interfering RNA (siRNA) or DNA into the cells (Kabilova et al., 2018; Kato et al., 2017) and successful results were obtained regarding the enhancing of cancer cell death and reduction of tumor growth causing efficient downregulation of specific proteins expressed in animal xenograft models (Kabilova et al., 2018). Another recent application was the synthesis of F-TLP based on polycationic lipid 1,26-bis(cholest-5-en-3β-yloxycarbonylamino)-7,11,16,20-tetraazahexacosan tetrahydrochloride (2X3), dioleoylphosphatidylethanolamine as lipid helper. This liposome was anchored with siRNA in different nitrogen to phosphate (N/P) ratios and, in vivo, the biodistribution of siRNA was characterized by high retention in the mouse body. Furthermore, the nanocarriers efficiently accumulated in the tumor (15 to 18% of total amount), and kidneys (71%) and were retained there for more than 24 h, causing efficient downregulation of p-glycoprotein expression in tumors, in N/P dependent manner (Kabilova et al., 2018).

Besides these findings described above, a known humanized monoclonal antibody that binds to FR-α (farletuzumab) which is current FDA on-going clinical trials phase II used in combination with CP plus PTX or CP plus pegylated liposomal DOX (NCT02289950). Previously it demonstrated a safe profile in Phase 1b clinical trial study in patients with platinum-sensitive epithelial ovarian cancer (Kim et al., 2016). Therefore, engineered lipid nanoparticles functionalized with farletuzumab have been highly challenging and could become a new approach for cancer treatment in the near future (Eloy, Petrilli, Trevizan, et al., 2017).

3.6   Liposomes Functionalized with Transferrin

Transferrin receptors (TfRs) are expressed at low levels in most normal human cells and tissues and overexpressed in malignant cells, which may be up to 100-fold higher than the average expression of normal cells. The TfRs, also known as TfR1, TfR2, and CD77, are cells membrane-associated type II glycoproteins involved in the regulation of cell growth and division process by cellular uptake of the iron. Transferrin binds the TfRs and both are internalized in clathrin-coated pits through receptor-mediated endocytosis and then, the iron is released into the cells from transferrin due to the decrease in pH endosome (Jhaveri et al., 2018).

Extracellular domains are similar for TfR1 and TfR2, but TfR2 shows 25-fold lower affinity to transferrin than TfR1. In addition, TfR2 is only expressed in hepatocytes and erythroid cells while TfR1 are abundant in hepatocytes, erythrocyte precursors and other tissues especially on cancer cells due the rapid dividing cellular process, but not in mature erythrocytes. A series of strategies have been reported to achieve targeting via TfRs such as using Tf, peptides, monoclonal antibodies or their fragments (H. H. Amin et al., 2017).

The use of TfR-targeted liposomes has been reported to deliver anticancer drugs such as oxaliplatin (OX), doxorubicin (DOX), 5-fluouracil (5-FU), resveratrol (RE), ceramide (CE), cisplatin (Cis), plasmid DNA and siRNA, into malignant cells, with promising results in the preclinical setting, enhancing drug cell uptake and drug efficacy (Belfiore et al., 2018; Zhang et al., 2017; Sriraman et al., 2016). For instance, a transferrin-targeted, RE-loaded liposome composed by egg Phosphatidylcholine, cholesterol, DSPE-PEG(2000), 1,2-dioleoyl-sn-glycero-3-phosphoethanolamine and cholesterol hemisuccinate showed favorable in vitro and in vivo efficacy, which warrant their further investigation for the treatment of glioblastoma. The authors concluded that the ability of resveratrol to arrest cells in the S-phase of the cell cycle, and selectively induce production of reactive oxygen species in cancer cells were probably responsible for its cytotoxic effects. The ability of transferrin receptor-targeted liposomes with RE to inhibit tumor growth and improve survival in mice was demonstrated. The formulation was significantly more cytotoxic and induced higher levels of apoptosis accompanied by activation of caspases 3/7 in glioblastoma cells when compared to free RE or untargeted liposomes. Percentages of animals surviving at day 25 were 60% for targeted-liposomes compared to only 20% for untargeted liposomes and free RE (Jhaveri et al., 2018).

Similar targeted liposome composition containing 5-FU induced apoptosis in colon cancer cells through activation of mitochondrial apoptosis pathways, indicating that transferrin-functionalized liposomes would provide a potential strategy to treat colon cancer by inducing apoptosis via mitochondria signaling pathway, consequently reducing the dose of the drug and causing fewer side-effects (Moghimipour et al., 2018).

There is an eminent scientific field exploring TfR as an excellent targeting marker for the treatment of cancer using monoclonal antibodies (Eloy, Petrilli, Trevizan, et al., 2017). In this way, the payload can be encapsulated within a cationic liposome surface-functionalized with an anti-TfR single-chain antibody fragment. Some studies have demonstrated preferential targeting and delivery of plasmid DNA, chemotherapeutic agents, small interfering RNAs and magnetic resonance imaging contrast agents and others. Kasper Bendix Johnsen and colleagues (2017) found that transferrin receptor-targeting increases the association between the immunoliposomes and primary endothelial cells in vitro, but this did not correlate with increased cargo transcytosis. Furthermore, the TfR-targeted immunoliposomes accumulated along the microvessels of the brains of rats. Noteworthy, the increased accumulation was correlated with both increased cargo uptake in the brain endothelium and subsequent cargo transport into the brain (Johnsen et al., 2017).

Dual ligand immunoliposomes for drug delivery to the brain using the anti-transferrin receptor monoclonal antibody (OX26MAb) and the anti-amyloid beta peptide antibody (19B8MAb), as nanocarriers of drugs for Alzheimer's disease therapy were also successfully reported. The authors concluded that they are only beginning to understand how different conditions affect the conjugation of antibodies to liposomes and their transport through the blood brain barrier. However, the obtained results showed that the cellular uptake of the immunoliposomes was substantially more efficient if an OX26MAb antibody is conjugated through the streptavidin-biotin complex instead of the maleimide group. Thus, it was demonstrated the importance of the conjugation method used to bind the antibody for efficient drug delivery to the brain (Loureiro et al., 2015).

The dual targeting in the same lipid nanocarriers was also proposed using DOX-loaded liposomes targeted with folic acid and transferrin in human cervical and ovarian carcinoma. As a result, the dual-targeted liposomes showed a sevenfold increase in cell association compared to either of the single-ligand targeted ones. In all cell lines studied, the dual-targeted liposomes showed significantly higher cytotoxic effects than the non-functionalized lipid nanoparticle, or single-ligand targeted liposomes. In the in vivo setting using xenograft nude mice model, it was found that both the dual-targeted as well as folic-acid targeted liposomes were able to equally inhibit tumor growth (Sriraman et al., 2016).

This improved drug delivery effect of dual ligand over single ligand TfR-targeting was also demonstrated by preclinical studies for daunorubicin and doxorubicin (chemotherapy) and plasmid DNA (gene therapy) using the rat model of lung cancer and brain glioma. Daunorubicin or doxorubicin-loaded liposomes surface-functionalized with transferrin and p-aminophenyl-α-D-manno-pyranoside or transferrin and cell-penetrating peptides demonstrated increased transfection efficiency than single ligand or non-ligand carriers in a mouse model of brain glioma. The results showed increased drug transport across the brain endothelial barrier, improved cellular uptake and increased survival compared to treatment with free drugs, single-ligand and non-ligand liposomes (Belfiore et al., 2018). The same was reported for plasmid DNA-loaded nanostructured lipid carrier surface-functionalized with both transferrin and hyaluronic acid in a mouse model of lung cancer (Zhang et al., 2017).

To date, few studies using the plasmid DNA SGT-53 and SGT-94 have progressed into human clinical trials. The SGT-53 is a complex composed of a wild-type p53 gene encapsulated in a liposome that is targeted to tumor cells by means of an anti-transferrin receptor single-chain antibody fragment attached to the outside of the liposome and the SGT-94 is a similar composition to deliver RB94, a tumor suppressor gene, a modified form of the retinoblastoma gene, RB110. RB94 has shown enhanced tumor suppression and tumor cell killing activity compared to RB110 in all tumor cell types studied, including bladder cancer cell lines. Moreover, RB94 has shown no toxicity to any normal human cells tested. Both SGT-53 and SGT-94 are sponsored by SynerGene Therapeutics, Inc. Complete Phase I study using SGT-53 (NCT00470613) and SGT-94 (NCT01517464) was designed to evaluate the safety and to establish a practically attainable and/or tolerable dose of these anti-cancer agents for use in further clinical trials. In addition, the clinical Phase Ib was designed to evaluate the safety and the effect of the SGT-53 in combination with docetaxel on tumor size or progression. Advanced studies of SGT-53 in children with refractory or recurrent solid tumors or combined SGT-53 plus gemcitabine/nab-paclitaxel for metastatic pancreatic cancer or phase II study of combined temozolomide and SGT-53 for treatment of recurrent glioblastoma are currently recruiting patients (ClinicalTrials.gov, 2018). These studies are summarized in Table 3.3.

Table 3.3   Current FDA on-going clinical trials (ClinicalTrials.gov, 2018)

Targeting ligands	Lipid nanoparticle	Clinical trial	Identification	Interventions	Conditions	Status
Transferrin	Liposome_	Phase II	NCT02340156	Temozolomide Genetic SGT-53	Recurrent gliobastoma	Recruiting
Transferrin	Liposome	Phase I	NCT02354547	Topotecan, cyclophosphamide Genetic sgt-53	Neoplasm	Recruiting
Transferrin	Liposome	Phase I	NCT02340117	Nab-paclitaxel, Gemcitabine, Genetic SGT-53	Metastatic pancreatic cancer	Recruiting
Transferrin	Liposome	Phase I	NCT00470613	Genetic SGT-53	Neoplasm	completed
Transferrin	Liposome	Phase I	NCT01517464	Genetic SGT-94	Neoplasm	completed
Folate	Liposome	Phase II	NCT02289950	Farletuzumab and doxorubicin	Ovarian cancer	Recruiting

3.7 Conclusion

Liposomes are the best-studied nanocarrier and have been available for over two decades in the clinic for the treatment of a variety of diseases, especially cancer. However, nonspecific drug delivery, despite the EPR effect of solid tumors, and resulting side effects, could be enhanced if targeting strategies are used. The literature has described a variety of targeting moieties which can be conjugated onto liposomal surface for active delivery, including the use of monoclonal antibodies and their fragments, such as the scFv fragment, peptides, including the cell penetrating peptides and tumor targeting peptides, and also folate and transferrin. Herein, we revised a variety of very recent papers on functionalized liposomes, and we found that there are many studies reporting successful in vitro and in vivo evaluation. While cell studies mostly indicated that functionalized liposomes are better uptaken by cells, in vivo studies demonstrated improved biodistribution and better anticancer activity in tumor xenografts. The preclinical successful evaluation motivated the conduction of some clinical trials, however no targeted liposome is still available in the market. It is expected that these delivery systems will move from bench to bedside after they prove safety, efficacy and are able to be industrially produced with reliability.

References

Accardo, Antonella, and Giancarlo Morelli. 2015. Peptide-Targeted Liposomes for Selective Drug Delivery: Advantages and Problematic Issues. *Biopolymers* 104(5): 462–479. doi:10.1002/bip.22678.

Alshaer, Walhan, Hervé Hillaireau, Juliette Vergnaud, Said Ismail, and Elias Fattal. 2015. Functionalizing Liposomes with Anti-CD44 Aptamer for Selective Targeting of Cancer Cells. *Bioconjugate Chemistry* 26(7): 1307–1313. doi:10.1021/bc5004313.

Alshaer, Walhan, Hervé Hillaireau, Juliette Vergnaud, Simona Mura, Claudine Deloménie, Félix Sauvage, Said Ismail, and Elias Fattal. 2018. Aptamer-Guided siRNA-Loaded Nanomedicines for Systemic Gene Silencing in CD-44 Expressing Murine Triple-Negative Breast Cancer Model. *Journal of Controlled Release* 271(October 2017). Elsevier: 98–106. doi:10.1016/j.jconrel.2017.12.022.

Amin, Hardik H., Nilesh M. Meghani, Chulhun Park, Van Hong Nguyen, Thao Truong Dinh Tran, Phuong Ha Lien Tran, and Beom Jin Lee. 2017. Fattigation-Platform Nanoparticles Using Apo-Transferrin Stearic Acid as a Core for Receptor-Oriented Cancer Targeting. *Colloids and Surfaces B: Biointerfaces* 159: 571–579. doi:10.1016/j.colsurfb.2017.08.014.

Amin, Mohamadreza, Mercedeh Mansourian, Gerben A. Koning, Ali Badiee, Mahmoud Reza Jaafari, and Timo L. M. Ten Hagen. 2015. Development of a Novel Cyclic RGD Peptide for Multiple Targeting Approaches of Liposomes to Tumor Region. *Journal of Controlled Release* 220: 308–315. doi:10.1016/j.jconrel.2015.10.039.

Apte, Anjali, Erez Koren, Alexander Koshkaryev, and Vladimir P. Torchilin. 2014. Doxorubicin in TAT Peptide-Modified Multifunctional Immunoliposomes Demonstrates Increased Activity against Both Drug-Sensitive and Drug-Resistant Ovarian Cancer Models. *Cancer Biology and Therapy* 15(1): 69–80. doi:10.4161/cbt.26609.

Ara, Mst. Naznin, Takashi Matsuda, Mamoru Hyodo, Yu Sakurai, Noritaka Ohga, Kyoko Hida, and Hideyoshi Harashima. 2014. Construction of an Aptamer Modified Liposomal System Targeted to Tumor Endothelial Cells. *Biological and Pharmaceutical Bulletin* 37(11): 1742–1749. doi:10.1248/bpb.b14-00338.

Atukorale, Prabhani U., Gil Covarrubias, Lisa Bauer, and Efstathios Karathanasis. 2017. Vascular Targeting of Nanoparticles for Molecular Imaging of Diseased Endothelium. *Advanced Drug Delivery Reviews* 113. Elsevier B. V.: 141–156. doi:10.1016/j.addr.2016.09.006.

Barbosa, Marcos V., Liziane O. F. Monteiro, Guilherme Carneiro, Andréa R. Malagutti, José M. C. Vilela, Margareth S. Andrade, Mônica C. Oliveira, Alvaro D. Carvalho-Junior, and Elaine A. Leite. 2015. Experimental Design of a Liposomal Lipid System: A Potential Strategy for Paclitaxel-Based Breast Cancer Treatment. *Colloids and Surfaces B: Biointerfaces* 136: 553–561. doi:10.1016/j.colsurfb.2015.09.055.

Barenholz, Yechezkel (Chezy). 2012. Doxil® - the First FDA Approved Nano-Drug: Lessons Learned. *Journal of Controlled Release* 160: 117–134.

Belfiore, Lisa, Darren N. Saunders, Marie Ranson, Kristofer J. Thurecht, Gert Storm, and Kara L. Vine. 2018. Towards Clinical Translation of Ligand-Functionalized Liposomes in Targeted Cancer Therapy: Challenges and Opportunities. *Journal of Controlled Release* 277: 1–13. doi:10.1016/j.jconrel.2018.02.040.

Belhadj, Zakia, Man Ying, Xie Cao, Xuefeng Hu, Changyou Zhan, Xiaoli Wei, Jie Gao, Xiaoyi Wang, Zhiqiang Yan, and Weiyue Lu. 2017. Design of Y-Shaped Targeting Material for Liposome-Based Multifunctional

Glioblastoma-Targeted Drug Delivery. *Journal of Controlled Release* 255: 132–141. doi:10.1016/j.jconrel.2017.04.006.

Bharti, Rashmi, Goutam Dey, Indranil Banerjee, Kaushik Kumar Dey, Sheetal Parida, B. N.Prashanth Kumar, Chandan Kanta Das, et al. 2017. Somatostatin Receptor Targeted Liposomes with Diacerein Inhibit IL-6 for Breast Cancer Therapy. *Cancer Letters* 388: 292–302. doi:10.1016/j.canlet.2016.12.021.

Boakye, Cedar H. A., Ketan Patel, and Mandip Singh. 2015. Doxorubicin Liposomes as an Investigative Model to Study the Skin Permeation of Nanocarriers. *International Journal of Pharmaceutics* 489(1–2): 106–116. doi:10.1016/j.ijpharm.2015.04.059.

Carlred, Louise, Vladana Vukojevic, Bjorn Johansson, Martin Schalling, Fredrik Hook, and Peter Sjovall. 2016. Imaging of Amyloid-Beta in Alzheimer's Disease Transgenic Mouse Brains with ToF-SIMS Using Immunoliposomes. *Biointerphases* 11(2): 02A312. doi:10.1116/1.4940215.

Chang, Minglu, Shanshan Lu, Fang Zhang, Tiantian Zuo, Yuanyuan Guan, Ting Wei, Wei Shao, and Guimei Lin. 2015. RGD-Modified pH-Sensitive Liposomes for Docetaxel Tumor Targeting. *Colloids and Surfaces B: Biointerfaces* 129. Elsevier B. V.: 175–182. doi:10.1016/j.colsurfb.2015.03.046.

Cheung, Anthony, Heather J. Bax, Debra H. Josephs, Kristina M. Ilieva, Giulia Pellizzari, James Opzoomer, Jacinta Bloomfield, et al. 2016. Targeting Folate Receptor Alpha for Cancer Treatment. *Oncotarget* 7(32): 52553–52574. doi:10.18632/oncotarget.9651.

ClinicalTrials.gov. 2018. Www.clinicaltrials.gov. 2018.

Darmostuk, Mariia, Silvie Rimpelova, Helena Gbelcova, and Tomas Ruml. 2014. Current Approaches in SELEX: An Update to Aptamer Selection Technology. *Biotechnology Advances* 33(6). Elsevier Inc.: 1141–1161. doi:10.1016/j.biotechadv.2015.02.008.

Deshpande, Pranali, Aditi Jhaveri, Bhushan Pattni, Swati Biswas, and Vladimir Torchilin. 2018. Transferrin and Octaarginine Modified Dual-Functional Liposomes with Improved Cancer Cell Targeting and Enhanced Intracellular Delivery for the Treatment of Ovarian Cancer. *Drug Delivery* 25(1): 517–532. doi:10.1080/10717544.2018.1435747.

Dissanayake, Shama, William A Denny, and Swarna Gamage. 2017. Recent Developments in Anticancer Drug Delivery Using Cell Penetrating and Tumor Targeting Peptides. *Journal of Controlled Release*

250: 62–76. http://www.sciencedirect.com/science/article/pii/S0168365917300603.

Dong, Sheng, Joshua Ding Wei Teo, Li Yan Chan, Chi-Lik Ken Lee, and Keitaro Sou. 2018. Far-Red Fluorescent Liposomes for Folate Receptor-Targeted Bioimaging. *ACS Applied Nano Materials*, 2: 1009–1013. doi:10.1021/acsanm.8b00084.

Dou, Xiao-Qian, Hua Wang, Jing Zhang, Fang Wang, Gui-Li Xu, Cheng-Cheng Xu, Huan-Hua Xu, Shen-Si Xiang, Jie Fu, and Hai-Feng Song. 2018. Aptamer – Drug Conjugate : Targeted Delivery of Doxorubicin in a HER3 Aptamer-Functionalized Liposomal Delivery System Reduces Cardiotoxicity. *International Journal of Nanomedicine* 13: 763–776.

Dou, Yannan, Kullervo Hynynen, and Christine Allen. 2017. To Heat or Not to Heat: Challenges with Clinical Translation of Thermosensitive Liposomes. *Journal of Controlled Release* 249: 63–73. doi:10.1016/j.jconrel.2017.01.025.

Eloy, Josimar O., Raquel Petrilli, Deise L. Chesca, Fabiano Saggioro, R. J. Lee, and J. M. Marchetti. 2017. Anti-HER2 Immunoliposomes for Co-Delivery of Paclitaxel and Rapamycin for Breast Cancer Therapy. *European Journal of Pharmaceutics and Biopharmaceutics* 115: 159–167. doi:10.1016/j.ejpb.2017.02.020.

Eloy, Josimar O., Raquel Petrilli, Lucas Noboru Fatori Trevizan, and Marlus Chorilli. 2017. Immunoliposomes: A Review on Functionalization Strategies and Targets for Drug Delivery. *Colloids and Surfaces B: Biointerfaces* 159: 454–467. doi:10.1016/j.colsurfb.2017.07.085.

Eloy, Josimar O., Raquel Petrilli, Renata F. V. Lopez, Robert J. Lee, and Robert J. Lee. 2015. Stimuli-Responsive Nanoparticles for siRNA Delivery. *Current Pharmaceutical Design* 21: 4131–4144.

Eloy, Josimar O., Raquel Petrilli, Robert W. Brueggemeier, Juliana Maldonado Marchetti, and Robert J. Lee. 2016. RapamycinLoaded Immunoliposomes Functionalized with Trastuzumab : A Strategy to Enhance Cytotoxicity to HER2 Positive Breast Cancer Cells. *Anti-Cancer Agents in Medicinal Chemistry* 17: 48–56. doi:10.2174/18715206166 66160526103432.

Farkhani, Samad Mussa, Alireza Valizadeh, Hadi Karami, Samane Mohammadi, Nasrin Sohrabi, and Fariba Badrzadeh. 2014. Cell Penetrating Peptides: Efficient Vectors for Delivery of Nanoparticles, Nanocarriers, Therapeutic and Diagnostic Molecules. *Peptides.* 57: 78–94. doi:10.1016/j.peptides.2014.04.015.

Fei, Weidong, Yan Zhang, Shunping Han, Jiaoyang Tao, Hongyue Zheng, Yinghui Wei, Jiazhen Zhu, Fanzhu Li, and Xuanshen Wang. 2017. RGD

Conjugated Liposome-Hollow Silica Hybrid Nanovehicles for Targeted and Controlled Delivery of Arsenic Trioxide against Hepatic Carcinoma. *International Journal of Pharmaceutics* 519: 250–262. doi: 10.1016/j. ijpharm.2017.01.031.

Fernandez, Marcos, Faiza Javaid, and Vijay Chudasama. 2018. Advances in Targeting the Folate Receptor in the Treatment/Imaging of Cancers. *Chemical Science* 9: 790–810. doi:10.1039/C7SC04004K.

Fisher, Richard K., Samuel I. Mattern-Schain, Michael D. Best, Stacy S. Kirkpatrick, Michael B. Freeman, Oscar H. Grandas, and Deidra J. H. Mountain. 2017. Improving the Efficacy of Liposome-Mediated Vascular Gene Therapy via Lipid Surface Modifications. *Journal of Surgical Research* 219: 136–144. doi:10.1016/j.jss.2017.05.111.

Fu, Han, Kairong Shi, Guanlian Hu, Yuting Yang, Qifang Kuang, Libao Lu, Li Zhang, et al. 2015. Tumor-Targeted Paclitaxel Delivery and Enhanced Penetration Using TAT-Decorated Liposomes Comprising Redox-Responsive Poly(ethylene Glycol). *Journal of Pharmaceutical Sciences* 104(3): 1160–1173. doi:10.1002/jps.24291.

Gazzano, Elena, Barbara Rolando, Konstantin Chegaev, Iris C Salaroglio, Joanna Kopecka, Isabella Pedrini, Simona Saponara, et al. 2018. Folate-Targeted Liposomal Nitrooxy-Doxorubicin : An E Ff Ective Tool against P- Glycoprotein-Positive and Folate Receptor-Positive Tumors. *Journal of Controlled Release* 270: 37–52.

Gholizadeh, Shima, Ganesh Ram R. Visweswaran, Gert Storm, Wim E. Hennink, Jan A. A. M. Kamps, and Robbert J. Kok. 2017. E-Selectin Targeted Immunoliposomes for Rapamycin Delivery to Activated Endothelial Cells. *International Journal of Pharmaceutics*, no. September. Elsevier: 0–1. doi:10.1016/j.ijpharm.2017.10.027.

Gómara, María José, Ignacio Pérez-Pomeda, José María Gatell, Victor Sánchez-Merino, Eloisa Yuste, and Isabel Haro. 2017. Lipid Raft-like Liposomes Used for Targeted Delivery of a Chimeric Entry-Inhibitor Peptide with Anti-HIV-1 Activity. *Nanomedicine: Nanotechnology, Biology, and Medicine* 13(2): 601–609. doi:10.1016/j.nano.2016.08.023.

Gregori, Maria, Mark Taylor, Elisa Salvati, Francesca Re, Simona Mancini, Claudia Balducci, Gianluigi Forloni, et al. 2017. Retro-Inverso Peptide Inhibitor Nanoparticles as Potent Inhibitors of Aggregation of the Alzheimer's Aβ Peptide. *Nanomedicine: Nanotechnology, Biology, and Medicine* 13(2): 723–732. doi:10.1016/j.nano.2016.10.006.

Gu, Zili, Minglu Chang, Yang Fan, Yanbin Shi, and Guimei Lin. 2017. NGR-Modified pH-Sensitive Liposomes for Controlled Release and Tumor Target Delivery of Docetaxel. *Colloids and Surfaces B: Biointerfaces*. Vol. 160. doi:10.1016/j.colsurfb.2017.09.052.

Guo, Bohong, Danqiao Xu, Xiaohong Liu, and Jun Yi. 2017. Enzymatic Synthesis and in vitro Evaluation of Folate-Functionalized Liposomes. *Drug Design, Development and Therapy* 11: 1839–1847.

Gupta, Nilesh, Fahad I. Al-Saikhan, Brijeshkumar Patel, Jahidur Rashid, and Fakhrul Ahsan. 2015. Fasudil and SOD Packaged in Peptide-Studded-Liposomes: Properties, Pharmacokinetics and Ex-Vivo Targeting to Isolated Perfused Rat Lungs. *International Journal of Pharmaceutics* 488(1–2): 33–43. doi:10.1016/j.ijpharm.2015.04.031.

Haeri, Azadeh, Sara Zalba, Timo L. M. ten Hagen, Simin Dadashzadeh, and Gerben A. Koning. 2016. EGFR Targeted Thermosensitive Liposomes: A Novel Multifunctional Platform for Simultaneous Tumor Targeted and Stimulus Responsive Drug Delivery. *Colloids and Surfaces B: Biointerfaces* 146: 657–669. doi:10.1016/j.colsurfb.2016.06.012.

Jhaveri, Aditi, Pranali Deshpande, Bhushan Pattni, and Vladimir Torchilin. 2018. Transferrin-Targeted, Resveratrol-Loaded Liposomes for the Treatment of Glioblastoma. *Journal of Controlled Release* 277: 89–101. doi:10.1016/j.jconrel.2018.03.006.

Jiao, Zhuomin, Yang Li, Hengyuan Pang, Yongri Zheng, and Yan Zhao. 2017. Pep-1 Peptide-Functionalized Liposome to Enhance the Anticancer Efficacy of Cilengitide in Glioma Treatment. *Colloids and Surfaces B: Biointerfaces* 158: 68–75. doi:10.1016/j.colsurfb.2017.03.058.

Johnsen, Kasper Bendix, Annette Burkhart, Fredrik Melander, Paul Joseph Kempen, Jonas Bruun Vejlebo, Piotr Siupka, Morten Schallburg Nielsen, Thomas Lars Andresen, and Torben Moos. 2017. Targeting Transferrin Receptors at the Blood-Brain Barrier Improves the Uptake of Immunoliposomes and Subsequent Cargo Transport into the Brain Parenchyma. *Scientific Reports* 7(1): 1–13. doi:10.1038/s41598-017-11220-1.

Kabilova, Tatyana O., Elena V. Shmendel, Daniil V. Gladkikh, Elena L. Chernolovskaya, Oleg V. Markov, Nina G. Morozova, Mikhail A. Maslov, and Marina A. Zenkova. 2018. Targeted Delivery of Nucleic Acids into Xenograft Tumors Mediated by Novel Folate-Equipped Liposomes. *European Journal of Pharmaceutics and Biopharmaceutics* 123: 59–70. doi:10.1016/j.ejpb.2017.11.010.

Kato, Tatsuya, Cheng S Jin, Hideki Ujiie, Daiyoon Lee, Kosuke Fujino, Hironobu Wada, Hsin-pei Hu, et al. 2017. Nanoparticle Targeted Folate Receptor 1-Enhanced Photodynamic Therapy for Lung Cancer. *Lung Cancer* 113: 59–68. doi:10.1016/j.lungcan.2017.09.002.

Kim, Hyunggun, Patrick H. Kee, Yonghoon Rim, Melanie R. Moody, Melvin E. Klegerman, Deborah Vela, Shao Ling Huang, David D. McPherson, and Susan T. Laing. 2015. Nitric Oxide-Enhanced Molecular Imaging of Atheroma Using Vascular Cellular Adhesion Molecule 1-Targeted Echogenic Immunoliposomes. *Ultrasound in Medicine and Biology* 41(6): 1701–1710. doi:10.1016/j.ultrasmedbio.2015.02.002.

Kim, Jung Seok, Dae Hwan Shin, and Jin Seok Kim. 2018. Dual-Targeting Immunoliposomes Using Angiopep-2 and CD133 Antibody for Glioblastoma Stem Cells. *Journal of Controlled Release* 269(November 2017). Elsevier: 245–257. doi:10.1016/j.jconrel.2017.11.026.

Kim, Kenneth H, Danijela Jelovac, Deborah K Armstrong, Benjamin Schwartz, Susan C Weil, Charles Schweizer, and Ronald D Alvarez. 2016. Phase 1b Safety Study of Farletuzumab, Carboplatin and Pegylated Liposomal Doxorubicin in Patients with Platinum- Sensitive Epithelial Ovarian Cancer. *Gynecologic Oncology* 140(2): 13–17. doi:10.1016/j.copsyc.2015.03.007.Current.

Krämer, Stefanie D., and H. Wunderli-Allenspach. 2003. No Entry for TAT(44–57) into Liposomes and Intact MDCK Cells: Novel Approach to Study Membrane Permeation of Cell-Penetrating Peptides. *Biochimica et Biophysica Acta - Biomembranes* 1609(2): 161–169. doi:10.1016/S0005-2736(02)00683-1.

Kuang, Huihui, Sook Hee Ku, and Efrosini Kokkoli. 2017. The Design of Peptide-Amphiphiles as Functional Ligands for Liposomal Anticancer Drug and Gene Delivery. *Advanced Drug Delivery Reviews* 110–111: 80–101. doi:10.1016/j.addr.2016.08.005.

Labidi, Soumaya, Nesrine Mejri, Aymen Lagha, Nouha Daoud, Houda El Benna, Mehdi Afrit, and Hamouda Boussen. 2016. Targeted Therapies in HER2-Overexpressing Metastatic Breast Cancer. *Breast Care* 11(6): 418–422. doi:10.1159/000452194.

Li, Fengqiao, Hao Mei, Yu Gao, Xiaodong Xie, Huifang Nie, and Tao Li. 2017. Biomaterials Co-Delivery of Oxygen and Erlotinib by Aptamer-Modi Fi Ed Liposomal Complexes to Reverse Hypoxia-Induced Drug Resistance in Lung Cancer. *Biomaterials* 145. Elsevier Ltd: 56–71. doi:10.1016/j.biomaterials.2017.08.030.

Li, Fengqiao, Hao Mei, Xiaodong Xie, Huijuan Zhang, Jian Liu, Tingting Lv, and Huifang Nie. 2017. Aptamer-Conjugated Chitosan-Anchored Liposomal Complexes for Targeted Delivery of Erlotinib to EGFR-Mutated Lung Cancer Cells. *The AAPS Journal* 19(3): 814–826. doi:10.1208/s12248-017-0057-9.

Li, Jin Ming, Wei Zhang, Hua Su, Yuan Yuan Wang, Cai Ping Tan, Liang Nian Ji, and Zong Wan Mao. 2015. Reversal of Multidrug Resistance in MCF-7/Adr Cells by Codelivery of Doxorubicin and BCL2 siRNA Using a Folic

Acid-Conjugated Polyethylenimine Hydroxypropyl-β-Cyclodextrin Nanocarrier. *International Journal of Nanomedicine* 10: 3147–3162. doi:10.2147/IJN.S67146.

Li, Lielie, Xueqin An, and Xiaojuan Yan. 2015. Folate-Polydiacetylene-Liposome for Tumor Targeted Drug Delivery and Fluorescent Tracing. *Colloids and Surfaces B: Biointerfaces* 134. Elsevier B. V.: 235–239. doi:10.1016/j.colsurfb.2015.07.008.

Li, Tianshu, Takuya Amari, Kentaro Semba, Tadashi Yamamoto, and Shinji Takeoka. 2017. Construction and Evaluation of pH-Sensitive Immunoliposomes for Enhanced Delivery of Anticancer Drug to ErbB2 over-Expressing Breast Cancer Cells. *Nanomedicine: Nanotechnology, Biology, and Medicine* 13(3). Elsevier Inc.: 1219–1227. doi:10.1016/j.nano.2016.11.018.

Li, Weijun, François Nicol, and Francis C. Szoka. 2004. GALA: A Designed Synthetic pH-Responsive Amphipathic Peptide with Applications in Drug and Gene Delivery. *Advanced Drug Delivery Reviews* 56(7): 967–985. doi:10.1016/j.addr.2003.10.041.

Liao, Zi Xian, Er Yuan Chuang, Chia Chen Lin, Yi Cheng Ho, Kun Ju Lin, Po Yuan Cheng, Ko Jie Chen, Hao Ji Wei, and Hsing Wen Sung. 2015. An AS1411 Aptamer-Conjugated Liposomal System Containing a Bubble-Generating Agent for Tumor-Specific Chemotherapy That Overcomes Multidrug Resistance. *Journal of Controlled Release* 208. Elsevier B. V.: 42–51. doi:10.1016/j.jconrel.2015.01.032.

Liu, Haoran, Junhua Mai, Jianliang Shen, Joy Wolfram, Zhaoqi Li, Guodong Zhang, Rong Xu, et al. 2018. A Novel DNA Aptamer for Dual Targeting of Polymorphonuclear Myeloid-Derived Suppressor Cells and Tumor Cells. *Theranostics* 8(1): 31–44. doi:10.7150/thno.21342.

Liu, Mei, Wei Li, Caroline A. Larregieu, Meng Cheng, Bihan Yan, Ting Chu, Hui Li, and Sheng Jun Mao. 2014. Development of Synthetic Peptide-Modified Liposomes with LDL Receptor Targeting Capacity and Improved Anticancer Activity. *Molecular Pharmaceutics* 11(7): 2305–2312. doi:10.1021/mp400759d.

Liu, Min Chen, Lin Liu, Xia Rong Wang, Wu Ping Shuai, Ying Hu, Min Han, and Jian Qing Gao. 2016. Folate Receptor-Targeted Liposomes Loaded with a Diacid Metabolite of Norcantharidin Enhance Antitumor Potency for H22 Hepatocellular Carcinoma Both in vitro and in vivo. *International Journal of Nanomedicine* 11: 1395–1412. doi:10.2147/IJN.S96862.

Liu, Ruiwu, Xiaocen Li, Wenwu Xiao, and Kit S Lam. 2017. Tumor-Targeting Peptides from Combinatorial Libraries. *Advanced Drug Delivery Reviews* 110: 13–37. doi:10.1080/10937404.2015.1051611.INHALATION.

Lopalco, Antonio, Annalisa Cutrignelli, Nunzio Denora, Angela Lopedota, Massimo Franco, and Valentino Laquintana. 2018. Transferrin Functionalized Liposomes Loading Dopamine HCl: Development and Permeability Studies across an in vitro Model of Human Blood–Brain Barrier. *Nanomaterials* 8(3): 178. doi:10.3390/nano8030178.

Loureiro, Joana A., Bárbara Gomes, Gert Fricker, Isabel Cardoso, Carlos A. Ribeiro, Cristiana Gaiteiro, Manuel A N Coelho, Maria do Carmo Pereira, and Sandra Rocha. 2015. Dual Ligand Immunoliposomes for Drug Delivery to the Brain. *Colloids and Surfaces B: Biointerfaces* 134: 213–219. doi:10.1016/j.colsurfb.2015.06.067.

Lozano, Neus, Zahraa S. Al-Ahmady, Nicolas S. Beziere, Vasilis Ntziachristos, and Kostas Kostarelos. 2015. Monoclonal Antibody-Targeted PEGylated Liposome-ICG Encapsulating Doxorubicin as a Potential Theranostic Agent. *International Journal of Pharmaceutics* 482(1–2). Elsevier B. V.: 2–10. doi:10.1016/j.ijpharm.2014.10.045.

Martinelli, E., R. De Palma, M. Orditura, F. De Vita, and F. Ciardiello. 2009. Anti-Epidermal Growth Factor Receptor Monoclonal Antibodies in Cancer Therapy. *Clinical and Experimental Immunology* 158(1): 1–9. doi:10.1111/j.1365-2249.2009.03992.x.

Meissner, Justyna M., Monika Toporkiewicz, Aleksander Czogalla, Lucyna Matusewicz, Kazimierz Kuliczkowski, and Aleksander F. Sikorski. 2015. Novel Antisense Therapeutics Delivery Systems: in vitro and in vivo Studies of Liposomes Targeted with Anti-CD20 Antibody. *Journal of Controlled Release* 220. The Authors: 515–528. doi:10.1016/j.jconrel.2015.11.015.

Miller, Kathy, Javier Cortes, Sara A Hurvitz, Ian E Krop, Debu Tripathy, Sunil Verma, Kaveh Riahi, et al. 2016. HERMIONE: A Randomized Phase 2 Trial of MM-302 plus Trastuzumab versus Chemotherapy of Physician's Choice plus Trastuzumab in Patients with Previously Treated, Anthracycline-Naïve, HER2-Positive, Locally Advanced/ metastatic Breast Cancer. *BMC Cancer* 16(1). BMC Cancer: 352. doi:10.1186/s12885-016-2385-z.

Miura, Naoya, Hidetaka Akita, Naho Tateshita, Takashi Nakamura, and Hideyoshi Harashima. 2017. Modifying Antigen-Encapsulating Liposomes with KALA Facilitates MHC Class I Antigen Presentation and Enhances Anti-Tumor Effects. *Molecular Therapy* 25(4): 1003–1013. doi:10.1016/j.ymthe.2017.01.020.

Moghimipour, Eskandar, Mohsen Rezaei, Zahra Ramezani, Maryam Kouchak, Mohsen Amini, Kambiz Ahmadi Angali, Farid Abedin Dorkoosh, and Somayeh Handali. 2018. Transferrin Targeted Liposomal

5-Fluorouracil Induced Apoptosis via Mitochondria Signaling Pathway in Cancer Cells. *Life Sciences* 194. Elsevier: 104–110. doi:10.1016/j. lfs.2017.12.026.

Moles, Ernest, Silvia Galiano, Ana Gomes, Miguel Quiliano, Cátia Teixeira, Ignacio Aldana, Paula Gomes, and Xavier Fernàndez-Busquets. 2017. ImmunoPEGliposomes for the Targeted Delivery of Novel Lipophilic Drugs to Red Blood Cells in a Falciparum Malaria Murine Model. *Biomaterials* 145: 178–191. doi:10.1016/j.biomaterials.2017.08.020.

Moles, Ernest, Kirsten Moll, Jun Hong Ch'ng, Paolo Parini, Mats Wahlgren, and Xavier Fernandez-Busquets. 2016. Development of Drug-Loaded Immunoliposomes for the Selective Targeting and Elimination of Rosetting Plasmodium Falciparum-Infected Red Blood Cells. *Journal of Controlled Release* 241: 57–67. doi:10.1016/j.jconrel.2016.09.006.

Moles, Ernest, Patricia Urbán, María Belén Jiménez-Díaz, Sara Viera-Morilla, Iñigo Angulo-Barturen, Maria Antònia Busquets, and Xavier Fernàndez-Busquets. 2015. Immunoliposome-Mediated Drug Delivery to Plasmodium-Infected and Non-Infected Red Blood Cells as a Dual Therapeutic/Prophylactic Antimalarial Strategy. *Journal of Controlled Release* 210: 217–229. doi:10.1016/j.jconrel.2015.05.284.

Monteiro, Liziane O. F., Renata S. Fernandes, Caroline M. R. Oda, Sávia C. Lopes, Danyelle M. Townsend, Valbert N. Cardoso, Mônica C. Oliveira, Elaine A. Leite, Domenico Rubello, and André L. B. de Barros. 2018. Paclitaxel-Loaded Folate-Coated Long Circulating and pH-Sensitive Liposomes as a Potential Drug Delivery System: A Biodistribution Study. *Biomedicine and Pharmacotherapy* 97: 489–495. doi:10.1016/j. biopha.2017.10.135.

Nahar, Kamrun, Shahriar Absar, Nilesh Gupta, Venkata Ramana Kotamraju, Ivan F. McMurtry, Masahiko Oka, Masanobu Komatsu, Eva Nozik-Grayck, and Fakhrul Ahsan. 2014. Peptide-Coated Liposomal Fasudil Enhances Site Specific Vasodilation in Pulmonary Arterial Hypertension. *Molecular Pharmaceutics* 11(12): 4374–4384. doi:10.1021/mp500456k.

Noble, Gavin T., Jared F. Stefanick, Jonathan D. Ashley, Tanyel Kiziltepe, and Basar Bilgicer. 2014. Ligand-Targeted Liposome Design: Challenges and Fundamental Considerations. *Trends in Biotechnology* 32(1): 32–45. doi:10.1016/j.tibtech.2013.09.007.

Normanno, Nicola, Antonella De Luca, Caterina Bianco, Luigi Strizzi, Mario Mancino, Monica R. Maiello, Adele Carotenuto, Gianfranco De Feo, Francesco Caponigro, and David S. Salomon. 2006. Epidermal Growth

Factor Receptor (EGFR) Signaling in Cancer. *Gene* 366(1): 2–16. doi:10.1016/j.gene.2005.10.018.

Ordóñez-Gutiérrez, Lara, Adrián Posado-Fernández, Davoud Ahmadvand, Barbara Lettiero, Linping Wu, Marta Antón, Orfeu Flores, Seyed Moein Moghimi, and Francisco Wandosell. 2016. Immuno PEG liposome-Mediated Reduction of Blood and Brain Amyloid Levels in a Mouse Model of Alzheimer's Disease Is Restricted to Aged Animals. *Biomaterials* 112. doi:http://dx.doi.org/10.1016/j.biomaterials.2016.07.027.

Park, Hanna, Dong Min Kim, Si Eun Baek, Keun Sik Kim, and Dong Eun Kim. 2015. Comparison of Drug Delivery Efficiency between Doxorubicin Intercalated in RNA Aptamer and One Encapsulated in RNA Aptamer-Conjugated Liposome. *Bulletin of the Korean Chemical Society* 36(10): 2494–2500. doi:10.1002/bkcs.10480.

Petazzi, Roberto Arturo, Andrea Gramatica, Andreas Herrmann, and Salvatore Chiantia. 2015. Time-Controlled Phagocytosis of Asymmetric Liposomes: Application to Phosphatidylserine Immunoliposomes Binding HIV-1 Virus-like Particles. *Nanomedicine: Nanotechnology, Biology, and Medicine* 11(8). Elsevier Inc.: 1985–1992. doi:10.1016/j.nano.2015.06.004.

Petrilli, Raquel, Josimar O. Eloy, Renata F. V. Lopez, and Robert J. Lee. 2016. Cetuximab Immunoliposomes Enhances Delivery of 5-FU to Skin Squamous Carcinoma Cells. *Anti-Cancer Agents in Medicinal Chemistry* 16. doi:10.2174/1871520616666160526110913.

Plourde, Kevin, Rabeb Mouna Derbali, Arnaud Desrosiers, Céline Dubath, Alexis Vallée-Bélisle, and Jeanne Leblond. 2017. Aptamer-Based Liposomes Improve Specific Drug Loading and Release. *Journal of Controlled Release* 251: 82–91. doi:10.1016/j.jconrel.2017.02.026.

Poh, Scott, Venkatesh Chelvam, Wilfredo Ayala-López, Karson S. Putt, and Philip S. Low. 2018. Selective Liposome Targeting of Folate Receptor Positive Immune Cells in Inflammatory Diseases. *Nanomedicine: Nanotechnology, Biology, and Medicine* 14(3). Elsevier Inc.: 1033–1043. doi:10.1016/j.nano.2018.01.009.

Rabenhold, Markus, Frank Steiniger, Alfred Fahr, Roland E. Kontermann, and Ronny Rüger. 2015. Bispecific Single-Chain Diabody-Immunoliposomes Targeting Endoglin (CD105) and Fibroblast Activation Protein (FAP) Simultaneously. *Journal of Controlled Release* 201. Elsevier B. V.: 56–67. doi:10.1016/j.jconrel.2015.01.022.

Ramana, Lakshmi Narashimhan, Shilpee Sharma, Swaminathan Sethuraman, Udaykumar Ranga, and Uma Maheswari Krishnan. 2015. Stealth Anti-CD4 Conjugated Immunoliposomes with Dual Antiretroviral

Drugs - Modern Trojan Horses to Combat HIV. *European Journal of Pharmaceutics and Biopharmaceutics* 89. Elsevier B. V.: 300–311. doi:10.1016/j.ejpb.2014.11.021.

Ramsey, Joshua D., and Nicholas H. Flynn. 2015. Cell-Penetrating Peptides Transport Therapeutics into Cells. *Pharmacology and Therapeutics* 154: 78–86. doi:10.1016/j.pharmthera.2015.07.003.

Riaz, Muhammad Kashif, Muhammad Adil Riaz, Xue Zhang, Congcong Lin, Ka Hong Wong, Xiaoyu Chen, Ge Zhang, Aiping Lu, and Zhijun Yang. 2018. Surface Functionalization and Targeting Strategies of Liposomes in Solid Tumor Therapy: A Review. *International Journal of Molecular Sciences* 19(1): 195–222. doi:10.3390/ijms19010195.

Saesoo, S., S. Sathornsumetee, P. Anekwiang, C. Treetidnipa, P. Thuwajit, S. Bunthot, W. Maneeprakorn, et al. 2018. Characterization of Liposome-Containing SPIONs Conjugated with Anti-CD20 Developed as a Novel Theranostic Agent for Central Nervous System Lymphoma. *Colloids and Surfaces B: Biointerfaces* 161. Elsevier B. V.: 497–507. doi:10.1016/j. colsurfb.2017.11.003.

Santiwarangkool, Sarochin, Hidekata Akita, Taichi Nakatani, Kenji Kusumoto, Hiroki Kimura, Masaru Suzuki, Masaharu Nishimura, Yusuke Sato, and Hideyoshi Harashima. 2017. PEGylation of the GALA Peptide Enhances the Lung-Targeting Activity of Nanocarriers That Contain Encapsulated siRNA. *Journal of Pharmaceutical Sciences* 106(9). Elsevier Ltd: 2420–2427. doi:10.1016/j.xphs.2017.04.075.

Sercombe, Lisa, Tejaswi Veerati, Fatemeh Moheimani, Sherry Y. Wu, Anil K. Sood, and Susan Hua. 2015. Advances and Challenges of Liposome Assisted Drug Delivery. *Frontiers in Pharmacology* 6 (DEC): 1–13. doi:10.3389/fphar.2015.00286.

Shi, Chenyang, Hui Cao, Wei He, Fei Gao, Yu Liu, and Lifang Yin. 2015. Novel Drug Delivery Liposomes Targeted with a Fully Human Anti-VEGF165 Monoclonal Antibody Show Superior Antitumor Efficacy in vivo. *Biomedicine and Pharmacotherapy* 73. Elsevier Masson SAS: 48–57. doi:10.1016/j.biopha.2015.05.008.

Shin, Dae Hwan, Sang Jin Lee, Jung Seok Kim, Jae Ha Ryu, and Jin Seok Kim. 2015. Synergistic Effect of Immunoliposomal Gemcitabine and Bevacizumab in Glioblastoma Stem Cell-Targeted Therapy. *Journal of Biomedical Nanotechnology* 11(11): 1989–2002. doi:10.1166/ jbn.2015.2146.

Siwowska, Klaudia, Raffaella Schmid, Susan Cohrs, Roger Schibli, and Cristina Müller. 2017. Folate Receptor-Positive Gynecological Cancer

Cells: in vitro and in vivo Characterization. *Pharmaceuticals* 10(3): 72. doi:10.3390/ph10030072.

Song, Xingli, Yi Ren, Jing Zhang, Gang Wang, Xuedong Han, Wei Zheng, and Linlin Zhen. 2015. Targeted Delivery of Doxorubicin to Breast Cancer Cells by Aptamer Functionalized DOTAP/DOPE Liposomes. *Oncology Reports* 34(4): 1953–1960. doi:10.3892/or.2015.4136.

Sriraman, Shravan Kumar, Giusseppina Salzano, Can Sarisozen, and Vladimir Torchilin. 2016. Anti-Cancer Activity of Doxorubicin-Loaded Liposomes Co-Modified with Transferrin and Folic Acid. *European Journal of Pharmaceutics and Biopharmaceutics* 105: 40–49.

Tai, Wanyi, Rubi Mahato, and Kun Cheng. 2010. The Role of HER2 in Cancer Therapy and Targeted Drug Delivery. *Journal of Controlled Release* 146(3). Elsevier B. V.: 264–275. doi:10.1016/j.jconrel.2010.04.009.

Ucar, Eser, Serap Teksoz, Cigdem Ichedef, Ayfer Yurt Kilcar, E Ilker Medine, Kadir Ari, Yasemin Parlak, B Elvan Sayit Bilgin, and Perihan Unak. 2017. Synthesis, Characterization and Radiolabeling of Folic Acid Modified Nanostructured Lipid Carriers as a Contrast Agent and Drug Delivery System. *Applied Radiation and Isotopes* 119: 72–79.

Wang, Jian, Zhitao Wu, Guoyu Pan, Junsheng Ni, Fangyuan Xie, Beige Jiang, Lixin Wei, Jie Gao, and Weiping Zhou. 2017. Enhanced Doxorubicin Delivery to Hepatocellular Carcinoma Cells via CD147 Antibody-Conjugated Immunoliposomes. *Nanomedicine: Nanotechnology, Biology, and Medicine.* Elsevier Inc., 1–12. doi:10.1016/j.nano.2017.09.012.

Wicki, Andreas, Reto Ritschard, Uli Loesch, Stefanie Deuster, Christoph Rochlitz, and Christoph Mamot. 2015. Large-Scale Manufacturing of GMP-Compliant Anti-EGFR Targeted Nanocarriers: Production of Doxorubicin-Loaded Anti-EGFR-Immunoliposomes for a First-in-Man Clinical Trial. *International Journal of Pharmaceutics* 484(1–2). Elsevier B. V.: 8–15. doi:10.1016/j.ijpharm.2015.02.034.

Williams, Jacob B, Clara M Buchanan, Ghaleb Husseini, and William G Pitt. 2017. Cytosolic Delivery of Doxorubicin from Liposomes to Multidrug-Resistant Cancer Cells via Vaporization of Perfluorocarbon Droplets. *Journal of Nanomedicine Research* 5(4): 0001221:10. doi:10.15406/jnmr.2017.05.00122.

Yang, Yanfang, Xiangyang Xie, Xueqing Xu, Xuejun Xia, Hongliang Wang, Lin Li, Wujun Dong, et al. 2016. Thermal and Magnetic Dual-Responsive Liposomes with a Cell-Penetrating Peptide-siRNA Conjugate for Enhanced and Targeted Cancer Therapy. *Colloids and Surfaces B: Biointerfaces* 146: 607–615. doi:10.1016/j.colsurfb.2016.07.002.

Yang, Yang, Yan Fang Yang, Xiang Yang Xie, Zhi Yuan Wang, Wei Gong, Hui Zhang, Ying Li, Fang Lin Yu, Zhi Ping Li, and Xing Guo Mei. 2015. Dual-Modified Liposomes with a Two-Photon-Sensitive Cell Penetrating Peptide and NGR Ligand for siRNA Targeting Delivery. *Biomaterials* 48: 84–96. doi:10.1016/j.biomaterials.2015.01.030.

Ying, Man, Qing Shen, Changyou Zhan, Xiaoli Wei, Jie Gao, Cao Xie, Bingxin Yao, and Weiyue Lu. 2016. A Stabilized Peptide Ligand for Multifunctional Glioma Targeted Drug Delivery. *Journal of Controlled Release* 243: 86–98. doi:10.1016/j.jconrel.2016.09.035.

Zalba, Sara, Ana M. Contreras, Azadeh Haeri, Timo L M Ten Hagen, I??igo Navarro, Gerben Koning, and Mar??a J. Garrido. 2015. Cetuximab-Oxaliplatin-Liposomes for Epidermal Growth Factor Receptor Targeted Chemotherapy of Colorectal Cancer. *Journal of Controlled Release* 210: 26–38. doi:10.1016/j.jconrel.2015.05.271.

Zhang, Bin, Yueying Zhang, and Dongmei Yu. 2017. Lung Cancer Gene Therapy: Transferrin and Hyaluronic Acid Dual Ligand-Decorated Novel Lipid Carriers for Targeted Gene Delivery. *Oncology Reports* 37(2): 937–944. doi:10.3892/or.2016.5298.

Zhen, Shuai, Yoichiro Takahashi, Shunichi Narita, Yi-chen Yang, and Xu Li. 2017. Targeted Delivery of CRISPR / Cas9 to Prostate Cancer by Modified gRNA Using a Flexible Aptamer-Cationic Liposome. *Oncotarget* 8(6): 9375–9387.

Zhou, Zhizhi, Mingying Liu, and Jiahuan Jiang. 2018. The Potential of Aptamers for Cancer Research. *Analytical Biochemistry*. Elsevier Inc. doi:10.1016/j.ab.2018.03.008.

Zuo, Tiantian, Yuanyuan Guan, Minglu Chang, Fang Zhang, Shanshan Lu, Ting Wei, Wei Shao, and Guimei Lin. 2016. RGD (Arg-Gly-Asp) Internalized Docetaxel-Loaded pH Sensitive Liposomes: Preparation, Characterization and Antitumoral Efficacy in vivo and in vitro. *Colloids and Surfaces B : Biointerfaces* 147: 90–99.

Zylberberg, Claudia, and Sandro Matosevic. 2016. Pharmaceutical Liposomal Drug Delivery: A Review of New Delivery Systems and a Look at the Regulatory Landscape. *Drug Delivery* 23(9): 3319–3329. doi:10.1080/ 10717544.2016.1177136.

Chapter 4

Liposomal Drug Delivery System and Its Clinically Available Products

Upendra Bulbake, Nagavendra Kommineni, and Wahid Khan

Department of Pharmaceutics, National Institute of Pharmaceutical Education and Research (NIPER), Hyderabad 500037, Telangana, India

mail4wahid@gmail.com

Liposomal drug delivery systems are extensively investigated clinically with many liposome-based formulations and technologies have already been utilized in the clinic. Liposomes are the first nanocarrier-based delivery system to make the transition from concept to clinical applications, and they are now a well-known technology platform with significant clinical acceptance. Liposomes are small, spherical vesicles with enclosed compartments separating an aqueous medium from phospholipid bilayer. Many drugs, including anticancer and antimicrobial agents, vaccines, genetic materials, etc., have been loaded into the aqueous or lipid phases of liposomes which alter their biodistribution profile and hence enhance their therapeutic indexes. A number of liposome-based products are on the market (Doxil®, Ambisome®, Epaxal®, Vyxeos™, etc.) and many more are in the pipeline (LEM-ETU, Endotag-I, Arikace, etc.). This

Handbook of Materials for Nanomedicine: Lipid-Based and Inorganic Nanomaterials
Edited by Vladimir Torchilin
Copyright © 2020 Jenny Stanford Publishing Pte. Ltd.
ISBN 978-981-4800-91-4 (Hardcover), 978-1-003-04507-6 (eBook)
www.jennystanford.com

chapter provides an overview of a liposomal drug carrier system, including its properties, classification, stability aspects, and method of preparation. Detailed insights about the formulation aspects of liposome-based product currently in clinic are also covered.

4.1 Introduction

The field of liposomal drug delivery system is rapidly expanding and its potentials have already been demonstrated by the many products on the market [1]. Liposomes, biodegradable and biocompatible nanocarriers, were discovered by A. Bangham in 1961 at the Babraham Institute, University of Cambridge [2]. Since then tremendous effort has been focused on the development of liposome as an efficient therapeutic tool against severe diseases including cancer, infectious disorders, vaccine delivery, gene delivery, etc. Several promising small-molecule drugs and genes previously considered less than useful due to problems of solubility, stability and toxicity can now be administered with the help of liposomes [3, 4]. Liposomes are formed by dispersing phospholipids in water and as a result forming one bilayer or a number of bilayers, each separated by aqueous phase which allow entrapping hydrophilic drugs into their aqueous space, while allowing hydrophobic drugs to be entrapped into the lipid bilayers (Fig. 4.1). The main structural components of liposomes are phospholipids/synthetic amphiphiles together with cholesterol, to control membrane permeability. Several methods have been utilized for the preparation of liposomes such as thin film hydration, ether and ethanol injection, freeze thawing, etc. Among them, the thin film hydration method is most widely used, in which components, i.e., phospholipids mixtures and/or drug, are dissolved in an organic solvent, then drying down them by rotary evaporation followed by addition of aqueous solvent to hydrate the film.

Currently, a good number of liposomal therapeutics are available for clinical use, approved by either the FDA in the United States (USFDA) or the European Medicines Agency (EMA) in the European Union. Doxil®, polyethylene glycol (PEG) functionalized liposomes loaded with doxorubicin, was the first approved liposomal product (FDA 1995) developed by Gabizon and Barenholz for Sequus

Pharmaceuticals, USA, for the treatment of HIV-associated Kaposi's sarcoma, advanced ovarian cancer, and multiple myeloma [5]. After Doxil®, other liposomal formulations are approved for various cancer treatment such as liposomal daunorubicin (DaunoXome®), liposomal vincristine (Marqibo®), etc., and most recent products include liposomal irinotecan (Onivyde™), and daunorubicin and cytarabine combination liposomes (Vyxeos®) were approved by the USFDA. Non-PEGylated liposomal doxorubicin (Myocet®) and liposomal mifamurtide (Mepact®) were approved by the EMA. Regulatory authorities also approved amphotericin B containing liposomes Amphotec® and Ambisome® in 1996 and 1997, respectively, for the treatment of serious fungal infections. Liposomal vaccine–based products, Epaxal® and Inflexal® V, are also clinically available which are manufactured by Crucell, Berna Biotech for vaccination against hepatitis and influenza viruses, respectively. The majority of these liposomes are not PEGylated, with the exception of Doxil® and Onivyde™ [6], which is surprising given the widely known advantages of small amounts of PEG have shown to confer to liposome delivery systems [7, 8]. Moreover, all of these liposomes are passively targeted, with no chemical or active based targeting moieties.

Considerable research has been done in the field of liposomes and many reviews on the biophysical aspects of liposome preparation, characterization, and optimization have been composed [9–12]. In this chapter, basic properties of liposomes and a comprehensive insight into the clinically available liposome formulations are covered in detail.

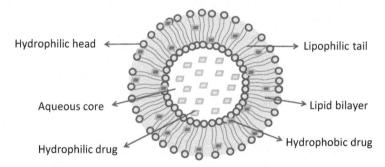

Figure 4.1 Schematic representation of liposome structure and its components.

4.2 Classification of Liposomes

Liposomes are classified either by the method of preparation or by the number of bilayers formed in the vesicle, or by size. The description of liposomes by the lamellarity and size is more common than by the method of their preparation. This chapter discusses the classification of liposomes on the basis of lamellarity.

4.2.1 Multilamellar Vesicles

Multilamellar Vesicle (MLV) liposomes have a size larger than 0.1 µm and consist of two or more lipid bilayers. Their method of preparation is simple, which includes thin film hydration method or hydration of lipids in excess of solvent. They are mechanically more stable on long-term storage. Because of the large size, they are cleared quickly by the reticulo-endothelial system (RES) cells and hence can be useful for targeting RES organs [13]. MLV have a larger trapped volume, i.e., amount of aqueous volume to lipid ratio. The drug entrapment into the liposomes can be enhanced by decreasing the hydration rate and by gentle mixing [14]. Encapsulation efficiency can also be enhanced by hydrating thin films of dry lipids [15]. Subsequent freeze drying and rehydration after mixing with the aqueous solvent (containing the drug) can form MLV with high encapsulation efficiency [16, 17].

4.2.2 Unilamellar Vesicles

Unilamellar vesicle (ULV) liposomes are spherical vesicles, composed by a single lipid bilayer of an amphiphilic lipid or mixture of those lipids, containing aqueous volume inside the chamber. ULVs are divided into two classes: small unilamellar vesicles (SUVs), with particle sizes up to 0.1 µm, and large unilamellar vesicles (LUVs), with particle sizes more than 0.1 µm up to a few microns.

4.2.2.1 Small unilamellar vesicles

Small unilamellar vesicles (SUVs) are smaller in size (generally less than 0.1 µm) in comparison to MLV and LUV, and have a single lipid bilayer. They have a low entrapped aqueous volume to lipid ratio and possess long circulation half-life. SUVs can be formulated by using solvent injection method (ether or ethanol injection methods) [18] or otherwise by reducing the size of MLV or LUV using extrusion or sonication process under an inert environment like nitrogen or Argon. The sonication can be done by using either a probe- or a bath-type sonicator. SUV can also be formed by passing MLV through a narrow opening under high pressure. SUVs are prone to aggregation and fusion if their surface charge is lower or negligible [19].

4.2.2.2 Large unilamellar vesicles

Large unilamellar vesicles (LUVs) consist of a single lipid bilayer and larger size generally greater than 0.1 μm. They possess good encapsulation efficiency, because they can keep a large volume of solvent in their cavity [20]. They contain high trapped volume therefore can be useful for entrapping hydrophilic drugs. LUV main advantage is that by using less amount of lipid large quantity of drug can be encapsulated. Like MLV, they are quickly cleared by RES cells, because of their larger size [13]. LUV can be prepared by different methods like detergent dialysis, ether injection, reverse phase evaporation, etc.

4.3 Basic Components in the Preparation of Liposomes

Liposomes are flexible in that their entire lipid membrane can be composed of either natural or synthetic phospholipids. Their properties can be modified entirely using suitable phospholipids. Their basic components are phospholipids and cholesterol as a stabilizer, other stabilizers also added sometimes to the mixture according to specific function of the liposome. Various types of lipids and their mixtures can be used to prepare a specific type of liposome.

4.3.1 Phospholipids

A phospholipid has both a hydrophobic and a hydrophilic component. Hence, it is called amphipathic molecule. A single phospholipid molecule has a phosphate head group on one end and two fatty acids chains side-by-side of that which makes the lipid "tails." The head is polar and hydrophilic due to negatively charged phosphate group. The phosphate groups, therefore, are attracted to the hydrophilic molecules in their environment. On the other hand, the lipid tails are nonpolar, hydrophobic, and uncharged. Hence, hydrophobic molecules are repelled by water. Some lipid tails contain saturated fatty acids and some contain unsaturated fatty acids. Phospholipid molecules as such are not water soluble, but rather, the molecules fused and align themselves in a planar bilayer form. Due to this planar arrangement hydrophilic portion of lipid bilayers has ability to interact with water. However, the hydrophobic portion will align away from water molecules.

Phospholipids are subdivided into four classes: phosphatidyl-choline, phosphatidylethanolamine, phosphatidylglycerol, and phosphatidylserine. These phospholipids possess the common features of fatty acids esterified to the 1 and 2 positions of the glycerol moiety with the phosphate group esterified to the 3 position.

Phosphatidylcholine, often referred to as lecithin, can be obtained from both synthetic and natural sources. It is readily extracted from soya bean and egg yolk but less readily from spinal cord and bovine heart. Phospholipids that contain the choline group are one of the most abundant lipids in nature. This class of phospholipids is most often used for liposome preparation. Phosphatidylcholine is very commonly used because it is relatively less costly, generally carries neutral charge, and is chemically inert. Another commonly used choline containing phospholipid is sphingomyelin or phosphoryl choline, which is readily extracted from nervous tissue. It is also a neutral molecule. Sphingomyelin hydrophobic chains are much more saturated than phosphatidylcholine. Sphingomyelins' transition phase temperature is also higher compared to the phosphatidylcholine phase transition temperature, near 37°C. This can bring heterogeneity in the bilayer membrane and generates domains in the bilayer membrane which can interact with adjacent sphingomyelin molecules and cholesterol.

Phosphatidylethanolamine head group is similar to phosphatidylcholines and directly attached hydrogen to the nitrogen of ethanolamine allows hydrogen-bonding interactions of neighboring molecules in the membrane. At neutral or low pH, the amino group is protonated and gives rise to neutral molecule, which prefers to produce hexagonal inverted lamellar structures only above the phase transition temperature.

In phosphatidylglycerol, the phosphate moiety esterifies the alcohol, which forms glycerol. Phosphatidylglycerol is found as a natural component of the lung surfactant of humans. It is synthesized by exchanging head group of a phosphatidylcholine-enriched phospholipid using the phospholipase D enzyme. Phosphatidylglycerol in the physiological pH range possesses a permanent negative charge.

Phosphatidylserine is a glycerophospholipid. It contains of two fatty acids linked by ester linkage to the glycerol first and second carbons and serine linked by a phosphodiester linkage to the glycerol third carbon. Phosphatidylserine is synthesized by base exchange reactions with phosphatidylethanolamine and phosphatidylcholine. The net charge on the phosphatidylserine head group is negative because of the negatively charged phosphate group.

4.3.2   Cholesterol

Cholesterol is one of the main components normally included into the lipid membrane of liposomes. It does not create the specific recognizable bilayer structure on its own. Cholesterol molecule orients itself between the phospholipid molecules with its hydroxyl group facing the water phase, the tricyclic ring sandwiched between the first few carbons of the fatty acyl chains, into the hydrocarbon core of the lipid bilayer. Cholesterol addition into the mixture stabilizes the liposomes, which in turn increases the phase transition temperature (T_c) of the bilayer membrane. Cholesterol addition also decreases the permeability of the membrane, hence keeping the liposome stable and intended drug entrapped [21]. Premature drug release is observed in liposomes without cholesterol and this is increased further when liposomes are confronted with high-density lipoproteins, which take up the phospholipids. Cholesterol is thought to obstruct this process, which in turn stabilizes the entire lipid membrane of the liposome [22].

4.4   Fundamental Properties of Liposomes

Liposomes are specific in their ability to incorporate drugs with different nature (hydrophilic or hydrophobic), which differs broadly in physicochemical properties such as bilayer fluidity, surface hydration, surface charge, and liposome size [23]. These properties have significant effects on the physical and/or biological performance of liposomes. Some major properties are discussed in the following subsections.

4.4.1 Fluidity of Lipid Bilayer

Lipid bilayer of liposome exhibits different physical states, a well-ordered arrangement or rigid gel phase below the phase transition temperature (T_c) and a disordered arrangement or fluid phase above T_c. At a temperature corresponding to T_c, the two phases coexist in equal proportions and maximum leakage of encapsulated drug from liposome is observed [24]. The fluidity of a liposome bilayer can be modified by choosing phospholipids with different T_c, which varies from −20 to 80°C depending upon the fatty acid chains length and nature (saturated or unsaturated). The liposome bilayer membranes composed of high-T_c lipids ($T_c > 37°C$) are less fluid at physiological temperature and less leaky, and liposomes composed of low-T_c lipids ($T_c < 37°C$) are more leaky at physiological temperatures. Incorporation of cholesterol also alters the fluidity of the liposome bilayer. Cholesterol in higher amounts (>30 mol%) can eliminate phase transition and also decrease the membrane permeability at a temperature $> T_c$, whereas incorporation of cholesterol in low concentration into the bilayer produces an increase in the membrane permeability [21].

4.4.2 Surface Charge

The surface charge on the liposome is an important property which influences the stability, biodistribution, and extent of liposome–cell interaction. On the basis of head group composition of the lipid and pH, liposomes may possess a neutral, positive, or negative charge on their surface. Lack of surface charge can cause higher liposomes aggregation, which in turn reduces their physical stability [25]. Liposomes with a neutral surface charge do not interact considerably with cells and the encapsulated drug mainly released extracellularly first and then enters into the cells [26]. After systemic administration, they have a minimum tendency to be cleared by cells of the reticulo-endothelial system (RES) and the maximum tendency to aggregate. Negatively charged liposomes are stable in dispersion and possess fewer tendencies to aggregate. Negatively charged liposomes show nonspecific cellular uptake and are taken up by the cell at a faster rate and to a higher extent compared to neutral liposomes [27]. Positively charged liposomes have been proposed to release their contents to cells by fusion with the cell membranes [28]. They are often utilized as a nucleic acid condensation reagent for nucleic acid delivery into the cells in gene therapy and possess higher tendency to interact with serum proteins, which results in enhanced uptake by the RES cells and eventual clearance by the spleen, lung, or liver. It has been shown that high doses of cationic liposomes produce varying degrees of tissue inflammation [29].

4.4.3 Liposome Size

The size of the liposome affects vesicle distribution and its clearance after systemic administration. The rate of liposome opsonized by RES increases with the size of the vesicles, i.e., small liposomes (≤ 0.1 μm) are cleared less rapidly and to a lower extent then large liposomes (>0.1 μm) [30]. Also, small liposomes show longer circulation half-life and increased accumulation in tumor tissue than large liposomes. Tumor capillaries are usually leakier than normal capillaries. They have large openings (up to 500 nm), through which liposomes can permeate and fluid can leak along with plasma proteins [31]. The extravasation of liposomes occurs due to the difference between intravascular hydrostatic and interstitial pressures [32].

4.4.4 Surface Steric Effect

Incorporation of small amounts of hydrophilic polymers (monosialoganglioside, polyethylene glycol (PEG), etc.) to the surface of the liposome bilayer membrane prevents the interaction of liposomes with blood components and reduces their aggregation. This strategy is generally called surface hydration or steric modification [10, 33]. This surface hydration offers steric hindrance to opsonin molecules adsorption on liposome which reduces the rate of liposome uptake by the RES cells (Fig. 4.2); hence, these sterically stabilized liposomes are more stable in biological environment and can show up to 10-fold greater circulation half-lives than conventional liposomes [34, 35].

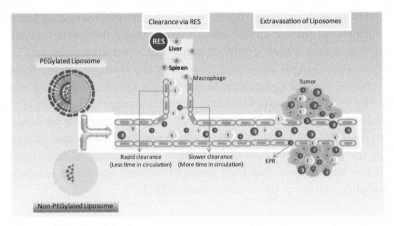

Figure 4.2 Pharmacokinetics of PEGylated liposomes and Non-PEGylated liposomes, adapted figure [1].

4.5 Limitations of Liposomes

4.5.1 Stability

Liposomal drug delivery systems must be sufficiently stable both chemically and physically so that they can be stored for longer periods of time, at least 1 year. Depending on their compositions, the liposomal drug delivery systems may have short shelf-lives partially because of physical and chemical instability.

Chemical instability may happen because of hydrolysis of four ester bonds and/or oxidation of unsaturated fatty acyl chains of liposome-forming phospholipids. Hydrolysis of ester bonds of phospholipids may be completely prohibited by water removal using freeze-drying process with a cryoprotectant. However, hydrophilic, low molecular weight, and non-bilayer interacting drugs containing liposomes may be stored as an aqueous dispersion because of the physical stability problems come across with freeze-drying of liposomes (e.g., loss of drug after rehydration and the tendency of increase in the average particle size). The storage temperature and pH of aqueous dispersion were found to be two most important parameters affecting hydrolysis of the phospholipid.

Therefore, to increase their stability, it is recommended that they should be stored at low temperatures (2–6°C, in a refrigerator) and at pH values close to neutral because phospholipids have their maximum stability at pH 6.5. Oxidative degradation can be reduced by preparation of liposomes under an oxygen free atmosphere, by addition of antioxidants, by use of high quality raw materials, and by storing at low temperatures which also reduces the rate of oxidation. Moreover, the partially saturated phospholipids for the preparation of the liposomes could be a better choice than polyunsaturated fatty acyl chains containing phospholipids [36].

Physical instability may be caused by changes in the mean particle size and size distribution because of vesicle fusion and aggregation, and entrapped drug loss due to leakage. Both of these processes (change in liposome size and drug leakage) alter the in vivo performance of the liposomal formulation and hence may affect the therapeutic effect of the drug. Changes in the mean particle size and size distribution of liposome suspension due to aggregation and fusion are strongly influenced by phospholipids' composition, medium pH, and composition [37]. Liposomes which not have a net electrical charge are less stable and prone to aggregation than electrically charged liposomes. Thus, incorporation of charged lipid in liposomal formulation can prevent or slow down aggregation. However, when charged lipid is used, variation of the pH at the bilayer and water interface from the bulk pH must be considered. The surface pH and the bulk pH difference can be large, particularly with a high amount of charged lipid in the bilayer and low ionic strength in the bulk aqueous solution. However, such pH difference can be useful when preparing liposomes containing drugs which are most stable at slightly alkaline pH [37]. The loss of entrapped, hydrophilic drugs because of leakage depends on the contents of the liposome, its size, and the physical state of the bilayer forming lipids (liquid crystalline or gel state). In general, leakage of drugs from liquid crystalline state liposomes is higher than gel state liposomes [38]. However, storage around T_c enhances the gel state bilayer permeability. Incorporation of suitable amount of cholesterol decreases the permeability of the bilayer [39].

4.5.2 Sterilization of Liposomal Formulations

Liposomal formulations are meant for parenteral use; hence, they must be sterile. There are five different sterilization methods available. They include steam, dry heat, gas, gamma radiation, and filtration. Among the different methods of sterilization, gamma-rays treatment damage the liposome membrane [40]. Lyophilized liposomes can be sterilized by ethylene oxide exposure, but the remains of the incompletely removed ethylene oxide and/or contamination caused due to it can be toxic [41]. The filtration method for sterilization can be used for SUVs of size <0.22 mm in size; however, for larger liposomes, this method is not suitable. Heat sterilization method has been used as a suitable alternative for certain types of MLVs and extruded liposomes without significant damage to their integrity [42]. Autoclaving method has been studied for the determination of leakage of various classes of drugs [43]. A marked leakage was seen with a hydrophilic compound, DOX, an amphipathic compound.

4.5.3 Encapsulation Efficiency

Encapsulation efficiency of liposomes is generally low. The drug at its therapeutic dose must be able to encapsulate inside the liposomes. Otherwise the quantity of lipids and other excipients of the liposomes may become toxic. Moreover, liposomal drugs pharmacokinetics can even be affected negatively. Hence, the entrapment method is of most importance. Active loading method is generally employed to improve the entrapment of certain drugs. In this method, uncharged drug (amphipathic weak acidic or basic drugs) that can cross the lipid bilayer membrane in uncharged form, but changes to the charged form once goes inside the liposome. This drug is now unable to come outside from the interior of the liposomes in the charged form. This effect can be established by creating a low pH environment inside the liposomes and suspending the liposomes in a neutral pH environment, which carries the drug [44]. However, the active loading method is not satisfactory for highly hydrophobic drugs like paclitaxel due to the low affinity of drug for the lipid bilayers [45].

4.6 Methods of Preparation of Liposomes

There are some conventional methods of liposome preparation. In all methods, the difference is in the step in which phospholipids are dissolved in organic solvents and drying down using rotary evaporator and then the film is redispersed in aqueous solvent [46].

4.6.1 Thin Film Hydration

This method first used by Bangham et al. [47] and it is still the commonly used method for the liposome preparation but having few limitations because of low encapsulation efficiency problem. In this method, liposomes are prepared by dissolving lipids in an organic solvent and then organic solvent is evaporated by thin film formation under vacuum. After complete organic solvent removal, the thin film lipid mixture is hydrated using aqueous media. The lipids suddenly swell and hydrate to form liposomes. This method yields heterogeneous sized MLVs over 1 µm in size.

4.6.2 Reverse Phase Evaporation

The reverse-phase evaporation (REV) method is a two-step procedure. A water/oil emulsion is prepared of lipids and media in excess organic solvent. Then organic phase removed under reduced pressure. The two phases (lipids and water) are usually homogenized by sonication or mechanical methods. Due to removal of the organic solvent, the phospholipid-coated water droplets come together to form a gel matrix. Further organic solvent removal, under reduced pressure, causes the gel matrix to form into a paste with smooth consistency. This paste contains suspension of LUVs [48]. Drug encapsulation efficiencies up to 50–60% can be achieved using this method [49]. This method can be used for both small- and large-molecule encapsulation. The main drawback of this method is the exposure of drug to organic solvents and to mechanical agitations.

4.6.3 Ethanol Injection

In 1973, the ethanol injection method was first described [50]. In this method, the lipid molecules precipitate and form bilayer planar fragments due to the immediate dilution of the ethanol in the aqueous media [51]. This then forms into liposomes, thereby encapsulating aqueous media. Batzri and Korn performed an experiment with a very low amount of lipid resulting in small liposomes and low encapsulation efficiency. The main advantage of the ethanol injection method is that a narrow distribution of small liposomes (< 100 nm) can be formed by simply injecting an ethanolic lipid solution in water, in one step, without sonication or extrusion [52].

4.6.4 Ether Injection

A lipids solution in diethyl ether or ether-methanol mixture is slowly injected to an aqueous solution of the drug to be encapsulated at 55–65°C or under reduced pressure. The removal of ether using vacuum results in the formation of the liposomes. The main disadvantages of this method are that the exposure of the drugs to be encapsulated to organic solvents at high temperature and the liposomes of heterogeneous size formed (70 to 200 nm) [18, 53]. An advantage of the ether injection method over the ethanol injection method is that the former method can be performed for extended periods to remove the solvent due to which a concentrated liposomal product forms with high encapsulation efficiencies.

4.6.5 Freeze Thaw Extrusion

This method is an extension of the conventional thin film method. In this method, first liposomes are formed by the film method and vortexed with the drug to be encapsulated until the whole film is dispersed and the resulting liposomes are frozen in slightly warm water and again vortexed. After two to three cycles of vortexing and freeze thaw, the sample is extruded three to four times. This is followed by six to seven freeze thaw cycles and additional eight to nine extrusions. These processes rupture and defuse SUVs during which the drug equilibrates between inside and outside of the liposome and liposomes themselves fuse and increase in size to form LUVs by extrusion technique.

4.6.6 Dehydration-Rehydration

Dehydration/hydration methods are based on the dried lipid films swelling upon exposure to an aqueous buffer. Because of the presence of buffer salts within the dehydrated film, an osmotic pressure gradient is generated which forces water between the bilayers, and the lamellae separate to produce liposomes. This process is frequently utilized to obtain mixtures of MLVs and ULVs. The proposed mechanism for liposome formation from a dehydrated film is the detachment [54] of membrane sheets, which eventually detach and close upon themselves and remain connected to the film. Mechanical agitation is required to detach the liposomes from the formed film.

4.7 Liposomal Products Available for Clinical Use

From the first liposomal product Doxil® approved in 1995 to the latest Vyxeos® in 2017, there are now many successful marketed liposomal products (Fig. 4.3) (Table 4.1). Most of them have to be administrated through the parenteral route due to the degradation of lipids in the gastrointestinal tract. However, some recent products such as Arikace® (see Section 4.4.2) can be subcutaneously injected or inhaled as aerosols. This section will introduce the approved liposomal products, while focusing more on their formulation aspects.

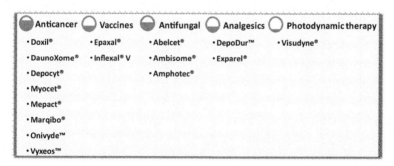

Figure 4.3   Liposome-based products in the management of various diseases.

Table 4.1   Marketed liposome-based products

Clinical products	Active agent	Lipid/lipid:drug molar ratio	Administration	Indication	Company/ approval year
Doxil®	Doxorubicin	HSPC:Cholesterol:PEG 2000-DSPE (56:39:5 molar ratio)	i.v.	Ovarian, breast cancer, Kaposi's sarcoma	Sequus Pharmaceuticals (1995)
DaunoXome®	Daunorubicin	DSPC and Cholesterol (2:1 molar ratio)	i.v.	AIDS-related Kaposi's sarcoma	NeXstar Pharmaceuticals (1996)
Depocyt®	Cytarabine/Ara-C	DOPC, DPPG, Cholesterol and Triolein	Spinal	Neoplastic meningitis	SkyPharma Inc. (1999)
Myocet®	Doxorubicin	EPC:Cholesterol (55:45 molar ratio)	i.v.	Combination therapy with cyclophosphamide in metastatic breast cancer	Elan Pharmaceuticals (2000)
Mepact®	Mifamurtide	DOPS:POPC (3:7 molar ratio)	i.v.	High-grade, resectable, non-metastatic osteosarcoma	Takeda Pharmaceutical Limited (2004)
Marqibo®	Vincristine	SM:Cholesterol (60:40 molar ratio)	i.v.	Acute lymphoblastic leukemia	Talon Therapeutics, Inc. (2012)

Clinical products	Active agent	Lipid/lipid:drug molar ratio	Administration	Indication	Company/ approval year
Onivyde™	Irinotecan	DSPC:MPEG-2000:DSPE (3:2:0.015 molar ratio)	i.v.	Combination therapy with fluorouracil and leucovorin in metastatic adenocarcinoma of the pancreas	Merrimack Pharmaceuticals Inc. (2015)
Vyxeos™	Daunorubicin and Cytarabine	DSPC:DSPG:Cholesterol (7:2:1 molar ratio)	i.v.	Acute myeloid leukemia	Jazz Pharmaceuticals Inc. (2017)
Epaxal®	Inactivated hepatitis A virus (strain RGSB)	DOPC:DOPE (75:25 molar ratio)	i.m.	Hepatitis A	Crucell, Berna Biotech (1993)
Inflexal® V	Inactivated hemaglutinine of Influenza virus strains A and B	DOPC:DOPE (75:25 molar ratio)	i.m.	Influenza	Crucell, Berna Biotech (1997)
Abelcet®	Amphotericin B	DMPC:DMPG (7:3 molar ratio)	i.v.	Invasive severe fungal infections	Sigma-Tau Pharmaceuticals (1995)

(*Continued*)

Table 4.1 *(Continued)*

Clinical products	Active agent	Lipid/lipid:drug molar ratio	Administration	Indication	Company/ approval year
Ambisome®	Amphotericin B	HSPC:DSPG:Cholesterol:Amphotericin B (2:0.8:1:0.4 molar ratio)	i.v.	Presumed fungal infections	Astellas Pharma (1997)
Amphotec®	Amphotericin B	Cholesteryl sulfate:Amphotericin B (1:1 molar ratio)	i.v.	Severe fungal infections	Ben Venue Laboratories Inc. (1996)
DepoDur™	Morphine sulfate	DOPC, DPPG, Cholesterol and Triolein	Epidural	Pain management	SkyPharma Inc. (2004)
Exparel®	Bupivacaine	DEPC, DPPG, Cholesterol and Tricaprylin	i.v.	Pain management	Pacira Pharmaceuticals, Inc. (2011)
Visudyne®	Verteporphin	Verteporphin:DMPC and EPG (1:8 molar ratio)	i.v.	Choroidal neovascularization	Novartis (2000)

HSPC, hydrogenated soy phosphatidylcholine; PEG, polyethylene glycol; DSPE, distearoyl-sn-glycero-phosphoethanolamine; DSPC, distearoylphosphatidylcholine; DOPC, dioleoylphosphatidylcholine; DPPG, dipalmitoylphosphatidylglycerol; EPC, egg phosphatidylcholine; DOPS, dioleoylphosphatidylserine; POPC, palmitoyloleoylphosphatidylcholine; SM, sphingomyelin; MPEG, methoxy polyethylene glycol; DMPC, dimyristoyl phosphatidylcholine; DMPG, dimyristoyl phosphatidylglycerol; DSPG, distearoylphosphatidylglycerol; DEPC, dierucoylphosphatidylcholine; DOPE, dioleoly-sn-glycero-phophoethanolamine; i.v, intravenous; i.m, intramuscular.

4.7.1 Liposomal Products Available in Market

4.7.1.1 Liposomal products in cancer treatment

4.7.1.1.1 *Doxil*®

Doxil® (stealth® liposomal DOX hydrochloride) injection is DOX hydrochloride incorporated into long circulating liposomes that contain surface-coated PEG chains. Doxil® is a product of Sequus Pharmaceuticals, USA, approved in 1995 as an i.v. injection for the treatment of HIV-associated Kaposi's sarcoma, advanced ovarian cancer, and multiple myeloma. Each single-dose vial of 10 mL and 25 mL contains 20 mg and 50 mg drug, respectively, and administered at an initial rate of 1 mg/min via i.v. infusion to minimize the risk of infusion related reactions.

Doxil® consists of a liquid dispersion of single lamellar vesicles with an approximate mean particle size in the range of 80–90 nm. The total lipid content of Doxil® is about 16 mg/mL and the DOX concentration is 2 mg/mL. Lipid components of Doxil® include the high T_c phospholipid, i.e., hydrogenated soy phosphatidylcholine (HSPC; T_c 52.5°C), cholesterol and distearoyl-phosphatidylethanolamine (DSPE) conjugated to PEG (*N*-carbamoylmethoxypolyethyleneglycol 2000 1,2-distearoyl-*sn*-glycerol-3 phosphoethanolamine sodium salt) in a molar ratio of 56:39:5 [55]. The HSPC and cholesterol ratio used provides a rigid bilayer at 37°C and below, promoting drug retention. DSPE is incorporated into the bilayer membrane of the liposomes which provides a stable anchor for the hydrophilic PEG chains (MW 2000) covalently bound to the ethanolamine head of DSPE and extending into the inner and outer water phase. DOX is encapsulated in the liposome by the remote loading approach (Fig. 4.4.) (first described by Barenholz [5]) using transmembrane gradient of ammonium sulfate. DOX is present in internal aqueous space of liposome (about 15,000 DOX molecules/liposome) at a drug-to-phospholipid ratio of approximately 150 μg/μmol in the presence of 155 mM ammonium sulfate and 200 μM deferoxamine mesylate. More than 90% of the drug is in the entrapped form present as a crystalline-like precipitate which is free from osmotic effects and therefore contributes to the stability of the entrapment [56]. The Doxil® liposomes are suspended in 10% sucrose.

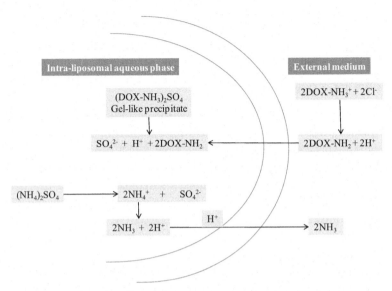

Figure 4.4 Remote loading approach of DOX into the intraliposomal aqueous phase. Liposomes are prepared at the desired concentration of ammonium sulfate. The gradient was formed by removing the ammonium sulfate from the external liposome medium. Intraliposomal NH_4^+ dissociates into NH_3 and H^+, NH_3 escape from the liposome and H^+ is retained in the liposome water phase. DOX HCl is added to the liposome dispersion at a temperature above the phase transition of the liposomal lipids. DOX, a cationic amphiphile, is present in equilibrium between an ionized and a non-ionized form. The latter form commutes across the liposome bilayer and becomes ionized once exposed to the internal H^+ environment, and forms a salt with the SO_4^{2-} anions, adapted figure [1].

Significant changes in the pharmacokinetic profile of DOX have been observed using Doxil®. The short distribution half-life of free DOX (~5 min) is remarkably changed in Doxil® which is cleared from plasma with a half-life of ~45 h. A slow plasma clearance (<0.1 L/h for Doxil® versus 45 L/h for free DOX) and a small volume of distribution (4 L for Doxil® versus 254 L for free DOX) are unique characteristics of Doxil®, as demonstrated in a pilot pharmacokinetic study [57]. In extensive studies with AIDS-related Kaposi's Sarcoma patients, Doxil® appears to be best therapy from the currently available therapies for this disease [58].

4.7.1.1.2 *DaunoXome®*

DaunoXome®, a daunorubicin (DNR) citrate liposomal injection, is a product of NeXstar Pharmaceuticals, USA approved in 1996 as single-dose vial for i.v. infusion for the treatment of HIV-associated Kaposi's sarcoma. DaunoXome® is an aqueous solution of DNR citrate encapsulated within neutrally charged liposomes composed of DSPC, cholesterol and DNR in a molar ratio of 10:5:1. DSPC and cholesterol are present in a molar ratio 2:1. The mean diameter is 45 nm and the total lipid to DNR ratio is 18.7:1. Specifically, DaunoXome® is the first liposomal product of its kind to be made solely of phospholipids [59]. Preparation of DNR into a citrate salt facilitates its encapsulation. The DNR liposome dispersion has pH between 4.9 and 6 and it is red and translucent. The vesicle membranes composed of DSPC:cholesterol lipids are below their T_c under physiological or ambient conditions. Therefore, these liposomes are more stable against the leakage of entrapped drug while stored under a broad range of conditions. Also, these DSPC:cholesterol liposomes are very well tolerated and have been given to patients at doses of over 2 g/m² in single injections without significant side effects. The particular liposomal formulation of DNR has been demonstrated to form liposomes with a exceptional physical stability, possibly preventing the entrapped DNR from rapid and extensive metabolization and minimizing protein binding [60].

Significant differences between the pharmacokinetic parameters of free DNR and DaunoXome® are observed in P1798 solid tumor and in plasma [61]. DaunoXome® produces 1 h plasma levels of DNR-equivalent (DNR fluorescent equivalents; parent drug plus fluorescent moieties) of 268 kg/mL, a 185-fold increase over free drug levels (1.4 kg/mL). DaunoXome® increases similar plasma AUC values, i.e., 227-fold over free drug. Tumor levels for DNR-equivalent at 1 h post administration of free drug were 9.6 μg/g and showed no significant increase after that. In contrast, DaunoXome® treatment produced DNR-equivalent accumulation within the tumor that continued through 8 h, peaking at 100 μg/g. The kinetics for this accumulation of DNR-equivalent followed an apparent first-order absorption process. Tissue DNR AUC ratios for DaunoXome® to free drug comparison showed that DaunoXome® produces greatest increase (10-fold) in tumor tissue [62].

4.7.1.1.3 *Depocyt®*

DepoCyt®, cytarabine/Ara-C liposomal injection, is a sterile suspension of the antimetabolite Ara-C, was approved by SkyePharm Inc., USA, in 1999 to improve the treatment of neoplastic meningitis (NM) through the sustained release of Ara-C. DepoCyt® increases the cerebrospinal fluid (CSF) half-life of Ara-C and hence decreases the dosing frequency and maintains sustained cytotoxic level over a period of 14 days [63]. The DepoCyt® liposomes are prepared by the DepoFoam™ technology, which offers a novel approach of sustained release drug delivery by encapsulation of drugs into multivesicular liposomes without modification in their molecular structure. DepoFoam™ particles are distinguished structurally from unilamellar vesicles and multilamellar vesicles in that each consists of microscopic spheroids with granular structure and single layered lipid particles composed of a honeycomb of numerous non-concentric vesicles containing the bounded drug. The particles are tens of microns in diameter and have large trapped aqueous volume, thereby affording delivery of large quantities of drugs in the encapsulated form in a small volume of injection. Each chamber is partitioned from the adjoining chambers by bilayer lipid membranes composed of synthetic analogs of naturally existing lipids (DOPC, DPPG, cholesterol, triolein, etc.). Extended release of drug attained because only outermost membranes breaks first of a multiple-layered non-concentric lipid vesicles result in release of entrapped drug to the external medium and the drug released from the internal vesicles results in a redistribution of drug inside the vesicles without drug release from the vesicles [64].

DepoCyt® liposomes contain Ara-C (50 mg), cholesterol (22 mg), triolein (6 mg), DOPC (28.5 mg), and DPPG (5 mg) per 5 mL of 0.9% preservative-free saline solution at a final pH of 5.5 to 8.5. In the DepoCyt® formulation, Ara-C is entrapped in the aqueous compartments of a matrix composed of above-mentioned phospholipids and the liposomes are spherical with size measuring about 20 μm in diameter. The residues of lipids used in the preparation of DepoCyt® liposomes are biodegradable and metabolized by the common metabolic pathways for phospholipids, triglycerides and cholesterol [65].

In phase I study, the AUC of free and encapsulated Ara-C after intraventricular administration of liposomal Ara-C was seen to be increased linearly with increasing dose [66]. The half-life of the drug in CSF after administration of a free Ara-C was 3.4 h. In comparison, following administering 50 mg of liposomal Ara-C (the phase II study recommended dose), the concentration of free Ara-C in CSF decreased biexponentially, with an 9.4 h initial half-life and 141 h terminal half-life. In this way, free Ara-C concentrations of >0.02 μg/mL were maintained in the lateral ventricles and lumbar sac for more than 14 days. This concentration is cytotoxic for practically all cancer cells when longer exposure to the drug is maintained [67].

4.7.1.1.4 *Myocet*®

Myocet®, non-PEGylated liposome formulation of DOX hydrochloride, was developed by Elan Pharmaceuticals, Princeton, USA, in 2000 to reduce the toxicity of DOX while maintaining its efficacy [68]. Combination of Myocet® and cyclophosphamide is indicated for the first line treatment of metastatic breast cancer in women. The product is available as a three-vial system: Myocet® DOX hydrochloride, Myocet® liposomes, and Myocet® buffer. Myocet® liposomes contain acidic EPC and cholesterol in a molar ratio of 55:45 and the DOX to total lipid ratio is approximately 0.27. At reconstitution, the DOX hydrochloride is dissolved in 20 ml of 0.9% sodium chloride for injection and slightly heated before a mixture of the liposomes and the buffer is added to form a liposome encapsulated DOX-citrate complex. The resulting reconstituted preparation of Myocet® contains 50 mg of DOX hydrochloride/ 25 mL of concentrate for liposomal dispersion for infusion (2 mg/mL). In Myocet®, by utilizing advanced encapsulation technology, DOX was incorporated into relatively small liposomes (150–250 nm) with more than 99% encapsulation efficiency and DOX to total lipid ratios surpass those predicted by passive encapsulation [69]. In this loading approach, sodium carbonate was added to an aqueous dispersion of liposomes in a proton-rich atmosphere (citrate buffer) creating a nearly neutral pH of 7.8 outside the liposome and acidic pH 4 inside the liposome. When DOX was added to these vesicles, its accumulation to the vesicle interior was determined by the acidic internal pH. Utilizing this process, DOX uptake surpasses theoretical predictions since the DOX forms a unique complex with citrate anions in the liposome interior. In addition to increased DOX loading, the complexation of DOX by citrate also reduces the rate of DOX leakage from these pH loaded liposomes [70].

The major difference between PEGylated (Doxil®) and Non PEGylated (Myocet®) DOX liposome is related to liposome-mediated changes in DOX plasma clearance rate. Even though some researchers attribute reduced DOX elimination rate to the use of PEGylated lipids, it is clear from numerous studies that the difference is due to changes in DOX release rates from the liposomes [71]. Myocet® releases more than 50% of its associated DOX within 1 h of i.v. injection and more than 90% of its encapsulated contents within 24 h. In contrast, liposomal formulations with saturated lipids comparable to the hydrogenated soya phosphatidylcholine/PEG lipid/cholesterol formulation release less than 10% of their entrapped DOX 24 h after i.v. administration [57].

In the preclinical and phase I studies, Myocet® was demonstrated to have a circulation half-life of about 3 times that of free DOX, although these results were derived from a wide range of plasma elimination half-lives [72]. Batist et al designed a multicentric clinical trial in which they compared Myocet® at 60 mg/m^2 and cyclophosphamide at 600 mg/m^2 with an identical dose of DOX and cyclophosphamide every 3 weeks in 297 women with metastatic breast cancer. Treatment was continued until progressive disease or prohibitive cardiac toxicity. There was no significant difference in response rate (43% versus 43%), median progression-free survival (5.2 months versus 5.5 months), or overall survival (21.2 months versus 16.4 months). Patients treated with Myocet® significantly showed reduced risk of cardiac toxicity, less mucositis, and absence of palmar plantar erythrodysesthesia while maintaining anti-tumor efficacy [73].

4.7.1.1.5 *Mepact*®

Mepact®, muramyl tripeptide phosphatidylethanolamine (MTP-PE) encapsulated liposome, was developed by Takeda Pharmaceutical Ltd. in 2004 for the treatment of osteosarcoma patients with postoperative combination chemotherapy in children, adolescents, and young adults who have gone through full macroscopic surgical resection. Mepact® has been designated as "orphan drug" by EMA. Mepact® may enhance the potential anti-tumor activity of macrophages. MTP-PE is the conjugate of muramyl tripeptide (MTP) and DPPE. It is a synthetic lipophilic derivative of muramyl dipeptide (MDP), a naturally component of bacterial cell walls. MTP-PE has common immunostimulant effects that are similar to those of natural MDP, with the addition of a prolonged circulation half-life in plasma and lower toxicity. Encapsulation of MTP-PE in the liposome (L-MTP-PE) greatly enhances the endocytosis and cell uptake of MTP-PE within the macrophage system and further prolongs its circulation half-life [74]. Mepact® liposomes are multi-lamellar vesicles (average diameter <100 nm) composed of synthetic phospholipids POPC and DOPS at a 7:3 molar ratio. 5 mg of liposomes contains 0.02 mg of MTP-PE (1:250 ratio). Each vial of Mepact® contains 1 g total lipids and 4 mg MTP-PE [75]. When MTP-PE is loaded into liposomes containing POPC and DOPS, the lipid particles bind the phosphatidyl serine receptors on macrophages. L-MTP-PE stimulates peptidoglycan recognition protein. This results in the initiation of cytotoxic functions [76]. Thus, Mepact® both inactive and active constituents target the lungs immune cells [77].

Preclinical results of Mepact® anti-tumor efficacy includes studies not only in rodents but also in dogs with hemangiosarcoma and osteosarcoma [78]. Dogs with spontaneous osteosarcoma showed significant improvement in survival in the L-MTP-PE treatment arms compared with placebo. A Phase III study demonstrated that the postoperative combination of Mepact® with other anti-tumor agents such as DOX, methotrexate, cisplatin, and ifosfamide improves survival in high-grade non-metastatic resectable osteosarcoma patients [79].

4.7.1.1.6   *Marqibo®*

Marqibo®, vincristine (VCR) sulfate liposome injection (VSLI), developed by Talon Therapeutics, Ltd., USA, was approved in 2012 for the treatment of advanced, relapsed, or refractory Philadelphia chromosome negative (Ph-) acute lymphoblastic leukemia (ALL) in an adolescent young adult population. Marqibo® was formulated to overcome the dosing and pharmacokinetic limitations of standard VCR. It is the encapsulation of VCR in aqueous interior of particular lipid vesicles called sphingosomes or optisomes [80]. Marqibo® sphingosomes are small (115 ± 25 nm in diameter), unilamellar vesicles made of SM and cholesterol in a molar ratio of 58:42. VSLI sphingosomes lipids' composition was specifically designed to facilitate the entrapment and retention of VCR, to prolong the circulation time of VCR, to slowly release the VCR in the tumor interstitium, and to increase extravasation into tumors. Also VSLI sphingosomes contribute to very low protein binding that results in a longer circulation time of the liposomes [81]. More than 90% of the drug is entrapped in the sphingosomes [81]. Approximately 18–40 % of encapsulated VCR was released in 24 h at 37°C in an in vitro study using human plasma [82]. Marqibo® single dose vial (31 mL) contains 5 mg VCR sulfate (0.16 mg/mL).

The rationale for the encapsulation of VCR in the sphingosomes is based on the potential of these vesicles to enhance the circulation half-life of the VCR and increase the accumulation of VCR at tumor sites. These VCR pharmacokinetic changes have been shown to be related with higher VCR accumulation at the tumor site and enhanced efficacy because cytotoxicity of VCR has been shown to be related to both duration of exposure and extracellular concentration [83]. Animal studies using VSLI have shown that loading of VCR in sphingosomes protects the VCR from the rapid elimination seen with VCR and enhances its circulation half-life, leading to significantly higher VCR concentrations in plasma for extended periods of time [82]. Marqibo® was effective and well tolerated when administer as a single agent in heavily pre-treated patients with relapsed or refractory non-Hodgkin's lymphomas at a 2.0 mg/m^2 dose administered intravenously over 1 h every 2 weeks [84].

4.7.1.1.7 Onivyde™

Onivyde™, irinotecan (IRI) liposome injection, is prescribed in combination with two other medicines, fluorouracil (5-FU) and leucovorin, for the treatment of metastatic pancreatic cancer patients previously treated with gemcitabine. Merrimack Pharmaceuticals Inc., USA, developed Onivyde™ in 2015. Onivyde™ liposomes are unilamellar lipid bilayer vesicles with an average diameter of 110 nm that encapsulates hydrophilic semisynthetic IRI hydrochloride trihydrate in a precipitated or gelated state as the sucrose octasulfate salt using a novel process known as intra-liposomal drug stabilization technology. This process utilizes intraliposomal trapping agents such as polyalkylammonium salt of a polymeric (polyphosphate) or nonpolymeric (sucrose octasulfate) highly charged multivalent anion. The process involves the formation of an drug-polyanion complex intraliposomally. Sucrose octasulfate is a high-charge density molecule having one strongly acidic negatively charged sulfate group. The triethylammonium component of the salt assists in drug loading as well, ensuring the charge neutrality of the liposome interior by allowing the efflux of cations accompanying the influx of the drug and possibly by formation of a self-perpetuating pH gradient to provide a driving force for progressive drug loading [85]. It was found that IRI hydrochloride trihydrate encapsulation in vesicles was quantitative up to 800 g IRI hydrochloride trihydrate/ mol phospholipid. The final molar ratio of drug:phospholipid is 1.36:1 for liposomes loaded at 800 g IRI hydrochloride trihydrate/ mol of phospholipid or 109,000 drug molecules/particle [86]. The Onivyde™ liposome is composed of DSPC, cholesterol and MPEG-2000-DSPE in the molar ratio of 3:2:0.015, which encapsulated more than 95% of the drug [87].

The pharmacokinetic parameters of nanoliposomal IRI in normal rats showed significantly longer circulation times for IRI encapsulated liposomes than free IRI formulation, particularly for liposomes loaded with intra-liposomal drug stabilization technology. Plasma half-lives for free IRI, IRI with triethylammonium polyphosphate and IRI with triethylammonium salts of sucrose octasulfate were 0.27, 6.80, and 10.7 h, respectively, as the free IRI drug was rapidly cleared from the circulation, with only 2% of the given dose detectable in the peripheral blood at 30 min after injection, and 35% of this residue found as carboxylate inactive form. In contrast to the pharmacokinetic values of free IRI, nanoliposomal IRI was associated with longer circulation half-lives and increased drug stability, with 23.2% of the given dose remaining in the plasma 24 h after administration, and no detectable conversion of the IRI prodrug to either its inactive form or its active metabolite at the specific time point [86]. Additional toxicity evaluations showed that nanoliposomal delivery of IRI resulted in a more than fourfold reduction of drug toxicity in mice, while the highest tolerated dose of triethylammonium salts of sucrose octasulfate nanoliposomal formulation was not achieved even after administration of the highest possible dose [86].

4.7.1.1.8 *Vyxeos*[TM]

Vyxeos[TM], daunorubicin (DNR) and cytarabine (Ara-C) liposome for injection, is a combination of DNR and Ara-C used for the treatment of acute high-risk myeloid leukemia (AML). Vyxeos[TM] is a recently approved (2017) product of Canada-based company, Jazz Pharmaceuticals. Vyxeos[TM] contains DNR and Ara-C in a molar ratio of 1:5. The 1:5 molar ratio of DNR:Ara-C has been shown to have additive/synergistic effects at killing leukemia cells in vitro and in murine models [88]. Phase III study data in elderly patients with secondary AML reveals greater median overall survival for Vyxeos[TM] (9.56 months) compared to the standard 7+3 regimen (Ara-C 100 mg/m^2/day by continuous infusion for 7 days and DNR 60 mg/m^2 on days 1, 2, and 3) (5.95 months) [89]. Vyxeos[TM] has received FDA breakthrough designation and may become the standard treatment for elderly patients with AML. Vyxeos[TM] liposomes are unilamellar bilayer vesicles with an average diameter of 100 nm prepared by the thin film hydration method. Vyxeos[TM] injection is supplied as a preservative-free, sterile, purple, lyophilized cake, in a single dose vial. Each single dose vial contains DNR (44 mg) and Ara-C (100 mg), and the following inactive ingredients: DSPC (454 mg), DSPG (132 mg), cholesterol (32 mg), copper gluconate (100 mg), triethanolamine (4 mg) and sucrose (2054 mg). The Vyxeos[TM] liposome is composed of DSPC, DSPG, and cholesterol in a 7:2:1 molar ratio [88].

The clinical data demonstrated a Vyxeos[TM] liposomal clearance of 0.108 L/h and 0.094 L/h and volume of distribution of 3.5 L and 4.6 L for DNR and Ara-C, respectively. Vyxeos[TM] liposomal DNR and Ara-C have terminal half-lives of 22.0 h and 33.9 h, respectively, than to the reported values of 18.5 h and 3.0 h for non-liposomal DNR and Ara-C, respectively. The AUCs for the maximum tolerated dose of 101 U/m^2 were 762.1 and 1990.6 mcg*hr/mL for DNR and Ara-C, respectively, which were 1000-fold greater than the reported non-liposomal values [90].

4.7.1.2 Liposomal products in viral infections

4.7.1.2.1 *Epaxal®*

Epaxal®, virosome-adjuvanted vaccine, is an aluminum-free liposomal vaccine for injection used to vaccinate against hepatitis A (HAV) (infectious jaundice) for adults and children from 1 year of age. Aluminum-free vaccines can be ideal for intradural or subcutaneous administration since they cause less local side effects than traditional aluminum-adsorbed vaccines [91]. Epaxal was the first product based on virosome technology manufactured by Crucell Berna Biotech, Switzerland, approved in 1993. Epaxal® product available as 0.5 ml prefilled syringes for i.m. injection containing at least 24 international units of HAV antigen. A pediatric product with half the adult dose, Epaxal® Junior, 0.25 ml, ≥ 12 IU HAV antigen, is also available. Virosomes are spherical unilamellar vesicles made up of phospholipids, PC, and PE, having mean diameter of 150 nm, and contain fusion active influenza glycoprotein hemagglutinin and neuraminidase intercalated into the bilayer [92, 93]. The Epaxal® virosomes lipids are DOPE and DOPC in a molar ratio of 25:75 [94]. Purified HAV and formalin-inactivated virions grown on MRC-5 human diploid cell culture are adsorbed to the surface of virosome. This constituent provides intrinsic adjuvant properties because the fusion active glycoprotein facilitates the delivery of the HAV antigen to immunocompetent cells. In fact hemagglutinin enables the virosome to bind to macrophage. The efficacy, immunogenicity, and tolerability of Epaxal® were demonstrated in a pediatric study: 20 seronegative toddlers (aged 12–16 months) and 80 seronegative children (aged 5–17 years) received a single i.m. dose with a subsequent booster dose at 12 months. At 4 weeks, 94% of the toddlers and 99% of the children had seroconverted. All the toddlers and 94% of children remained seroprotected at 12 months and showed a strong antibody rise after the booster dose [95].

4.7.1.2.2 *Inflexal® V*

Inflexal® V, liposomal adjuvanted vaccine for active immunization against influenza virus, was developed by using virosomal technology by the Crucell Berna Biotech, Switzerland, and it is only adjuvanted influenza vaccine available for all age groups. Inflexal® V is a parenterally given trivalent virosome influenza vaccine consists of three monovalent virosome pools, each formed with one influenza strain's specific HA and NA glycoproteins [96]. The influenza strains selection are dependent on the yearly recommendations of the WHO [97]. Inflexal® V production process is similar to Epaxal®, surface antigens neuraminidase and hemagglutinin of the influenza virus are integrated into PC bilayer liposomes, yielding unilamellar virosomes with an mean diameter of approximately 150 nm [96]. The particulate structure and the function of the surface hemagglutinin protein is to bind to the cell receptor, mediate pH-dependent fusion with endosomes and stimulate the immune system which are responsible for the adjuvant function [98]. Inflexal® V composed of 10% envelope lipids DOPC:DOPE in a 75:25 molar ratio [99] and 15 μg hemagglutinin each of influenza A and B virus strains. Inflexal® V contains no formaldehyde or thiomersal and is fully biodegradable. The purity of the vaccine is reflected in its very low ovalbumin content, which is an indication of the amount of residual egg protein. In a study of 453 healthy children aged 6–71 months, Inflexal® V fulfilled the EMA criteria given for adults for seroprotection, seroconversion, and increases in geometric mean titer for all the three influenza strains. The vaccine was found to be safe and well tolerated. Furthermore, Inflexal® V demonstrated significantly greater immunogenicity in children over the comparator split influenza vaccine used in the study for the A/H1N1 strain [100].

4.7.1.3 Liposomal products in fungal infections

4.7.1.3.1 *Abelcet®*

Abelcet®, Amphotericin B (AmB) lipid complex injection, developed by Sigma-Tau Pharmaceuticals in 1995 for the treatment of serious fungal infections in patients who are refractory to or intolerant of the conventional AmB therapy. Abelcet® is a dispersion in sodium chloride (0.9%; w/v) of very fine particles of AmB complexed in a 1:1 molar ratio with a mixture of the phospholipids DMPC and

DMPG in a 7:3 molar ratio. Unlike liposomal AmB, Abelcet® does not possess a spherical structure. Indeed, freeze-fracture electron micrographs have shown Abelcet® to possess a ribbon-like structure (1 to 10 mm diameter), resulting in a higher molar ratio of AmB in the formulation compared with liposomal/colloidal preparations [101, 102]. Despite the high amount of AmB in the formulation, there is small spontaneous release of AmB into the circulation immediately after infusion. Pharmacokinetic studies showed that there is deposition in the RES [103]. Release of this deposition at local sites of infection would be due to the action of lipases enzymes. These lipases might derive at least partially from the infecting yeasts themselves, also from inflammatory cells present at the infection site [104]. The combined effects of RES organ deposition, localized release, and high amounts of AmB within the complex might explain its enhanced efficacy than with conventional AmB formulations [105].

A single-dose pharmacokinetic study of ^{14}C-labeled AmB, loaded in Abelcet® and AmB, showed a reduced peak concentration in serum and AUC for Abelcet®. At 24 h following administration, there was accumulation of Abelcet® especially in the liver and the spleen, which are often sites of invasive fungal infections. The Abelcet® concentration in the lungs was higher than AmB at 1 h and 3 h following administration, but immediately diminished to comparable levels with AmB by 12 h following administration [103]. Abelcet® efficacy was demonstrated in immunocompetent and immunosuppressed murine models of fungal infection with aspergillus, blastomycosis, and cryptococcus; the effective dose for survival of 50% of infected animals was 1–5-fold higher for Abelcet® than conventional AmB [106].

4.7.1.3.2 *Ambisome*®

Ambisome®, liposomal AmB, is non-pyrogenic lyophilized liposomes for i.v. infusion, developed by Astellas Pharma, USA, and approved in 1997 for use as empirical treatment for recognized fungal infections in patients with febrile neutropenia and for aspergillosis, cryptococcosis, and candidiasis infections refractory to AmB. This was the first drug approved for this indication in the USA. Ambisome® is a unilamellar liposomal product of AmB and is a true liposomal formulation. It is different from the lipid-complexed

AmB formulations in that it consists entirely of uniform small spherical unilamellar lipid vesicles less than 100 nm in diameter. AmB is incorporated tightly within the liposomal bilayer membrane through the formation of a noncovalent charged complex between the positively charged mycosamine in AmB and the negatively charged DSPG, and hydrophobic interactions with the cholesterol components of the membrane in Ambisome®. The lipid bilayer consists of hydrogenated soy PC, cholesterol, DSPG, and AmB in a molar ratio of 2:1:0.8:0.4 [107]. These lipids have high T_c and were chosen to prepare a formulation that would be stable at 37°C. Other excipients such as antioxidants, disodium succinate hydrate and a-tocopherol, and sucrose as an isotonic agent are also present in the formulation.

Gondal et al. compared the pharmacokinetics of Ambisome® with conventional formulation of AmB. Ambisome® (1 and 5 mg/kg) and conventional formulation of AmB (1 mg/kg) were administered in mice. The C_{max} was much higher following Ambisome® than conventional formulation of AmB. After Ambisome® (1 and 5 mg/kg), C_{max} values were 8 and 50 mg/L, respectively, while after conventional formulation of AmB (1 mg/kg), C_{max} was 1.5 mg/L. Ambisome® disappeared more slowly from serum than conventional formulation of AmB, which due to the small, rigid, spherical liposomes are taken up less rapidly by the RES. The first- and second-phase half-lives of Ambisome® were 1.6 and 17 h, respectively, for the 1 mg/kg dose and 1.6 and > 24 h for 5 mg/kg dose, whereas those for conventional formulation of AmB were 0.5 and 11 h, respectively [108, 109].

4.7.1.3.3 *Amphotec*®

Amphotec®, colloidal dispersion of cholesteryl sulfate and AmB, is a sterile, pyrogen-free, lyophilized powder for i.v. administration. Ben Venue Laboratories, Bedford, USA, developed Amphotec® in 1996 for the treatment of patients with invasive aspergillosis where renal impairment or unacceptable toxicity prohibits the use of AmB deoxycholate in therapeutic doses, and in invasive aspergillosis patients where previous AmB deoxycholate therapy has failed. Amphotec® consists of AmB and cholesteryl sulfate (metabolite of cholesterol) in a molar ratio of 1:1 in a highly organized structure. Two molecules of AmB bind to two molecules of cholesteryl sulfate and form a tetramer with both a hydrophilic and a hydrophobic portion. These tetramers form a disk-type structure by aggregating into spiral arms which is approximately 120 nm in diameter [110]. After administration, Amphotec® remains largely intact and is rapidly cleared from the circulation by the macrophage phagocyte system cells, predominately by Kupfer cells of the liver. Consequently, less drug is available for binding to low density lipoproteins and, as a result, less is delivered to the kidney due to which reduced acute toxicity, hemolysis, and lipoprotein binding occur [111]. In a Phase I study, Amphotec® was tolerated at higher doses compared to conventional AmB, with a similar pattern of acute side effects, but less renal toxicity [112]. Later clinical studies focused on patients who were intolerant or unresponsive to the conventional AmB therapy. These clinical studies, which used different doses up to 8 mg/kg/day, showed high response rates against serious fungal infections that included aspergillosis, coccidioidomycosis, and candidiasis [113, 114].

4.7.1.4 Liposomal products in pain management

4.7.1.4.1 *DepoDur™*

DepoDur™, morphine sulfate extended-release liposomal formulation, is a sterile dispersion of liposomes intended for single-dose administration by the epidural route at the lumber level. DepoDur™ is a unique drug developed by SkyePharma, San Diego, approved in 2004 for the treatment of postoperative pain or prior to major surgery or following clamping the umbilical cord during cesarean section. DepoDur™ is designed for extended release of the drug up to 48 h. DepoDur™ is a multivesicular liposomal dispersion of morphine sulfate developed by Depofoam™ technology (discussed above in Section 4.7.1.1.3). The mean diameter of the DepoDur™ liposomes ranges from 17 to 23 μm [115]. DepoDur™ liposomes are dispersed in a 0.9% sodium chloride solution. DepoDur™ each vial contains morphine sulfate at a concentration of 10 mg/mL. Other phospholipids and inactive ingredients are DOPC (4.2 mg/mL), cholesterol (3.3 mg/mL) DPPG (0.9 mg/mL) tricaprylin (0.3 mg/mL) and triolein (0.1 mg/mL). A preclinical study in dogs showed maximum CSF levels for DepoDur™ and standard epidural morphine. The maximum concentration for DepoDur™ was one-third compared to standard epidural morphine. The time to attain maximum levels in the CSF for DepoDur™ was 3 h (10 mg) and 11 h (30 mg), compared with standard epidural morphine at 5 min (5 mg). The CSF levels of DepoDur™ returned to 0 at 120 h (10 mg) and 144 h (30 mg) compared to 24 to 48 h for the standard epidural morphine [116].

4.7.1.4.2 *Exparel®*

Exparel®, bupivacaine extended release liposome, is a non-pyrogenic, sterile, white to off-white preservative-free aqueous dispersion of multivesicular liposomes. Pacira Pharmaceuticals, Inc., USA, developed Exparel® in 2011, and it is indicated for administration into the surgical site to produce postsurgical local analgesia. Exparel® is developed for extended release of the drug up to 72 h. It is a multivesicular liposomal formulation of bupivacaine developed by Depofoam™ technology. It consists of multivesicular liposomes with average diameter in the range of 24–31 μm [117]. Exparel® contains a novel phospholipid ingredient, dierucoylphosphatidylcholine (DEPC) 8.2 mg/mL, which is specific to this formulation and has not previously been included in other DepoFoam™ technology–based approved products, e.g., DepoCyt® and DepoDur™ [118]. Each single vial contains bupivacaine at a concentration of 13.3 mg/mL. Other phospholipids and inactive ingredients are cholesterol (4.7 mg/mL) DPPG (0.9 mg/mL) and tricaprylin (2.0 mg/mL). Davidson et al. conducted pharmacokinetic study on human volunteers, in which they compared s.c. injection of 20 mL of 2% Exparel® versus 20 mL of 0.5% conventional bupivacaine. No difference in the C_{max} between the two groups was found: 0.87±0.45 versus 0.83±0.34 in conventional bupivacaine and Exparel® groups, respectively, despite a fourfold higher bupivacaine dose and a 9.8-fold increase in the terminal half-life displayed by the Exparel® group (131±58 versus 1294±860 min in conventional bupivacaine and Exparel® groups, respectively). The time for peak level increased sevenfold in the Exparel® group than the group administered conventional bupivacaine, which was attributed to the extended release of bupivacaine from the Exparel® [119].

4.7.1.5 Liposomal products in photodynamic therapy

4.7.1.5.1 *Visudyne®*

Visudyne®, verteporfin (VPF) liposomal injection, is a first light activated drug used in photodynamic therapy or verteporfin therapy approved by the regulatory authorities. Novartis AG, Switzerland, developed Visudyne® in 2000 for the treatment of choroidal neovascularization (CNV) secondary to age-related macular degeneration (AMD). VPF therapy has provided the means to treat patients for whom no other therapy has been proven effective in large-scale randomized clinical trials. Visudyne® contains VPF, is a synthetic chlorin-like porphyrin which shows strong absorption peaks between 650 and 700 nm, and also possesses strong hydrophobic characteristic. Verteporfin activation by red laser light in the flowing blood of the eye causes its site-specific activity in the treatment of wet macular degeneration [120]. Visudyne® liposome is a unilamellar vesicle consisting of DMPC, egg phosphatidyl glycerol (EPG), and ascorbyl palmitate. The molar ratio of VPF and mixture of phospholipids is approximately 1:8.0 and the median size of the liposomes are between 150 and 300 nm. Use of these fluid lipids enables fast release of VPF in the circulation and the hydrophobic nature of VPF prompts redistribution to blood proteins in vivo [121]. Higher hydrophobicity of VPF resulted in 100% encapsulation efficiency of loaded into the liposome. Biodistribution studies for aqueous VPF and liposomal VPF showed significantly higher accumulation of VPF in tumor tissue with liposomal VPF compared with aqueous VPF. Elimination rates were found to be approximately equivalent, i.e., circulation half-lives of 16.9 h for aqueous VPF and 16.1 h for liposomal VPF. In a biological assay in tumor-bearing mice, when PDT was administered 3 h following i.v. administration of liposomal VPF and aqueous VPF formulation, the liposomal VPF was found superior to the aqueous VPF formulation. In vitro plasma distribution studies showed the distribution of free VPF and liposomal VPF as 49.1%±2.6% and 91.1%±0.3%, respectively [122].

4.7.2 Liposomal Products in Various Phases of Clinical Trials

There has been much research in the field of liposomes, and many liposomal-based products have been approved by the regulatory authorities for various clinical trials. Table 4.2 highlights various liposomal products undergoing clinical trial investigation along with their current development phase, composition, route of administration, and indication.

Table 4.2 Liposome-based products in various phases of clinical trials

Products name/code	Active agent	Lipid composition	Administration	Indication	Company
		Phase I clinical trials			
LEM-ETU	Mitoxantrone	DOPC, cholesterol and cardiolipin	i.v.	Various cancers	NeoPharm Labs Ltd.
Liposomal Grb-2	Antisense oligodeoxynucleotide growth factor receptor bound protein 2 (Grb-2)	Unknown	i.v.	Hematologic malignancies	Bio-Path holdings
INX-0125	Vinorelbine tartrate	Cholesterol and sphingomyelin	i.v.	Advanced solid tumors	Inex Pharmaceuticals
INX-0076	Topotecan	Cholesterol and sphingomyelin	i.v.	Advanced solid tumors	Inex Pharmaceuticals
TKM-080301	PLK1 siRNA	Unique LNP technology (formerly referred to as stable nucleic acid-lipid particles or SNALP)	Hepatic intra-arterial administration	Neuroendocrine tumors	Tekmira Pharmaceuticals
Atu027	PKN3 siRNA	AtuFECT01	i.v.	Pancreatic cancer	Silence Therapeutics

Products name/code	Active agent	Lipid composition	Administration	Indication	Company
2B3-101	Doxorubicin	Glutathione PEGylated liposomes	i.v.	Solid tumors	2-BBB therapeutic
MTL-CEBPA	CEBPA siRNA	SMARTICLES® liposomal nanoparticles	i.v.	Liver cancer	MiNA Therapeutics
ATI-1123	Docetaxel	Protein stabilizing liposomes (PSL™)	i.v.	Solid tumors	Azaya therapeutic
LiPlaCis	Cisplatin	The lipid composition of the LiPlasomes is tailored to be specifically sensitive to degradation by the sPLA2 enzyme	i.v.	Advanced solid tumors	Oncology Venture
MCC-465	Doxorubicin	DPPC, cholesterol and maleimidated palmitoyl phosphatidyl ethanolamine; immunoliposomes tagged with PEG and the F(ab⃞)2 fragment of human monoclonal antibody GAH	i.v.	Metastatic stomach cancer	Mitsubishi Tanabe Pharma Corporation

(Continued)

Table 4.2 *(Continued)*

Products name/code	Active agent	Lipid composition	Administration	Indication	Company
SGT-53	p53 gene	Cationic lipids complexed with plasmid DNA encoding wild-type p53 tumor suppressor protein	i.v.	Various solid tumors	SynerGene Therapeutics
Alocrest	Vinorelbine	Sphingomyelin/cholesterol (OPTISOME™)	i.v.	Breast and lung cancers	Spectrum Pharmaceuticals
Phase II clinical trials					
Aroplatin	Platinum analogue cis-(trans- R,R-1,2-diaminocyclohexane) bis (neodecanoato) platinum (II)	DMPC and DMPG	i.v.	Metastatic colorectal cancer	Agenus Inc.
Liposomal annamycin	Semi-synthetic doxorubicin analogue annamycin	DMPC and DMPG	i.v.	Relapsed or refractory acute myeloid leukemia	Aronex Pharmaceuticals
SPI-077	Cisplatin	Soybean phosphatidylcholine, cholesterol	i.v.	Lung, head and neck cancer	Alza Corporation

Products name/code	Active agent	Lipid composition	Administration	Indication	Company
OSI-211	Lurtotecan	HSPC and cholesterol	i.v.	Ovarian, head and neck cancer	OSI Pharmaceuticals
S-CKD602	Potent topoisomerase I inhibitor	Phospholipids covalently bound to mPEG	i.v.	Cancer	Alza Corporation
LE-SN38	Irinotecan's active metabolite	DOPC, cholesterol and cardiolipin	i.v.	Advanced colorectal cancer	NeoPharm Labs Ltd.
LEP-ETU	Paclitaxel	DOPC, cholesterol and cardiolipin	i.v.	Cancer	NeoPharm Labs Ltd.
Endotag-I	Paclitaxel	DOTAP: DOPC: Paclitaxel	i.v.	Breast and pancreatic cancers	Medigene
Atragen	All-trans retinoic acid	DMPC and soybean oil	i.v.	Hormone-resistant prostate cancer; renal cell carcinoma and acute myelogenous leukemia	Aronex Pharmaceuticals

(Continued)

Table 4.2 *(Continued)*

Products name/code	Active agent	Lipid composition	Administration	Indication	Company
		Phase III clinical trials			
Arikace	Amikacin	DPPC and cholesterol	Aerosol delivery	Lung infections	Transave Inc.
Stimuvax	Tecemotide	Cholesterol, DMPG, DPPC	s.c.	Non-small cell lung cancer	Oncothyreon Inc.
T4N5 liposomal lotion	T4 endonuclease V	Egg lecithin	Topical	Xeroderma pigmentosum	AGI Dermatics Inc.
Liprostin	Prostaglandin E-1 (PGE-1)	Unknown	i.v.	Restenosis after angioplasty	Endovasc Inc.
ThermoDox	Doxorubicin	DPPC, Myristoyl stearyl phosphatidylcholine and DSPE-N-[amino(polyethylene glycol)-2000]	i.v.	Hepatocellular carcinoma and also recurring chest wall breast cancer	Celsion
Lipoplatin	Cisplatin	DPPG, soy phosphatidyl choline, mPEG-distearoyl phosphatidylethanolamine lipid conjugate and cholesterol	i.v.	Non-small cell lung cancer	Regulon Inc.

4.8 Conclusions

Liposomes have fully grown as a drug delivery system. Marketed formulations to fight fatal diseases, such as cancer, are available. The advantage of liposomal drug carriers systems over other colloidal drug carrier systems is their high versatility in terms of their in vivo physico-chemical behavior. This is a result of their versatile molecular structure and a capability to form bilayers with various characteristics. Well-defined structures with many functional properties are possible: long-circulating liposomes (PEGylated liposomes; Doxil®), conventional liposomes (Vyxeos™), liposome for vaccination (Epaxal®), temperature-sensitive liposomes (ThermoDox), cationic-charged liposomes (EndoTAG-1), etc. In summary, based on the pharmaceutical uses and clinically available formulations, we can state that liposomes as drug carrier systems have definitely established their position in modern drug delivery systems.

References

1. Bulbake U, Doppalapudi S, Kommineni N, Khan W. Liposomal formulations in clinical use: An updated review. *Pharmaceutics*. 2017; 9(2): 12.

2. Bangham A, Standish MM, Watkins JC. Diffusion of univalent ions across the lamellae of swollen phospholipids. *Journal of Molecular Biology*. 1965; 13(1): 238-IN27.

3. Torchilin VP. Multifunctional nanocarriers. *Advanced Drug Delivery Reviews*. 2012; 64: 302–315.

4. Bulbake U, Jain A, Khan W. Nanocarriers as non-viral vectors in gene delivery application. In *Multifunctional Nanocarriers for Contemporary Healthcare Applications*: IGI Global. 2018; p. 357–380.

5. Barenholz YC. Doxil®—the first FDA-approved nano-drug: Lessons learned. Journal of controlled release. 2012; 160(2): 117–134.

6. Chang T, Shiah H, Yang C, Yeh K, Cheng A, Shen B, Wang Y, Yeh C, Chiang N, Chang J. Phase I study of nanoliposomal irinotecan (PEP02) in advanced solid tumor patients. *Cancer Chemotherapy and Pharmacology*. 2015; 75(3): 579–586.

7. Otsuka H, Nagasaki Y, Kataoka K. PEGylated nanoparticles for biological and pharmaceutical applications. *Advanced Drug Delivery Reviews.* 2012; 64: 246–255.

8. Gref R, Minamitake Y, Peracchia MT, Trubetskoy V, Torchilin V, Langer R. Biodegradable long-circulating polymeric nanospheres. *Science.* 1994; 263(5153): 1600–1603.

9. Sriraman SK, Torchilin VP. Recent advances with liposomes as drug carriers. *Advanced Biomaterials and Biodevices.* 2014: 79–119, DOI: 10.1002/9781118774052.ch3.

10. Torchilin VP. Recent advances with liposomes as pharmaceutical carriers. *Nature Reviews Drug Discovery.* 2005; 4(2): 145–160.

11. Lasic DD. Novel applications of liposomes. *Trends in Biotechnology.* 1998; 16(7): 307–321.

12. Drummond DC, Meyer O, Hong K, Kirpotin DB, Papahadjopoulos D. Optimizing liposomes for delivery of chemotherapeutic agents to solid tumors. *Pharmacological Reviews.* 1999; 51(4): 691–744.

13. Sharma A, Sharma US. Liposomes in drug delivery: Progress and limitations. *International Journal of Pharmaceutics.* 1997; 154(2): 123–140.

14. Olson F, Hunt C, Szoka F, Vail W, Papahadjopoulos D. Preparation of liposomes of defined size distribution by extrusion through polycarbonate membranes. *Biochimica et Biophysica Acta (BBA)-Biomembranes.* 1979; 557(1): 9–23.

15. Barenholz Y, Gibbes D, Litman B, Goll J, Thompson T, Carlson F. A simple method for the preparation of homogeneous phospholipid vesicles. *Biochemistry.* 1977; 16(12): 2806–2810.

16. Ohsawa T, Miura H, Harada K. A novel method for preparing liposome with a high capacity to encapsulate proteinous drugs: Freeze-drying method. *Chemical and Pharmaceutical Bulletin.* 1984; 32(6): 2442–2445.

17. Kirby CJ, Gregoriadis G. A simple procedure for preparing liposomes capable of high encapsulation efficiency under mild conditions. *Liposome Technology.* 1984; 1: 19–27.

18. Deamer D, Bangham A. Large volume liposomes by an ether vaporization method. *Biochimica et Biophysica Acta (BBA)-Biomembranes.* 1976; 443(3): 629–634.

19. Hamilton R, Goerke J, Guo L, Williams MC, Havel RJ. Unilamellar liposomes made with the French pressure cell: A simple preparative

and semiquantitative technique. *Journal of Lipid Research.* 1980; 21(8): 981–992.

20. Vemuri S, Rhodes C. Preparation and characterization of liposomes as therapeutic delivery systems: A review. *Pharmaceutica Acta Helvetiae.* 1995; 70(2): 95–111.

21. Corvera E, Mouritsen O, Singer M, Zuckermann M. The permeability and the effect of acyl-chain length for phospholipid bilayers containing cholesterol: Theory and experiment. *Biochimica et Biophysica Acta (BBA)-Biomembranes.* 1992; 1107(2): 261–270.

22. Kirby C, Gregoriadis G. The effect of the cholesterol content of small unilamellar liposomes on the fate of their lipid components in vivo. *Life Sciences.* 1980; 27(23): 2223–2230.

23. Straubinger RM, Sharma A, Murray M, Mayhew E. Novel Taxol formulations: Taxol-containing liposomes. *Journal of the National Cancer Institute Monographs.* 1993(15): 69–78.

24. Risbo J, Jorgensen K, Sperotto MM, Mouritsen OG. Phase behavior and permeability properties of phospholipid bilayers containing a short-chain phospholipid permeability enhancer. *Biochimica et Biophysica Acta (BBA)-Biomembranes.* 1997; 1329(1): 85–96.

25. Sharma A, Straubinger RM. Novel taxol formulations: Preparation and characterization of taxol-containing liposomes. *Pharmaceutical Research.* 1994; 11(6): 889–896.

26. Sharma A, Straubinger NL, Straubinger RM. Modulation of human ovarian tumor cell sensitivity to N-(phosphonacetyl)-L-aspartate (PALA) by liposome drug carriers. *Pharmaceutical Research.* 1993; 10(10): 1434–1441.

27. Lee RJ, Low PS. Folate-mediated tumor cell targeting of liposome-entrapped doxorubicin in vitro. *Biochimica et Biophysica Acta (BBA)-Biomembranes.* 1995; 1233(2): 134–144.

28. Felgner JH, Kumar R, Sridhar C, Wheeler CJ, Tsai YJ, Border R, Ramsey P, Martin M, Felgner PL. Enhanced gene delivery and mechanism studies with a novel series of cationic lipid formulations. *Journal of Biological Chemistry.* 1994; 269(4): 2550–2561.

29. Scheule RK, George JAS, Bagley RG, Marshall J, Kaplan JM, Akita GY, Wang KX, Lee ER, Harris DJ, Jiang C. Basis of pulmonary toxicity associated with cationic lipid-mediated gene transfer to the mammalian lung. *Human Gene Therapy.* 1997; 8(6): 689–707.

30. Harashima H, Sakata K, Funato K, Kiwada H. Enhanced hepatic uptake of liposomes through complement activation depending on the size of liposomes. *Pharmaceutical Research.* 1994; 11(3): 402–406.

31. Jain RK. Transport of molecules across tumor vasculature. *Cancer and Metastasis Reviews.* 1987; 6(4): 559–593.

32. Yuan F, Leunig M, Huang SK, Berk DA, Papahadjopoulos D, Jain RK. Mirovascular permeability and interstitial penetration of sterically stabilized (stealth) liposomes in a human tumor xenograft. *Cancer Research.* 1994; 54(13): 3352–3356.

33. Allen T, Chonn A. Large unilamellar liposomes with low uptake into the reticuloendothelial system. *FEBS Letters.* 1987; 223(1): 42–46.

34. Allen T, Hansen C, Martin F, Redemann C, Yau-Young A. Liposomes containing synthetic lipid derivatives of poly (ethylene glycol) show prolonged circulation half-lives in vivo. *Biochimica et Biophysica Acta (BBA)-Biomembranes.* 1991; 1066(1): 29–36.

35. Klibanov AL, Maruyama K, Beckerleg AM, Torchilin VP, Huang L. Activity of amphipathic poly (ethylene glycol) 5000 to prolong the circulation time of liposomes depends on the liposome size and is unfavorable for immunoliposome binding to target. *Biochimica et Biophysica Acta (BBA)-Biomembranes.* 1991; 1062(2): 142–148.

36. Grit M, Crommelin DJ. Chemical stability of liposomes: Implications for their physical stability. *Chemistry and Physics of Lipids.* 1993; 64(1–3): 3–18.

37. Lichtenberg D, Barenholz Y. Liposomes: Preparation, characterization, and preservation. *Methods of Biochemical Analysis.* 1988; 33: 337–462.

38. Crommelin D, Van Bommel E. Stability of liposomes on storage: Freeze dried, frozen or as an aqueous dispersion. *Pharmaceutical Research.* 1984; 1(4): 159–163.

39. Deamer DW, Bramhall J. Permeability of lipid bilayers to water and ionic solutes. *Chemistry and Physics of Lipids.* 1986; 40(2–4): 167–188.

40. Ianzini F, Guidoni L, Indovina P, Viti V, Erriu G, Onnis S, Randaccio P. Gamma-irradiation effects on phosphatidylcholine multilayer liposomes: Calorimetric, NMR, and spectrofluorimetric studies. *Radiation Research.* 1984; 98(1): 154–166.

41. Ratz H, Freise J, Magerstedt P, Schaper A, Preugschat W, Keyser D. Sterilization of contrast media (Isovist) containing liposomes by ethylene oxide. *Journal of Microencapsulation.* 1989; 6(4): 485–492.

42. KIKUCHI H, CARLssoN A, YACHI K, HIROTA S. Possibility of heat sterilization of liposomes. *Chemical and Pharmaceutical Bulletin.* 1991; 39(4): 1018–1022.

43. Zuidam NJ, Lee SS, Crommelin DJ. Sterilization of liposomes by heat treatment. *Pharmaceutical Research.* 1993; 10(11): 1591–1596.

44. Choucair A, Lim Soo P, Eisenberg A. Active loading and tunable release of doxorubicin from block copolymer vesicles. *Langmuir.* 2005; 21(20): 9308–9313.

45. Sharma A, Mayhew E, Bolcsak L, Cavanaugh C, Harmon P, Janoff A, Bernacki RJ. Activity of paclitaxel liposome formulations against human ovarian tumor xenografts. *International Journal of Cancer.* 1997; 71(1): 103–107.

46. Mozafari MR. Liposomes: An overview of manufacturing techniques. *Cellular and Molecular Biology Letters.* 2005; 10(4): 711.

47. Bangham A, Standish MM, Watkins J. Diffusion of univalent ions across the lamellae of swollen phospholipids. *Journal of Molecular Biology.* 1965; 13(1): 238-IN27.

48. Szoka F, Papahadjopoulos D. Procedure for preparation of liposomes with large internal aqueous space and high capture by reverse-phase evaporation. *Proceedings of the National Academy of Sciences.* 1978; 75(9): 4194–4198.

49. Papahadjopoulos D, Vail W, Jacobson K, Poste G. Cochleate lipid cylinders: Formation by fusion of unilamellar lipid vesicles. *Biochimica et Biophysica Acta (BBA)-Biomembranes.* 1975; 394(3): 483–491.

50. Batzri S, Korn ED. Single bilayer liposomes prepared without sonication. *Biochimica et Biophysica Acta (BBA)-Biomembranes.* 1973; 298(4): 1015–1019.

51. Lasic D. Mechanisms of liposome formation. *Journal of Liposome Research.* 1995; 5(3): 431–441.

52. Stano P, Bufali S, Pisano C, Bucci F, Barbarino M, Santaniello M, Carminati P, Luisi PL. Novel Camptothecin Analogue (Gimatecan)-Containing Liposomes Prepared by the Ethanol Injection Method. *Journal of Liposome Research.* 2004; 14(1–2): 87–109.

53. Schieren H, Rudolph S, Finkelstein M, Coleman P, Weissmann G. Comparison of large unilamellar vesicles prepared by a petroleum ether vaporization method with multilamellar vesicles: ESR, diffusion and entrapment analyses. *Biochimica et Biophysica Acta (BBA)-General Subjects.* 1978; 542(1): 137–153.

54. Lasic DD. *Liposomes: From Physics to Applications*: Elsevier Science Ltd; 1993.

55. Working P, Dayan A. Pharmacological-toxicological expert report. CAELYX.(Stealth liposomal doxorubicin HCl). *Human & Experimental Toxicology.* 1996; 15(9): 751–785.

56. Lasic D, Frederik P, Stuart M, Barenholz Y, McIntosh T. Gelation of liposome interior A novel method for drug encapsulation. *FEBS Letters.* 1992; 312(2–3): 255–258.

57. Gabizon A, Catane R, Uziely B, Kaufman B, Safra T, Cohen R, Martin F, Huang A, Barenholz Y. Prolonged circulation time and enhanced accumulation in malignant exudates of doxorubicin encapsulated in polyethylene-glycol coated liposomes. *Cancer Research.* 1994; 54(4): 987–992.

58. Harrison M, Tomlinson D, Stewart S. Liposomal-entrapped doxorubicin: An active agent in AIDS-related Kaposi's sarcoma. *Journal of Clinical Oncology.* 1995; 13(4): 914–920.

59. Allen TM, Martin FJ. Advantages of liposomal delivery systems for anthracyclines. *Seminars in Oncology.* 2004; 31(6 Suppl 13): 5–15:.

60. Petre CE, Dittmer DP. Liposomal daunorubicin as treatment for Kaposi's sarcoma. *International Journal of Nanomedicine.* 2007; 2(3): 277.

61. Forssen EA. The design and development of DaunoXome® for solid tumor targeting in vivo. *Advanced Drug Delivery Reviews.* 1997; 24(2–3): 133–150.

62. Forssen EA, Coulter DM, Proffitt RT. Selective in vivo localization of daunorubicin small unilamellar vesicles in solid tumors. *Cancer Research.* 1992; 52(12): 3255–3261.

63. Kim S, Chatelut E, Kim JC, Howell SB, Cates C, Kormanik PA, Chamberlain MC. Extended CSF cytarabine exposure following intrathecal administration of DTC 101. *Journal of Clinical Oncology.* 1993; 11(11): 2186–2193.

64. Mantripragada S. A lipid based depot (DepoFoam® technology) for sustained release drug delivery. *Progress in Lipid Research.* 2002; 41(5): 392–406.

65. Murry DJ, Blaney SM. Clinical pharmacology of encapsulated sustained-release cytarabine. *Annals of Pharmacotherapy.* 2000; 34(10): 1173–1178.

66. Chamberlain MC, Kormanik P, Howell SB, Kim S. Pharmacokinetics of intralumbar DTC-101 for the treatment of leptomeningeal metastases. *Archives of Neurology.* 1995; 52(9): 912–917.

67. Frei E, Bickers JN, Hewlett JS, Lane M, Leary WV, Talley RW. Dose schedule and antitumor studies of arabinosyl cytosine (NSC 63878). *Cancer Research.* 1969; 29(7): 1325–1332.

68. Balazsovits J, Mayer L, Bally M, Cullis P, McDonell M, Ginsberg R, Falk R. Analysis of the effect of liposome encapsulation on the vesicant properties, acute and cardiac toxicities, and antitumor efficacy of doxorubicin. *Cancer Chemotherapy and Pharmacology*. 1989; 23(2): 81–86.

69. Madden TD, Harrigan PR, Tai LC, Bally MB, Mayer LD, Redelmeier TE, Loughrey HC, Tilcock CP, Reinish LW, Cullis PR. The accumulation of drugs within large unilamellar vesicles exhibiting a proton gradient: A survey. *Chemistry and Physics of Lipids*. 1990; 53(1): 37–46.

70. Li X, Cabral-Lilly D, Janoff A, Perkins W. Complexation of internalized doxorubicin into fiber bundles affects its release rate from liposomes. *Journal of Liposome Research*. 2000; 10(1): 15–27.

71. Gabizon A, Martin F. Polyethylene glycol-coated (pegylated) liposomal doxorubicin. *Drugs*. 1997; 54(4): 15–21.

72. Cowens J, Creaven P, Greco W, Brenner D, Tung Y, Ostro M, Pilkiewicz F, Ginsberg R, Petrelli N. Initial clinical (phase I) trial of TLC D-99 (doxorubicin encapsulated in liposomes). *Cancer Research*. 1993; 53(12): 2796–2802.

73. Batist G, Ramakrishnan G, Rao CS, Chandrasekharan A, Gutheil J, Guthrie T, Shah P, Khojasteh A, Nair MK, Hoelzer K. Reduced cardiotoxicity and preserved antitumor efficacy of liposome-encapsulated doxorubicin and cyclophosphamide compared with conventional doxorubicin and cyclophosphamide in a randomized, multicenter trial of metastatic breast cancer. *Journal of Clinical Oncology*. 2001; 19(5): 1444–1454.

74. Maeda M, Knowles RD, Kleinerman ES. Muramyl tripeptide phosphatidylethanolamine encapsulated in liposomes stimulates monocyte production of tumor necrosis factor and interleukin-1 in vitro. *Cancer Communications*. 1991; 3(10–11): 313–321.

75. Pahl JH, Kwappenberg KM, Varypataki EM, Santos SJ, Kuijjer ML, Mohamed S, Wijnen JT, van Tol MJ, Cleton-Jansen A-M, Egeler RM. Macrophages inhibit human osteosarcoma cell growth after activation with the bacterial cell wall derivative liposomal muramyl tripeptide in combination with interferon-γ. *Journal of Experimental & Clinical Cancer Research*. 2014; 33(1): 27.

76. Galligioni E, Favaro D, Santarosa M, Quaia M, Spada A, Freschi A, Alberti D. Induction and maintenance of monocyte cytotoxicity during treatment with liposomes containing muramyl tripeptide despite tachyphylaxis to the cytokine response. *Clinical Cancer Research*. 1995; 1(5): 493–499.

77. Anderson P, Tomaras M, McConnell K. Mifamurtide in osteosarcoma--A practical review. *Drugs of today (Barcelona, Spain: 1998)*. 2010; 46(5): 327–337.

78. MacEwen EG, Kurzman ID, Rosenthal RC, Smith BW, Manley PA, Roush JK, Howard PE. Therapy for osteosarcoma in dogs with intravenous injection of liposome-encapsulated muramyl tripeptide. Oxford University Press; 1989.

79. Luetke A, Meyers PA, Lewis I, Juergens H. Osteosarcoma treatment– where do we stand? A state of the art review. *Cancer Treatment Reviews*. 2014; 40(4): 523–532.

80. Webb M, Harasym T, Masin D, Bally M, Mayer L. Sphingomyelin-cholesterol liposomes significantly enhance the pharmacokinetic and therapeutic properties of vincristine in murine and human tumour models. *British Journal of Cancer*. 1995; 72(4): 896–904.

81. Johnston MJ, Semple SC, Klimuk SK, Edwards K, Eisenhardt ML, Leng EC, Karlsson G, Yanko D, Cullis PR. Therapeutically optimized rates of drug release can be achieved by varying the drug-to-lipid ratio in liposomal vincristine formulations. *Biochimica et Biophysica Acta (BBA)-Biomembranes*. 2006; 1758(1): 55–64.

82. Krishna R, Webb MS, Onge GS, Mayer LD. Liposomal and nonliposomal drug pharmacokinetics after administration of liposome-encapsulated vincristine and their contribution to drug tissue distribution properties. *Journal of Pharmacology and Experimental Therapeutics*. 2001; 298(3): 1206–1212.

83. Boman NL, Cullis PR, Mayer LD, Bally MB, Webb MS. Liposomal vincristine: The central role of drug retention in defining therapeutically optimized anticancer formulations. *Long Circulating Liposomes: Old Drugs, New Therapeutics*: Springer; 1998. pp. 29–49.

84. Sarris A, Hagemeister F, Romaguera J, Rodriguez M, McLaughlin P, Tsimberidou A, Medeiros L, Samuels B, Pate O, Oholendt M. Liposomal vincristine in relapsed non-Hodgkin's lymphomas: Early results of an ongoing phase II trial. *Annals of Oncology*. 2000; 11(1): 69–72.

85. Haran G, Cohen R, Bar LK, Barenholz Y. Transmembrane ammonium sulfate gradients in liposomes produce efficient and stable entrapment of amphipathic weak bases. *Biochimica et Biophysica Acta (BBA)-Biomembranes*. 1993; 1151(2): 201–215.

86. Drummond DC, Noble CO, Guo Z, Hong K, Park JW, Kirpotin DB. Development of a highly active nanoliposomal irinotecan using a novel intraliposomal stabilization strategy. *Cancer Research*. 2006; 66(6): 3271–3277.

87. Hong K, Drummond DC, Kirpotin D. Liposomes useful for drug delivery. Google Patents; 2016.

88. Tardi P, Johnstone S, Harasym N, Xie S, Harasym T, Zisman N, Harvie P, Bermudes D, Mayer L. In vivo maintenance of synergistic cytarabine: Daunorubicin ratios greatly enhances therapeutic efficacy. *Leukemia Research*. 2009; 33(1): 129–139.

89. Cassidy S, Syed BA. Acute myeloid leukaemia drugs market. Nature Publishing Group; 2016.

90. Nikanjam M, Capparelli E, Lancet JE, Kolitz JE, Louie AC, Schiller GJ. *Enhanced cytarabine and daunorubicin population pharmacokinetics when administered as cpx-351: A novel liposomal formulation not requiring dose reduction for mild renal or hepatic dysfunction. Am Soc Hematology.* 2016; 128: 3955.

91. Clarke PD, Adams P, Ibáñez R, Herzog C. Rate, intensity, and duration of local reactions to a virosome-adjuvanted vs. an aluminium-adsorbed hepatitis A vaccine in UK travellers. *Travel Medicine and Infectious Disease*. 2006; 4(6): 313–318.

92. Glück R, Moser C, Metcalfe IC. Influenza virosomes as an efficient system for adjuvanted vaccine delivery. *Expert Opinion on Biological Therapy*. 2004; 4(7): 1139–1145.

93. Glück R. Adjuvant activity of immunopotentiating reconstituted influenza virosomes (IRIVs). *Vaccine*. 1999; 17(13–14): 1782–1787.

94. Zylberberg C, Matosevic S. Pharmaceutical liposomal drug delivery: A review of new delivery systems and a look at the regulatory landscape. *Drug Delivery*. 2016; 23(9): 3319–3329.

95. Riedemann S, Reinhardt G, Frösner G, Glück R, Herzog C. Universal seroconvertion and low rate of side effects after one dose of the new hepatitis A vaccine EPAXAL, even when given to children less than 2 years of age. *Schweiz Med Wochenschr*. 1996; 126: 339.

96. Mischler R, Metcalfe IC. Inflexal® V a trivalent virosome subunit influenza vaccine: Production. *Vaccine*. 2002; 20: B17–B23.

97. Gerdil C. The annual production cycle for influenza vaccine. *Vaccine*. 2003; 21(16): 1776–1779.

98. Moser C, Amacker M, Kammer AR, Rasi S, Westerfeld N, Zurbriggen R. Influenza virosomes as a combined vaccine carrier and adjuvant system for prophylactic and therapeutic immunizations. *Expert Review of Vaccines*. 2007; 6(5): 711–721.

99. Glück R, Metcalfe I. New technology platforms in the development of vaccines for the future. *Vaccine*. 2002; 20: B10–B6.

100. Kanra G, Marchisio P, Feiterna-Sperling C, Gaedicke G, Lazar H, Durrer P, KÜrsteiner O, Herzog C, Kara A, Principi N. Comparison of immunogenicity and tolerability of a virosome-adjuvanted and a split influenza vaccine in children. *The Pediatric Infectious Disease Journal.* 2004; 23(4): 300–306.

101. Janoff A, Boni L, Popescu M, Minchey S, Cullis PR, Madden T, Taraschi T, Gruner S, Shyamsunder E, Tate M. Unusual lipid structures selectively reduce the toxicity of amphotericin B. *Proceedings of the National Academy of Sciences.* 1988; 85(16): 6122–6126.

102. Madden T, Janoff A, Cullis P. Incorporation of amphotericin B into large unilamellar vesicles composed of phosphatidylcholine and phosphatidylglycerol. *Chemistry and Physics of Lipids.* 1990; 52(3–4): 189–198.

103. Olsen SJ, Swerdel MR, Blue B, Clark JM, Bonner DP. Tissue distribution of amphotericin B lipid complex in laboratory animals. *Journal of Pharmacy and Pharmacology.* 1991; 43(12): 831–835.

104. Janoff A, Perkins W, Saletan S, Swenson C. Amphotericin B lipid complex (ABLC™): A molecular rationale for the attenuation of amphotericin B related toxicities. *Journal of Liposome Research.* 1993; 3(3): 451–471.

105. Gates C, Pinney R. Amphotericin B and its delivery by liposomal and lipid formulations. *Journal of Clinical Pharmacy and Therapeutics.* 1993; 18(3): 147–153.

106. Clark JM, Whitney RR, Olsen SJ, George RJ, Swerdel MR, Kunselman L, Bonner DP. Amphotericin B lipid complex therapy of experimental fungal infections in mice. *Antimicrobial Agents and Chemotherapy.* 1991; 35(4): 615–621.

107. Adler-Moore JP, Proffitt RT. Development, characterization, efficacy and mode of action of AmBisome, a unilamellar liposomal formulation of amphotericin B. *Journal of Liposome Research.* 1993; 3(3): 429–450.

108. Gondal JA, Swartz RP, Rahman A. Therapeutic evaluation of free and liposome-encapsulated amphotericin B in the treatment of systemic candidiasis in mice. *Antimicrobial Agents and Chemotherapy.* 1989; 33(9): 1544–1548.

109. De Marie S, Janknegt R, Bakker-Woudenberg I. Clinical use of liposomal and lipid-complexed amphotericin B. *Journal of Antimicrobial Chemotherapy.* 1994; 33(5): 907–916.

110. Guo LS, Fielding RM, Lasic DD, Hamilton RL, Mufson D. Novel antifungal drug delivery: Stable amphotericin B-cholesteryl sulfate discs. *International Journal of Pharmaceutics.* 1991; 75(1): 45–54.

111. Working P. Amphotericin B colloidal dispersion. *Chemotherapy.* 1999; 45(Suppl. 1): 15–26.

112. Sanders SW, Buchi K, Goddard M, Lang J, Tolman K. Single-dose pharmacokinetics and tolerance of a cholesteryl sulfate complex of amphotericin B administered to healthy volunteers. *Antimicrobial Agents and Chemotherapy.* 1991; 35(6): 1029–1034.

113. White MH, Anaissie EJ, Kusne S, Wingard JR, Hiemenz JW, Cantor A, Gurwith M, Du Mond C, Mamelok RD, Bowden RA. Amphotericin B colloidal dispersion vs. amphotericin B as therapy for invasive aspergillosis. *Clinical Infectious Diseases.* 1997; 24: 635–642.

114. Bowden RA, Cays M, Gooley T, Mamelok RD, van Burik J-A. Phase I study of amphotericin B colloidal dispersion for the treatment of invasive fungal infections after marrow transplant. *Journal of Infectious Diseases.* 1996; 173(5): 1208–1215.

115. Alam M, Hartrick CT. Extended-release epidural morphine (DepoDur™): An old drug with a new profile. *Pain Practice.* 2005; 5(4): 349–353.

116. Yaksh TL, Provencher JC, Rathbun ML, Myers RR, Powell H, Richter P, Kohn FR. Safety assessment of encapsulated morphine delivered epidurally in a sustained-release multivesicular liposome preparation in dogs. *Drug Delivery.* 2000; 7(1): 27–36.

117. Angst MS, Drover DR. Pharmacology of drugs formulated with DepoFoam™. *Clinical Pharmacokinetics.* 2006; 45(12): 1153–1176.

118. Richard BM, Rickert DE, Newton PE, Ott LR, Haan D, Brubaker AN, Cole PI, Ross PE, Rebelatto MC, Nelson KG. Safety evaluation of EXPAREL (DepoFoam bupivacaine) administered by repeated subcutaneous injection in rabbits and dogs: Species comparison. *Journal of Drug Delivery.* 2011; 2011: 467429.

119. Davidson EM, Barenholz Y, Cohen R, Haroutiunian S, Kagan L, Ginosar Y. High-dose bupivacaine remotely loaded into multivesicular liposomes demonstrates slow drug release without systemic toxic plasma concentrations after subcutaneous administration in humans. *Anesthesia & Analgesia.* 2010; 110(4): 1018–1023.

120. Bressler NM. Verteporfin therapy of subfoveal choroidal neovascularization in age-related macular degeneration: Two-year results of a randomized clinical trial including lesions with occult with no classic choroidal neovascularization—verteporfin in photodynamic therapy report 2. *American Journal of Ophthalmology.* 2002; 133(1): 168–169.

121. Strong HA, Levy J, Huber G, Fsadni M. Vision through photodynamic therapy of the eye. Google Patents; 1998.

122. Richter AM, Waterfield E, Jain AK, Canaan AJ, Allison BA, Levy JG. Liposomal delivery of a photosensitizer, benzoporphyrin derivative monoacid ring A (BPD), to tumor tissue in a mouse tumor model. *Photochemistry and Photobiology*. 1993; 57(s1): 1000–1006.

Chapter 5

Solid Lipid Nanoparticles

Karsten Mäder

Martin Luther University Halle-Wittenberg, Institute of Pharmacy,
Kurt-Mothes-Str. 3, D-06120 Halle (Saale), Germany
karsten.maeder@pharmazie.uni-halle.de

5.1 Introduction: History and Concept of SLN

Nanosized drug delivery systems have been developed to overcome one or several of the following problems: (i) low and highly variable drug concentrations after peroral administration due to poor absorption, rapid metabolism and elimination, (ii) poor drug solubility, which excludes i.v. injection of an aqueous drug solution, and (iii) drug distribution to other tissues combined with high toxicity (e.g., cancer drugs). Several systems, including micelles, liposomes, polymer nanoparticles, nanoemulsions, and nanocapsules have been established. Since the 1990s, solid lipid nanodispersions (SLN) have attracted increased attention. It is the aim of this chapter to discuss the general features of these systems with respect to manufacturing and performance.

In the past, solid lipids have been mainly used for rectal and dermal applications. In the beginning of the 1980s, Speiser and

Handbook of Materials for Nanomedicine: Lipid-Based and Inorganic Nanomaterials
Edited by Vladimir Torchilin
Copyright © 2020 Jenny Stanford Publishing Pte. Ltd.
ISBN 978-981-4800-91-4 (Hardcover), 978-1-003-04507-6 (eBook)
www.jennystanford.com

coworkers developed solid lipid microparticles (by spray drying) [1] and "nanopellets for peroral administration" [2]. These Nanopellets were produced by dispersion of melted lipids with high-speed mixers or ultrasound. The manufacturing process was unable to reduce all particles to the submicron size. A considerable amount of microparticles was present in the samples. This might not be a serious problem for peroral administration, but it excludes an intravenous injection. "Lipospheres," described by Domb, are close related systems [3–5]. They are also produced by means of high shear mixing or ultrasound and also often contain considerable amounts of microparticles.

The quality of the SLN has been significant improved by the use of HPH in the early 1990s [6–8]. Higher shear forces and a better distribution of the energy force more effective particle disruption, compared with high shear mixing and ultrasound. Dispersions obtained by this HPH are called solid lipid nanoparticles (SLN™). Most SLN dispersions produced by HPH are characterized by an average particle size below 500 nm and a low microparticle content. Other production procedures are based on the use of organic solvents HPH/solvent evaporation [9] or on dilution of microemulsions [10, 11]. The ease and efficacy of manufacturing lead to an increased interest in SLN. Furthermore, it has been claimed that SLN combine the advantages yet without inheriting the disadvantages of other colloidal carriers [12, 13]. Proposed advantages include the following:

- possibility of controlled drug release and drug targeting
- increased drug stability
- high drug pay load
- feasibility to incorporate lipophilic and hydrophilie drugs
- no biotoxicity of the carrier
- avoidance of organic solvents
- no problems with respect to large-scale production and sterilization

However, during the past years, some of these claims have been questioned and it became evident that SLN are rather complex systems which possess not only advantages but also serious limitations.

5.2 Ingredients and Production of Solid Lipid Nanoparticles

5.2.1 General Ingredients

General ingredients include solid lipids, emulsifier(s), and water. The term lipid is used generally in a very broad sense and includes triglycerides (e.g., tristearine, hard fat), partial glycerides, pegylated lipids, fatty acids (stearic acid), steroids (e.g., cholesterol), and waxes (e.g., cetylpalmitate). All classes of emulsifiers (with respect to charge and molecular weight) have been used to stabilize the lipid dispersion. The most frequently used compounds include different kinds of poloxamer, polysorbates, lecithin, and bile acids. It has been found that the combination of emulsifiers might prevent particle agglomeration more efficiently.

Unfortunately, poor attention has been given by most investigators to the physicochemical properties of the lipid. Fatty acids, partial glycerides, and other polar lipids are able to interact with water to much a greater extent, compared with long chain triglycerides (e.g., they might form liquid crystalline phases). Polar lipids will have much more interaction with stabilizers (e.g., formation of mixed micelles), while more lipophilic lipids will show phase segregation. The author strongly suggests to follow the proposal by Small and to classify lipids according to their interactions with water [14].

5.2.2 SLN Preparation

5.2.2.1 High shear homogenization and ultrasound

High shear homogenization and ultrasound are dispersion techniques which were initially used for the production of solid lipid nanodispersions [123]. Both methods are widespread and easy to handle. However, dispersion quality is often poor due to the presence of microparticles. Furthermore, metal contamination has to be considered if ultrasound is used. Ahlin et al. used a rotor-stator homogenizer to produce SLN from different lipids, including trimyristin, tripalmitin, tristearin, partial glycerides (Witepsol®W35, Witepsol®H35) and glycerol tribehenate (Compritol 888) by melt emulsification [15]. They investigated the influence of different process parameters, including emulsification time, stirring rate and cooling conditions on the particle size and the zeta potential. Poloxamer 188 was used as steric stabilizer. For Witepsol®W35 dispersions, the following parameters were found to produce the best SLN quality: stirring 8 min at 20000 rpm, the optimum cooling conditions 10 min at 5000 rpm at room temperature. In contrary, the best conditions for Dynasan 116 dispersions were 10 min emulsification at 25 000 rpm and 5 min of cooling at 5000 rpm in cool water (T = 16°C). An increased stirring rate did not significantly decrease the particle size, but improved the polydispersity index slightly. No general rule can be derived from differences in the established optimum emulsification and cooling conditions. In most cases, average particle sizes in the range of 100–200 nm were obtained in this study.

5.2.3   High-Pressure Homogenization

HPH has emerged as a very reliable and the most powerful technique for the preparation of SLN. HPH has been used for many years for the production of nanoemulsions for parenteral nutrition. In most cases, scaling up represents zero or limited problems. High-pressure homogenizers push a liquid with high pressure 100–3000 bar) through a narrow gap (in the range of few microns). The fluid accelerates on a very short distance to very high velocities. The high shear stress disrupts the particles down to the submicron range. Typical lipid contents are in the range of 5% to 10%. Even higher lipid concentrations (up to 40%) have been homogenized to lipid nanodispersions [16]. Two general approaches of the homogenization step, the hot and the cold homogenization techniques, can be used for the production of SLN [17, 18]. In both cases, a preparatory step involves the drug incorporation into the bulk lipid by dissolving the drug in the lipid melt.

5.2.4   Hot Homogenization

The hot homogenization is carried out at temperatures above the melting point of lipid. Therefore, it is in fact the homogenization of an emulsion. A pre-emulsion of the drug-loaded lipid melt and the aqueous emulsifier phase (same temperature) is obtained by high-shear mixing device (Ultraturrax). The quality of the pre-emulsion is very important for the final product quality. In general, higher temperatures result in lower particle sizes due to the decrease of the viscosity of the inner phase [19]. However, high temperatures may also increase the degradation rate of the drug and the carrier. The homogenization step can be repeated several times. It should be kept in mind, however, that HPH increases the temperature of the sample (approximately 10°C for 500 bar [20]). In most cases, 3 to 5 homogenization cycles at 500 to 1500 bar are sufficient. Increasing the homogenization pressure or the number of cycles often results in an increase of the particle size due to particle coalescence, which occurs as a result of the high kinetic energy of the particles [21]. It is important to note that the primary product of the hot homogenization is a nanoemulsion due to the liquid state of the lipid. Solid particles are expected to be formed by the following cooling of the sample to room temperature, or to temperatures below. Due to the small particle size and the presence of emulsifiers, lipid crystallization may be highly retarded and the sample may remain as a supercooled melt for several months [22].

5.2.5   Cold Homogenization

Cold homogenization has been developed to overcome the following three problems of the hot homogenization technique:

(1) temperature-induced drug degradation
(2) drug distribution into the aqueous phase during homogenization
(3) complexity of the crystallization step of the nanoemulsion, leading to several modifications and/or supercooled melts

The first preparatory step for cold homogenization is the same as in the hot homogenization procedure and includes the solubilization of the drug in the melt of the bulk lipid. However, the following steps differ. The drug-containing melt is rapidly cooled. The high cooling rate favors a homogenous distribution of the drug within the lipid matrix. The solid, drug-containing lipid is milled to microparticles. Typical particle sizes obtained by means of ball or mortar milling are in the range of 50 to 100 microns. Low temperatures increase the fragility of the lipid, and therefore favor particle disruption. The solid lipid microparticles are suspended in a chilled emulsifier solution. The pre-emulsion is subjected to HPH at or below room temperature. An effective temperature control and regulation is needed in order to ensure the unmolten state of the lipid due to the increase in temperature during homogenization [20]. In general, compared with hot homogenization, larger particle sizes and a broader size distribution are observed in cold homogenized samples [23]. A modified version of this technique has been published by the group of Müller-Goymann. They dispersed a solid 1:1 lecithin–hard fat mixture (described as solid reversed micelles) in Tween containing water using high-pressure homogenization [24].

5.2.5.1 SLN prepared by solvent emulsification/evaporation

Solvent emulsification/evaporation processes adapt techniques which have been previously used for the production of polymeric micro- and nanoparticles. The solid lipid is dissolved in a water-immiscible organic solvent (e.g., cyclohexane, or chloroform) that is emulsified in an aqueous phase. Upon evaporation of the solvent, a nanoparticle dispersion is formed by precipitation of the lipid in the aqueous medium. Westesen prepared nanoparticles of tripalmitate by dissolving the triglyceride in chloroform [25]. This solution was emulsified into an aqueous phase by high-pressure homogenization. The organic solvent was removed from the emulsion by evaporation and/or reduced pressure. The mean particle size ranges from approximately 30 to 100 nm depending on the lecithin/co-surfactant blend. Particles with very small diameters (30 nm) were obtained by using bile salts as co-surfactants. Comparable small particle size distributions were not achievable by melt emulsification of similar composition. The mean particle size depends on the concentration of the lipid in the organic phase. Very small particles could only be obtained with low fat loads (5%w) related to the organic solvent. With increasing lipid content, the efficacy of the homogenization declines due to the higher viscosity of the dispersed phase.

5.2.5.2 SLN preparations by solvent injection

The solvent injection method has been developed by Fessi to produce polymer nanoparticles [26]. Nanoparticles were only produced with solvents which distribute very rapidly into the aqueous phase (e.g., ethanol, acetone, DMSO), while larger particle sizes were obtained with more lipophilic solvents. According to Fessi, the particle size is critically determined by the velocity of the distribution processes and only water miscible solvents can be used. The solvent injection method can also be used for the production of solid lipid nanoparticles [27, 28]. However, the method is limited to lipids which dissolve in the polar organic solvent. Advantages of the method are the avoidance of elevated temperatures and high shear stress. However, the lipid concentration in the primary suspension will be less compared with high-pressure-homogenization. Furthermore, the use of organic solvents clearly represents a drawback of the method.

5.2.5.3 SLN preparations by dilution of microemulsions or liquid crystalline phases

SLN preparation techniques which are based on the dilution of microemulsions have been developed by Gasco and coworkers. Unfortunately, there is no common agreement within the scientific community about the definition of a microemulsion. One part of the scientific community understands under microemulsions high fluctuating systems which can be regarded as critical solutions, and therefore do not contain an inner and outer phase. This model has been confirmed by self-diffusion NMR studies of Lindman [29]. In contrast, Gasco and other scientists understand microemulsions as two systems composed of an inner and outer phase (e.g., O/W-microemulsions). They are made by stirring an optical transparent mixture at 65–70°C, typically composed of a low melting lipid fatty acid (e.g., stearic acid), emulsifier (e.g., polysorbate 20, polysorbate 60, soy phosphatidylcholin, taurodeoxycholic acid sodium salt), co-emulsifiers (e.g., Butanol, Na-monooctylphosphate), and water. The hot microemulsion is dispersed in cold water (2–3°C) under stirring. Typical volume ratios of the hot microemulsion to the cold water are in the range of 1:25 to 1:50. The dilution process is critically determined by the composition of the microemulsion. According to the literature, the droplet structure is already contained in the microemulsion, and therefore, no energy is required to achieve submicron particle sizes [30, 31]. The temperature gradient and the pH-value determine the product quality in addition to the composition of the microemulsion. High temperature gradients facilitate rapid lipid crystallization and prevent aggregation [32, 33]. Due to the dilution step, lipid contents which are achievable are considerably lower, compared with the HPH based formulations. Another disadvantage includes the use of organic solvents. Dahms and Seidel describe a similar approach to produce SLN. A hot liquid crystalline phase (instead of a microemulsion) is diluted in cold water to yield a solid lipid nanodispersion [34]. This approach avoids the use of high-pressure homogenization and organic solvents, and therefore represents an interesting opportunity.

5.2.6 Further Processing

5.2.6.1 Sterilization

Sterility is required for parenteral formulations. Dry or wet heat, filtration, gamma-irradiation, chemical sterilization, and aseptic production are general, opportunities to achieve sterility. The sterilization should not change the properties of the sample with respect to physical and chemical stability and the drug release kinetics. Sterilization by heat is a reliable procedure which is most commonly used. It was also applied for Liposomes [35, 36]. Steam sterilization will cause the formation of an oil in water emulsion, due to the melting of the lipid particles. The formation of SLN requires recrystallization of the lipids. Concerns are related to temperature-induced changes of the physical and chemical stability. The correct choice of the emulsifier is of significant importance for the physical stability of the sample at high temperatures. Increased temperatures will affect the mobility and the hydrophilicity of all emulsifiers, but to a different extent. Schwarz found that Lecithin is preferable to Poloxamer for steam sterilization, as only a minor increase in the particle size and the number of microparticles was observed after steam sterilization [37, 38]. An increase in particle size for Poloxamer 188-stabilized Compritol-SLN was observed after steam sterilization. It was found that a decrease of the sterilization temperature from 121 to 110°C can reduce sterilization-induced particle aggregation to a large extent. This destabilization can be attributed to the decreased steric destabilization of the Poloxamer. It is well known for PEG-based emulsifiers that increased temperatures lead to dehydratization of the ethylenoxide chains, pointing to a decrease of the thickness of the protecting layer. It has been demonstrated by 1H-NMR spectroscopy on Poloxamer-stabilized lipid nanoparticles, that even a moderate temperature increase from RT to 37°C decreases the mobility of the ethylenoxide chains on the particle surface [39]. Results of Freitas et al. indicate that the lowering of the lipid content (to 2%), and the surface modification of the glass vials and nitrogen purging might prevent the particle growth to a large extent and avoid gelation [40] Further studies of Cavalli et al. [41] and Heiati [42] demonstrate the possibility of steam sterilization of drug-loaded SLN.

Filtration sterilization of dispersed systems requires very high pressure and is not applicable to particles >200 nm. As most SLN particles are close to this size, filtration is of no practical use, due to the clocking of the filters. Few studies investigated the possibility of gamma-sterilization. It must be kept in mind that free radicals are formed during gamma-sterilization in all samples, due to the high energy of the gamma-rays. These radicals may recombine with no modification of the sample or undergo secondary reactions which might lead to chemical modifications of the sample. The degree of sample degradation depends on the general chemical reactivity and the molecular mobility and the presence of oxygen. It is therefore not surprising that chemical changes of the lipid bilayer components of liposomes were observed after gamma-irradiation [43] Schwarz investigated the impact of different sterilization techniques (steam sterilization or gamma-sterilization) on SLN characteristics [37, 38]. In comparison to lecithin-stabilized systems, Poloxamer-stabilized SLN were less stable after steam sterilization. However, this difference was not detected for gamma-sterilized samples. Compared with steam sterilization at 121°C, the increase in particle size after gamma-irradiation was lower, but comparable to steam sterilized samples at 110°C. Unfortunately, most investigators did not search for steam sterilization or irradiation-induced chemical degradation products. It should be kept in mind that degradation does not always cause increased particle sizes. In contrast, the formation of species like lysophosphatides or free fatty acids could even preserve small particle sizes, but might cause toxicological problems (e.g., by hemolysis). Further studies with more focus on chemical degradation products are clearly necessary to permit valid statements of the possibilities of SLN sterilization.

5.2.6.2 Drying by lyophilization, nitrogen purging and spray drying

SLN are thermodynamic unstable systems, and therefore, particle growth has to be minimized. Furthermore, SLN ingredients and incorporated drugs are often unstable and might be hydrolyzed or oxidized. The transformation of the aqueous SLN-suspension in a dry, redispersible powder is therefore often a necessary step to ensure storage stability of the samples. Lyophilization is widely used and is a promising way to increase chemical and physical SLN stability over extended periods of time. Lyophilization also offers principle possibilities for SLN incorporation into pellets, tablets, or capsules. Two additional transformations are necessary which might be the source of additional stability problems. The first transformation, from aqueous dispersion to powder, involves the freezing of the sample and the evaporation of water under vacuum. Freezing might cause stability problems due to the freezing out effect which results in the changes of the osmolarity and the pH. The second transformation, resolubilization, involves situations at least in its initial stages which favor particle aggregation (i.e., low water and high particle content, high osmotic pressure).

The protective effect of the surfactant can be compromised by lyophilization [44]. It has been found that the lipid content of the SLN dispersion should not exceed 5%, so as to prevent an increase in the particle size. Direct contact of lipid particles is decreased in diluted samples. Furthermore, diluted SLN dispersions will also have higher sublimation velocities and a higher specific surface area [45].The addition of cryoprotectors (e.g., sorbitol, mannose, trehalose, glucose, polyvinylpyrrolidon) will be necessary to decrease SLN aggregation and to obtain a better redispersion of the dry product. Schwarz et al. investigated the lyophilization of SLN in detail [46]. Best results were obtained with the cryoprotectors glucose, mannose, maltose, and trehalose in the concentration range between 10% and 15%. The observations come into line with the results of the studies on liposome lyophilization, which indicated that trehalose was the most sufficient substance to prevent liposome fusion and the leakage of the incorporated drug [47]. Encouraging results obtained with unloaded SLN cannot predict the quality of drug-loaded lyophilizates. Even low concentrations of 1% tetracain or etomidat caused a significant increase in particle size, excluding an intravenous administration [46]. Siekmann and Westesen investigated the lyophilization of tripalmitate-SLN using glucose, sucrose, maltose, and trehalose as cryoprotective agents [48]. Handshaking of redispersed samples was an insufficient method, but bath sonification produced better results. Average particle sizes of all lyophilized samples with cryoprotective agents were 1.5 to 2.4 times higher than the original dispersions. One-year storage caused increased particle sizes of 4 to 6.5 times compared with the original dispersion. In contrast to the lyophilizates, the aqueous dispersions of tyloxapol/phospholid-stabilized tripalmitate SLN exhibited remarkable storage stability. The instability of the SLN lyophilizates can be explained by the sintering of the particles. TEM pictures of tripalmitate SLN show an anisometrical, platelet-like shape of the particles. Lyophilization changes the properties of the surfactant layer due to the removal of water, and increases the particle concentration which favors particle aggregation. Increased particle sizes after lyophilization (2.1 to 4.9 times) were also reported by Cavalli [41]. Heiati compared the influence of four cryoprotectors (i.e., trehalose, glucose, lactose and mannitol) on the particle size of azidothymidine palmitate-loaded SLN Iyophilizates [42]. In agreement to other reports, trehalose was found to be the most effective cryoprotectant. The freezing procedure will affect the crystal structure and the properties of the lyophilizate. Literature data suggest that the freezing process needs to be optimized to a particular sample size. Schwarz recommended rapid freezing in liquid nitrogen [46]. In contrast, other researchers observed the best results after a slow freezing process [49]. Again, best results were obtained with samples of low lipid content and with the cryoprotector trehalose. Slow freezing in a deep freeze (−70°C) was superior to rapid cooling in liquid nitrogen. Furthermore, introduction of an additional thermal treatment of the frozen SLN dispersion (2 h at −22°C; followed by 2 h temperature decrease to −40°C) was found to improve the quality of the lyophilizate. Lyophilization has also been used to stabilize retinoic acid-loaded SLN [50].

An interesting alternative to lyophilization has been suggested by Gasco's group. Drying with a nitrogen stream at low temperatures of 3 to 10°C has been found to be superior [51]. Compared with lyophilization, the advantages of this process are the avoidance of freezing and the energy efficiency resulting from the higher vapor pressure of water. Spray drying has been scarcely for SLN drying, although it is cheaper compared with lyophilization. Freitas obtained a redispersible powder with this method, which meets the general requirements of i.v.-injections, with regard to the particle size and the selection of the ingredients [52]. Spray drying might potentially cause particle aggregation due to high temperatures, shear forces, and partial melting of the particles. Freitas recommends the use of lipids with high melting points >70°C to avoid sticking and aggregation problems. Furthermore, the addition of carbohydrates and low lipid contents favor the preservation of the colloidal particle size in spray drying.

5.3 SLN Structure and Characterization

The characterization of SLN is a necessity but also a great challenge. Lipid characterization itself is not trivial, as the statement by Laggner shows [53]: "Lipids and fats, as soft condensed material in general, are very complex systems, which not only in their static structures but also with respect to their kinetics of supramolecular formation. Hysteresis phenomena or supercooling can gravely complicate the task of defining the underlying structures and boundaries in a phase diagram." This is especially true for lipids in the colloidal size range. Therefore, possible artifacts caused by sample preparation (removal of emulsifier from particle surface by dilution, induction of crystallization processes, changes of lipid modifications) should be kept in mind. For example, the contact of the SLN dispersion with new surfaces (e.g., a syringe needle) might induce lipid crystallization or modification, and sometimes result in the spontaneous transformation of the low viscous SLN-dispersion into a viscous gel. The most important parameters of SLN include particle size and shape, the kind of lipid modification and the degree of crystallization, and the surface charge. Photon correlation spectroscopy (PCS) and Laser Diffraction (LD) are the most powerful techniques for routine measurements of particle size. It should be kept in mind that both methods are not "measuring" particle sizes. Rather, they detect light scattering effects which are used to calculate particle sizes. For example, uncertainties may result from nonspherical particle shapes. Platelet structures commonly occur during lipid crystallization [54] and are very often described in the SLN literature [55–59]. The influence of the particle shape on the measured size is discussed by Sjöström [55]. Further difficulties arise both in PCS and LD measurements for samples which contain several populations of different size. Therefore, additional techniques might be useful. For example, light microscopy is recommended although it is not sensitive to the nanometer size range. It gives a fast indication about the presence and the character of microparticles. Electron microscopy provides, in contrast to PCS and LD, direct information on the particle shape [57, 58]. Atomic force microscopy (AFM) has attracted increasing attention. A cautionary note applies to the use of AFM in the field of nanoparticles, as an immobilization of the SLN by solvent removal is required to assess their shape by the AFM tip. This procedure is likely to cause substantial changes of the molecular structure of the particles. Zur Mühlen demonstrated the ability of AFM to image the morphological structure of SLN [60]. The sizes of the visualized particles are of the same magnitude, compared with the results of PCS measurements. AFM investigations revealed the disk-like structure of the particles. Dingler investigated cetylpalmitate SLN (stabilized by polyglycerol methylglucose distearate, Tego Care 450) by electron microscopy and AFM and found an almost spherical form of the particles [61]. The usefulness of cross flow field-flow-fractionation (FFF) for the characterization of colloidal lipid nanodispersions has also been demonstrated [58]. Lipid nanodispersions with constant lipid content, but different ratios of liquid and solid lipids did show similar particle sizes in dynamic light scattering. However, retention times in FFF were remarkably dissimilar due to the different particle shapes (i.e., spheres vs. platelets). Anisotropic particles such as platelets will be constrained by the cross flow much more heavily compared with the spheres of similar size. The very high anisometry of SLN particles has been confirmed by electron microscopy, where very thin particles of 15 nm thickness and the length of several hundred nanometers became visible.

The measurement of the zeta potential allows predictions about the storage stability of colloidal dispersions [62]. In general, particle aggregation is less likely to occur for charged particles (i.e., high zeta potential) due to electric repulsion. However, this rule cannot strictly apply to systems which contain steric stabilizers, because the adsorption of steric stabilizers will decrease the zeta potential due to the shift in the shear plane of the particle.

Particle size analysis is just one aspect of SLN quality. The same attention has to be paid on the characterization of lipid crystallinity and modification, because these parameters are strongly correlated with drug incorporation, viscosity and release rates. Thermodynamic stability and lipid packing density increase, and drug incorporation rates decrease in the following order:

supercooled melt < α-modification < β'-modification < β-modification

In general, it has been found that melting and crystallization processes of nanoscaled material can differ considerable from that of the bulk material [63]. The thermodynamic properties of material having small nanometer dimensions can be considerably different, compared with the material in bulk form (e.g., the reduction of melting point). This occurs because of the tremendous influence of the surface energy.

This statement is also valid for SLN, where lipid crystallization and modification changes might be highly retarded due to the small size of the particles and the presence of emulsifiers [64]. Moreover, crystallization might not occur at all. It has been shown that samples which were previously described as SLN (solid lipid particles) were in fact supercooled melts (liquid lipid droplets) [65]. The impact of the emulsifier on SLN lipid crystallization has been shown by Bunjes [66]. The same group demonstrated also a size-dependent melting of SLN [67].

Differential scanning calorimetry (DSC) and X-ray scattering are most commonly applied to assess the status of the lipid. DSC uses the fact that different lipid modifications possess different melting points and melting enthalpies. By means of X-ray scattering, it is possible to assess the length of the long and short spacings of the lipid lattice. It is highly recommended to measure the SLN dispersion themselves, because solvent removal will often lead to modification changes. Sensitivity problems and long measurement times of conventional X-ray sources might be overcome by synchrotron irradiation [64]. In addition, this method permits to conduct time resolved experiments and allows the detection of intermediate states of colloidal systems which will be non-detectable by conventional X-ray methods [53].

It was shown that SLN might form superstructures by parallel alignment of SLN platelets. These reversible particle self-assemblies were observed by Illing et al. in tripalmitin dispersions when the lipid concentration exceeds 40 mg/g. Higher lipid concentrations did enhance particle self-assembly. The tendency to form self-assemblies has been found to depend on the particle shape, the lipid, and the surfactant concentration [68]. Infrared and Raman Spectroscopy are useful tools to investigate structural properties of lipids and they might give complementary information to X-ray and DSC [54]. Raman measurements on SLN show that the arrangement of lipid chains of SLN dispersions changes with storage [69].

Rheometry might be particularly useful for the characterization of the viscoelastic properties of SLN dispersions. The rheological properties are important with respect to the dermatological use of SLN, but they also provide useful information about the structural features of SLN dispersions and their storage dependency.

Studies of Lippacher show that the SLN dispersion possess higher elastic properties than emulsions of comparable lipid content [70–72]. Furthermore, a sharp increase of the elastic module is observed at a certain lipid content. This point indicates the transformation from a low viscous lipid dispersion to an elastic system with a continuous network of lipid nanocrystals. Illing and Unruh did compare the rheological properties of trimyristic, tripalmitic, and tristearic SLN suspensions. The results indicate that the viscosity of triglyceride suspensions increases with the lipid chain length and an increased anisotropy of the particles [73]. Souto et al. used rheology to study the influence of SLN addition on the rheological properties of hydrogels [74]. The co-existence of additional colloidal structures (micelles, liposomes, mixed micelles, nanodispersed liquid crystalline phases, supercooled melts, drug-nanoparticles) has to be taken into account for all SLN dispersions. Unfortunately, many investigators neglect this aspect, although the total amount of surface-active compounds is often comparable to the total amount of the lipid. The characterization and quantification are serious challenges due to the similarities in size. In addition, the sample preparation will modify the equilibrium of the complex colloidal system. Dilution of the original SLN dispersion with water might cause the removal of surfactant molecules from the particle surface and induce further changes such as crystallization or the changes of the lipid modifications. It is therefore highly desirable to use methods which are sensitive to the simultaneous detection of different colloidal species, which do not require preparatory steps such as Raman-, NMR-, and ESR (EPR) spectroscopy.

NMR active nuclei of interest are 1H, 13C, 19F, and 35P. Due to the different chemical shifts, it is possible to attribute the NMR signals to particular molecules or their segments. For example, lipid methyl protons give signals at 0.9 ppm, while protons of the polyethylenglycole chains give signals at 3.7 ppm. Simple 1H-NMR spectroscopy permits an easy and rapid detection of supercooled melts, due to the low linewidths of the lipid protons [69, 75–77]. This method is based on the different proton relaxation times in the liquid and semisolid/solid state. Protons in the liquid state give sharp signals with high signal amplitudes, while semisolid/ solid protons give very broad or invisible NMR signals under these circumstances. NMR has been used to characterize calixarene SLN [78] and hybrid lipid particles (NLC), which are composed of liquid and solid lipids [59]. Protons from solid lipids are not detected by standard NMR, but they can be visualized by solid state NMR. A drawback of solid state NMR is the rapid spinning of the sample that might cause artifacts. A paper describes the use of this method to monitor the distribution of Q10 in lipid matrices [79]. Unfortunately, the authors did use "drying of the sample to constant weight" as a preparatory step, which will cause significant changes of the sample characteristics. ESR requires the addition of paramagnetic spin probes to investigate SLN dispersions. A large variety of spin probes is commercially available. The corresponding ESR spectra give information about the microviscosity and micropolarity. ESR permits the direct, repeatable, and non-invasive characterization of the distribution of the spin probe between the aqueous and the lipid phase [80]. Experimental results demonstrate that storage-induced crystallization of SLN leads to an expulsion of the probe out of the lipid into the aqueous phase [81]. Furthermore, using an ascorbic acid reduction assay, it is possible to monitor the time scale of the exchange between the aqueous and the lipid phase [59]. The transfer rates of molecules between SLN and liposomes or cells have been determined by ESR [82].

5.4 The "Frozen Emulsion Model" and Alternative SLN Models

Lipid nanoemulsions are composed of a liquid oily core and a surfactant layer (e.g., lecithin). They are widely used for the parenteral delivery of poorly soluble drugs [83–85]. The original idea of SLN was to achieve a controlled release of incorporated drugs by increasing the viscosity of the lipid matrix. Therefore, it is not surprising that in original model, SLN is being described as "frozen emulsions" (see Fig. 5.1, left and middle) [86, 87]. However, lipids are known to crystallize very frequently in anisotropic platelet shapes [54]. Sjöström et al. described in 1995 that the particle shape of cholesterylacetate SLN did strongly depend on the emulsifier [55]. Platelet shaped particles have been detected for lecithin-stabilized particles, while PEG-20-sorbitanmonolaurate-stabilized particles preserved their spherical shape. Anisotropic particles have been found in numerous other SLN dispersions [56–59]. Based on the experimental results, a platelet shaped SLN model can be proposed as an alternative (see Fig. 5.1, right).

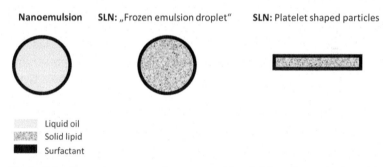

Nanoemulsion   **SLN:** „Frozen emulsion droplet"   **SLN:** Platelet shaped particles

Liquid oil
Solid lipid
Surfactant

Figure 5.1 General structure of a nanoemulsion (left), and proposed models for SLN: Frozen emulsion droplet model (middle) and platelet shaped SLN model (right).

In the year 2000, Westesen questioned the frozen emulsion droplet model with the following statement [88]: "Careful physicochemical characterization has demonstrated that these lipid-based nanosuspensions (solid lipid nanoparticles) are not just emulsions with solidified droplets. During the development process of these systems, interesting phenomena have been observed, such as gel formation on solidification and upon storage, unexpected dynamics of polymorphic transitions, extensive annealing of nanocrystals over significant periods of time, stepwise melting of particle fractions in the lower-nanometer-size range, drug expulsion from the carrier particles on crystallization and upon storage, an extensive supercooling."

Her comment highlights the complex behavior and changes of SLN dispersions. In addition, the presence of competing colloidal structures (e.g., micelles, liposomes, mixed micelles, nanodispersed liquid crystalline phases, supercooled melts and drug-nanoparticles) should be considered. Additional colloids might have an impact on very different aspects, including the correct measurement of particle size, drug incorporation and toxicity. A study shows that the cell toxicity of the SLN dispersion was reduced by dialysis due to the removal of water soluble components [89].

5.5 Nanostructured Lipid Carriers

Nanostructured lipid carriers (NLC) have been proposed as a new SLN generation with improved characteristics [90]. The general idea behind the system is to improve the poor drug loading capacity of SLN by "mixing solid lipids with spatially incompatible lipids leading to special structures of the lipid matrix" [91], while still preserving controlled release features of the particles. Three different types of NLC have been proposed (NLC 1: The imperfect structured type, NLC II: The structureless type and NLC III: The multiple type). Unfortunately, these structural proposals have not been supported by experimental data. They assume a spherical shape and they are not compatible with lipid platelet structures. For example, NLC III structures should contain small oily drop lets in a solid lipid sphere (Fig. 5.2, left). Detailed analytical examination of NLC systems by Jores et al. demonstrate that "nanospoon" structures are formed, in which the liquid oil adheres on the solid surface of a lipid platelet (Fig. 5.2, right).

Jores et al. concluded that "neither SLN nor NLC lipid nanoparticles showed any advantage with respect to incorporation rate or retarded accessibility to the drug, compared with conventional nanoemulsions. The experimental data concludes that NLCs are not spherical solid lipid particles with embedded liquid droplets, but rather, they are solid platelets with oil present between the solid platelet and the surfactant layer." Very similar structures have been found on Q10-loaded SLN by Bunjes et al. [92].

Liquid oil
Solid lipid
Surfactant

Figure 5.2 Proposed NLC III structure (modified after [91]) and experimental determined "nanospoon" structure described by Jores et al. (side view of particle) [58, 59].

5.6 Drug Localization and Release

Proposed advantages of SLN, compared with nanoemulsions, include increased protection capacity against drug degradation and controlled release possibilities due to the solid lipid matrix. The general low capacity of crystalline structures to accommodate foreign molecules is a strong argument against the proposed rewards.

It is therefore necessary to distinguish between drug association and drug incorporation. Drug association means that the drug is associated with the lipid, but it might be localized in the surfactant layer or between the solid lipid and the surfactant layer (similar to the oil in Fig. 5.2, right). Drug incorporation would mean the distribution of the drug within the lipid matrix. Another limiting aspect comes from the fact that the platelet structure of SLN, which is found in many systems, leads to a tremendous increase in surface area and the shortening of the diffusion lengths. Furthermore, additional colloid structures present in the sample are alternatives for drug localization the SLN for drug incorporation as it was pointed out by Westesen [88]: "The estimation of drug distribution is difficult for dispersions consisting of more than one type of colloidal particle. Depending on the type of stabilizer and on the concentration ratio of stabilizer to matrix material significant numbers of particles such as liposomes and/or (mixed) micelles may coexist with the expected type of particles."

The detailed investigation of drug localization is very difficult and only a few studies exist. Parelectric spectroscopy has been used to investigate the localization of glucocorticoids. The results indicate that the drug molecules are attached to the particle surface, but not incorporated into the lipid matrix. With Betamethasonvalerate, the loading capacity of the particle surface was clearly below the usual concentration of 0.1% [93]. Lukowski used Energy Dispersive X-ray Analysis and found that the drugs triamcinolone, dexamethasone, and chloramphenicol are partially stored at the surface of the individual nanoparticles [94]. The importance of the emulsifier is reflected in a study from Danish scientists [95]. They produced gamma-cyhalothrin (GCH)-loaded lipid micro- and nanoparticles. GCH had only limited solubility in the solid lipid and was expulsed during storage. The appearance of GCH crystals was strongly dependent from the solubility of the GCH in the emulsifier solutions. Emulsifier with high GCH solubility provoked rapid crystal growth. This observation is in accordance with a mechanism of crystal growth according to Ostwald ripening. Slovenian scientist found that ascorbylpalmitate was more resistant against oxidation in non-hydrogenated soybean lecithin liposomes, compared with SLN [96]. It shows that liposomes might have a higher protection capacity compared with SLN.

Fluorescence and ESR methods have been used by Jores et al. to monitor the microenvironment and the mobility of model drugs. The results indicate that even highly lipophilic compounds are pushed into a polar environment during lipid crystallization. Therefore, the incorporation capacity of SLN is very poor for most molecules [69]. A nitroxide reduction assay gave results in accordance with the results of the distribution. Compared with nanoemulsions, nitroxides were more accessible in SLN and NLC to ascorbic acid, localized in the aqueous environment. Therefore, nanoemulsions were more protective than SLN and NLC systems. Fluorescence and Raman spectroscopy gave also evidence that curcumin was not incorporated in the lipid matrix of SLN, but was localized in the surfactant layer [97].

Drug release from SLN and NLC could be controlled by either the diffusion of the drug or the erosion of the matrix. The original idea was to achieve a controlled release of SLN due to the slowing down of drug diffusion to the particle surface. This idea is, however, questionable due to drug expulsion during lipid crystallization. In addition, very short diffusion lengths in nanoscaled delivery systems lead to short diffusion times, even in highly viscous or solid matrices. In most cases, the delivery of the drug will be controlled by the slow dissolution rate in the aqueous environment. Drug release rate will be highly dependent on the presence of further solubilizing colloids (e.g., micelles), which are able to work as a shuttle for the drug and the presence or absence of a suitable acceptor compartment. Many investigators studied only the release in buffer media. A controlled release pattern under such conditions is not surprising, as it is caused by low solubilization kinetics due to the poor solubility of the drug according to the Noyes-Whitney equitation. In vivo, acceptor compartments will be present (e.g., lipoproteins, membranes) and will speed up release processes significantly. Whenever possible, drug-loaded SLN should be compared with nano-suspensions to separate the general features of the drug and the influence of the lipid matrix. Results by Kristl et al. indicate that lipophilic nitroxides diffuse between SLN and liposomes. The diffusion kinetics was strongly dependent on the nitroxide structure. In contrast, uptake of nitroxides in cells was similar between lipophilic nitroxides, suggesting endocytosis as the main mechanism [82]. The detailed mechanisms of drug release in vivo are poorly understood. In vitro data by different groups demonstrate that SLN are degraded by lipases [98–101]. Degradation by lipase depends on the lipid and strongly on the surfactant. Steric stabilization (e.g., by poloxamer) of SLN and NLC are less accessible because lipase needs an interface for activation. It is also known that highly crystalline lipids are poorly degraded by lipase. The data by Heider et al demonstrate that the commonly used method of titration might underestimate the degree of lipid digestion even with backtitration [100].

5.7 Administration Routes and in vivo Data

SLN and NLC can be administrated at different routes, including peroral, dermal, intravenous, and pulmonal. Peroral administration of SLN could enhance the drug absorption and modify the absorption kinetics. Despite the fact that in most of the SLN preparations the drug will be associated but not incorporated in the lipid, SLN might have advantages due to enhanced lymphatic uptake, enhanced bioadhesion or increased drug solubilization by SLN lipolysis products such as fatty acids and monoglycerides. A serious challenge represents the preservation of the colloidal particle size in the stomach, where low pH values and high ionic strengths favor agglomeration and particle growth. Zimmermann and Müller studied the stability of different SLN formulations in artificial gastric juice [102]. The main findings of this study are that (i) some SLN dispersions preserve their particle size under acidic conditions, and (ii) there is no general lipid and surfactant which are superior to others. The particular interactions between lipid and stabilizer are determining the robustness of the formulation. Therefore, the suitable combination of ingredients has to be determined on a case-by-case basis. Several animal studies show increased absorption of poorly soluble drugs. The efficacy of orally administrated triptolide-free drug and triptolide-loaded SLN have compared in the carrageenan-induced rat paw edema by Mei et al. [103]. Their results suggest that SLN can enhance the anti-inflammatory activity of triptolide and decrease triptolide-induced hepatotoxicity. The usefulness of SLN to increase the absorption of the poorly soluble drug all-trans retinoic acid has been shown by Hu et al. on rats [104]. Gasco's group investigated the uptake and distribution of tobramycin-loaded SLN in rats [105, 106]. They observed an increased uptake into the lymph, which causes prolonged drug residence times in the body of the animals. Furthermore, AUC and clearance rates did depend on the drug load. The same group described also enhanced absorption of idarubicin-loaded solid lipid nanoparticles (IDA-SLN), in comparison to the drug solution. Furthermore, the authors described that SLN were able to pass the blood-brain barrier and concluded that duodenal administration of IDA-SLN modifies the pharmacokinetics and tissue distribution of idarubicin [107].

Parenteral administration of SLN is of great interest too. To avoid the rapid uptake of the SLN by the RES system after i.v. injection, stealth SLN particles have been developed by the adoption of the stealth concept from liposomes and polymer nanoparticles. Reports indicate that Doxorubicin-loaded stealth SLN circulate for long period of time in the blood and change the tissue distribution [108]. Therefore, SLN could be alternatives to marketed stealth-liposomes, which can decrease the heart toxicity of this drug due to changed biodistribution. Long circulation times have also been observed for Poloxamer-stabilized SLN with Paclitaxel [109]. A recent study by Stelzner et al. shows the high promise of SLN based adjuvants for immunization [110].

The dermal application is of particular interest and it might be considered as the main application of SLN [111]. SLN have occlusive properties which are related to the solid structure of the lipid [112]. Human in vivo results of the group of Müller demonstrate that SLN can improve skin hydration and viscoelasticity [113]. SLN have also UV protection capacity due to their reflection of UV light [114]. Furthermore, data by Schäfer-Korting suggest SLN can be used to decrease drug side effects due to SLN mediated drug targeting to particular skin layers [115].

Further reports describe additional applications of SLN as well as gene delivery [116], delivery to the eye [117], pulmonary delivery [118, 119], and drug targeting of anticancer drugs [120]. Studies of the different groups also propose the use of SLN for brain targeting to deliver MRI contrast agents [121] or antitumor drugs [122, 123].

5.8 Summary and Outlook

SLN and NLC are now investigated by many scientists worldwide. In contradiction to early proposals, they certainly do not combine all the advantages of the other colloidal drug carriers and avoid the disadvantages of them. SLN are complex colloidal dispersions, not just "frozen emulsions." SLN dispersions are very susceptible to the sample history and storage conditions. Disadvantages of SLN include gel formation on solidification and upon storage, unexpected dynamics of polymorphic transitions, extensive annealing of nanocrystals over significant periods of time, stepwise melting of particle fractions in the lower-nanometer-size range, drug expulsion from the carrier particles on crystallization and upon storage, and extensive supercooling. The anisotropic shape of many SLN dispersions increases the surface area significantly, decreases the diffusion lengths to the surface and changes the rheological behavior dramatically (e.g., gel formation). Furthermore, the presence of alternative colloidal structures (micelles, liposomes) has to be considered to contribute to drug localization. In most cases, the drug will be associated with the lipid and not incorporated. Studies demonstrate that SLN and NLC might have no advantages compared with submicron emulsions, with regard to protection from the aqueous environment.

On the other hand, animal data suggest that SLN can change the pharmacokinetics and the toxicity of drugs. In many cases, drug incorporation might not be required and drug association with the lipid can be sufficient for lymphatic uptake. Clearly, more detailed studies are necessary to get a deeper understanding of the in vivo fate of these carriers [124]. Whenever possible, SLN and NLC systems should be compared directly with alternative nanosized carriers (e.g., liposomes, nanoemulsions, nanosuspensions) to evaluate their true potential.

References

1. Eldern T, Speiser P, and Hincal A (1991) Optimization of spray-dried and congealed lipid micropellets and characterization of their surface morphology by scanning electron microscopy. *Pharm Res* 8: 47–54.

2. Speiser P (1990) Lipidnanopellets als Trägersystem für Arzneimittel zur peroralen Anwendung, European Patent EP 0167825.

3. Domb AJ (1993) Lipospheres for controlled delivery of substances. United States Patent No. 5188837.

4. Domb AJ (1995) Long acting injectable oxytetracycline-lipposphere formulation. *Int J Pharm* 124: 271–278.

5. Domb AJ (1993) Lipposphere parenteral delivery system. *Proc Intl Symp Control Rel Bioact Mater* 20: 346–347.

6. Siekmann Band Westesen K (1992) Submicron-sized parenteral carrier systems based on solid lipids. *Pharm Pharmacol Lett* 1: 123–126.

7. Müller RH, Lucks JS (1996) Arzneistoffträger aus festen Lipidteilchen. Feste Lipid- nanosphären (SLN). European Patent No. 0605497.

8. Müller RH, Mehnert W, Lucks JS, Schwarz C, zur Mühlen A, Weyhers H, Freitas C and Rühl D (1995) Solid lipid nanoparticles (SLN): An alternative colloidal carrier system for controlled drug delivery. *Eur J Pharm Biopharm* 41: 62–69.

9. Sjöström B and Bergenstahl B (1992) Preparation of submicron drug particles in lecithin- stabilized o/w emulsions. I. Model studies of the precipitation of cholesteryl acetate. *Int J Pharm* 88: 53–62.

10. Cavalli R, Caputo O and Gasco MR (1993) Solid lipospheres of doxorubicin and idaru bicin. *Int J Pharm* 89: R9–RI2.

11. Gasco MR (1993) Method for producing solid lipid microspheres having a narrow size distribution. United States Patent No. 5250236.

12. Müller RH (1997) *Pharmazeutische Technologie, Moderne Arzneiformen.* Wiss. Verlagsges. Stuttgart.

13. Müller RH and Runge SA (1998) Solid lipid nanoparticles (SLN) for controlled drug delivery, in *Submicron Emulsions in Drug Targeting and Delivery,* Benita S (ed.), Harwood Academic Publishers.

14. Small D (1986) *Handbook of Lipids. The Physical Chemistry of Lipids: From Alkanes to Phospholipids.* Plenum Press, New York.

15. Ahlin P, Kristl J, and Smid-Kobar J (1998) Optimization of procedure parameters and physical stability of solid lipid nanoparticles in dispersions. *Acta Pharm* 48: 257–267.

16. Lippacher A, Müller RH, and Mäder K (2000) Investigation on the viscoelastic properties of lipid based colloidal drug carriers. *Int J Pharm* 196: 227–230.

17. zur Mühlen A and Mehnert W (1998) Drug release and release mechanism of prednisolone loaded solid lipid nanoparticles. *Pharmazie* 53:552–555.

18. zur Mühlen A, Schwarz C, and Mehnert W (1998) Solid lipid nanoparticles (SLN) for controlled drug delivery: Drug release and release mechanism. *Eur J Pharm Biopharm* 45: 149–155.

19. Lander R, Manger W, Scouloudis M, Ku A, Davis C, and Lee A (2000) Gaulin homogenization: A mechanistic study. *Biotechnol Prog* 16: 80–85.

20. Jahnke S (1998) The theory of high pressure homogenization, in *Emulsions and Nanosuspensions for the Formulation of Poorly Soluble Drugs,* Müller RH, Benita S, and Böhm B (eds), Medpharm Scientific Publishers: Stuttgart, pp. 177–200.

21. Siekmann B and Westesen K (1994) Melt-homogenized solid lipid nanoparticles stabilized by the nonionic surfactant tyloxapol, 1. Preparation and particle size determination. *Pharm Pharmacol Lett* 3: 194–197.

22. Bunjes H, Siekmann B, and Westesen K (1998), Emulsions of supercooled melts: A novel drug delivery system, in *Submicron*

Emulsions in *Drug Targeting and Delivery,* Benita S (ed), Harwood Academic Publishers.

23. zur Mühlen A (1996) Feste Lipid-Nanopartikel mit prolongierter Wirkstoffliberation: Herstellung, Langzeitstabilität, Charakterisierung, Freisetzungsverhalten und -mechanismen, PhD thesis, Free University of Berlin.

24. Friedrich l and Müller-Goymann CC (2003) Characterization of solidified reverse micellar solutions (SRMS) and production development of SRMS-based nanosuspensions. *Eur J Pharm Biopharm* 56: 111–119.

25. Siekmann B and Westesen K (1996) Investigations on solid lipid nanoparticles prepared by precipitation in o/w emulsions. *Eur J Pharm Biopharm* 43: 104–109.

26. Fessi H, Puisieux F, Ammoury N, and Benita S (1989) Nanocapsule formation by interfacial polymer deposition following solvent displacement. *Int J Pharm* 55: R1–R4.

27. Hu FQ, Yuan H, Zhang HH, and Fang M (2002) Preparation of solid lipid nanoparticles with clobetasol propionate by a novel solvent diffusion method in aqueous system and physicochemical characterization. *Int J Pharm* 239: 121–128

28. Schubert MA and Müller-Goymann CC (2003) Solvent injection as a new approach for manufacturing lipid nanoparticles: Evaluation of the method and process parameters. *Eur J Pharm Biopharm* 55: 125–131.

29. Danielsson l and Lindman B (1981) The definition of microemulsion. *Coll Surf* B 3: 391–392.

30. Gasco MR (1997) Solid lipid nanospheres from warm micro-emulsions. *Pharma Technol Eur* 52–58.

31. Boltri L, Canal T, Esposito PA, and Carli F (1993) Lipid nanoparticles: Evaluation of some critical formulation parameters. *Proc Intl Symp Control Rel Bioact Mater* 20: 346–347.

32. Cavalli R, Marengo E, Rodriguez L, and Gasco MR (1996) Effect of some experimental factors on the production process of solid lipid nanoparticles. *Eur J Pharm Biopharm* 43: 110–115.

33. Gasco MR, Morel S, and Carpigno R (1992) Optimization of the incorporation of desoxycortisone acetate in lipospheres. *Eur J Pharm Biopharm* 38: 7–10.

34. Dahms G and Seidel FH (2004) Method for the preparation of solid-lipid nanoparticles (SLNs) without high pressure homogenizer for

pharmaceutical, cosmetic and food applications. German Patent application DE 2003-10312763 20030321.

35. Zuidam NT, Lee SS L, and Crommelin DJA (1992) Sterilization of liposomes by heat treatment. *Pharm Res* 10: 1591–1596.

36. Lukyanov AN and Torchilin VP (1994) Autoclaving of liposomes. *J Microencap* 11: 669–672.

37. Schwarz C and Mehnert W (1995) Sterilization of drug-free and tetracaine-loaded solid lipid nanoparticles (SLN). Proc 1st World Meeting APGI/APV, Budapest, 485–486.

38. Schwarz C, Freitas C, Mehnert W, and Müller RH (1995) Sterilization and physical stability of drug-free and etomidate-loaded solid lipid nanoparticles. *Proc Intl Symp Control Rel Bioact Mat* 22: 766–767.

39. Liedtke S, Jores K, Mehnert W, and Mäder K (2000) Possibilities of non-invasive physicochemical characterisation of colloidal drug carriers, *27th Intl Symp Control Rel Bioact Mater* Vol. 27, Controlled Release Society, Paris, 1088–1089.

40. Freitas C (1998) Feste Lipid-Nanopartikel (SLN): Mechanismen der physikalischen Destabilisierung und Stabilisierung. PhD thesis, Free University of Berlin.

41. Cavalli R, Caputo O, Carlotti ME, Trotta M, Scarnecchia C, and Gasco MR (1997) Sterilization and freeze-drying of drug-free and drug-loaded solid lipid nanoparticles. *Int J Pharm* 148: 47–54.

42. Heiati H, Tawashi R and Phillips NC (1998) Drug retention and stability of solid lipid nanoparticles containing azidothymidine palmitate after autoclaving, storage and lyophilisation. *J Microencap* 15: 173–184.

43. Sculier JP, Coune A, Brassine C, Laduron C, Atassi G, Ruysschert GM, and Fruhling J (1986) Intravenous infusion of high doses of liposomes containing NSC 251635, a water insoluble cytostatic agent. A pilot study with pharmacokinetic data. *J Clin Oncol* 4: 789–797.

44. Rupprecht H (1993) Physikalisch-chemische Grundlagen der Gefriertrocknung, in Essig D and Oschmann R (eds), *Lyophilization.* Paperback APV, Band 35, Wissenschaftliche Verlagsgesellschaft mbH, Stuttgart, pp. 13–38.

45. Pikal MI, Shah S, Roy ML, and Putman R (1990) The secondary drying stage of freeze drying: Drying kinetics as a function of temperature and chamber pressure. *Int J Pharm* 60: 203–217.

46. Schwarz C and Mehnert W (1997) Freeze-drying of drug-free and drug-loaded solid lipid nanoparicles. *Int J Pharm* 157: 171–179.

47. Crowe LM, Crowe JH, Rudolph A, Womersley C, and Appel L (1985) Preservation of freeze-dried liposomes by trehalose. *Arch Biochem Biophys* 242: 240–247.

48. Siekmann B and Westesen K (1994) Melt-homogenized solid lipid nanoparticles stabilized by the nonionic surfactant tyloxapol, II. Physicochemical characterization and lyophilisation. *Pharm Pharmacol Lett* 3: 225–228.

49. Zimmermann E, Müller RH, and Mäder K (2000) Influence of different parameters on reconstitution of lyophilized SLN. *Int J Pharm* 196: 211–213.

50. Lim SI, Lee MK, and Kim CK (2004) Altered chemical and biological activities of all-trans retinoic acid incorporated in solid lipid nanoparticle powders. *J Control Rel* 100: 53–61.

51. Marengo E, Cavalli R, Rovero G, and Gasco MR (2003) Scale-up and optimization of an evaporative drying process applied to aqueous dispersions of solid lipid nanoparticles. *Pharm Dev Techn* 8: 299–309.

52. Freitas C and Müller RH (1998) Spray-drying of solid lipid nanoparticles (SLNTM). *Eur J Pharm Biopharm* 46: 145–151.

53. Laggner P (1999) X-ray diffraction of lipids, in *Spectral Properties of Lipids*, Hamilton RJ, and Cast J (eds.), Sheffield Academic Press.

54. Garti N and Sato K (eds.) (1998) *Crystallization and Polymorphism of Fats and Fatty Acids*, Marcel Dekker; New York and Basel.

55. Sjöström B, Kaplun A, Talmon Y, and Cabane B (1995) Structures of nanoparticles prepared from oil in water emulsions. *Pharm Res* 12: 39–48.

56. Siekmann B and Westesen K (1992) Sub-micron sized parenteral carrier systems based on solid lipid, *Pharm Pharmacol Lett* 1: 123–126.

57. Illing A, Unruh T, and Koch MHJ (2004) Investigation on particle self-assembly in solid lipid-based colloidal drug carrier systems. *Pharm Res* 21: 592–597.

58. Jores K, Mehnert W, Drechsler M, Bunjes H, Johann C, and Mäder K (2004) Investigations on the structure of solid lipid nanoparticles (SLN) and oil-loaded solid lipid nanoparticles by photon correlation spectroscopy, field-flow fractionation and transmission electron microscopy. *J Control Rel* 95: 217–227.

59. Jores K, Mehnert W, and Mäder K (2003) Physicochemical investigations on solid lipid nanoparticles (SLN) and on oil-loaded

solid lipid nanoparticles: A NMR- and ESR-study. *Pharm Res* 20: 1274–1283.

60. zur Mühlen A, zur Mühlen E, Niehus H, and Mehnert W (1996) Atomic force microscopy studies of solid lipid nanoparticles. *Pharm Res* 13: 1411–1416.

61. Dingler A, Blum RP, Niehus H, Müller RH, and Gohla S (1999) Solid lipid nanoparticles (SLN™/LipopearlsTM): A pharmaceutical and cosmetic carrier for the application of vitamin E in dermal products. *J Microencap* 16: 751–767.

62. Müller RH (1996) *Zetapotential und Partikelladung: Kurze Theorie, praktische Meßdurchführung, Dateninterpretation.* Wissenschaftliche Verlagsgesellschaft Stuttgart.

63. Lai SL, Guo JY, Petrova V, Ramanath G, and Allen LH (1996) Size-dependent melting properties of small tin particles: Nanocalorimetric measurements. *Phys Rev Lett* 77: 99–103.

64. Westesen K, Siekmann B, and Koch MHJ (1993) Investigations on the physical state of lipid nanoparticles by synchrotron radiation X-ray diffraction. *Int J Pharm* 93: 189–199.

65. Westesen K and Bunjes H (1995) Do nanoparticles prepared from lipids solid at room temperature always possess a solid matrix? *Int J Pharm* 115: 129–131.

66. Bunjes H, Koch MHJ, and Westesen K (2003) Influence of emulsifiers on the crystallization of solid lipid nanoparticles. *J Pharm Sci* 92: 1509–1520.

67. Bunjes H, Koch MHJ, and Westesen K (2000) Effect of particle size on colloidal solid triglycerides. *Langmuir* 16: 5234–5241.

68. Illing A, Unruh T, and Koch MHJ (2004) Investigation on particle self-assembly in solid lipid-based colloidal drug carrier systems. *Pharm Res* 21: 592–597.

69. Jores K (2004) Lipid nanodispersions as drug carrier systems: A physicochemical characterization, Thesis, University of Halle (http://sundoc.bibliothek.uni-halle.de/diss-online/04/04H310/prom.pdf).

70. Lippacher A, Müller RH, and Mäder K (2004) Liquid and semisolid SLN dispersions for topical application: Rheological characterization. *Eur J Pharm Biopharm* 58: 561–567.

71. Lippacher A, Müller RH, and Mäder K (2001) Preparation of semisolid drug carriers for topical application based on solid lipid nanoparticles. *Int J Pharm* 214: 9–12.

72. Lippacher A, Müller RH, and Mäder K (2002) Semisolid SLN dispersions for topical application: Influence of formulation and production parameters on viscoelastic properties. *Eur J Pharm Biopharm* 53: 155–160.

73. Illing A and Unruh T (2004) Investigation on the flow behavior of dispersions of solid triglyceride nanoparticles. *Int J Pharm* 284: 123–131.

74. Souto EB, Wissing SA, Barbosa CM, and Müller RH (2004) Evaluation of the physical stability of SLN and NLC before and after incorporation into hydrogel formulations. *Eur J Pharm Biopharm* 58: 83–90.

75. Westesen K and Siekmann B (1997) Investigation of the gel formation of phospholipids stabilized solid lipid nanoparticles. *Int J Pharm* 151: 35–45.

76. Bunjes H, Westesen K, and Koch MHJ (1996) Crystallization tendency and polymorphic transitions in triglyceride nanoparticles. *Int J Pharm* 129: 159–173.

77. Zimmermann E, Liedtke S, Müller RH, and Mäder K (1999) 1H-NMR as a method to characterize colloidal carrier systems, *Proc Int Symp Control Rel Bioact Mater* 26: 591–592.

78. Ahlin P, Kristl, J Pecar S, Strancar J, and Sentjurc M (2003) The effect of lipophilicity of spin-labeled compounds on their distribution in solid lipid nanoparticle dispersions studied by electron paramagnetic resonance. *J Pharm Sci* 92: 58–66.

79. Wissing, S A, Müller RH, Manthei L, and Mayer C (2004) Structural characterization of Q10-loaded solid lipid nanoparticles by NMR spectroscopy. *Pharm Res* 21: 400–405.

80. Ahlin P, Kristl J, Pecar S, Strancar J, and Sentjurc M (2003) The effect of lipophilicity of spin-labeled compounds on their distribution in solid lipid nanoparticle dispersions studied by electron paramagnetic resonance. *J Pharm Sci* 92: 58–66.

81. Liedtke S, Zimmermann E, Müller RH, and Mäder K (1999) Physical characterisation of solid lipid nanoparticles (SLN™). *Proc Intl Symp Control Rel Bioact Mater* 26: 595–596.

82. Kristl J, Volk B, Ahlin P, Gombac K, and Sentjurc M (2003) Interactions of solid lipid nanoparticles with model membranes and leukocytes studied by EPR. *Int J Pharm* 256: 133–140.

83. Müller RH and Heinemann S (1994) Fat emulsions for parenteral nutrition IV: Lipofundin MCT/LCT regimens for total parenteral nutrition (TPN) with high electrolyte load. *Int J Pharm* 107: 121–132.

84. Klang SH, Parnas M, and Benita S (1998) Emulsions as drug carriers: Possibilities, limitations and future perspectives, in *Emulsions and Nanosuspensions for the Formulation of Poorly Soluble Drugs,* Müller, RH, Benita S, and Böhm B (eds), Medpharm Scientific Publishers, Stuttgart, pp. 31–65.

85. Davis SS, Washington C, West P, and Illum L (1987) Lipid emulsions as drug delivery systems. *Ann NY Acad Sci* 507: 75–88.

86. Müller RH and Runge SA (1998) Solid lipid nanoparticles (SLN) for controlled drug delivery, in *Submicron Emulsions in Drug Targeting and Delivery,* Benita S (ed), Harwood Academic Publishers; Amsterdam, pp. 219–234.

87. Mehnert W and Mäder K (2001) Solid lipid nanoparticles: Production, characterization and applications. *Adv Drug Del Rev* 47: 165–196.

88. Westesen K (2000) Novel lipid-based colloidal dispersions as potential drug administration systems. Expectations and reality. *Coll Polym Sci* 278: 608–618.

89. Heydenreich AY, Westmeier R, Pedersen N, Poulsen HS, and Kristensen HG (2003) Preparation and purification of cationic solid lipid nanospheres: Effects on particle size, physical stability, and cell toxicity. *Int J Pharm* 254: 83–87.

90. Wissing SA, Kayser O, and Müller RH (2004) Solid lipid nanoparticles for parenteral drug delivery. *Adv Drug Del Rev* 56: 1257–1272.

91. Müller RH, Radtke M, and Wissing SA (2002) Nanostructured lipid matrices for improved microencapsulation of drugs. *Int J Pharm* 242: 121–128.

92. Bunjes H, Drechsler M, Koch MHJ, and Westesen K (2001) Incorporation of the model drug ubidecarenone into solid lipid nanoparticles. *Pharm Res* 18: 287–293.

93. Sivaramakrishnan R, Nakamura C, Mehnert W, Korting HC, Kramer KD, and Schäfer-Korting M (2004) Glucocorticoid entrapment into lipid carriers: Characterization by parelectric spectroscopy and influence on dermal uptake. *J Control Rel* 97: 493–502.

94. Lukowski G and Kasbohm J (2001) Energy Dispersive X-ray Analysis of loaded solid lipid nanoparticles. Proceedings: *28th International Symposium on Controlled Release of Bioactive Materials and 4th Consumer & Diversified Products Conference,* San Diego, CA, United States, pp. 516–517.

95. Frederiksen HK, Kristensen HG, and Pedersen M (2003) Solid lipid microparticle formulations of the pyrethroid gamma-cyhalothrin-

incompatibility of the lipid and the pyrethroid and biological properties of the formulations. *J Control Rel* 86: 243–252.

96. Kristl J, Volk B, Gasperlin M, Sentjurc M, and Jurkovic P (2003) Effect of colloidal carriers on ascorbyl palmitate stability. *Eur J Pharm Sci* 19: 181–189.

97. Noack A, Hause G, and Mäder K. (2012) Physicochemical characterization of curcuminoid-loaded solid lipid nanoparticles. *Int J Pharm* 423: 440–451.

98. Olbrich C, Kayser O and Müller RH (2002) Lipase degradation of Dynasan 114 and 116 solid lipid nanoparticles (SLN): Effect of surfactants, storage time and crystallinity. *Int J Pharm* 237: 119–128.

99. Olbrich C, Kayser O and Müller RH (2002) Enzymatic Degradation of Dynasan 114 SLN: Effect of Surfactants and Particle Size. *J Nanopar Res* 4: 121–129.

100. Noack A and Mäder K: (2012) In vitro digestion of curcuminoid-loaded lipid nanoparticles. *J Nanopar Res* 14, 1113.

101. Heider M, Hause G, and Mäder K (2016) Does the commonly used pH-stat method with back titration really quantify the enzymatic digestibility of lipid drug delivery systems? A case study on solid lipid nanoparticles (SLN). *Eur J Pharm Biopharm* 109: 194–205.

102. Zimmermann E and Müller RH (2001) Electrolyte- and pH-stabilities of aqueous solid lipid nanoparticle (SLN) dispersions in artificial gastrointestinal media. *Eur J Pharm Biopharm* 52: 203–210.

103. Mei Z, Li X, Wu Q, Hu S, and Yang X (2005) The research on the anti-inflammatory activity and hepatotoxicity of triptolide-loaded solid lipid nanoparticle. *Pharmacol Res* 51: 345–351.

104. Hu LD, Tang X, and Cui FD (2004) Solid lipid nanoparticles (SLNs) to improve oral bioavailability of poorly soluble drugs. *J Pharm Pharmacol* 56: 1527–1535.

105. Bargoni A. Cavalli R, Zara GP, Fundaro A, Caputo O and Gasco MR (2001) Transmucosal transport of tobramycin incorporated in solid lipid nanoparticles (SLN) after duodenal administration to rats. Part 11. Tissue distribution. *Pharmacol Res* 43: 497–502.

106. Cavalli R, Bargoni A, Podio V, Muntoni E, Zara GP, and Gasco MR (2003) Duodenal administration of solid lipid nanoparticles loaded with different percentages of tobramycin. *J Pharm Sci* 92: 1085–1094.

107. Zara CP, Bargoni A, Cavalli R, Fundaro A, Vighetto D, and Casco MR (2002) Pharmacokinetics and tissue distribution of idarubicin-loaded

solid lipid nanoparticles after duodenal administration to rats. *J Pharm Sci* 91: 1324–1333.

108. Zara CP, Cavalli R, Bargoni A, Fundaro A, Vighetto D, and Casco MR (2002) Intravenous administration to rabbits of non-stealth and stealth doxorubicin-loaded solid lipid nanoparticles at increasing concentrations of stealth agent: Pharmacokinetics and distribution of doxorubicin in brain and other tissues. *J Drug Targ* 10: 327–335.

109. Chen D, Lu W, Yang T, Li J, and Zhang Q (2002) Preparation and characterization of long-circulating solid lipid nanoparticles containing paclitaxel. *Yixueban* 34: 57–60.

110. Stelzner JJ, Behrens M, Behrens SE, and Mäder K. (2018) Squalene containing solid lipid nanoparticles, a promising adjuvant system for yeast vaccines. *Vaccine* 36: 2314–2320.

111. Müller RH, Radtke M, and Wissing SA (2002) Solid lipid nanoparticles (SLN) and nanostructured lipid carriers (NLC) in cosmetic and dermatological preparations. *Adv Drug Del Rev* 54(Suppl1): S131–S155.

112. Wissing SA and Müller RH (2002) The influence of the crystallinity of lipid nanoparticles on their occlusive properties. *Int J Pharm* 242: 377–379.

113. Wissing SA and Müller RH (2003) The influence of solid lipid nanoparticles on skin hydration and viscoelasticity: *in vivo* study. *Eur J Pharm Biopharm* 56: 67–72.

114. Wissing SA and Müller RH (2001) Solid lipid nanoparticles (SLN): A novel carrier for UV blockers. *Pharmazie* 56: 783–786.

115. Maia CS, Mehnert W, Schaller M, Korting HC, Gysler A, Haberland A, and Schäfer-Korting M (2002) Drug targeting by solid lipid nanoparticles for dermal use. *J Drug Targ* 10: 489–495.

116. Rudolph C, Schillinger U, Ortiz, A, Tabatt K, Plank C, Müller RH, and Rosenecker J (2004) Application of novel solid lipid nanoparticle (SLN)-gene vector formulations based on a dimeric HIV-1 TAT-peptide in vitro and in vivo. *Pharm Res* 21: 1662–1669.

117. Casco MR, Zara CP, and Saettone MF (2004) Pharmaceutical compositions suitable for the treatment of ophthalmic diseases. Patent application WO2004039351.

118. Videira MA, Botelho MF, Santos AC, Couveia LF, Pedroso De Lima JJ, and Almeida AJ (2002) Lymphatic uptake of pulmonary delivered radiolabelled solid lipid nanoparticles. *J Drug Targ* 10: 607–613.

119. Weber S, Zimmer A, and Pardeike J (2014) Solid lipid nanoparticles (SLN) and nanostructured lipid carriers (NLC) for pulmonary application: A review of the state of the art. *Eur J Pharm Biopharm* 86: 7–22.

120. Stevens PJ, Sekido M, and Lee RJ (2004) Synthesis and evaluation of a hematoporphyrin derivative in a folate receptor-targeted solid-lipid nanoparticle formulation. *Anticancer Res* 24: 161–165.

121. Peira E, Marzola P, Podio V, Aime S, Sbarbati A, and Casco MR (2003) In vitro and in vivo study of solid lipid nanoparticles loaded with superparamagnetic iron oxide. *J Drug Targ* 11: 19–24.

122. Wang JX, Sun X, and Zhang ZR (2002) Enhanced brain targeting by synthesis of 3′,5′-dioctanoyl-5-fluoro-2′-deoxyuridine and incorporation into solid lipid nanoparticles. *Eur J Pharm Biopharm* 54: 285–290.

123. Zara CP, Cavalli R, Bargoni A, Fundaro A, Vighetto D, and Gasco MR (2002) Intravenous administration to rabbits of non-stealth and stealth doxorubicin-loaded solid lipid nanoparticles at increasing concentrations of stealth agent: Pharmacokinetics and distribution of doxorubicin in brain and other tissues. *J Drug Targ* 10: 327–335.

124. Mu H and Holm R (2018) Solid lipid nanocarriers in drug delivery: Characterisation and design. *Expert Opin Drug Deliv* doi: 10.1080/17425247.2018.1504018.

Chapter 6

Lipoproteins for Biomedical Applications: Medical Imaging and Drug Delivery

Pratap C. Naha,[a] **Stephen E. Henrich,**[b,c,d,e] **David P. Cormode,**[a,f,g] **and C. Shad Thaxton** [b,c,d,e]

[a]*Department of Radiology, University of Pennsylvania, 3400 Spruce St, 1 Silverstein, Philadelphia, Pennsylvania 19104, USA*

[b]*Department of Urology, Northwestern University, Chicago, Illinois, USA*

[c]*Simpson Querrey Institute for Bionanotechnology, Northwestern University, Chicago, Illinois, USA*

[d]*International Institute for Nanotechnology, Northwestern University, Chicago, Illinois, USA*

[e]*Robert H. Lurie Comprehensive Cancer Center, Northwestern University, Chicago, Illinois, USA*

[f]*Department of Bioengineering, University of Pennsylvania, 3400 Spruce St, 1 Silverstein, Philadelphia, Pennsylvania 19104, USA*

[g]*Department of Cardiology, University of Pennsylvania, 3400 Spruce St, 1 Silverstein, Philadelphia, Pennsylvania 19104, USA*

david.cormode@uphs.upenn.edu, cthaxton003@md.northwestern.edu

The development of nanoparticle delivery vehicles for drugs and contrast-generating media is an important field. Many artificial

Handbook of Materials for Nanomedicine: Lipid-Based and Inorganic Nanomaterials
Edited by Vladimir Torchilin
Copyright © 2020 Jenny Stanford Publishing Pte. Ltd.
ISBN 978-981-4800-91-4 (Hardcover), 978-1-003-04507-6 (eBook)
www.jennystanford.com

systems have been reported for these applications, such as liposomes, micelles, polymeric nanoparticles, and iron oxides. However, natural nanoparticles such as lipoproteins have significant advantages for these applications. Lipoproteins are an endogenous group of nanoparticles whose main role is the transport of fats between tissues. They are biocompatible, biodegradable and are naturally targeted to cell types of interest for imaging or drug delivery. Furthermore, there is now an expansive body of work that illustrates how lipoproteins can carry a plethora of payloads such as fluorophores, gadolinium chelates, radionuclides, nanocrystals, drugs, or nucleic acids. This chapter reviews the roles of lipoproteins as delivery vehicles. First, background on the different lipoprotein classes (i.e., HDL, LDL, VLDL, and chylomicrons) is provided. Next, we cover the applications of lipoproteins as contrast agents for MRI, CT, PET, and fluorescence imaging. Last, the literature on lipoprotein drug or nucleic acid carriers and their roles in treating cancers and cardiovascular disease are discussed. Overall, the goal of this chapter is to give the reader an understanding of the basics of the field of lipoprotein-based delivery vehicles.

6.1 Introduction

Novel biomedical technologies can improve patient health via improved diagnoses, early disease detection, and with safer or more effective treatments. Early and more precise disease diagnosis allows selection of the most appropriate therapies and enhances the likelihood of successful treatment. Non-invasive medical imaging technologies, such as magnetic resonance imaging (MRI), computed tomography (CT), positron emission tomography (PET), variants of mammography such as dual energy mammography (DEM) and ultrasound (US) are currently used in the clinic for the diagnosis of many diseases and conditions [1–8]. The use of contrast agents with these imaging techniques can better identify abnormal tissue or can allow disease classification [9].

Nanoparticles have been the focus of intense interest as contrast agents for imaging techniques such as MRI [1, 7, 10, 11], CT [2–4, 12, 13], DEM [5, 6], US [8, 14–16] and photoacoustics (PA) [17–20].

Nanoparticles possess several advantages as contrast agents or as a drug delivery vehicle, i.e., their synthesis usually requires very few steps, their size can often be tuned in a wide range, they can have long circulation half-lives, can be efficiently targeted to specific tissues or organs, and multiple drugs and imaging probes can be integrated into the same nanoparticles [19, 21]. The nanomedicine market is growing very rapidly, and it is projected to reach up to $350 billion (USD) by 2025 [22].

Many nanoparticle-based drug and contrast-generating material delivery vehicles have been proposed. These systems are frequently based on polymers, such as poly lactic-co-glycolic acid (PLGA) [23–29] or poly lactic acids (PLA) [30–32]. These polymers are biodegradable, biocompatible, and are Food and Drug Administration (FDA) approved. In addition to polymeric drug delivery nanoparticles, liposomes are also FDA approved and can carry drug or imaging probes to the disease sites. Liposomes are lipid bilayers that encapsulate an aqueous core. To date, numerous formulations based on liposomes such as Doxil, Lipodox, DaunoXome, Marqibo, Ambisome, Abelcet, Amphotec, Depocyt, Visudyne, DepoDur, Epaxal, and Inflexal V are in the clinic [33].

Lipoproteins are a set of natural nanoparticles whose main function is the transport of fats in the body. Similar to liposomes, lipoproteins possess interesting properties as a drug and imaging probe delivery systems. Lipoproteins are endogenous, non-immunogenic, can carry contrast-generating materials (such as nanocrystals, quantum dots, and fluorophores), drugs and nucleic acids [34–40]. More interestingly, lipoproteins can be passively targeted to macrophages, low-density lipoprotein receptor (LDLr) or scavenger receptor type BI (SR-BI) [41–43]. Also, lipoproteins can be re-directed to other targets for imaging and therapy [44, 45]. Based on their physico-chemical properties, i.e., size and surface composition, lipoprotein-based nanoparticles can have long circulation half-lives compared to non-lipoprotein nanoparticles [46, 47]. In this chapter, we discuss the different lipoprotein types, their role in the human body, and their biomedical applications as delivery vehicles for contrast agents and drugs.

6.2   Lipoproteins

Lipoproteins are composed of proteins (apolipoproteins), phospholipids, cholesterol esters and triacylglycerols, and are classified mainly into four subtypes, i.e., chylomicrons, very low-density lipoprotein (VLDL), low-density lipoprotein (LDL), and high-density lipoprotein (HDL). The structure of each type of lipoprotein is similar and each of them is spherical in shape when mature (nascent HDL can be discoidal). The surface of lipoproteins consists of an integrated mixture of apolipoproteins and amphiphilic lipids (mostly phospholipids and unesterified cholesterol), with a core of neutral lipids (triacylglycerols, cholesteryl esters and small amounts of unesterified cholesterol, etc.). On the other hand, lipoproteins differ in their size, lipid composition, major apolipoprotein, function and density (Table 6.1) [48]. Chylomicrons are the largest type of lipoprotein, while HDL is the smallest (Fig. 6.1) [49]. Chylomicrons are synthesized mostly in the intestine, range in size from 80 to 1200 nm (Table 6.1) [49], and are responsible for transportation of lipids from the intestinal lumen to the liver [50]. Chylomicrons also act as a carrier for lipid-based drugs, transporting them to the lymphatic system [50].

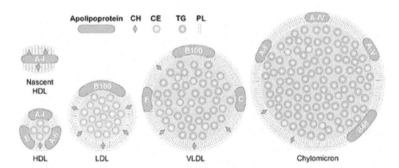

Figure 6.1 Schematic representations of the compositions of different lipoproteins. Figure reproduced with permission from [49].

VLDL is synthesized via a two-step process in the hepatocytes of the liver. First, microsomal triglyceride transfer protein (MTP) transfers lipids to apolipoprotein B (apoB) during its translation. Next, fusion occurs between apoB-containing precursor particles and triglyceride droplets to form mature VLDL [51]. The main role of VLDL is in the transport of endogenous lipids. VLDL is converted to intermediate density lipoprotein (IDL) after interactions with lipoprotein lipase, and then to LDL [52]. The LDLr is expressed in many tissues and causes LDL internalization via receptor mediated endocytosis. The internalized LDL degrades in lysosomes releasing cholesterol to the cell [53].

Table 6.1   Properties and major role of lipoproteins

Lipoprotein	Size (nm)	Major apolipoprotein	Density (g/dl)	Major role
Chylomicron	80–1200	B-48, A-I, A-II, C, E	< 0.95	Transport of dietary fats
VLDL	35–80	B-100, C, E	0.95–1.006	Transport of fats from the liver
IDL	27–30	B-100, C, E	1.006–1.019	Transport of fats from the liver
LDL	22–27	B-100	1.019–1.063	Delivery of fats to peripheral tissues
HDL	7–13	A-I, A-II, C, E	1.063–1.25	Removal of cholesterol from peripheral tissues

Source: Data reused with the permission from [48].

HDL is the smallest lipoprotein, and is largely produced in the liver and intestine. Its protein component mostly consists of apolipoprotein AI (apoA-I) and apolipoprotein A-II (apoA-II) [54]. ApoA-I is an activator of lecithin-cholesterol acyltransferase (LCAT). LCAT transfers acyl chains from phospholipids to cholesterol. This releases mono-acyl phospholipids, and concentrates cholesterol from both tissues and other lipoproteins [55]. HDL is classified into five sub-types based on shape, size, composition, density, and surface charge, i.e., HDL_{2a}, HDL_{2b}, HDL_{3a}, HDL_{3b}, and HDL_{3c} [55, 56]. HDL cholesterol is known as "good cholesterol," since high levels of HDL in the blood are associated with lower risk of cardiovascular disease [57]. On the other hand, LDL cholesterol is known as "bad cholesterol" since high levels of LDL in the blood are associated with higher risk of cardiovascular disease [58].

As mentioned above, lipoproteins play major roles in cardiovascular diseases. For example, high LDL levels are associated with greater risk of myocardial infarction, ischemic heart disease, and ischemic stroke [58]. Consequently, several drug technologies have been developed to reduce plasma LDL levels, as presented in Table 6.2, and their mechanisms of action of lowering LDL are presented in Fig. 6.2. For example, a class of small molecule–based drugs known as statins decrease plasma levels of LDL via inhibition of an enzyme called 3-hydroxy-3-methylglutaryl coenzyme A reductase (HMGCR), which is responsible for cholesterol synthesis in the liver (Fig. 6.2) [58, 59]. As a result, LDL receptor expression increases in the liver, enhances the clearance of LDL from the blood and the risk of cardiovascular disease decreases [60, 61]. Another approach has been to use antibodies against proprotein convertase subtilisin/kexin type 9 (PCSK9), which is a protein that binds to the LDLr to signal for its degradation. Anti-PCSK9 antibodies prevent PCSK9 from binding to LDLr, therefore increasing LDLr levels and reducing plasma LDL levels (Fig. 6.2) [58, 62].

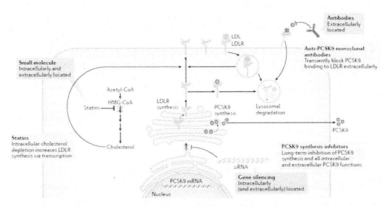

Figure 6.2 Schematic representation of the drug technologies used or proposed to reduce plasma LDL cholesterol levels. Reproduced with permission from [58].

Table 6.2 Summary of drug technologies reducing lipids and lipoprotein level in the plasma

Parameters	Drug technologies			
	Small molecules	Antibodies	Gene silencing with antisense oligonucleotides	Gene silencing with small interfering RNA
Chemical Structure	Organic compound	Protein	Single-stranded RNA	Double-stranded RNA
Mass (kDa)	<1	~150	~12	~21
Mechanism of action	Blocks enzyme or receptor in cells	Blocks protein in plasma	Blocks gene mRNA transcripts in cell	Blocks gene mRNA transcripts in cell
Potential for off-target adverse effects	High likelihood of off-target, non-tissue-specific effects	Low, given high specificity for target	Low, given high specificity for target with third-generation agents	Low, given high specificity for target
Immunogenicity	Low	High	High	High
Efficacy	50% reduction in LDL-cholesterol levels	60% reduction in LDL-cholesterol levels	90% reduction in lipoprotein(a) levels	50% reduction in LDL-cholesterol levels
Variation of within-person drug response	High	High	Low	Low
Half-life	Days	Weeks	Months	>1 year
Administration route	Oral	Subcutaneous	Subcutaneous	Subcutaneous
Dosing frequency	Daily	Weekly to twice monthly	Monthly	Twice yearly
Targets	Proteins in ng to µg	Proteins in µg to mg	Lipoproteins in g	Lipoproteins in g

Source: Data reused with the permission from [58].

On the other hand, LDL receptors are over expressed in certain cancer types, so these tumors have higher affinity toward LDL than normal tissue [63]. Therefore, LDL could be an attractive drug or contrast agent delivery vehicle for these tumors.

As mentioned above, HDL cholesterol is known as good cholesterol, and high levels of HDL reduce risk of cardiovascular disease. Furthermore, it has very high affinity towards macrophages due to their expression of an HDL receptor, i.e., scavenger receptor type BI (SR-BI) [64, 65]. Therefore, HDL could be an attractive drug delivery vehicle for macrophage-associated diseases such as inflammation, cancers and atherosclerotic plaque. In addition, there is considerable interest in approaches that could be used to elevate HDL levels. Pharmacological approaches have been trialed, such as cholesterylester transferase protein (CETP) inhibitors [66]. In addition, efforts to raise HDL cholesterol levels have been made by direct injections of HDL derived substances. For example, several HDL-based nanoparticles such as apoA-I Milano, CER-001 and CSL-112 are in clinical trials [67, 68]. ApoA-I Milano is a variant of apoA-I which is associated with decreased risk of cardiovascular disease [69]. A clinical trial reported that HDL nanoparticles reconstituted with recombinant apoA-I Milano (referred to as ETC-216 or MDCO-216) significantly reduced atherosclerosis burden in coronary arteries, when administered weekly for five weeks [70]. A clinical trial of an HDL nanoparticle reconstituted with wild type apoA-I (referred to as CSL-112) did not reduce atherosclerotic plaque burden compared to placebo; however, it improved plaque characterization index and coronary scores [71]. The same formulation was tested in type-2 diabetes patients, where it was found to reduce platelet aggregation, and therefore it could be an effective therapeutic strategy for reducing vascular complications for type-2 diabetic patients [72].

Lipoproteins have been proposed as delivery vehicle for drugs, nucleic acids, and imaging probes. For example, an LDL lowering statin, simvastatin, has been found to also be a potent anti-inflammatory drug and a recent study showed that simvastatin loaded HDL nanoparticles efficiently reduced inflammation in atherosclerotic plaques [40]. Another study showed that a synthetic HDL nanoparticle inhibits toll-like receptor 4 (TLR4) inflammatory responses via scavenging and neutralizing lipopolysaccharide (LPS) toxins [73]. HDL-based nanoparticles have been proposed as a delivery vehicle for nucleic acids [74–76], peptides [77–79] and drugs [40, 80–82]. In addition, HDL nanoparticles have been shown to be promising imaging probes for CT [83], MRI [84, 85], PET [86, 87] and fluorescence imaging [35, 42]. Similarly, LDL has been shown to be an efficient drug and imaging probe delivery system as well [41, 88].

Lipoproteins have the following advantages as a platform for drug or imaging probe delivery:

1. They have long circulation half-lives.
2. They are endogenous, biocompatible, and biodegradable.
3. They are non-immunogenic.
4. It is possible to target to macrophage-associated diseases, such as cancers, inflammations and atherosclerotic plaque and other targets.
5. It is possible to integrate multiple agents (i.e., drugs, peptides, nucleic acids or imaging probes, etc.) into the core or at the surface.

However, the major disadvantage with this type of delivery system is that by loading them with drugs or contrast-generating materials, characteristics such as particle size, stability, safety, and degradability may be compromised. Therefore, loading of drugs and contrast agents into lipoprotein-based delivery systems needs to be optimized on a case-by-case basis.

6.3 Lipoprotein-Based Contrast Agents

6.3.1 High-Density Lipoprotein Contrast Agents

6.3.1.1 Labeled native HDL contrast agents

HDL has been proposed as a delivery vehicle for contrast agents in many studies. The first report on HDL as an imaging probe examined the pharmacokinetics of HDL. In this study, the authors labeled the protein component of HDL with I-125, and positron emission tomography (PET) imaging was used for imaging HDL in a mouse model of atherosclerosis [89]. The authors found that around 30% of the injected dose is still in the systemic circulation at 24 h post injection. Biodistribution data revealed that most of the HDL accumulated in the liver, heart, kidney, aorta, and lungs. Interestingly, they found that HDL was accumulated in atherosclerotic plaques in abdominal aortas using autoradiography [89].

6.3.1.2 Reconstituted HDL contrast agents

Reconstituted HDL has been adapted as contrast agents for several imaging modalities, such as CT, MRI, fluorescence, and PET [42, 74, 90–95]. To reconstitute HDL, it is first separated into its proteins and lipid components, which are then reassembled in combination with other materials such as contrast-generating materials or drugs. This approach allows the integration of payloads into the lipid layer or the hydrophobic core. HDL reconstitution has traditionally been done by incubation of the proteins with lipid vesicles; however, a group led by Fayad, Mulder, and Fisher recently reported HDL synthesis using microfluidic chip technology [96]. In addition, this group has reported many studies on HDL as contrast agents for medical imaging applications. The first of these studies focused on HDL as a contrast agent for MR and fluorescence imaging [90, 92]. In this report, the authors labeled HDL using lipids whose headgroup included either a gadolinium chelate to produce MRI contrast or a fluorophore for fluorescence-based techniques. After injections of this agent into apoE KO mice on a high fat diet (a model of atherosclerosis), enhanced MR signal was found in arterial plaques due to accumulation of these gadolinium labeled HDL nanoparticles. Confocal microscopy and histopathology revealed that the labeled HDL accumulated in the macrophage-rich region of the plaque. This group developed several techniques to label HDL with gadolinium chelates to create stronger MR contrast [97], and explored the use of an apoE-based lipopeptide incorporated into the lipid layer of HDL, which resulted in higher uptake in macrophages via binding to LDLr expressed on macrophages [98]. As a result, stronger MR contrast was achieved in vivo compared to the HDL without lipopeptide [98]. Other researchers also reported that HDL labeled with gadolinium is an effective MR contrast agent [99, 100]. A recent study showed that HDL labeled with europium produced MRI contrast via chemical exchange saturation transfer (CEST) [101].

Duivenvoorden et al. reported an HDL-based nanoparticle that acted as a contrast agent for both MR and fluorescence imaging, as well as delivered a statin drug payload, rendering these nanoparticles a theranostic [40]. The HDL was labeled with gadolinium chelates and a fluorophore (Cy 5.5) in the lipid layer as well as either an anti-inflammatory drug (simvastatin) or DiR (another lipophilic fluorophore) loaded into the core. Schematic depictions and TEM images of statin loaded HDL and Gd-Cy5.5-statin-HDL nanoparticles are shown in Fig. 6.3A. Gd-Cy5.5-statin-HDL was injected intravenously into apoE KO mice and MR imaging was performed before and 24 h after injection. The imaging data revealed that stronger MR signal was found in the plaque region of mouse aortas after injection of the agent (Fig. 6.3B). Similarly, fluorescence imaging revealed that HDL nanoparticles preferentially accumulated in the plaque region of the aorta compared to controls (Fig. 6.3C), indicating that this HDL-based agent could effectively deliver drug to atherosclerotic plaque.

In addition to MR and fluorescence imaging, reconstituted HDL has also been reported as a probe for PET imaging by the Reiner group [86]. In this study, they conjugated Zr-89 (a radioisotope for PET imaging) to HDL either via phospholipids or apoA1. This agent was injected into breast tumor–bearing mice intravenously, and the results revealed that around 17% of injected dose accumulated in the tumor [86].

6.3.1.3 Nanocrystal loaded HDL contrast agents

Interestingly, nanocrystals can be encapsulated in the hydrophobic core of HDL. Several studies have reported iron oxide or gold nanoparticles encapsulated in HDL as contrast agents for MRI or CT, respectively.[35, 42, 83, 95] Multiple components such as nanocrystals, gadolinium chelates and fluorophores could be incorporated into HDL simultaneously for multimodal imaging applications (Fig. 6.3D) [102]. Cormode et al. reported injections of gold nanoparticle-loaded HDL (Au-HDL) into atherosclerotic mice. At 24 h post-injection, the authors found significantly higher CT attenuation in atherosclerotic plaques (Fig. 6.3E) compared to control groups. Similar results were observed with MRI and fluorescence imaging when injected with iron oxide-HDL and quantum dot-HDL, respectively [42]. Another study showed that with quantum dots in the core, and with Cy 5.5-labeled lipids, HDL could be used to study the bio-interactions of HDL with other lipoproteins and cells via Förster resonance energy transfer (FRET) [94].

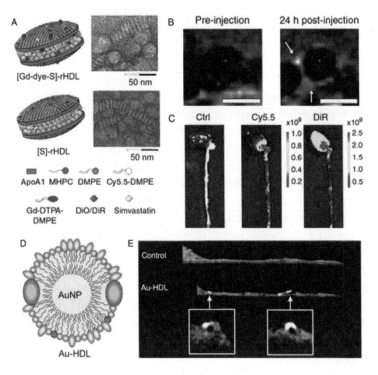

Figure 6.3 HDL-based contrast agents for MRI, fluorescence imaging and CT. (A) Schematic representations of HDL nanoparticles loaded with combinations of gadolinium, dyes or statin, and TEM of these nanoparticles. (B) In vivo MR images of an atherosclerotic mouse acquired pre-injections and 24 h post-injection with [Gd-dye-S]-HDL. (C) Ex vivo fluorescence imaging of aortas from atherosclerotic mouse injected with saline or [Gd-dye-S]-HDL. (D) Schematic representation of gold nanoparticle loaded HDL (Au-HDL). E) Ex vivo CT imaging of the aortas from atherosclerotic mice injected with Au-HDL or saline. A, B, and C reproduced with permission from [40]. D and E reproduced with permission from Cormode et al., 2008 [42].

6.3.1.4   Re-routed HDL contrast agents

Several studies have shown that HDL nanoparticles can be re-directed to other targets rather than its natural targets via modification with peptides, small molecules or proteins [44, 78, 103, 104]. In a study by Chen et al., it was shown that after conjugation of arginine-glycine-aspartic acid (RGD) peptides to the apoA1 component of HDL (RGD-HDL) it efficiently targeted the $\alpha_v\beta_3$-integrin, since RGD-HDL specifically bound to endothelial cells expressing the $\alpha_v\beta_3$-integrin [44]. In this study, the authors labeled RGD-HDL with gadolinium for MR and a NIR fluorophore (DiR) for fluorescence imaging. In vivo imaging revealed that RGD-HDL nanoparticles accumulated in tumors, as stronger MR contrast was observed compared to untargeted HDL nanoparticles. Confocal microscopy on tumor tissue samples confirmed the localization of RGD-HDL in the endothelial cells at 1 h post-injection [44]. Another study from this group showed that gadolinium labeled HDL nanoparticles functionalized with the collagen-specific EP3533 peptide can visualize intra-plaque macrophages and collagen content in atherosclerotic plaque with MRI [103].

Folic acid (FA) can be used as a targeting moiety for tumors that express folate receptors (FR) on cell surfaces. Corbin et al. developed a folate conjugated HDL nanoparticle platform to target an ovarian cancer that over expresses the folate receptor. Successful targeting of these nanoparticles was determined using fluorescence imaging [104]. Similarly, HDL conjugated with endothelial growth factor (EGF) can be rerouted to the tumors expressing the EGF receptor [105]. A recent study by Ding et al. showed that RGD peptide conjugated HDL can efficiently deliver siRNA to the tumor cells expressing $\alpha_v\beta_3$-integrin, as demonstrated by fluorescence imaging of Cy5 labeled HDL [78].

6.3.2 Low-Density Lipoprotein–Based Imaging Probes

6.3.2.1 Labeled native LDL contrast agents

Similar to HDL, LDL has been proposed as contrast agent for nuclear imaging and various radionuclei such as I-123, I-125, In-111, F-19 and Tc-99m have been used to label LDL [89, 106–109]. For example, oxidized LDL (ox-LDL) was labeled with Tc-99m to compare its pharmacokinetics with native LDL [110]. The authors found that ox-LDL was rapidly cleared from the circulation compared to the native LDL, as most of the ox-LDL was accumulated in the liver. However, they found that ox-LDL accumulated in the plaque (carotid artery) significantly more than the controls, indicating that LDL-based contrast agents could be used for imaging atherosclerotic plaque.

LDL has also been explored as a contrast agent for fluorescence imaging [88, 111–113]. In a study by Li et al., native LDL was labeled with DiI, a lipophilic fluorophore, and used for imaging tumors that over express LDLr [113]. In this study, the authors used two different tumor models, i.e., B16 and HepG2, and found that LDL-DiI is more homogenously distributed in HepG2 tumors than B16 tumors, indicating a differential pattern of LDLr expression within these tumors.

Numerous studies have reported the use of LDL as a MRI contrast agent. For example, LDL was labeled with manganese ions [114], whose five unpaired electrons render it highly paramagnetic, and therefore can be used as an MRI contrast agent [115, 116]. Mn-LDL was incubated with a murine monocyte cell line and after 24 h of incubation, pellets of these cells were scanned with MRI. A significant difference in signal was observed in the cells incubated with Mn-LDL compared to controls, which indicates the potential of Mn-LDL as a MRI contrast agent.

A notable study on LDL receptor (LDLr) imaging with MRI was reported by Corbin et al. [117]. The authors labeled LDL with a gadolinium chelate (Gd-DTPA-SA-LDL), as schematically depicted in Fig. 6.4A [88]. After labeling, Gd-DTPA-SA-LDL was incubated with cells that do and do not over express LDLr. They found that more Gd-DTPA-SA-LDL was internalized in HepG2 cells (+ve LDLr) compared to CHO cells (-ve LDLr). Next, they performed in vivo imaging using a HepG2 tumor–bearing mouse model. They found that MR signal in the liver and tumor post-injection increased as a function of time compared to pre-scans, due to accumulation of Gd-DTPA-SA-LDL in the tumors and liver (Fig. 6.4B).

In addition, Li et al. reported another LDL MRI contrast agent variant [118]. In this study, the authors labeled LDL with PTIR267, a gadolinium chelate derivative conjugated to a fluorophore. This agent was injected intravenously into B16 tumor–bearing mice. In vivo imaging data revealed that stronger MR signals were found in the tumors of the mice injected with gadolinium chelate labeled LDL compared to unlabeled LDL. More recently, LDL labeled with gadolinium chelates have been studied in a mouse model of atherosclerosis using MRI [119, 120]. Lowell et al. labeled LDL with gadolinium (Gd-LDL), and then injected it into atheroma mice. MR imaging was performed at 24 and 48 h post injection. The images revealed that very strong signal developed in the plaque region at 48 h post-injection, compared to controls [120].

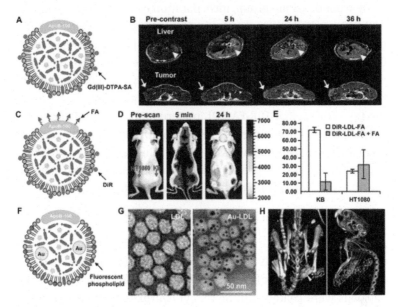

Figure 6.4 LDL-based contrast agents for MRI, fluorescence imaging and CT. (A). Schematic representation of Gd labeled LDL (Gd-LDL). (B) In vivo MR imaging of a tumor-bearing mouse, pre- and post-injection with Gd-LDL. Arrowheads point to the liver and the arrows indicate the tumor. (C) Schematic representation of DiR-LDL-FA. (D) Pre- and post-injection of DiR-LDL-FA in vivo fluorescence imaging of a mouse bearing both KB and HT1080 tumors. (E). Fluorescence intensity of tumor extracts from an in vivo competition-inhibition assay, where free FA was used to block the FR mediated endocytic pathway of DiR-LDL-FA. (F). Schematic representation of LDL loaded gold nanoparticle (Au-LDL). (G). TEM of LDL and Au-LDL nanoparticles. (H). In vivo CT imaging of B16-F10 tumor–bearing mouse injected with Au-LDL. Au-LDL accumulated in the tumor (left image) and in the liver (right image) as indicated in yellow. Figure reproduced with permission from Zhu and Xia, 2017 [88].

6.3.2.2 Re-routed LDL contrast agents

LDL has been re-directed to targets other than LDLr. For example, Zheng et al. developed a folic acid conjugated LDL (FA-LDL) contrast agent for imaging tumors that over express FR at their surface [45].

In this study, authors conjugated folic acid to the lysine residue of the protein component (apoB) of LDL, and labeled it with Dil for fluorescence imaging. Via in vitro studies, they showed that FA-LDL was specifically directed to cells that over express FR. In another study from the same group, incorporation of DiR into FA-LDL (DiR-FA-LDL) resulted in a contrast agent for in vivo fluorescence imaging of tumors overexpressing FR (Fig. 6.4C) [121]. In vitro experiments demonstrated that DiR-FA-LDL was targeted to cells that over express FR. In vivo optical imaging was done with mice bearing two types of tumors that either did or did not express FR. The resulting images showed that DiR-FA-LDL preferentially accumulated in the tumor (KB) that overexpressed FR (Fig. 6.4D), since the fluorescence intensity in those tumors was significantly higher than in the control tumors. An in vivo competition-inhibition experiment confirmed the specificity of DiR-FA-LDL to the KB tumor (Fig. 6.4E).

6.3.2.3 Nanocrystal loaded LDL contrast agents

Allijn et al. reported a method for loading hydrophobically coated inorganic nanocrystals (for example, gold, iron oxide or quantum dots) and hydrophobic fluorophores into LDL [41]. The hydrophobic payloads were coated with myristoyl-hydroxy-phosphatidylcholine (MHPC) prior to loading into LDL cores. The resulting MHPC micelles were sonicated with native LDL, which led to the payload being transferred into the core of LDL. A schematic depiction of gold cores encapsulated in LDL is shown in Fig. 6.4F [88]. Transmission electron micrographs show the loading of gold nanoparticles within LDL (Fig. 6.4G). In vitro experiments with B16-F10 cells (a melanoma cell line that overexpresses LDLr) showed that significant amounts of Au-LDL nanoparticles were internalized. Furthermore, in vivo CT imaging revealed enhanced contrast in mice bearing B16-F10 tumors injected with Au-LDL (Fig. 6.4H).

6.3.2.4 Reconstituted LDL contrast agents

Due to the difficulty of the reconstitution process for LDL, this technique has not been used widely for making LDL-based contrast agents; however, a few studies have been reported. For example, reconstitution was used to load LDL with iodinated triglyceride to create a CT contrast agent [122]. LDL was isolated from human plasma, then iodinated triglyceride was loaded into the core of LDL according to the Krieger method [123]. Iodinated triglyceride loaded LDL was incubated with HepG2 cells, which were subsequently scanned as cell pellets with a CT scanner. They found that HepG2 cell pellets incubated with LDL loaded iodinated triglyceride produced nearly twofold higher CT contrast than control cells or cells incubated with free iodinated triglyceride.

6.3.3 Very Low-Density Lipoprotein–Based Contrast Agents

VLDL-based contrast agents have only been rarely explored. In one study, VLDL was labeled with I-123 (VLDL-I-123), this agent was injected intravenously into rabbits, i.e., New Zealand white rabbit (normal rabbits) and homozygous Watanabe-heritable hyperlipidemic (WHHL) rabbits (rabbits with low expression of LDLr in their liver) [124]. Radioactivity in the different organs, i.e., liver and heart was recorded using a gamma camera. Very little radioactivity was detected in the liver of WHHL rabbits compared to normal rabbits. This study found that LDLr expression in different tissues could be detected via nuclear imaging using this VLDL-based contrast agent.

6.3.4 Chylomicron-Based Contrast Agents

Similar to VLDL, chylomicrons are also rarely explored as contrast agents. A few studies reported chylomicrons labeled with radioactive materials or gold nanoparticles; however, imaging experiments were not performed in those studies [125, 126]. Bruns et al. developed a chylomicron-based nanoparticle platform, in which authors loaded quantum dots and iron oxide nanocrystals in the core [127]. Quantum dots provided contrast for fluorescence imaging, whereas iron oxides provided contrast for MR imaging. The authors employed this chylomicron-like particle to study chylomicron metabolism using fluorescence imaging and MRI.

6.4 Drug Delivery with Lipoproteins

6.4.1 Drug Delivery with High-Density Lipoprotein

6.4.1.1 Properties of HDL in drug delivery

In addition to their widely recognized role in reverse cholesterol transport, HDL plays major roles as systemic delivery agents. In the native setting, HDL species (which are 7–13 nm in diameter) harbor and deliver a diverse set of cargoes including circulating miRNAs [128], endogenous lipids [55], proteins [129, 130], and small hydrophobic metabolites and signaling molecules. This versatility has made HDL extremely attractive as a drug delivery platform, particularly for cancer and inflammation. One property of HDLs which makes them ideal candidates for delivery of anti-cancer agents is their ability to target tumor cells. While a majority of chemotherapeutics are indiscriminately uptaken by healthy and malignant tissues, HDLs are capable of targeting tumor tissue both actively and passively. First, nanoparticles such as HDL are often preferentially uptaken by the leaky, fenestrated vasculature of tumor tissue. Due to impaired lymphatic drainage and the resultant elevated hydrostatic pressures, particles of this size are also more likely to be retained in tumor tissues once they have extravasated. These phenomena together are known as the enhanced permeability and retention (EPR) effect [131], although the strength of this effect is variable between tumor types as well as between tumors of the same type [131, 132]. Since both HDL and LDL are within the right size range (i.e., ~5–200 nm) and have long circulation half-lives, they can deliver substantial drug payloads via the EPR effect. Beyond the EPR effect, HDL also may actively target tumors via specific HDL receptors, as mentioned above [133–137]. Cargo delivered by HDLs to tumor cells also avoids the endolysosomal pathway [65, 138], which offers a distinct advantage over many competing delivery systems, especially for nucleic acid cargo [74], as most drug-loaded nanoparticles which enter the endolysosomal pathway undergo degradation in lysosomes before interacting with the intended target.

One of the hallmark structural features of lipoproteins is a dynamic, hydrophobic core [139, 140]. This core is capable of expanding and contracting throughout the process of cargo loading and unloading, making it well suited for efficient, iterative delivery of hydrophobic cargo. It is estimated that ~40% of new therapeutics are poorly water soluble, which severely limits the quantity of drug that can be administered [141]. Efficient loading of these poorly soluble drugs into the hydrophobic core of HDLs can often be accomplished, enabling higher achievable doses and increased circulating half lives. In sum, HDL is an attractive drug delivery vehicle for anti-cancer therapy for at least four major reasons: (1) passive targeting of tumor tissue via the EPR effect, (2) active targeting of tumor tissue through upregulated HDL receptors such as SR-B1, and (3) a hydrophobic core which is well suited to load and deliver lipophilic drugs which would otherwise be poorly soluble, and (4) a non-endocytic cargo delivery pathway.

6.4.1.2 Exploiting the targeting properties of native HDL and other lipoproteins

While synthetic HDL mimics and rHDL formulations have each been developed for drug delivery, some recent approaches have instead exploited native circulating lipoproteins to enhance drug stability and delivery. As a noteworthy example of this approach, a recent article describes novel drug conjugates which were shown to be efficiently loaded into circulating lipoproteins upon systemic injection [81], significantly enhancing circulating half life and delivery to target tissues. The group tested a number of squalene-modified compounds, which when injected intravenously exhibited strong association with either HDL or LDL, and imparted enhanced targeting of the drugs to SR-B1 or LDLr expressing cells. Some squalene-modified compounds showed preferential binding to HDL (e.g., adenosine and Cy5.5) while others showed preferential association with LDL (e.g., gemcitabine). Radiolabeled squalene-modified gemcitabine was found to penetrate malignant tissues in mouse tumor xenografts, with twofold greater uptake in the LDLr high-expressing breast cancer line MDA-MB-231 than the LDLr low-expressing breast cancer line MCF-7. These uptake experiments were performed in mice fed on a high-fat diet in order to increase circulating LDL [81].

6.4.1.3   rHDL for drug delivery in cancer

Many groups have investigated rHDL or discoidal HDL as delivery systems for anti-cancer therapy. rHDL is often loaded with a hydrophobic compound and may be altered with additional functional moieties to enhance targeting or facilitate an auxiliary therapeutic effect. Crosby et al. recently published a report detailing the synthesis of a chimeric rHDL nanodisk consisting of an α-CD20 single chain variable fragment antibody conjugated to human apoA-1 for lymphoma [142]. The rHDL-based chimera was shown to bind tightly to the surface of CD20-positive lymphoma cells and off-load encapsulated hydrophobic compounds which were efficiently delivered to the intracellular space. Another recent study describes the development of a dual therapeutic platform directed against breast cancer using rHDL [143]. The group first formulated an amphiphilic complex of p53 plasmid DNA using a cationic lipid dimethyldioctadecylammonium bromide (DODAB), and subsequently encapsulated this complex and the hydrophobic chemotherapy drug paclitaxel in rHDL to form stable particles, ~150–200 nm in diameter. This platform was shown to induce robust expression of p53 in vitro and in vivo, and inhibit tumor growth by ~65% in an orthotopic mouse model of breast cancer. A similar dual formulation of rHDL with p53 plasmid DNA and anti-angiogenic therapy was previously reported for use in bladder cancer [144]. Xiong et al. recently investigated ultrasound-stimulated drug delivery using infrared dye-loaded rHDL in an orthotopic model of breast cancer. Here, ultrasound stimulation was shown to significantly approved drug accumulation in tumor tissues over unstimulated controls [145].

6.4.1.4   rHDL for drug delivery in cardiovascular and neurodegenerative disease

Beyond cancer, rHDL have demonstrated significant promise as delivery vehicles for cardiovascular disease. As mentioned above, rHDL were loaded with simvastatin and shown to inhibit inflammation in atherosclerotic plaques in an apoE knockout mouse model of atherosclerosis [146]. In a more recent report, Tang et al. demonstrated that statin-loaded rHDL is able to attenuate progression of inflammation in atherosclerotic plaques by inhibiting plaque macrophage proliferation [147]. In this study, a two-step regimen was designed and implemented which involved a 1-week induction therapy with statin-loaded rHDLs followed by an 8-week oral statin therapy regimen. This two-step approach was shown to dramatically reduce plaque size by 43% after the 1st week, while reducing macrophage prevalence by 65%. These reductions were also maintained over the subsequent 8 weeks.

rHDL-based delivery systems have recently expanded their application to neurodegenerative disease in a series of papers demonstrating striking results in pre-clinical models of Alzheimer's. Song et al. recently reported that a bio-inspired rHDL-ApoE3 nanostructure was capable of crossing the blood–brain barrier, binding amyloid-beta monomer and oligomer with high affinity, and facilitating amyloid beta degradation by glial cells. The particle was able to rescue the memory deficits and associated neurologic changes in a mouse model of Alzheimer's disease [148]. An even more recent publication from this group reported a multi-modal therapeutic formulation of these particles, which incorporated monosialotetrahexosylganglioside (GM1) as a targeting moiety, and loaded the neuroprotective peptide NAP into the particle core [149]. The NAP-loaded GM1-rHDL particles exhibited stronger neuroprotective effects and memory loss rescue than either NAP or GM1-rHDL alone, and no evidence of cytotoxicity was found.

6.4.1.5 Synthetic HDL mimetics for drug delivery

Synthetic HDL mimetics have exhibited significant promise as drug delivery agents for cardiovascular disease. A recent article describes a novel HDL-like nanoparticle (HDL NP) which delivers nitric oxide as therapy for vascular disease (Fig. 6.5) [150]. Specifically, a new gold nanoparticle-based HDL NP was developed which incorporates an S-nitrosylated phospholipid onto the nanoparticle surface (SNO HDL NP). This particle thus combines the vasoprotective functions of HDL with the vasoprotective functions of nitric oxide in a single therapeutic platform. While nitric oxide's beneficial effects on vascular function are well demonstrated [151–154], its short half life [155] and lack of targeting are significant limitations to its translation to therapy. The SNO HDL NP formulation overcomes both of these limitations by using a naturally targeted agent (HDL) alongside a stable nitric oxide-functionalized lipid donor functionalized on the particle surface. The particle was shown to reduce ischemia/reperfusion injury in a mouse kidney transplant model, and also reduce atherosclerotic plaque burden in a mouse model of atherosclerosis (Fig. 6.5D,E) [150]. Another recent report described the development of a polymer-based HDL mimetic for atherosclerosis. In this work, a microfluidic technique was employed to assemble particles using poly(lactic-*co*-glycolic) acid, phospholipids, and apoA-1. These particles were 30–90 nm in diameter, and were shown to efficiently efflux cholesterol, and target atherosclerotic plaques in vivo in ApoE knockout mice [156].

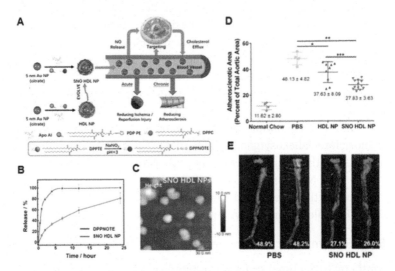

Figure 6.5 Nitric oxide-delivering HDL mimetic nanoparticles (SNO HDL NP) reduce atherosclerotic burden. (A). Schematic of the synthesis and downstream application of a hybrid nitric oxide-carrying HDL nanoparticle platform which incorporates an S-nitrosylated lipid assembled on the particle surface. (B). In vitro nitric oxide release profiles for SNO HDL NP vs. the S-nitrosylated lipid, 1,2-dipalmitoyl-*sn*-glycero-3-phosphonitrosothioethanol (DPPNOTE) alone at 37°C. (C). Atomic force microscopy of SNO HDL NPs revealed spherical particles ~13 nm in diameter. (D). Quantification of atherosclerotic plaque burden in Apo E knockout mice that had been on a high fat diet or normal diet for 12 weeks, and treated as indicated 3×/week for an additional 6 weeks. (E). Representative images of aortas from mice treated with PBS, or SNO HDL NPs. Red color indicates atherosclerotic plaques. Reproduced with permission from [150].

The above-mentioned particle platforms highlight the diversity of approaches available when using HDL as a drug delivery system.

These bio-inspired delivery vehicles range from 7–200 nm in diameter, use both inorganic and organic materials as core scaffolds, deliver drugs in the lipophilic core of the particle as well as on the particle surface, and have achieved significant pre-clinical success in both cardiovascular disease and malignancy.

6.4.2   Drug Delivery with Low-Density Lipoprotein

LDL has garnered significant attention as a delivery vehicle for both cancer and atherosclerosis, particularly due to the expression of LDLr on the surface of tumor cells and plaque-resident foam macrophages. Similar to HDL, LDL is also capable of both active and passive targeting of tumor tissue. Several tumor types exhibit upregulated expression of LDLr to enhance delivery of cholesterol and cholesteryl esters to fuel malignancy, including pancreatic ductal carcinoma [157] and glioblastoma [158, 159]. Furthermore, because LDL is larger in diameter than HDL (Table 6.1), its hydrophobic core has more volume for the encapsulation of hydrophobic drugs and is particularly well suited for this purpose.

A recent study investigated the use of ultrasound stimulation to enhance blood–brain barrier penetration of lipoprotein nanoparticles [160], which could therefore be of interest for neurodegenerative disease or CNS malignancies such as glioblastoma. LDL nanoparticles were formulated using LDL isolated from human serum that was subsequently subjected to a core replacement process in which the neuroprotective agent docosahexaenoic acid (DHA) was loaded into the particle core. Healthy rats underwent ultrasound pulsation of localized brain regions prior to being injected intravenously with fluorescent dye-labeled, DHA-loaded LDL nanoparticles. Rats exposed to ultrasound pulsation demonstrated significantly greater uptake of LDL nanoparticles and DHA than unstimulated controls. Moreover, there were no signs of CNS inflammation in rats exposed to ultrasound pulsation, indicating that ultrasound pulsation could be a safe and effective means to enhance blood–brain barrier penetration of lipoprotein-based nanoparticle therapeutics and imaging agents [160]. The same group recently reported the use of DHA-loaded LDL nanoparticles for a different clinical target, namely liver cancer. The group demonstrated that hepatic artery injection of DHA-loaded

LDL particles reduced growth of orthotopic liver tumors in rats by disrupting redox balance in tumor cells [161].

Recent work from Zhang et al. reports the synthesis of a doxorubicin-loaded LDL peptide-DNA hybrid nanomaterial for anti-cancer therapy [162]. The approach sought to exploit the structural advantages of duplex DNA, namely to enable intercalation of doxorubicin into the material prior to delivery. To stabilize the DNA, several cationic polymers were used along with an LDLr-targeting peptide to enhance tumor targeting of the delivery system. This particle formulation increased tumor cell death and intracellular ROS production in vitro, and reduced intracellular elimination of doxorubicin [162].

Several groups have recently reported new LDL-based drug delivery systems using polysaccharides to assemble hybrid nanoparticles. Zhou et al. reported the development of a pectin-LDL hybrid nanogel for oral delivery of curcumin [163]. In this work, the plant-based polysaccharide pectin was complexed with LDL in a pH- and heat-induced assembly process. FTIR spectra demonstrated that both hydrophobic and electrostatic interactions comprised the driving forces for nanogel assembly. In a separate study, fabrication of a delivery vehicle using chitosan-LDL nanoparticles to encapsulate doxorubicin was demonstrated. Chitosan and LDL were combined in a one-step synthesis to obtain ~150–200 nm particles which were subsequently loaded with doxorubicin, achieving greater than 50% encapsulation efficiency [164]. The loaded particles were efficiently taken up by gastric cancer cells and successfully induced apoptosis.

6.4.3 Drug Delivery with Very Low-Density Lipoprotein

VLDL has not often been used for applications in drug delivery. Two likely reasons for this are that VLDL has a relatively short circulation half life (3–5 h) [165], and that VLDL undergoes a large number of transformations in the systemic circulation. Over its maturation process, VLDL acquires and loses several distinct classes of apolipoproteins, and undergoes a series of interactions with other serum lipoproteins and lipid metabolizing enzymes, which makes it unwieldy as a delivery vehicle. However, VLDL may be an important player with respect to the distribution and delivery of free drugs in the systemic circulation. A recent article published by Yamamoto et al. investigated 42 randomly selected drugs for their associations with different lipoprotein serum fractions after oral administration in mice. Of the drugs tested, more than 20% were detected primarily in lipoprotein fractions. A lipoprotein lipase inhibitor was then used to increase levels of circulating VLDL in the blood stream, and drug concentrations were re-assessed. While the concentration of well-soluble or albumin-bound drugs did not significantly change, the concentration of drugs primarily found in lipoprotein fractions increased, which was likely due to increased VLDL levels [166].

6.4.4 Drug Delivery with Chylomicrons

Chylomicrons are poor candidates as drug delivery systems due to their short half lives and heterogeneity [167, 168]. However, the oral bioavailability of some drugs can be effectively modulated by altering intestinal lymphatic transport of the drug via chylomicrons. Chylomicron flow can be pharmacologically inhibited by agents such as cycloheximide or colchicine [169]. Hence, the relative importance of chylomicron-mediated lymphatic transport from enterocytes to the systemic circulation with respect to a drug's oral bioavailability can be examined by using such an inhibitor prior to oral administration. Recently, one group demonstrated that for self-emulsifying drug delivery systems with >25% oil fraction, the lymphatic transport route via chylomicrons is a significant contributor to oral bioavailability of the immunosuppressive drug sirolimus [170].

6.5 Nucleic Acid Delivery with Lipoproteins

Therapeutic oligonucleotides have remarkable promise in cancer, cardiovascular disease, and beyond. RNA interference (RNAi) specifically has proven to be a robust mechanism for disease modulation via targeted knockdown of disease-relevant gene expression. RNAi can be mediated by several different RNA sub-types, including short interfering RNAs (siRNA), short hairpin RNAs (shRNA), and microRNAs (miRNA). While these RNAs can each exhibit potent knockdown of target mRNA expression, they are inherently unstable in their naked form and therefore unviable for most clinical applications [171, 172]. However, enhanced stability of RNAi therapeutics can be achieved through a number of mechanisms (reviewed here) [173]. One mechanism relevant to this review is the conjugation of cholesterol or cholesteryl esters to the 5' or 3' end of the antisense strand of siRNA. This modification allows nucleic acids to bind albumin or lipoproteins in the systemic circulation, which prolongs their circulating half life and improves stability [174–176]. In vivo, cholesterol-modified siRNAs have been shown to primarily target the liver, due to binding of the RNA to circulating HDL, and subsequent metabolism and recycling of HDL in the liver via SR-B1 [176, 177].

In 2011, Vickers et al. discovered that native HDL isolated from human serum contains high quantities of miRNAs, and that these miRNAs exhibit improved stability over naked miRNAs [128]. By contrast, there was comparatively little miRNA observed in the LDL fraction of serum samples from both healthy donors and those with cardiovascular disease, suggesting that HDLs may be better suited to RNA delivery. This discovery prompted significant interest in using HDL and HDL mimetics as RNAi delivery systems. There have primarily been two categories of approaches to use HDL for RNAi delivery, namely the use of reconstituted HDL or synthetic HDL mimetics.

Synthetic HDL mimetics for siRNA delivery have employed one of two core materials, gold nanoparticles or HDL-mimicking peptide-phospholipid scaffolds. First, bio-inspired HDL nanoparticles have been synthesized using a 5 nm gold nanoparticle core. The fabrication of these materials is carried out by functionalizing the surface of the 5 nm gold particle with apoA-1 and phospholipids. The resultant HDL NPs are size, surface chemistry, and functional mimics of native mature, spherical HDL, containing 2–4 copies of apoA-1 per particle and an outer phospholipid layer consisting of 1,2-dipalmitoyl-*sn*-glycero-3-phosphocholine (DPPC) [178]. These particles have been shown to interact strongly with SR-B1, as well as ABCA1 and ABCG1 to induce cholesterol efflux from immune cells (especially myeloid cells) [179, 180], neoplastic cells [179], and other cell types in vitro and in vivo. These particles have been shown to be effective for nucleic acid delivery, using both anti-sense DNA and siRNA. HDL NPs were able to efficiently absorb cholesterylated anti-sense DNA onto their surface and modulate gene expression in SR-B1 expressing cells in vivo [74].

A recent article reported the development of a hybrid nanoparticle platform, which combined a bio-inspired spherical HDL mimetic with a lipid-RNA complex for siRNA delivery (Fig. 6.6) [181]. The cationic lipid 1,2-dioleoyl-3-trimethylammonium-propane (DOTAP) was employed to complex anionic ssRNA and anionic HDL mimetic particles together into multi-lamellar RNA delivery vehicles ~100 nm in diameter. In vitro studies demonstrated that these particles were efficiently taken up by multiple cancer cell lines via specific targeting of the HDL receptor SR-B1. Moreover, the particles were found to be highly resistant to nuclease degradation, and mediated potent knockdown of the expression of androgen receptor in vitro in prostate cancer cells. In vivo knockdown efficacy and sharp inhibition of tumor growth was also demonstrated in prostate cancer xenograft mouse model [181]. Since the initial publication of this approach, the platform has been further tested and proven efficacious using several different gene targets in mouse models of ovarian cancer [182, 183].

While the above-mentioned synthetic HDL mimetic RNAi delivery agents used an inorganic or hybrid core material, organic core materials such as peptides have also been used as scaffolds. Lin et al. recently reported a peptide-phospholipid–based HDL mimetic nanomaterial which was used to encapsulate and deliver cholesterol-modified siRNA [184]. While the group had previously demonstrated in vitro efficacy for this platform, the report details potent knockdown in vivo of bcl-2 protein in KB tumor-bearing mice and inhibition of tumor growth. In a separate report, one group attempted to alter the natural targeting properties of HDLs for SR-B1 in order to avoid accumulation of siRNA in the liver and adrenal glands. The group used a cyclic RGD peptide-modified apoA-1 to encapsulate cholesterol-modified siRNA. This "re-routed" HDL nanoparticle for siRNA delivery demonstrated knockdown of the Pokemon gene in vivo in tumor-bearing mice, tumor growth inhibition, and extended survival [78]. Additionally, a platform previously highlighted in section 4 of this chapter using a single chain variable fragment antibody-apoA-1 fusion protein targeted against CD20+ mantle cell lymphoma cells has also been used to deliver siRNA [185] in addition to hydrophobic drugs.

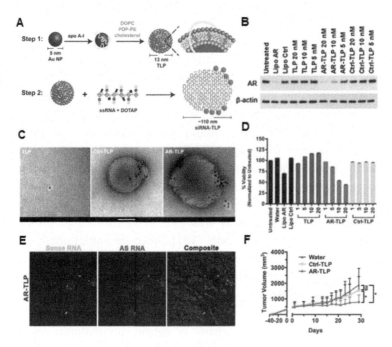

Figure 6.6 A synthetic, bio-inspired siRNA delivery system using cationic lipids formulated with HDL mimetic nanoparticles. (A). Schematic of the stepwise fabrication process for assembly of siRNA templated lipoprotein nanoparticles (siRNA-TLPs). In Step 1, TLPs are synthesized using a 5 nm gold nanoparticle scaffold, apoA-1, and phospholipids. In Step 2, TLPs are mixed with single-strand RNA (ssRNA), complement strands of a siRNA duplex, and complexed with 1,2-dioleoyl-3-trimethylammonium-propane (DOTAP). (B). Western blot for androgen receptor (AR) of lysate obtained from LNCaP prostate cancer cells treated with siRNA-TLPs, scrambled RNA control TLPs, and lipofectamine controls. (C). Transmission electron microscopy (TEM) images of HDL mimetic TLPs prior to formulation with the cationic lipid DOTAP, formulation of TLPs with DOTAP and scrambled RNA control, and functional siRNA-TLPs. (D). Cell viability of LNCaP prostate cancer cells treated with siRNA-TLPs and controls. (E). LNCaP tumor uptake of sense (Cy3) and antisense (Cy5) labeled siRNA sequences assembled with siRNA-TLPs 24 h after a single systemic administration assessed with confocal microscopy. Tumor tissues were counterstained with DAPI (blue). (F). Tumor volume measurements of LNCaP prostate cancer xenografts in athymic nude mice upon treatment with water, siRNA-TLP therapy, or scrambled RNA TLP control. Reproduced with permission from [76].

Ding et al. recently published an rHDL formulation with cholesterol-modified siRNA as anti-angiogenic therapy for breast cancer. The group demonstrated both in vitro and in vivo efficacy, finding that the particles decreased VEGF expression, reduced intratumor microvessel formation, and inhibited breast tumor xenograft growth [186]. Another group recently reported apoA and apoE-based rHDL formulations for liver-targeted delivery of cholesterol-modified siRNA [187]. ApoE-based particles were found to be significantly more effective than apoA-based delivery vehicles for liver targeting. A separate study used a similar rHDL formulation with cholesterol-modified siRNA for liver cancer. Here, in vivo knockdown studies were conducted and efficient knockdown of Pokemon and Bcl-2 protein were observed along with tumor growth inhibition [188].

While HDL-based delivery systems have attracted much interest in recent years due to the discovery that HDLs are natural carriers of miRNAs, LDL-based nucleic acid delivery systems have also been explored. A hybrid LDL-chitosan nanoparticle formulation was recently described by Zhu et al. for co-delivery of siRNA and doxorubicin to hepatocytes. While this report did not demonstrate in vivo knockdown efficacy, the platform substantially reduced siRNA uptake and degradation by macrophages in vitro, and demonstrated marked accumulation in tumor tissue in an orthotopic mouse model of hepatocellular carcinoma [189].

Finally, in a recent report lipopeptide nanoparticles were used as siRNA delivery vehicles and further studied when administered in combination with various apolipoproteins [190]. The group found that the efficacy and cytotoxicity of lipopeptide nanoparticles were largely unaffected by co-administration with most apolipoprotein sub-types; however, administration with apoE greatly enhanced luciferase knockdown in vitro in HeLa cells. Moreover, knockdown in vivo in hepatocytes was significantly diminished in apoE knockout mice compared to wild type controls. This suggests that some nanoparticle-based siRNA systems may benefit greatly by complexing with lipoproteins ex vivo during vehicle fabrication or in vivo after administration [190].

6.6 Conclusions

The examples given above confirm that lipoproteins are highly promising vehicles for the delivery of drugs, nucleic acids and contrast-generating media. In the past decade, new synthetic methods have been reported, new payloads trialed and impressive new therapeutic results published. Zr-89 labeled HDL is currently in clinical trials [95]. Overall, the field is in a position of strength. During the next decade, we expect to see continued activity in the area with new groups joining the field, additional clinical trials and lipoprotein-based nanoparticles being used for novel applications.

References

1. D. R. Arifin, D. A. Kedziorek, Y. Fu, K. W. Chan, M. T. McMahon, C. R. Weiss, D. L. Kraitchman, J. W. Bulte, Microencapsulated cell tracking, *NMR Biomed.*, 26 (2013) 850–859.

2. A. L. Bernstein, A. Dhanantwari, M. Jurcova, R. Cheheltani, P. C. Naha, T. Ivanc, E. Shefer, D. P. Cormode, Improved sensitivity of computed tomography towards iodine and gold nanoparticle contrast agents via iterative reconstruction methods, *Sci. Rep.*, 6 (2016) 26177.

3. A. L. Brown, P. C. Naha, V. Benavides-Montes, H. I. Litt, A. M. Goforth, D. P. Cormode, synthesis, x-ray opacity, and biological compatibility of ultra-high payload elemental bismuth nanoparticle x-ray contrast agents, *Chem. Mater.*, 26 (2014) 2266–2274.

4. D. P. Cormode, S. Si-Mohamed, D. Bar-Ness, M. Sigovan, P. C. Naha, J. Balegamire, F. Lavenne, P. Coulon, E. Roessl, M. Bartels, M. Rokni, I. Blevis, L. Boussel, P. Douek, Multicolor spectral photon-counting computed tomography: In vivo dual contrast imaging with a high count rate scanner, *Sci. Rep.*, 7 (2017) 4784.

5. R. Karunamuni, P. C. Naha, K. C. Lau, A. Al-Zaki, A. V. Popov, E. J. Delikatny, A. Tsourkas, D. P. Cormode, A. D. Maidment, Development of silica-encapsulated silver nanoparticles as contrast agents intended for dual-energy mammography, *Eur. Radiol.*, 26 (2016) 3301–3309.

6. P. C. Naha, K. C. Lau, J. C. Hsu, M. Hajfathalian, S. Mian, P. Chhour, L. Uppuluri, E. S. McDonald, A. D. Maidment, D. P. Cormode, Gold silver alloy nanoparticles (GSAN): An imaging probe for breast cancer screening with dual-energy mammography or computed tomography, *Nanoscale*, 8 (2016) 13740–13754.

7. P. C. Naha, A. A. Zaki, E. Hecht, M. Chorny, P. Chhour, E. Blankemeyer, D. M. Yates, W. R. Witschey, H. I. Litt, A. Tsourkas, D. P. Cormode, Dextran coated bismuth-iron oxide nanohybrid contrast agents for computed tomography and magnetic resonance imaging, *J. Mater. Chem. B Mater. Biol. Med.*, 2 (2014) 8239–8248.

8. N. Teraphongphom, P. Chhour, J. R. Eisenbrey, P. C. Naha, W. R. Witschey, B. Opasanont, L. Jablonowski, D. P. Cormode, M. A. Wheatley, Nanoparticle loaded polymeric microbubbles as contrast agents for multimodal imaging, *Langmuir*, 31 (2015) 11858–11867.

9. M. F. Kircher, J. K. Willmann, Molecular body imaging: MR imaging, CT, and US. part I. principles, *Radiology*, 263 (2012) 633–643.

10. K. Liu, L. Dong, Y. Xu, X. Yan, F. Li, Y. Lu, W. Tao, H. Peng, Y. Wu, Y. Su, D. Ling, T. He, H. Qian, S. H. Yu, Stable gadolinium based nanoscale lyophilized injection for enhanced MR angiography with efficient renal clearance, *Biomaterials*, 158 (2018) 74–85.

11. D. Ni, W. Bu, E. B. Ehlerding, W. Cai, J. Shi, Engineering of inorganic nanoparticles as magnetic resonance imaging contrast agents, *Chem. Soc. Rev.*, 46 (2017) 7438–7468.

12. Q. Y. Cai, S. H. Kim, K. S. Choi, S. Y. Kim, S. J. Byun, K. W. Kim, S. H. Park, S. K. Juhng, K. H. Yoon, Colloidal gold nanoparticles as a blood-pool contrast agent for X-ray computed tomography in mice, *Invest. Radiol.*, 42 (2007) 797–806.

13. T. L. McGinnity, O. Dominguez, T. E. Curtis, P. D. Nallathamby, A. J. Hoffman, R. K. Roeder, Hafnia (HfO_2) nanoparticles as an X-ray contrast agent and mid-infrared biosensor, *Nanoscale*, 8 (2016) 13627–13637.

14. C. Hernandez, L. Nieves, A. C. de Leon, R. Advincula, A. A. Exner, The role of surface tension in gas nanobubble stability under ultrasound, *ACS Appl Mater Interfaces*, 10 (2018) 9949–9956.

15. W. Tang, Z. Yang, S. Wang, Z. Wang, J. Song, G. Yu, W. Fan, Y. Dai, J. Wang, L. Shan, G. Niu, Q. Fan, X. Chen, Organic semiconducting photoacoustic nanodroplets for laser-activatable ultrasound imaging and combinational cancer therapy, *ACS Nano*, 12 (2018) 2610–2622.

16. O. Perlman, I. S. Weitz, S. Sivan, H. Abu-Khalla, M. Benguigui, Y. Shaked, H. Azhari, Copper oxide loaded PLGA nanospheres: Towards a multifunctional nanoscale platform for ultrasound based imaging and therapy, *Nanotechnology*, 29 (2018) 185102.

17. R. Cheheltani, R. M. Ezzibdeh, P. Chhour, K. Pulaparthi, J. Kim, M. Jurcova, J. C. Hsu, C. Blundell, H. I. Litt, V. A. Ferrari, H. R. Allcock, C. M. Sehgal, D. P. Cormode, Tunable, biodegradable gold nanoparticles as

contrast agents for computed tomography and photoacoustic imaging, *Biomaterials*, 102 (2016) 87–97.

18. C. Xu, F. Chen, H. F. Valdovinos, D. Jiang, S. Goel, B. Yu, H. Sun, T. E. Barnhart, J. J. Moon, W. Cai, Bacteria-like mesoporous silica-coated gold nanorods for positron emission tomography and photoacoustic imaging-guided chemo-photothermal combined therapy, *Biomaterials*, 165 (2018) 56–65.

19. M. Wang, K. Deng, W. Lu, X. Deng, K. Li, Y. Shi, B. Ding, Z. Cheng, B. Xing, G. Han, Z. Hou, J. Lin, Rational design of multifunctional Fe@ gamma-Fe$_2$O$_3$ @H-TiO$_2$ nanocomposites with enhanced magnetic and photoconversion effects for wide applications: From photocatalysis to imaging-guided photothermal cancer therapy, *Adv. Mater.*, 30 (2018) 1706747.

20. J. Liu, X. Cai, H. C. Pan, A. Bandla, C. K. Chuan, S. Wang, N. Thakor, L. D. Liao, B. Liu, Molecular engineering of photoacoustic performance by chalcogenide variation in conjugated polymer nanoparticles for brain vascular imaging, *Small*, 14 (2018) 1703732.

21. D. P. Cormode, P. C. Naha, Z. A. Fayad, Nanoparticle contrast agents for computed tomography: A focus on micelles, *Contrast Media Mol. Imaging*, 9 (2014) 37–52.

22. https://www. grandviewresearch.com/press-release/global-nanomedicine-market (accessed on 22nd February 2018).

23. A. J. Mieszawska, A. Gianella, D. P. Cormode, Y. Zhao, A. Meijerink, R. Langer, O. C. Farokhzad, Z. A. Fayad, W. J. Mulder, Engineering of lipid-coated PLGA nanoparticles with a tunable payload of diagnostically active nanocrystals for medical imaging, *Chem. Commun.*, 48 (2012) 5835–5837.

24. A. J. Mieszawska, Y. Kim, A. Gianella, I. van Rooy, B. Priem, M. P. Labarre, C. Ozcan, D. P. Cormode, A. Petrov, R. Langer, O. C. Farokhzad, Z. A. Fayad, W. J. Mulder, Synthesis of polymer-lipid nanoparticles for image-guided delivery of dual modality therapy, *Bioconjug. Chem.*, 24 (2013) 1429–1434.

25. E. R. Swy, A. S. Schwartz-Duval, D. D. Shuboni, M. T. Latourette, C. L. Mallet, M. Parys, D. P. Cormode, E. M. Shapiro, Dual-modality, fluorescent, PLGA encapsulated bismuth nanoparticles for molecular and cellular fluorescence imaging and computed tomography, *Nanoscale*, 6 (2014) 13104–13112.

26. P. C. Naha, V. Kanchan, P. K. Manna, A. K. Panda, Improved bioavailability of orally delivered insulin using Eudragit-L30D coated PLGA microparticles, *J. Microencapsul*, 25 (2008) 248–256.

27. M. Di Francesco, R. Primavera, D. Romanelli, R. Palomba, R. C. Pereira, T. Catelani, C. Celia, L. Di Marzio, M. Fresta, D. Di Mascolo, P. Decuzzi, Hierarchical microplates as drug depots with controlled geometry, rigidity and therapeutic efficacy, *ACS Appl. Mater. Interfaces*, 10 (2018) 9280–9289.

28. M. E. Halwes, K. M. Tyo, J. M. Steinbach-Rankins, H. B. Frieboes, Computational modeling of antiviral drug diffusion from poly(lactic-co-glycolic-acid) fibers and multicompartment pharmacokinetics for application to the female reproductive tract, *Mol. Pharm.*, 15 (2018) 1534–1547.

29. H. Kim, L. Niu, P. Larson, T. A. Kucaba, K. A. Murphy, B. R. James, D. M. Ferguson, T. S. Griffith, J. Panyam, Polymeric nanoparticles encapsulating novel TLR7/8 agonists as immunostimulatory adjuvants for enhanced cancer immunotherapy, *Biomaterials*, 164 (2018) 38–53.

30. P. C. Naha, V. Kanchan, A. K. Panda, Evaluation of parenteral depot insulin formulation using PLGA and PLA microparticles, *J. Biomater. Appl.*, 24 (2009) 309–325.

31. M. Gai, M. A. Kurochkin, D. Li, B. Khlebtsov, L. Dong, N. Tarakina, R. Poston, D. J. Gould, J. Frueh, G. B. Sukhorukov, In-situ NIR-laser mediated bioactive substance delivery to single cell for EGFP expression based on biocompatible microchamber-arrays, *J. Control. Release*, 276 (2018) 84–92.

32. E. Graham-Gurysh, K. M. Moore, A. B. Satterlee, K. T. Sheets, F. C. Lin, E. M. Bachelder, C. R. Miller, S. D. Hingtgen, K. M. Ainslie, Sustained delivery of doxorubicin via acetalated dextran scaffold prevents glioblastoma recurrence after surgical resection, *Mol. Pharm.*, 15 (2018) 1309–1318.

33. B. S. Pattni, V. V. Chupin, V. P. Torchilin, New developments in liposomal drug delivery, *Chem. Rev.*, 115 (2015) 10938–10966.

34. D. P. Cormode, P. A. Jarzyna, W. J. M. Mulder, Z. A. Fayad, Modified natural nanoparticles as contrast agents for medical imaging, *Adv. Drug Deliv. Rev.*, 62 (2010) 329–338.

35. T. Skajaa, D. P. Cormode, E. Falk, W. J. Mulder, E. A. Fisher, Z. A. Fayad, High-density lipoprotein-based contrast agents for multimodal imaging of atherosclerosis, *Arterioscler Thromb. Vasc. Bio.*, 30 (2010) 169–176.

36. M. G. Damiano, R. K. Mutharasan, S. Tripathy, K. M. McMahon, C. S. Thaxton, Templated high density lipoprotein nanoparticles as potential therapies and for molecular delivery, *Adv. Drug Del. Rev.*, 65 (2013) 649–662.

37. R. Kannan Mutharasan, L. Foit, C. Shad Thaxton, High-density lipoproteins for therapeutic delivery systems, *J. Mater. Chem. B*, 4 (2016) 188–197.

38. H. Huang, W. Cruz, J. Chen, G. Zheng, Learning from biology: Synthetic lipoproteins for drug delivery, *Wiley Interdiscip Rev. Nanomed. Nanobiotechnol.*, 7 (2015) 298–314.

39. L. Song, H. Li, U. Sunar, J. Chen, I. Corbin, A. G. Yodh, G. Zheng, Naphthalocyanine-reconstituted LDL nanoparticles for in vivo cancer imaging and treatment, *Int. J. Nanomed.*, 2 (2007) 767–774.

40. R. Duivenvoorden, J. Tang, D. P. Cormode, A. J. Mieszawska, D. Izquierdo-Garcia, C. Ozcan, M. J. Otten, N. Zaidi, M. E. Lobatto, S. M. van Rijs, B. Priem, E. L. Kuan, C. Martel, B. Hewing, H. Sager, M. Nahrendorf, G. J. Randolph, E. S. Stroes, V. Fuster, E. A. Fisher, Z. A. Fayad, W. J. Mulder, A statin-loaded reconstituted high-density lipoprotein nanoparticle inhibits atherosclerotic plaque inflammation, *Nat. Commun.*, 5 (2014) 3065.

41. I. E. Allijn, W. Leong, J. Tang, A. Gianella, A. J. Mieszawska, F. Fay, G. Ma, S. Russell, C. B. Callo, R. E. Gordon, E. Korkmaz, J. A. Post, Y. M. Zhao, H. C. Gerritsen, A. Thran, R. Proksa, H. Daerr, G. Storm, V. Fuster, E. A. Fisher, Z. A. Fayad, W. J. M. Mulder, D. P. Cormode, Gold nanocrystal labeling allows low-density lipoprotein imaging from the subcellular to macroscopic level, *ACS Nano*, 7 (2013) 9761–9770.

42. D. P. Cormode, T. Skajaa, M. M. van Schooneveld, R. Koole, P. Jarzyna, M. E. Lobatto, C. Calcagno, A. Barazza, R. E. Gordon, P. Zanzonico, E. A. Fisher, Z. A. Fayad, W. J. Mulder, Nanocrystal core high-density lipoproteins: A multimodality contrast agent platform, *Nano Lett.*, 8 (2008) 3715–3723.

43. K. C. Vickers, P. Sethupathy, J. Baran-Gale, A. T. Remaley, Complexity of microRNA function and the role of isomiRs in lipid homeostasis, *J. Lipid Res.*, 54 (2013) 1182–1191.

44. W. Chen, P. A. Jarzyna, G. A. van Tilborg, V. A. Nguyen, D. P. Cormode, A. Klink, A. W. Griffioen, G. J. Randolph, E. A. Fisher, W. J. Mulder, Z. A. Fayad, RGD peptide functionalized and reconstituted high-density lipoprotein nanoparticles as a versatile and multimodal tumor targeting molecular imaging probe, *FASEB J.*, 24 (2010) 1689–1699.

45. G. Zheng, J. Chen, H. Li, J. D. Glickson, Rerouting lipoprotein nanoparticles to selected alternate receptors for the targeted delivery of cancer diagnostic and therapeutic agents, *PNAS*, 102 (2005) 17757–17762.

46. M. E. Davis, Z. G. Chen, D. M. Shin, Nanoparticle therapeutics: An emerging treatment modality for cancer, *Nat. Rev. Drug Discov.*, 7 (2008) 771–782.

47. S. Eisenberg, H. G. Windmueller, R. I. Levy, Metabolic fate of rat and human lipoprotein apoproteins in the rat, *J. Lipid Res.*, 14 (1973) 446–458.

48. C. S. Thaxton, J. S. Rink, P. C. Naha, D. P. Cormode, Lipoproteins and lipoprotein mimetics for imaging and drug delivery, *Adv. Drug Deliv. Rev.*, 106 (2016) 116–131.

49. D. A. Bricarello, J. T. Smilowitz, A. M. Zivkovic, J. B. German, A. N. Parikh, Reconstituted lipoprotein: A versatile class of biologically-Inspired nanostructures, *ACS Nano*, 5 (2011) 42–57.

50. J. A. Yanez, S. W. Wang, I. W. Knemeyer, M. A. Wirth, K. B. Alton, Intestinal lymphatic transport for drug delivery, *Adv. Drug Deliv. Rev.*, 63 (2011) 923–942.

51. G. S. Shelness, J. A. Sellers, Very-low-density lipoprotein assembly and secretion, *Curr. Opin. Lipidol.*, 12 (2001) 151–157.

52. M. Merkel, R. H. Eckel, I. J. Goldberg, Lipoprotein lipase: Genetics, lipid uptake, and regulation, *J. Lipid Res.*, 43 (2002) 1997–2006.

53. J. Babiak, L. L. Rudel, 2 Lipoproteins and atherosclerosis, Baillière's, *J. Clin. Endocrinol. Metab.*, 1 (1987) 515–550.

54. K. M. McMahon, L. Foit, N. L. Angeloni, F. J. Giles, L. I. Gordon, C. S. Thaxton, Synthetic high-density lipoprotein-like nanoparticles as cancer therapy, *Cancer Treat. Res.*, 166 (2015) 129–150.

55. A. Kontush, M. Lhomme, M. J. Chapman, Unraveling the complexities of the HDL lipidome, *J. Lipid Res.*, 54 (2013) 2950–2963.

56. R. S. Rosenson, H. B. Brewer, Jr., M. J. Chapman, S. Fazio, M. M. Hussain, A. Kontush, R. M. Krauss, J. D. Otvos, A. T. Remaley, E. J. Schaefer, HDL measures, particle heterogeneity, proposed nomenclature, and relation to atherosclerotic cardiovascular events, *Clin. Chem.*, 57 (2011) 392–410.

57. D. J. Gordon, B. M. Rifkind, High-density lipoprotein--the clinical implications of recent studies, *N. Engl. J. Med.*, 321 (1989) 1311–1316.

58. B. G. Nordestgaard, S. J. Nicholls, A. Langsted, K. K. Ray, A. Tybjaerg-Hansen, Advances in lipid-lowering therapy through gene-silencing technologies, *Nat. Rev. Cardiol.*, 15 (2018) 261–272.

59. J. Shepherd, S. M. Cobbe, I. Ford, C. G. Isles, A. R. Lorimer, P. W. MacFarlane, J. H. McKillop, C. J. Packard, Prevention of coronary heart disease with pravastatin in men with hypercholesterolemia. West of

Scotland Coronary Prevention Study Group, *N. Engl. J. Med.*, 333 (1995) 1301–1307.

60. Therapeutic response to lovastatin (mevinolin) in nonfamilial hypercholesterolemia. A multicenter study. The Lovastatin Study Group II, *JAMA*, 256 (1986) 2829–2834.

61. D. W. Bilheimer, S. M. Grundy, M. S. Brown, J. L. Goldstein, Mevinolin and colestipol stimulate receptor-mediated clearance of low density lipoprotein from plasma in familial hypercholesterolemia heterozygotes, *PNAS*, 80 (1983) 4124–4128.

62. K. J. Moore, I. J. Goldberg, Emerging roles of PCSK9: More than a one-trick pony, *Arter. Thromb. Vasc. Biol.*, 36 (2016) 211–212.

63. R. A. Firestone, Low-density lipoprotein as a vehicle for targeting antitumor compounds to cancer cells, *Bioconjug. Chem.*, 5 (1994) 105–113.

64. R. S. Rosenson, H. B. Brewer, Jr., W. S. Davidson, Z. A. Fayad, V. Fuster, J. Goldstein, M. Hellerstein, X. C. Jiang, M. C. Phillips, D. J. Rader, A. T. Remaley, G. H. Rothblat, A. R. Tall, L. Yvan-Charvet, Cholesterol efflux and atheroprotection: Advancing the concept of reverse cholesterol transport, *Circulation*, 125 (2012) 1905–1919.

65. N. Sabnis, M. Nair, M. Israel, W. J. McConathy, A. G. Lacko, Enhanced solubility and functionality of valrubicin (AD-32) against cancer cells upon encapsulation into biocompatible nanoparticles, *Int. J. Nanomed.*, 7 (2012) 975–983.

66. A. H. Mohammadpour, F. Akhlaghi, Future of cholesteryl ester transfer protein (CETP) inhibitors: A pharmacological perspective, *Clin. Pharmacokinet*, 52 (2013) 615–626.

67. B. A. Kingwell, M. J. Chapman, A. Kontush, N. E. Miller, HDL-targeted therapies: Progress, failures and future, *Nat. Rev. Drug Discov.*, 13 (2014) 445–464.

68. B. R. Krause, A. T. Remaley, Reconstituted HDL for the acute treatment of acute coronary syndrome, *Curr. Opin. Lipidol.*, 24 (2013) 480–486.

69. G. Franceschini, C. R. Sirtori, A. Capurso, 2nd, K. H. Weisgraber, R. W. Mahley, A-IMilano apoprotein. Decreased high density lipoprotein cholesterol levels with significant lipoprotein modifications and without clinical atherosclerosis in an Italian family, *J. Clin. Invest.*, 66 (1980) 892–900.

70. S. E. Nissen, T. Tsunoda, E. M. Tuzcu, P. Schoenhagen, C. J. Cooper, M. Yasin, G. M. Eaton, M. A. Lauer, W. S. Sheldon, C. L. Grines, S. Halpern, T. Crowe, J. C. Blankenship, R. Kerensky, Effect of recombinant ApoA-I

Milano on coronary atherosclerosis in patients with acute coronary syndromes: A randomized controlled trial, *JAMA*, 290 (2003) 2292–2300.

71. P. G. Lerch, V. Fortsch, G. Hodler, R. Bolli, Production and characterization of a reconstituted high density lipoprotein for therapeutic applications, *Vox Sang*, 71 (1996) 155–164.

72. A. C. Calkin, B. G. Drew, A. Ono, S. J. Duffy, M. V. Gordon, S. M. Schoenwaelder, D. Sviridov, M. E. Cooper, B. A. Kingwell, S. P. Jackson, Reconstituted high-density lipoprotein attenuates platelet function in individuals with type 2 diabetes mellitus by promoting cholesterol efflux, *Circulation*, 120 (2009) 2095–2104.

73. L. Foit, C. S. Thaxton, Synthetic high-density lipoprotein-like nanoparticles potently inhibit cell signaling and production of inflammatory mediators induced by lipopolysaccharide binding Toll-like receptor 4, *Biomaterials*, 100 (2016) 67–75.

74. K. M. McMahon, R. K. Mutharasan, S. Tripathy, D. Veliceasa, M. Bobeica, D. K. Shumaker, A. J. Luthi, B. T. Helfand, H. Ardehali, C. A. Mirkin, O. Volpert, C. S. Thaxton, Biomimetic high density lipoprotein nanoparticles for nucleic acid delivery, *Nano Lett.*, 11 (2011) 1208–1214.

75. D. L. Michell, K. C. Vickers, HDL and microRNA therapeutics in cardiovascular disease, *Pharmacol. Ther.*, 168 (2016) 43–52.

76. K. M. McMahon, M. P. Plebanek, C. S. Thaxton, Properties of native high-density lipoproteins inspire synthesis of actively targeted in vivo sirna delivery vehicles, *Adv. Funct. Mater.*, 26 (2016) 7824–7835.

77. H. J. Park, R. Kuai, E. J. Jeon, Y. Seo, Y. Jung, J. J. Moon, A. Schwendeman, S. W. Cho, High-density lipoprotein-mimicking nanodiscs carrying peptide for enhanced therapeutic angiogenesis in diabetic hindlimb ischemia, *Biomaterials*, 161 (2018) 69–80.

78. Y. Ding, Y. Han, R. Wang, Y. Wang, C. Chi, Z. Zhao, H. Zhang, W. Wang, L. Yin, J. Zhou, Rerouting native HDL to predetermined receptors for improved tumor-targeted gene silencing therapy, *ACS Appl. Mater. Interfaces*, 9 (2017) 30488–30501.

79. A. Alaarg, M. L. Senders, A. Varela-Moreira, C. Perez-Medina, Y. Zhao, J. Tang, F. Fay, T. Reiner, Z. A. Fayad, W. E. Hennink, J. M. Metselaar, W. J. M. Mulder, G. Storm, A systematic comparison of clinically viable nanomedicines targeting HMG-CoA reductase in inflammatory atherosclerosis, *J. Control. Release*, 262 (2017) 47–57.

80. A. R. Neves, J. F. Queiroz, S. A. C. Lima, S. Reis, Apo E-functionalization of solid lipid nanoparticles enhances brain drug delivery: Uptake

mechanism and transport pathways, *Bioconjug. Chem.*, 28 (2017) 995–1004.

81. D. Sobot, S. Mura, S. O. Yesylevskyy, L. Dalbin, F. Cayre, G. Bort, J. Mougin, D. Desmaele, S. Lepetre-Mouelhi, G. Pieters, B. Andreiuk, A. S. Klymchenko, J. L. Paul, C. Ramseyer, P. Couvreur, Conjugation of squalene to gemcitabine as unique approach exploiting endogenous lipoproteins for drug delivery, *Nat. Commun.*, 8 (2017) 15678.

82. J. S. Rink, W. Sun, S. Misener, J. J. Wang, Z. J. Zhang, M. R. Kibbe, V. P. Dravid, S. S. Venkatraman, C. S. Thaxton, Nitric oxide-delivering high-density lipoprotein-like nanoparticles as a biomimetic nanotherapy for vascular disease, *ACS Appl. Mater. Interfaces*, 10 (2018) 6904–6916.

83. D. P. Cormode, E. Roessl, A. Thran, T. Skajaa, R. E. Gordon, J. P. Schlomka, V. Fuster, E. A. Fisher, W. J. M. Mulder, R. Proksa, Z. A. Fayad, Atherosclerotic plaque composition: Analysis with multicolor CT and targeted gold nanoparticles, *Radiology*, 256 (2010) 774–782.

84. D. P. Cormode, K. C. Briley-Saebo, W. J. Mulder, J. G. Aguinaldo, A. Barazza, Y. Ma, E. A. Fisher, Z. A. Fayad, An ApoA-I mimetic peptide high-density-lipoprotein-based MRI contrast agent for atherosclerotic plaque composition detection, *Small*, 4 (2008) 1437–1444.

85. D. P. Cormode, J. C. Frias, Y. Ma, W. Chen, T. Skajaa, K. Briley-Saebo, A. Barazza, K. J. Williams, W. J. Mulder, Z. A. Fayad, E. A. Fisher, HDL as a contrast agent for medical imaging, *Clin. Lipidol.*, 4 (2009) 493–500.

86. C. Perez-Medina, J. Tang, D. Abdel-Atti, B. Hogstad, M. Merad, E. A. Fisher, Z. A. Fayad, J. S. Lewis, W. J. Mulder, T. Reiner, PET Imaging of tumor-associated macrophages with 89Zr-labeled high-density lipoprotein nanoparticles, *J. Nucl. Med.*, 56 (2015) 1272–1277.

87. C. Perez-Medina, T. Binderup, M. E. Lobatto, J. Tang, C. Calcagno, L. Giesen, C. H. Wessel, J. Witjes, S. Ishino, S. Baxter, Y. Zhao, S. Ramachandran, M. Eldib, B. L. Sanchez-Gaytan, P. M. Robson, J. Bini, J. F. Granada, K. M. Fish, E. S. Stroes, R. Duivenvoorden, S. Tsimikas, J. S. Lewis, T. Reiner, V. Fuster, A. Kjaer, E. A. Fisher, Z. A. Fayad, W. J. Mulder, In vivo PET imaging of HDL in multiple atherosclerosis models, *JACC Cardiovasc. Imaging*, 9 (2016) 950–961.

88. C. Zhu, Y. Xia, Biomimetics: Reconstitution of low-density lipoprotein for targeted drug delivery and related theranostic applications, *Chem. Soc. Rev.*, 46 (2017) 7668–7682.

89. A. Shaish, G. Keren, P. Chouraqui, H. Levkovitz, D. Harats, Imaging of aortic atherosclerotic lesions by (125)I-LDL, (125)I-oxidized-LDL, (125)I-HDL and (125)I-BSA, *Pathobiology*, 69 (2001) 225–229.

90. J. C. Frias, K. J. Williams, E. A. Fisher, Z. A. Fayad, Recombinant HDL-like nanoparticles: A specific contrast agent for MRI of atherosclerotic plaques, *J. Am. Chem. Soc.*, 126 (2004) 16316–16317.

91. Q. Song, M. Huang, L. Yao, X. Wang, X. Gu, J. Chen, J. Chen, J. Huang, Q. Hu, T. Kang, Z. Rong, H. Qi, G. Zheng, H. Chen, X. Gao, Lipoprotein-based nanoparticles rescue the memory loss of mice with Alzheimer's disease by accelerating the clearance of amyloid-beta, *ACS Nano*, 8 (2014) 2345–2359.

92. J. C. Frias, Y. Ma, K. J. Williams, Z. A. Fayad, E. A. Fisher, Properties of a versatile nanoparticle platform contrast agent to image and characterize atherosclerotic plaques by magnetic resonance imaging, *Nano Lett.*, 6 (2006) 2220–2224.

93. T. Skajaa, D. P. Cormode, P. A. Jarzyna, A. Delshad, C. Blachford, A. Barazza, E. A. Fisher, R. E. Gordon, Z. A. Fayad, W. J. Mulder, The biological properties of iron oxide core high-density lipoprotein in experimental atherosclerosis, *Biomaterials*, 32 (2011) 206–213.

94. T. Skajaa, Y. Zhao, D. J. van den Heuvel, H. C. Gerritsen, D. P. Cormode, R. Koole, M. M. van Schooneveld, J. A. Post, E. A. Fisher, Z. A. Fayad, C. de Mello Donega, A. Meijerink, W. J. Mulder, Quantum dot and Cy5.5 labeled nanoparticles to investigate lipoprotein biointeractions via Forster resonance energy transfer, *Nano Lett.*, 10 (2010) 5131–5138.

95. W. J. M. Mulder, M. M. T. van Leent, M. Lameijer, E. A. Fisher, Z. A. Fayad, C. Perez-Medina, High-density lipoprotein nanobiologics for precision medicine, *Acc. Chem. Res.*, 51 (2018) 127–137.

96. Y. Kim, F. Fay, D. P. Cormode, B. L. Sanchez-Gaytan, J. Tang, E. J. Hennessy, M. Ma, K. Moore, O. C. Farokhzad, E. A. Fisher, W. J. M. Mulder, R. Langer, Z. A. Fayad, Single step reconstitution of multifunctional high-density lipoprotein-derived nanomaterials using microfluidics, *ACS Nano*, 7 (2013) 9975–9983.

97. K. C. Briley-Saebo, S. Geninatti-Crich, D. P. Cormode, A. Barazza, W. J. M. Mulder, W. Chen, G. B. Giovenzana, E. A. Fisher, S. Aime, Z. A. Fayad, High-relaxivity gadolinium-modified high-density lipoproteins as magnetic resonance imaging contrast agents, *J. Phys. Chem. B*, 113 (2009) 6283–6289.

98. W. Chen, E. Vucic, E. Leupold, W. J. Mulder, D. P. Cormode, K. C. Briley-Saebo, A. Barazza, E. A. Fisher, M. Dathe, Z. A. Fayad, Incorporation of an apoE-derived lipopeptide in high-density lipoprotein MRI contrast agents for enhanced imaging of macrophages in atherosclerosis, *Contrast Media Mol. Imaging*, 3 (2008) 233–242.

99. C. E. Carney, I. L. Lenov, C. J. Baker, K. W. MacRenaris, A. L. Eckermann, S. G. Sligar, T. J. Meade, Nanodiscs as a modular platform for multimodal MR-optical imaging, *Bioconjug. Chem.*, 26 (2015) 899–905.

100. A. B. Sigalov, Nature-inspired nanoformulations for contrast-enhanced in vivo MR imaging of macrophages, *Contrast Media Mol. Imaging*, 9 (2014) 372–382.

101. Q. Wang, S. Chen, Q. Luo, M. Liu, X. Zhou, A europium-lipoprotein nanocomposite for highly-sensitive MR-fluorescence multimodal imaging, *RSC Adv.*, 5 (2015) 1808–1811.

102. J. Rieffel, U. Chitgupi, J. F. Lovell, Recent advances in higher-order, multimodal, biomedical imaging agents, *Small*, 11 (2015) 4445–4461.

103. W. Chen, D. P. Cormode, Y. Vengrenyuk, B. Herranz, J. E. Feig, A. Klink, W. J. M. Mulder, E. A. Fisher, Z. A. Fayad, Collagen-specific peptide conjugated HDL nanoparticles as MRI contrast agent to evaluate compositional changes in atherosclerotic plaque regression, *JACC Cardiovasc Imaging*, 6 (2013) 373–384.

104. I. R. Corbin, K. K. Ng, L. Ding, A. Jurisicova, G. Zheng, Near-infrared fluorescent imaging of metastatic ovarian cancer using folate receptor-targeted high-density lipoprotein nanocarriers, *Nanomedicine (London)* 8 (2013) 875–890.

105. Z. Zhang, J. Chen, L. Ding, H. Jin, J. F. Lovell, I. R. Corbin, W. Cao, P. C. Lo, M. Yang, M. S. Tsao, Q. Luo, G. Zheng, HDL-mimicking peptide-lipid nanoparticles with improved tumor targeting, *Small*, 6 (2010) 430–437.

106. J. Pietzsch, R. Bergmann, K. Rode, C. Hultsch, B. Pawelke, F. Wuest, J. van den Hoff, Fluorine-18 radiolabeling of low-density lipoproteins: A potential approach for characterization and differentiation of metabolism of native and oxidized low-density lipoproteins in vivo, *Nucl. Med. Biol.*, 31 (2004) 1043–1050.

107. J. M. Rosen, S. P. Butler, G. E. Meinken, T. S. Wang, R. Ramakrishnan, S. C. Srivastava, P. O. Alderson, H. N. Ginsberg, Indium-111-labeled LDL: A potential agent for imaging atherosclerotic disease and lipoprotein biodistribution, *J. Nucl. Med.*, 31 (1990) 343–350.

108. H. N. Ginsberg, S. J. Goldsmith, S. Vallabhajosula, Noninvasive imaging of 99mtechnetium-labeled low density lipoprotein uptake by tendon xanthomas in hypercholesterolemic patients, *Arteriosclerosis*, 10 (1990) 256–262.

109. H. Sinzinger, H. Bergmann, J. Kaliman, P. Angelberger, Imaging of human atherosclerotic lesions using 123I-low-density lipoprotein, *Eur. J. Nucl. Med. Mol. Imaging*, 12 (1986) 291–292.

110. L. Iuliano, A. Signore, S. Vallabajosula, A. R. Colavita, C. Camastra, G. Ronga, C. Alessandri, E. Sbarigia, P. Fiorani, F. Violi, Preparation and biodistribution of 99m technetium labelled oxidized LDL in man, *Atherosclerosis*, 126 (1996) 131–141.

111. G. Zheng, H. Li, M. Zhang, S. Lund-Katz, B. Chance, J. D. Glickson, Low-density lipoprotein reconstituted by pyropheophorbide cholesteryl oleate as target-specific photosensitizer, *Bioconjug. Chem.*, 13 (2002) 392–396.

112. S. P. Wu, I. Lee, P. P. Ghoroghchian, P. R. Frail, G. Zheng, J. D. Glickson, M. J. Therien, Near-infrared optical imaging of B16 melanoma cells via low-density lipoprotein-mediated uptake and delivery of high emission dipole strength tris[(porphinato)zinc(II)] fluorophores, *Bioconjug. Chem.*, 16 (2005) 542–550.

113. H. Li, Z. Zhang, D. Blessington, D. S. Nelson, R. Zhou, S. Lund-Katz, B. Chance, J. D. Glickson, G. Zheng, Carbocyanine labeled LDL for optical imaging of tumors, *Acad. Radiol.*, 11 (2004) 669–677.

114. L. M. Mitsumori, J. L. Ricks, M. E. Rosenfeld, U. P. Schmiedl, C. Yuan, Development of a lipoprotein based molecular imaging MR contrast agent for the noninvasive detection of early atherosclerotic disease, *Int. J. Card. Imaging*, 20 (2004) 561–567.

115. E. M. Gale, S. Mukherjee, C. Liu, G. S. Loving, P. Caravan, Structure-redox-relaxivity relationships for redox responsive manganese-based magnetic resonance imaging probes, *Inorg. Chem.*, 53 (2014) 10748–10761.

116. B. Drahoš, I. Lukeš, É. Tóth, Manganese(II) complexes as potential contrast agents for MRI, *Eur. J. Inorg. Chem.*, 2012 (2012) 1975–1986.

117. I. R. Corbin, H. Li, J. Chen, S. Lund-Katz, R. Zhou, J. D. Glickson, G. Zheng, Low-density lipoprotein nanoparticles as magnetic resonance imaging contrast agents, *Neoplasia*, 8 (2006) 488–498.

118. H. Li, B. D. Gray, I. Corbin, C. Lebherz, H. Choi, S. Lund-Katz, J. M. Wilson, J. D. Glickson, R. Zhou, MR and fluorescent imaging of low-density lipoprotein receptors, *Acad. Radiol.*, 11 (2004) 1251–1259.

119. Y. Yamakoshi, H. Qiao, A. N. Lowell, M. Woods, B. Paulose, Y. Nakao, H. Zhang, T. Liu, S. Lund-Katz, R. Zhou, LDL-based nanoparticles for contrast enhanced MRI of atheroplaques in mouse models, *Chem. Commun.*, 47 (2011) 8835–8837.

120. A. N. Lowell, H. Qiao, T. Liu, T. Ishikawa, H. Zhang, S. Oriana, M. Wang, E. Ricciotti, G. A. FitzGerald, R. Zhou, Y. Yamakoshi, Functionalized low-density lipoprotein nanoparticles for in vivo enhancement of

atherosclerosis on magnetic resonance images, *Bioconjug. Chem.*, 23 (2012) 2313–2319.

121. J. Chen, I. R. Corbin, H. Li, W. Cao, J. D. Glickson, G. Zheng, Ligand conjugated low-density lipoprotein nanoparticles for enhanced optical cancer imaging in vivo, *J. Am. Chem. Soc.*, 129 (2007) 5798–5799.

122. M. L. Hill, I. R. Corbin, R. B. Levitin, W. Cao, J. G. Mainprize, M. J. Yaffe, G. Zheng, In vitro assessment of poly-iodinated triglyceride reconstituted low-density lipoprotein: Initial steps toward CT molecular imaging, *Acad. Radiol.*, 17 (2010) 1359–1365.

123. M. Krieger, [34] Reconstitution of the hydrophobic core of low-density lipoprotein, in: *Methods in Enzymology*, Academic Press (1986) pp. 608–613.

124. M. Huettinger, J. R. Corbett, W. J. Schneider, J. T. Willerson, M. S. Brown, J. L. Goldstein, Imaging of hepatic low density lipoprotein receptors by radionuclide scintiscanning in vivo, *PNAS*, 81 (1984) 7599–7603.

125. A. Bowler, T. G. Redgrave, J. C. Mamo, Chylomicron-remnant clearance in homozygote and heterozygote Watanabe-heritable-hyperlipidaemic rabbits is defective. Lack of evidence for an independent chylomicron-remnant receptor, *Biochem. J.*, 276 (Pt 2) (1991) 381–386.

126. N. Xu, L. Zhou, R. Odselius, A. Nilsson, Uptake of radiolabeled and colloidal gold-labeled chyle chylomicrons and chylomicron remnants by rat platelets in vitro, *Arter. Thromb. Vasc. Biol.*, 15 (1995) 972–981.

127. O. T. Bruns, H. Ittrich, K. Peldschus, M. G. Kaul, U. I. Tromsdorf, J. Lauterwasser, M. S. Nikolic, B. Mollwitz, M. Merkel, N. C. Bigall, S. Sapra, R. Reimer, H. Hohenberg, H. Weller, A. Eychmuller, G. Adam, U. Beisiegel, J. Heeren, Real-time magnetic resonance imaging and quantification of lipoprotein metabolism in vivo using nanocrystals, *Nat. Nanotechnol.*, 4 (2009) 193–201.

128. K. C. Vickers, B. T. Palmisano, B. M. Shoucri, R. D. Shamburek, A. T. Remaley, MicroRNAs are transported in plasma and delivered to recipient cells by high-density lipoproteins, *Nat. Cell Biol.*, 13 (2011) 423–433.

129. W. S. Davidson, R. A. Silva, S. Chantepie, W. R. Lagor, M. J. Chapman, A. Kontush, Proteomic analysis of defined HDL subpopulations reveals particle-specific protein clusters: Relevance to antioxidative function, *Arter. Thromb. Vasc. Biol.*, 29 (2009) 870–876.

130. A. S. Shah, L. Tan, J. L. Long, W. S. Davidson, Proteomic diversity of high density lipoproteins: Our emerging understanding of its importance in lipid transport and beyond, *J. Lipid Res.*, 54 (2013) 2575–2585.

131. H. Maeda, H. Nakamura, J. Fang, The EPR effect for macromolecular drug delivery to solid tumors: Improvement of tumor uptake, lowering of systemic toxicity, and distinct tumor imaging in vivo, *Adv. Drug Deliv. Rev.*, 65 (2013) 71–79.

132. V. Torchilin, Tumor delivery of macromolecular drugs based on the EPR effect, *Adv. Drug Deliv. Rev.*, 63 (2011) 131–135.

133. C. G. Leon, J. A. Locke, H. H. Adomat, S. L. Etinger, A. L. Twiddy, R. D. Neumann, C. C. Nelson, E. S. Guns, K. M. Wasan, Alterations in cholesterol regulation contribute to the production of intratumoral androgens during progression to castration-resistant prostate cancer in a mouse xenograft model, *Prostate*, 10 (2010) 390–400.

134. P. M. Cruz, H. Mo, D. W. McConathy, N. A. Sabnis, A. G. Lacko, The role of cholesterol metabolism and cholesterol transport in carcinogenesis: A review of scientific findings, relevant to future cancer therapeutics, *Front. Pharmacol.*, 4 (2013) 119.

135. E. Thysell, I. Surowiec, E Hornberg, S. Crnalic, A. Widmark, A. I. Johansson, P. Stattin, A. Bergh, T. Moritz, H. Antti, P. Wikstrom. Metabolomic characterization of human prostate cancer bone metastases reveals increased levels of cholesterol, *PLOS One*, 5 (2010) e14175.

136. Y. Zheng, Y. Liu, H. Jin, S. Pan, Y. Qian, C. Huang, Y. Zeng, Q. Luo, M. Zeng, Z. Zhang, Scavenger receptor B1 is a potential biomarker of human nasopharyngeal carcinoma and its growth is inhibited by HDL-mimetic nanoparticles, *Theranostics*, 3 (2013) 477–486.

137. J. Li, J. Wang, M. Li, L. Yin, X.-A. Li, T.-G. Zhang, Up-regulated expression of scavenger receptor class B type 1 (SR-B1) is associated with malignant behaviors and poor prognosis of breast cancer, *Pathol. Res. Pract.*, 212 (2016) 555–559.

138. K. K. Ng, J. F. Lovell, G. Zheng, Lipoprotein-inspired nanoparticles for cancer, *Acc. Chem. Res.*, 44 (2011) 1105–1113.

139. S. Eisenberg, Lipoproteins and lipoprotein metabolism. A dynamic evaluation of the plasma fat transport system, *Klinische Wochenschrift*, 61 (1983) 119–132.

140. R. Prassl, Human low density lipoprotein: The mystery of core lipid packing, *J. Lipid Res.*, 52 (2011) 187–188.

141. O. C. Farokhzad, R. Langer, Impact of nanotechnology on drug delivery, *ACS Nano*, 3 (2009) 16–20.

142. N. M. Crosby, M. Ghosh, B. Su, J. A. Beckstead, A. Kamei, J. B. Simonsen, B. Luo, L. I. Gordon, T. M. Forte, R. O. Ryan, Anti-CD20 single chain

variable antibody fragment–apolipoprotein A-I chimera containing nanodisks promote targeted bioactive agent delivery to CD20-positive lymphomas, *Biochem. Cell Biol.*, 93 (2015) 343–350.

143. W. Wang, K. Chen, Y. Su, J. Zhang, M. Li, J. Zhou, Lysosome-Independent intracellular drug/gene codelivery by lipoprotein-derived nanovector for synergistic apoptosis-inducing cancer-targeted therapy, *Biomacromolecules*, 19 (2018) 438–446.

144. Q. Ouyang, Z. Duan, G. Jiao, J. Lei, A biomimic reconstituted high-density-lipoprotein-based drug and p53 gene co-delivery system for effective antiangiogenesis therapy of bladder cancer, *Nanoscale Res. Lett.*, 10 (2018) 965.

145. F. Xiong, S. Nirupama, S. R. Sirsi, A. Lacko, K. Hoyt, Ultrasound-stimulated drug delivery using therapeutic reconstituted high-density lipoprotein nanoparticles, *Nanotheranostics*, 1 (2017) 440–449.

146. R. Duivenvoorden, J. Tang, D. P. Cormode, A. J. Mieszawska, D. Izquierdo-Garcia, C. Ozcan, M. J. Otten, N. Zaidi, M. E. Lobatto, S. M. V. Rijs, B. Priem, E. L. Kuan, C. Martel, B. Hewing, H. Sager, M. Nahrendorf, G. J. Randolph, E. S. G. Stroes, V. Fuster, E. A. Fisher, Z. A. Fayad, W. J. M. Mulder, A statin-loaded reconstituted high-density lipoprotein nanoparticle inhibits atherosclerotic plaque inflammation, *Nat. Commun.*, 5 (2014) 3065.

147. J. Tang, M. E. Lobatto, L. Hassing, S. V. D. Staay, S. M. V. Rijs, C. Calcagno, M. S. Braza, S. Baxter, F. Fay, B. L. Sanchez-Gaytan, R. Duivenvoorden, H. B. Sager, Y. M. Astudillo, W. Leong, S. Ramachandran, G. Storm, C. Pérez-Medina, T. Reiner, D. P. Cormode, G. J. Strijkers, E. S. G. Stroes, F. K. Swirski, M. Nahrendorf, E. A. Fisher, Z. A. Fayad, W. J. M. Mulder, Inhibiting macrophage proliferation suppresses atherosclerotic plaque inflammation, *Sci. Adv.*, 1 (2015) e1400223.

148. Q. Song, M. Huang, L. Yao, X. Wang, X. Gu, J. Chen, J. Chen, J. Huang, Q. Hu, T. Kang, Z. Rong, H. Qi, G. Zheng, H. Chen, X. Gao, Lipoprotein-based nanoparticles rescue the memory loss of mice with Alzheimer's disease by accelerating the clearance of amyloid-beta, *ACS Nano*, 8 (2014) 2345–2359.

149. M. Huang, M. Hu, Q. Song, H. Song, J. Huang, X. Gu, X. Wang, J. Chen, T. Kang, X. Feng, D. Jiang, G. Zheng, H. Chen, X. Gao, GM1-modified lipoprotein-like nanoparticle: Multifunctional nanoplatform for the combination therapy of Alzheimer's disease, *ACS Nano*, 9 (2015) 10801–10816.

150. J. S. Rink, W. Sun, S. Misener, J.-J. Wang, Z. J. Zhang, M. R. Kibbe, V. P. Dravid, S. Venkatraman, C. S. Thaxton, Nitric oxide-delivering high-density lipoprotein-like nanoparticles as a biomimetic nanotherapy

for vascular diseases, *ACS Appl. Mater Interfaces*, 10 (2018) 6904–6916.

151. U. Forstermann, Nitric oxide and oxidative stress in vascular disease, *Pflugers Arch.*, 459 (2010) 923–939.

152. J. O. Lundberg, M. T. Gladwin, E. Weitzberg, Strategies to increase nitric oxide signalling in cardiovascular disease, *Nat. Rev. Drug Discov.*, 14 (2015) 623.

153. U. Förstermann, N. Xia, H. Li, Roles of vascular oxidative stress and nitric oxide in the pathogenesis of atherosclerosis, *Circ. Res.*, 120 (2017) 713–735.

154. A. Sharma, S. Sellers, N. Stefanovic, C. Leung, S. M. Tan, O. Huet, D. J. Granville, M. E. Cooper, J. B. D. Haan, P. Bernatchez, Direct endothelial nitric oxide synthase activation provides atheroprotection in diabetes-accelerated atherosclerosis, *Diabetes*, 64 (2015) 3937–3950.

155. J. F. Quinn, M. R. Whittaker, T. P. Davis, Delivering nitric oxide with nanoparticles, *J. Control. Release*, 205 (2015) 190–205.

156. B. L. Sanchez-Gaytan, F. Fay, M. E. Lobatto, J. Tang, M. Ouimet, Y. Kim, S. E. M. V. D. Staay, S. M. V. Rijs, B. Priem, L. Zhang, E. A. Fisher, K. J. Moore, R. Langer, Z. A. Fayad, W. J. M. Mulder, HDL-mimetic PLGA nanoparticle to target atherosclerosis plaque macrophages, *Bioconjug. Chem.*, 26 (2015) 443–451.

157. F. Guillaumond, G. Bidaut, M. Ouaissi, S. Servais, V. Gouirand, O. Olivares, S. Lac, L. Borge, J. Roques, O. Gayet, M. Pinault, C. Guimaraes, J. Nigri, C. Loncle, M.-N. Lavaut, S. Garcia, A. Tailleux, B. Staels, E. Calvo, R. Tomasini, J. L. Iovanna, S. Vasseur, Cholesterol uptake disruption, in association with chemotherapy, is a promising combined metabolic therapy for pancreatic adenocarcinoma, *PNAS*, 112 (2015) 2473–2478.

158. D. Guo, F. Reinitz, M. Youssef, C. Hong, D. Nathanson, D. Akhavan, D. Kuga, A. N. Amzajerdi, H. Soto, S. Zhu, I. Babic, K. Tanaka, J. Dang, A. Iwanami, B. Gini, J. DeJesus, D. D. Lisiero, T. T. Huang, R. M. Prins, P. Y. Wen, H. I. Robins, M. D. Prados, L. M. DeAngelis, I. K. Mellinghoff, M. P. Mehta, C. D. James, A. Chakravarti, T. F. Cloughesy, P. Tontonoz, P. S. Mischel, An LXR agonist promotes glioblastoma cell death through inhibition of an EGFR/AKT/SREBP-1/LDLR–dependent pathway, *Cancer Discov.*, 1 (2011) 442–456.

159. M. Nikanjam, E. A. Blakely, K. A. Bjornstad, X. Shu, T. F. Budinger, T. M. Forte, Synthetic nano-low density lipoprotein as targeted drug delivery vehicle for glioblastoma multiforme, *Int. J. Pharm.*, 328 (2007) 86–94.

160. R. S. Mulik, C. Bing, M. Ladouceur-Wodzak, I. Munaweera, R. Chopra, I. R. Corbin, Localized delivery of low-density lipoprotein docosahexaenoic acid nanoparticles to the rat brain using focused ultrasound, *Biomaterials*, 83 (2016) 257–268.

161. X. Wen, L. Reynolds, R. S. Mulik, S. Y. Kim, T. V. Treuren, L. H. Nguyen, H. Zhu, I. R. Corbin, Hepatic arterial infusion of low-density lipoprotein docosahexaenoic acid nanoparticles selectively disrupts redox balance in hepatoma cells and reduces growth of orthotopic liver tumors in rats, *Gastroenterology*, 150 (2016) 488–498.

162. N. Zhang, J. Tao, H. Hua, P. Sun, Y. Zhao, Low-density lipoprotein peptide-combined DNA nanocomplex as an efficient anticancer drug delivery vehicle, *Eur. J. Pharm. Biopharm.*, 94 (2015) 20–29.

163. M. Zhou, T. Wang, Q. Hu, Y. Luo, Low density lipoprotein/pectin complex nanogels as potential oral delivery vehicles for curcumin, *Food Hydrocoll.*, 57 (2016) 20–29.

164. J. Tian, S. Xu, H. Deng, X. Song, X. Li, J. Chen, F. Cao, B. Li, Fabrication of self-assembled chitosan-dispersed LDL nanoparticles for drug delivery with a one-step green method, *Int. J. Pharm.*, 517 (2017) 25–34.

165. K. G. Parhofer, P. Hugh, R. Barrett, D. M. Bier, G. Schonfeld, Determination of kinetic parameters of apolipoprotein B metabolism using amino acids labeled with stable isotopes, *J. Lipid Res.*, 32 (1991) 1311–1323.

166. H. Yamamoto, T. Takada, Y. Yamanashi, M. Ogura, Y. Masuo, M. Harada-Shiba, H. Suzuki, VLDL/LDL acts as a drug carrier and regulates the transport and metabolism of drugs in the body, *Sci. Rep.*, 7 (2017) 633.

167. A. Gustafson, Chylomicron metabolism, *J. Internal Med.*, 179 (1966) 29–32.

168. S. M. Grundy, H. Y. I. Mok, Chylomicron clearance in normal and hyperlipidemic man, *Metabolism*, 25 (1976) 1225–1239.

169. A. Dahan, A. Hoffman, Evaluation of a chylomicron flow blocking approach to investigate the intestinal lymphatic transport of lipophilic drugs, *Eur. J. Pharm. Sci.*, 24 (2005) 381–388.

170. M. Sun, X. Zhai, K. Xue, L. Hu, X. Yang, G. Li, L. Si, Intestinal absorption and intestinal lymphatic transport of sirolimus from self-microemulsifying drug delivery systems assessed using the single-pass intestinal perfusion (SPIP) technique and a chylomicron flow blocking approach: Linear correlation with oral bioavailabilities in rats, *Eur. J. Pharm. Sci.*, 43 (2011) 132–140.

171. S. Gao, F. Dagnaes-Hansen, E. J. B. Nielsen, J. Wengel, F. Besenbacher, K. A. Howard, J. Kjems, The effect of chemical modification and nanoparticle formulation on stability and biodistribution of siRNA in mice, *Mol. Ther.*, 17 (2009) 1225–1233.

172. R. P. Hickerson, A. V. Vlassov, Q. Wang, D. Leake, H. Ilves, E. Gonzalez-Gonzalez, C. H. Contag, B. H. Johnston, R. L. Kaspar, Stability study of unmodified siRNA and relevance to clinical use, *Oligonucleotides*, 18 (2008) 345–354.

173. S. H. Ku, S. D. Jo, Y. K. Lee, K. Kim, S. H. Kim, Chemical and structural modifications of RNAi therapeutics, *Adv. Drug Del. Rev.*, 104 (2016) 16–28.

174. C. Wolfrum, S. Shi, K. N. Jayaprakash, M. Jayaraman, G. Wang, R. K. Pandey, K. G. Rajeev, T. Nakayama, K. Charrise, E. M. Ndungo, T. Zimmermann, V. Koteliansky, M. Manoharan, M. Stoffel, Mechanisms and optimization of, *Nat. Biotechnol.*, 25 (2007) 1149.

175. J. Soutschek, A. Akinc, B. Bramlage, K. Charisse, R. Constien, M. Donoghue, S. Elbashir, A. Geick, P. Hadwiger, J. Harborth, M. John, V. Kesavan, G. Lavine, R. K. Pandey, T. Racie, K. G. Rajeev, I. Röhl, I. Toudjarska, G. Wang, S. Wuschko, D. Bumcrot, V. Koteliansky, S. Limmer, M. Manoharan, H.-P. Vornlocher, Therapeutic silencing of an endogenous gene by systemic administration of modified siRNAs, *Nature*, 432 (2004) 173.

176. C. Lorenz, P. Hadwiger, M. John, H. P. Vornlocher, C. Unverzagt, Steroid and lipid conjugates of siRNAs to enhance cellular uptake and gene silencing in liver cells, *Bioorg. Med. Chem. Lett.*, 14 (2004) 4975–4977.

177. M. K. Bijsterbosch, E. T. Rump, R. Vrueh, R. Dorland, R. Veghel, K. L. Tivel, E. A. L. Biessen, T. Berkel, M. Manoharan, Modulation of plasma protein binding and in vivo liver cell uptake of phosphorothioate oligodeoxynucleotides by cholesterol conjugation, *Nucleic Acids Res.*, 28 (2000) 2717–2725.

178. C. S. Thaxton, W. L. Daniel, D. A. Giljohann, A. D. Thomas, C. A. Mirkin, Templated spherical high density lipoprotein nanoparticles, *J. Am. Chem. Soc.*, 131 (2009) 1384–1385.

179. M. P. Plebanek, R. K. Mutharasan, O. Volpert, A. Matov, J. C. Gatlin, C. S. Thaxton, Nanoparticle targeting and cholesterol flux through scavenger receptor Type B-1 inhibits cellular exosome uptake, *Sci. Rep.*, 5 (2015) 15724.

180. M. P. Plebanek, D. Bhaumik, P. J. Bryce, C. S. Thaxton, Scavenger receptor type B1 and lipoprotein nanoparticle inhibit myeloid-derived suppressor cells, *Mol. Cancer Ther.*, 17 (2018) 686–697.

181. K. M. McMahon, M. P. Plebanek, C. S. Thaxton, Properties of native high-density lipoproteins inspire synthesis of actively targeted in vivo siRNA delivery vehicles, *Adv. Funct. Mater.*, 26 (2016) 7824–7835.

182. A. E. Murmann, K. M. McMahon, A. Haluck-Kangas, N. Ravindran, M. Patel, C. Y. Law, S. Brockway, J. J. Wei, C. S. Thaxton, M. E. Peter, Induction of DISE in ovarian cancer cells in vivo, *Oncotarget*, 8 (2017) 84643–84658.

183. A. E. Murmann, Q. Q. Gao, W. E. Putzbach, M. Patel, E. T. Bartom, C. Y. Law, B. Bridgeman, S. Chen, K. M. McMahon, C. S. Thaxton, M. E. Peter, Small interfering RNAs based on huntingtin trinucleotide repeats are highly toxic to cancer cells, *EMBO Rep*, 19 (2018) e45336.

184. Q. Lin, J. Chen, H. Jin, K. K. Ng, M. Yang, W. Cao, L. Ding, Z. Zhang, G. Zheng, Efficient systemic delivery of siRNA by using high-density lipoprotein-mimicking peptide lipid nanoparticles, *Nanomedicine (London)*, 7 (2012) 1813–1825.

185. J. B. Simonsen, B. Su, M. Ghosh, J. Beckstead, T. M. Forte, R. O. Ryan, Utilizing a reconfigured Hdl particle to target and deliver sirna to mantle cell lymphoma cells, *Biophys. J.*, 106 (2014) 188a.

186. Y. Ding, Y. Wang, J. Zhou, X. Gu, W. Wang, C. Liu, X. Bao, C. Wang, Y. Li, Q. Zhang, Direct cytosolic siRNA delivery by reconstituted high density lipoprotein for target-specific therapy of tumor angiogenesis, *Biomaterials*, 35 (2014) 7214–7227.

187. T. Nakayama, J. S. Butler, A. Sehgal, M. Severgnini, T. Racie, J. Sharman, F. Ding, S. S. Morskaya, J. Brodsky, L. Tchangov, V. Kosovrasti, M. Meys, L. Nechev, G. Wang, C. G. Peng, Y. Fang, M. Maier, K. G. Rajeev, R. Li, J. Hettinger, S. Barros, V. Clausen, X. Zhang, Q. Wang, R. Hutabarat, N. V. Dokholyan, C. Wolfrum, M. Manoharan, V. Kotelianski, M. Stoffel, D. W. Y. Sah, Harnessing a physiologic mechanism for siRNA delivery with mimetic lipoprotein particles, *Mol. Ther.*, 20 (2012) 1582–1589.

188. Y. Ding, W. Wang, M. Feng, Y. Wang, J. Zhou, X. Ding, X. Zhou, C. Liu, R. Wang, Q. Zhang, A biomimetic nanovector-mediated targeted cholesterol-conjugated siRNA delivery for tumor gene therapy, *Biomaterials*, 33 (2012) 8893–8905.

189. Q.-L. Zhu, Y. Zhou, M. Guan, X.-F. Zhou, S.-D. Yang, Y. Liu, W.-L. Chen, C.-G. Zhang, Z.-Q. Yuan, C. Liu, A.-J. Zhu, X.-N. Zhang, Low-density lipoprotein-coupled N-succinyl chitosan nanoparticles co-delivering siRNA and doxorubicin for hepatocyte-targeted therapy, *Biomaterials*, 35 (2014) 5965–5976.

190. Y. Dong, K. T. Love, J. R. Dorkin, S. Sirirungruang, Y. Zhang, D. Chen, R. L. Bogorad, H. Yin, Y. Chen, A. J. Vegas, C. A. Alabi, G. Sahay, K. T. Olejnik, W. Wang, A. Schroeder, A. K. R. Lytton-Jean, D. J. Siegwart, A. Akinc, C. Barnes, S. A. Barros, M. Carioto, K. Fitzgerald, J. Hettinger, V. Kumar, T. I. Novobrantseva, J. Qin, W. Querbes, V. Koteliansky, R. Langer, D. G. Anderson, Lipopeptide nanoparticles for potent and selective siRNA delivery in rodents and nonhuman primates, *PNAS*, 111 (2014) 3955–3960.

Chapter 7

Exosomes in Cancer Disease, Progression, and Drug Resistance

Taraka Sai Pavan Grandhi,[a,b] Rajeshwar Nitiyanandan,[c] and Kaushal Rege[d]

[a]*Biomedical Engineering, Arizona State University, Tempe, Arizona 85287-6106, USA*
[b] *Current address: Advanced Assays Group, Genomics Institute of the Novartis Research Foundation, La Jolla, California 92121, USA*
[c]*Biological Design, Arizona State University, Tempe, Arizona 85287-6106, USA*
[d]*Chemical Engineering, Arizona State University, Tempe, Arizona 85287-6106, USA*
sgrandhi@gnf.org

7.1 Introduction

Exosomes are cell-type–dependent extracellular vesicles, 30–150 nm in dimension, that play a critical role as mediators of communication between different cell types in the body [1]. Exosomes consist of a lipid bilayer enclosing cargo of DNA, RNA and/or proteins and are directed to a specific destination [2–6] by means of proteins displayed on their surface [7]. These extracellular vesicles are produced by all cells of the body [8–10] and were previously considered as cellular waste or debris without a significant biological role. However, recent

Handbook of Materials for Nanomedicine: Lipid-Based and Inorganic Nanomaterials
Edited by Vladimir Torchilin
Copyright © 2020 Jenny Stanford Publishing Pte. Ltd.
ISBN 978-981-4800-91-4 (Hardcover), 978-1-003-04507-6 (eBook)
www.jennystanford.com

studies have shed light on the diverse roles exosomes and their contents play in multiple disease biologies [11]. Exosomal delivery of cargo for communication is now considered a key means of cell–cell communication in addition to that mediated by cell–cell contact and soluble molecules [12]. Unsurprisingly, these messengers play an important role in the growth and metastasis of cancer cells [13–16], including in actively preparing distant metastatic sites for organotropic cancer progression [7, 15, 17]. Exosomes have also been demonstrated to play a role in active modulation of the microenvironment [18–21]. These roles have led to the emergence of exosomes as promising biomarkers for cancer disease progression, metastases, and recurrence [22–27]. In this chapter, we review the roles of exosomes in cancer growth and progression (metastases) and their promise as biomarkers and immunotherapeutics in cancer. Details on the methods of isolation and purification of exosomes and use of exosomes as nanocarriers of nucleic acids for cancer therapy are described in our previous review [28].

7.2 Exosome Biogenesis

During the process of endocytosis, the invagination of cell surface plasma membrane and budding into the cell gives rise to endocytic vesicles that fuse to generate early endosomes and mature into late endosomes. Exosomes are created by the inward budding of the late endosomal limiting membrane [10] (Fig. 7.1). Inward budding of the endosomal limiting membrane generates intraluminal endosomal vesicles (ILVs) that are loaded with cargo destined for future extracellular release. The late endosome containing ILVs (termed multivesicular body or MVBs) fuses with the plasma membrane, releasing its content into the extracellular space [29]. The ILVs once released into the extracellular space are considered to be exosomes [10, 29] (Fig. 7.1).

Molecular mechanisms controlling the biogenesis, secretion and uptake of exosomes remain to be fully understood. Some of the well-known mechanisms of cargo loading into the exosomes include ESCRT system [30–32], tetraspanin-directed mechanisms [33] and lipid-dependent mechanisms [34]. ESCRT system (endosomal sorting complex required for transport) best known for MVB biogenesis was initially considered only effective in recognizing and sorting ubiquitinated cargo into ILVs destined for lysosomal fusion and degradation. However, Baietti et al. [35] showed the ability of syndecan heparan sulfate proteoglycans and syntenin (cytoplasmic adaptor protein of syndecans) to recruit ESCRTs for exosome biogenesis. It was shown that syntenin interacts with ESCRTs via binding to ALIX (an ESCRT-III-binding protein) to package cargo and trigger vesicle formation destined to be within exosomes. Other proteins such as CHMP4 and AAA+ ATPase VPS4 critical for MVB biogenesis that interact directly with ALIX were shown to remain conserved in this process. ALIX adaptor protein has been shown to bind and sort other proteins such as RNA-binding protein Argonuate 2 (Ago2), RNA-binding protein SYNCRIP which are essential for miRNA processing and sorting into exosomes [36, 37]. These studies indicate ALIX adaptor protein as a shuttle for sorting and loading cytoplasmic proteins into exosomes [35].

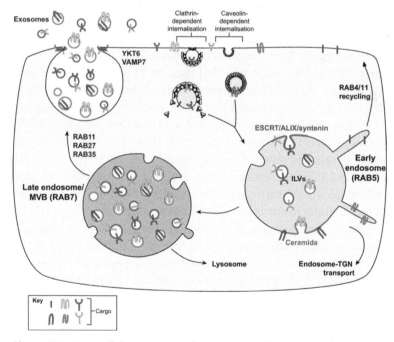

Figure 7.1 Intracellular processes for exosome biogenesis and secretion. Reproduced from [123] with permission from Copyright Clearance Center and the publisher (The Company of Biologists).

Tetraspanins (CD63, CD9, CD81, and CD82) are integral membrane protein components that are highly enriched in exosomes. They organize into tetraspanin-enriched domains by interacting with lipids, other cytosolic and transmembrane proteins [33, 38]. Cytoplasmic and transmembrane protein interactome of tetraspanins is considered another major source of protein loading into the exosomes [33]. Sphingosine and fatty acid containing ceramides have been shown to be essential in exosome biogenesis [39]. It is well known that ceramides can induce stable and spontaneous curvature in lipid bilayer membranes (liposomes), and play an essential role in inward budding of late endosomal limiting membrane for ILV generation [39]. Kajimoto et al. [34] showed that continuous metabolization of ceramide into sphingosine 1-phosphate (S1P) activates inhibitory G protein–coupled S1P receptors, essential for successful cargo sorting into ILVs in HeLa cells; S1P signaling was shown to play a role in the sorting of the tetraspanin CD63 protein into exosomes [34].

Apart from proteins, mRNA, miRNA, and dsDNA have been shown to be trafficked via exosomes between different target cells [40]. Adaptor protein ALIX was shown to bind miRNA and Ago2, allowing their sorting into exosomes [36]. Lavello et al. [36] showed that ALIX/Ago2 immunoprecipitates obtained from human liver stem-like cells (HLSCs) contained significant amounts of miRNA. Further ALIX knockdown significantly reduced the amount of miRNAs present in the HLSC-derived exosomes. Other proteins abundantly found in exosomes such as heterogeneous ribonucleoprotein A2B1 (hnRNPA2B1) and Annexin-2 have also been suggested to play critical roles in the sorting of mRNAs and miRNAs into exosomes due to their affinity towards specific ribonucleotides [41–44].

While the systems described in the previous paragraphs elucidate cargo sorting into the ILVs, trafficking of MVBs towards the cell membrane for fusion and exosomal release is equally essential. Rab family of small GTPases have been known to control endosomal and secretory vesicle trafficking within the cell [45]. Rab GTPases associate reversibly with MVB membrane and regulate the sequential steps in its trafficking to the cell surface for exosomal release. Rab 4 and 5 associate with early endosomes whereas Rab 7 and 9 associate with late endosomes. Rabs also interact with specific effector proteins such as SNARE proteins (soluble N-ethylmaleimide-sensitive fusion protein-attachment protein receptor) to drive vesicle docking and fusion to targeted membranes [46, 47]. Specifically, Rab27a was shown to regulate MVBs fusion with plasma membrane to release exosomes via interaction with a SNARE protein (Syntaxin 1a) [48, 49] (Fig. 7.1). Further, Rabs are also responsible for deciding the fate of the MVBs. For example, Vanlandingham et al. [50] found that replacement of Rab5 with Rab7 caused MVBs fusion to lysosomes resulting in degradation of MVB cargo. Once released from target cells, exosomes traffic to their target cell type to be uptaken and processed. Exosome uptake by target cells is mostly thought to be via receptor-mediated endocytosis, non-specific endocytosis and membrane fusion, governed microenvironmental conditions such as low pH, etc. While progress has been made in understanding the exosomal biogenesis and its homing to the target, factors that initiate their formation still remains unknown and are understudied.

7.3 Exosomes in Cancer Progression

The ability to accurately traffic and target diverse cargo consisting of proteins, DNA and/or RNA to a specific cell type makes exosomes very valuable for cancer cells to modify their microenvironment. While genetic and epigenetic mutations initiate oncogenic transformation of a particular cell type, cancer progression and metastases critically depend on a cooperative tumor microenvironment, replete with an immunosuppressive phenotype. Studies over the past decade have shown the seminal role cancer cell–derived exosomes play in modifying immediate and distant microenvironments into pro-tumorigenic niches. In addition, exosomes released from cells within the pro-tumorigenic microenvironment (e.g., cancer-associated fibroblasts) can further fuel the growth and development of the disease. Collectively, these studies have shown an integral role of exosomes in cancer progression and metastases.

7.3.1 Impact of Cancer Cell–Derived Exosomes on Stromal Cells

Solid tumors of epithelial origin contain several different cells including fibroblasts/stromal cells [51–53], endothelial cells [21], pericytes [54], tissue-resident stem cells [55, 56], and macrophages [57, 58] in the microenvironment. These cells are actively converted into pro-tumorigenic phenotypes for promoting cancer progression and metastases. Exosomes released from cancer cells can actively convert stromal cells/mesenchymal stem cells in the vicinity into myofibroblast-like cells with cancer-associated fibroblast (CAFs) phenotype [59–61]. Chowdhury et al. [60] showed that the exosomes derived from DU145 prostate cancer cells could modify bone marrow mesenchymal stem cells (BM-MSCs) into pro-tumorigenic and pro-angiogenic myofibroblasts. Exosomal TGF-β derived from DU145 cells was shown to convert the BM-MSCs into a-SMA positive myofibroblasts. Treatment of BM-MSCs with cancer cell–derived exosomes triggered the release of VEGF-A and HGF and upregulated the mRNA expression of pro-tumorigenic proteins MMP-3, MMP-13, and SerpinA1 in the transformed myofibroblast cells. Incubation of DU145 prostate cancer cells with conditioned media (CM)-derived from exosome transformed BM-MSCs caused higher proliferation, motility, and viability compared to treatment with untransformed CM control. Spheroids generated from co-cultures of DU145 prostate cancer cells and transformed myofibroblasts showed increased invasion of cancer cells out of the spheroids when embedded into 3D Matrigel columns compared to cancer cells from single cell spheroids. Weber et al. [62] also reported the role of exosomal TGF-β in the conversion of stromal cells into cancer-associated fibroblasts in mesothelioma lung cancer. Their results showed that exosomal TGF-β–induced conventional SMAD-dependent actin remodeling and upregulation a-SMA expression in the recipient cells. The exosomal components also caused SMAD-independent changes such as deposition of pericellular coat composed of hyaluronic acid, which likely promoted tumor progression and dissemination.

Exosomes derived from bone marrow–derived tumors were shown to be able to re-engineer their surroundings into pro-tumorigenic niches. Paggetti et al. [61] showed that exosomes from chronic lymphocytic leukemia (CLL) cells were able to modify stromal cells into a phenotype similar to cancer-associated fibroblasts. They showed that CLL-derived exosomes were actively incorporated into the endothelial and mesenchymal stem cells, which induced a phenotype resembling cancer-associated fibroblasts. CLL exosomes were highly enriched in miRNAs such as miR-21, miR-155, miR-

146a, miR-148a, and let-7g and anti-apoptotic proteins, angiogenic factors, RNA processing proteins and heat shock proteins which were hypothesized to induce transformation of the target cells. Exosomes were shown to activate several kinases in the target cells within an hour of incorporation and initiate nuclear factor (NF)-κB signaling which led to major changes in target cell survival, migration, proliferation and RNA expression. A strong CAF gene signature comprising upregulation of CXCL1, IL6, IL34, CCL2, ICAM1, and MMP1 genes was observed in the target cells followed by significant upregulation in levels of cytokines, chemokines and proangiogenic factors. As multiple CAF-mediated functions are dependent on (NF)-κB signaling, activation of (NF)-κB signaling pathways in target cells by exosomes strongly supports their role in microenvironment transformation.

7.3.2 Impact of Cancer Cell–Derived Exosomes on Endothelial Cells

Cancer cell–derived exosomes can also significantly impact the endothelial cells in the tumor vicinity [63]. Schillaci et al. [64] showed that the exosomes derived from highly metastatic colon cancer cells (SW620Exos) were able to induce higher endothelial hyperpermeability compared to a non-metastatic colon cancer cell line (SW480Exos). Higher hyperpermeability could aid in metastatic progression of the cells due to improper tight junctions. Compared to exosomes derived from a non-metastatic cell line, those derived from the metastatic cancer cells induced significantly higher hyperpermeability in a confluent HUVEC cell layer in vitro as determined using FITC-Dextran. Exosomes derived from the metastatic cell line caused cytosolic delocalization of the cell surface tight junction protein VE-cadherin, and other adherens junction (AJ) proteins, β-catenin and p120-catenin. Proteomic analyses showed ~ 150 proteins that were significantly different between the exosomes derived from SW480Exos and SW620Exos cells. Exosomes derived from the metastatic cell line were enriched in proteins related to cytoskeletal transformation, actin and microtubule remodeling. Exosomes were enriched in thrombin, a RhoA activator responsible for cytoskeletal changes in the target cells, and treating the monolayer of endothelial cells with thrombin resulted in similar levels of hyperpermeability as observed with the SW620Exos cells [64].

Sun et al. [65] showed that exosomes derived from glioma cancer stem cells could induce angiogenesis in the brain endothelial cells (BECs) via the miR-21/VEGF signaling axis. Exosomes derived from the glioma cells contained microRNA 21 and VEGF and triggered significantly higher expression of phosphorylated VEGFR2, endothelial cell mobility and tube forming capability in BECs compared to the scrambled controls. Zhang et al. [63] also found an increase in VEGF expression in HUVEC cells after treatment with 786-0 renal cell-derived exosomes indicating their potential role in inducing angiogenesis in cancer microenvironment.

7.3.3 Impact of Cancer Cell–Derived Exosomes on Antigen-Presenting Cells and Other Immune Cells

Dendritic cells, monocytes, and macrophages constitute the mononuclear phagocyte system which are largely responsible for constantly surveying the body for any signs of external or internal disease. These white blood cells, responsible for engulfing foreign substances and cellular debris, identify diseased and infected cells based on the surface expression of proteins that distinguish them from healthy counterparts. Upon internalization and digestion of the diseased cell proteins, these cells have the ability to present the antigen to the naïve T-cells initiating an immune response. Cancer cells evade immune detection by creating a microenvironment that allows their progression and metastases, and exosomes play an important role in the development of an immune-suppressive microenvironment. Salimu et al. [66] showed that DU145 prostate cancer cell–derived exosomes immunosuppress dendritic cells by inducing the expression of CD73 protein, which results in the inhibition of proinflammatory cytokines TNFα and IL12 production. Exosomal prostaglandin E2 (PGE2) was found as the potential driver of CD73 induction and subsequent inhibition of TNFα and IL12 production. Immunosuppressed dendritic cells were unable to elicit an immune response as measured by the impaired CD8$^+$ T cell responses. Cancer cell–derived exosomal PGE2 was also shown to impact differentiation of myeloid cells in the bone marrow. Xiang et al. [67]. showed that murine breast carcinoma-derived exosomes are taken up by myeloid cells in the bone marrow which induce their differentiation towards immunosuppressive myeloid-derived suppressor cell (MDSCs) phenotype. The authors showed that the PGE2 and TGF-β in the exosomes were responsible for the accumulation of the (CD11b⁺ Gr-1⁺) MDSCs. Addition of anti-PGE2 and anti-TGFβ antibodies significantly reduced the accumulation of (CD11b⁺ Gr-1⁺) MDSC cells. Immunosuppressive MDSCs have been implicated in recruiting CD4⁺ Foxp3⁺ CD25⁺ regulatory T cells that create an immunosuppressive environment in the tumor by actively suppressing the proliferation of CD4⁺ and CD8⁺ T cells [68].

Tumor-derived exosomes have also been shown to directly induce the expansion of regulatory T cells and induce apoptosis in tumor reactive activated CD8⁺ T cells. Wieckowski et al. [69] showed that tumor-derived microvesicles, derived from sera of melanoma and SCHNN (head and neck squamous cell carcinoma) patients, induced apoptosis in activated primary CD8⁺ T cells. FasL (Fas ligand), TRAIL and programmed death-1 ligand present on microvesicles surface were hypothesized to induce apoptosis in the tumor-reactive CD8⁺ T cells. The level of FasL on the microvesicle surface from patient sera correlated with their apoptosis-inducing activity and disease activity in the patients. Further, tumor-derived microvesicles were also shown to expand the immunosuppressive CD4⁺CD25^{high}FOXP3⁺ Treg population via interaction between MV-derived B7-H2 and ICOS⁺CD4⁺ T cells.

Cancer cell–derived exosomes were also shown to modify macrophages in the secondary sites away from the tumor. Chow et al. [70] showed that breast cancer-derived exosome–induced NF-kB activation in distant macrophages that led to the secretion of proinflammatory cytokines such as IL-6, TNF-α, GCSF, and CCL2 via TLR2 engagement. In vitro–produced breast cancer cell–derived exosomes, when injected into immunocompromised mice, were efficiently internalized by macrophages present in the lung and brain which created an immunosuppressive environment for the eventual metastasis of the cancer cells. Tumor cell-derived exosomes were also shown to impact and mislead platelets to initiate thrombin generation and enabling subsequent cancer progression [71]. These studies point to the role that cancer exosomes play in modifying the immune system to support tumor growth and expansion.

7.3.4 Tumor-Derived Exosomes Regulate Organotypic Metastases

In addition to modifying their immediate vicinity, the impact of tumor-derived exosomes is evident during organotypic metastases. A fundamental characteristic of cancer is to metastasize to different organs from its primary source and exosomes were recently shown to have a major role in the process. Primary tumor-derived exosomes have been shown to carry crucial cargo to targets cells within metastatic organ to initiate the formation of pre-metastatic niche [7, 17] and prepare the site for cancer cell arrival. Hoshimo et al. [7] showed that exosomal integrin composition determines their homing and organotypic fates. In this seminal study, the authors showed that the exosomes derived from lung-, liver-, and brain-tropic breast (MDA-MB-231) and pancreatic cancer cells (BxPC-3 and HPAF-II) preferentially bind and fuse with the cells from their target organ, such as liver Kupffer cells, lung fibroblasts and epithelial cells and brain endothelial cells. Specifically, exosomal integrins $ITG\beta_4$ and $ITG\beta_5$ were implicated in lung and liver tropism. Further, upon binding these target cells, the exosomal contents were shown to modify these cells to initiate pre-metastatic niche formation for eventual arrival of the cancer cells. Unbiased gene expression analysis of liver resident Kupffer cells treated with either BxPC-3-LiT exosomes (from liver tropic BxPC-3 pancreatic cancer cells) or BxPC-3-LiT $ITG\beta_5$KD exosomes (controls) for two weeks identified over 900 genes upregulated in response to the exosome treatment. Cell migration genes were most prominently upregulated (twofold for 221 genes; fourfold for 42 genes) followed by pro-inflammatory genes such as S100 (S100A8 and S100P). Several S100 genes (S100A4, -A6, -A10, -A11, -A13, and -A16) were also upregulated in 4175-lung tropic tumor (4175-LuT) exosome-educated lung WI-38 fibroblasts and in human bronchial epithelial cells (HBEpCs) compared to those educated with control exosomes (4175-LuT $ITG\beta_4$KD exosomes) (Fig. 7.2). Western blot analysis confirmed upregulation of phosphorylated Src in response to treatment with $ITG\beta_4$ containing exosomes. Phosphorylated Src plays a crucial role in S100A4 expression and subsequent pro-inflammatory cascade.

Their results showed the critical role played by exosomal integrins in homing to distant metastatic sites and directly influencing the resident cells towards facilitating tumor metastases.

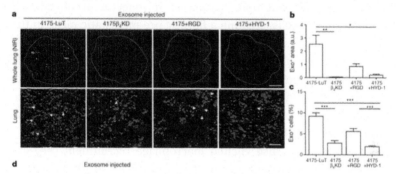

Figure 7.2 Exosomes from lung tropic MDA-MB-231 sub-line (4175-LuT cells) exhibit organ tropism and localize to lungs. (a) Top, Near-infrared whole-lung imaging of 4175-LuT- or control 4175β_4KD-derived exosomes loaded with PKH26 red dye, or 4175-LuT-derived exosomes pre-incubated with RGD or HYD-1 blocking peptides loaded with PKH26 red dye. Bottom, fluorescence microscopy. Arrowheads indicate exosome foci. (b) Quantification of exosome-positive areas from the whole-lung images from top section of (a). (c) Immunofluorescence quantification of exosome-positive cells from bottom section of (a) Data are mean±s.e.m. *$P<0.05$; **$P<0.01$; ***$P<0.001$ by one-way ANOVA (b, c, e); ***$P<0.001$ by two-tailed Student's t-test (d). The image has been reproduced from Hoshino et al. [7] with permission from Springer Nature.

Liu et al. [17] showed that the tumor exosomal RNAs (derived from melanoma and Lewis lung cancer cells) activate TLR3 on alveolar epithelial cells and recruit neutrophils to promote pre-metastatic niche formation. Tumor exosomal RNA content and not tumor RNA content activated lung epithelial TLR3 and induced chemokine production (CXCL1, CXCL2, CXCL5, and CXCL12), recruited neutrophils and promoted the establishment of pre-metastatic niche. Tumor exosomal RNA was shown to be involved in both expression and activation of lung epithelial TLR3. While the tumor exosomal RNA induced the expression of TLR3 by activation of NF-kB and MAPK pathways in the lung epithelial cells, small nuclear RNA content within the tumor exosomes was shown to activate TLR3.

Costa-Silva et al. [15] elucidated the role of pancreatic cancer exosomes in conditioning the host liver and initiating hepatic pre-metastatic niche formation for liver metastases. Pancreatic ductal adenocarcinoma (PDAC) cell–derived exosomes were shown to induce expression and secretion of TGF-β in liver resident Kupffer cells and significant upregulation of fibronectin production by hepatic stellate cells, creating a fibrotic pre-metastatic niche in the liver. Exosomal cargo of macrophage inhibitory factor (MIF) was shown to induce the release of TGF-β by liver resident Kupffer cells that in turn promoted deposition of fibronectin by the hepatic stellate cells. Control non-tumor cell–derived exosomes failed to produce similar changes in the host liver microenvironment. Further, the local production of TGF-β from Kupffer cells after education by PDAC-derived exosomes also elicited macrophage recruitment (F4/80+ macrophages) from the host bone marrow, aiding the pre-metastatic niche formation. The authors further showed that TGFβ was upregulated in Kupffer cells during the early PanIN stage, suggesting a role of PDAC-derived exosomes during the pre-tumoral stages of liver pre-metastatic niche formation. Macrophage inhibitory factor (MIF) was found to be significantly elevated in pancreatic cancer patients with progression of disease after diagnosis, indicating the clinical relevance of these findings. Tumor-derived exosomes have now been implicated in organotropic metastases of multiple other cancers including melanoma metastases to bone marrow and lymph nodes [16, 72], breast cancer metastases to brain [73] and renal cancer metastases to lung [74]. These studies point to the role of exosomes as critical orchestrators of metastatic progression of primary cancers.

7.3.5 Activated Tumor Microenvironment–Derived Exosomes Accelerate Cancer Progression

While cancer cell–derived exosomes modify their microenvironment, exosomes derived from the modified tumor microenvironment components can also alter cancer cells into a more metastatic and invasive phenotype [75, 76]. This vicious cycle of cross communication between the different cells of the microenvironment eventually drives the cancer progression. For example, it has been shown that exosomes derived from cancer-associated fibroblasts in esophageal cancer contain microRNA species including miR-33a and miR-326, which play an important role in tumor progression by influencing cell–cell junctions, endocytosis, and cell adhesions [77]. Zhao et al. [76] showed that the exosomes derived from patient-derived cancer-associated fibroblasts (CAFs) modulate prostate cancer cell metabolism by inhibiting oxidative phosphorylation and increasing glycolysis and glutamine-dependent carboxylation in cancer cells. Alternatives to oxidative phosphorylation often help in cancer cell survival under nutrient deprivation and nutrient stressed conditions [78]. CAF-derived exosomes contain high amounts of lactate and acetate which help replenish the TCA cycle metabolites and also act as source of lipids. Ultra-high-performance liquid chromatography of patient CAF–derived exosomes showed high levels of arginine, alanine, valine, serine, glutamine, glutamate, proline, threonine, asparagine, and leucine, all of which can provide an "off-the-shelf" pool of metabolite cargo that can help cancer cells evade nutrient stress and nutrient deprivation (Figs. 7.3 and 7.4). Even senescent fibroblasts were shown to induce tumor cell proliferation [79]. Takasugi et al. [79] showed that senescent fibroblasts release exosomes containing EphA2 bind to ephrin-A1 on MCF-7 breast cancer cells inducing their proliferation. Exosomes derived from pre-senescent cells failed to induce similar level of proliferation as those from senescent cells due to the lack of EphA2 within the pre-senescent exosomes.

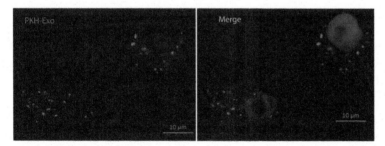

Figure 7.3 Cancer-associated fibroblast-derived exosomes carry metabolite cargo to cancer cells. Representative fluorescence image shows CAFs exosomes were uptaken by prostate cancer cells. Prostate cancer cells were incubated with PKH26 (red dye)-labeled CAFs exosomes for 3 h. Blue, cell nuclei; Red, PKH-Exo. The image has been reproduced from Zhao et al. [76] with permission from *eLife* one under Creative Commons Attribution License.

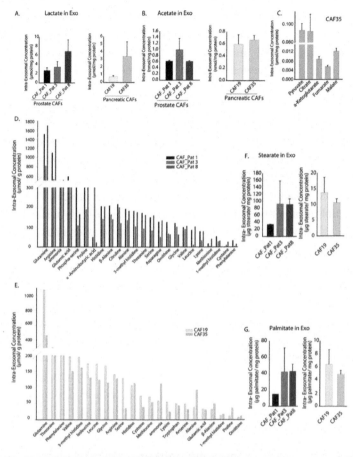

Figure 7.4 Cancer-associated fibroblast-derived exosomes carry metabolite cargo to cancer cells. Prostate and pancreatic cancer patient fibroblasts were shown to carry cargo crucial for cancer cell metabolism. Intra-exosomal lactate (A) and acetate (B) concentrations were measured in exosomes isolated from three prostate and two pancreatic CAFs using enzymatic assays. Intra-exosomal metabolites were extracted by methanol/chloroform method and protein concentration was used for normalization. ($n = 3$). (C) TCA cycle metabolites, including pyruvate, citrate, α-ketoglutarate, fumarate, and malate were measured using GC-MS in exosomes isolated from pancreatic patient CAF35. ($n = 3$). (D, E) Amino acids were measured using ultra-high-performance liquid chromatography (UPLC) inside CAF-derived exosomes (CDEs) (prostate CAFs: [D]; pancreatic CAFs: [E]). Significant levels of amino acids were detected inside CDEs. ($n = 3$). (F-G) Stearate and palmitate were detected at high levels using GC-MS inside pancreatic and prostate CDEs ($n \geq 3$). Data in (A–C), (F–G) are expressed as mean ± SEM. The image and description has been reproduced from Zhao et al. [76] with permission from *eLife* one under Creative Commons Attribution License.

Exosomes derived from other components of the tumor microenvironment such as platelets, endothelial cells and macrophages also participate in the cross-communication to enhance tumor growth. Platelet-derived exosomes have been shown to promote tumor growth and angiogenesis [80]. Wieczorek et al. [81] showed that exosomes/microvesicles derived from platelets transformed and activated multiple lung cancer cells (A549, CRL2066, CRL 2062, HTB 183, HTB 177) into more carcinogenic phenotype. The exosomes/microvesicles were shown to transfer integrin CD41 to the lung cancer cells which activated MAP kinase and increased the expression of membrane type 1-matrix metalloproteinase. In A549 lung cancer cells, the exosomes stimulated proliferation, upregulate cyclin expression and increase trans-Matrigel invasion. Further, exosomal contents stimulated mRNA expression of angiogenic factors such as VEGF, MMP-9, IL-8, and HGF. Further, intravenous injection of platelet-derived exosomes/microvesicles covered Lewis lung carcinoma cells (LLC) into syngeneic mice resulted in significantly more metastatic lung loci compared to animals injected with bare LLC cells. These studies and findings indicate the complex and intricate roles exosomes derived from tumor microenvironment can play in tumor progression and metastases.

7.4 Exosomes in Cancer Drug Resistance

Many cancer patients initially respond to therapeutics but relapse to a resistant form of the disease [82, 83]. Three frequently described pathways for exosome-mediated tumor drug resistance mechanisms have been described: exosome-mediated transfer of miRNAs to tumor cells, drug removal via exosome pathways, and neutralization of antibody-based drugs. Richards et al. [75] showed that exosomes derived from cancer-associated fibroblasts (CAFs) regulate survival and proliferation of pancreatic cancer cells after chemotherapeutic insult. The authors showed that CAFs exposed to gemcitabine, a chemotherapeutic drug approved for pancreatic cancer, significantly increase the production of exosomes containing miR-146a and Snail mRNA. These exosomes were shown to induce chemoresistance-inducing factor Snail in recipient epithelial cells and promote proliferation and drug resistance. Treatment of pancreatic cancer-associated CAFs with an inhibitor of exosome release GW4869 reduced exosome release from CAFs and sensitized pancreatic cancer cells to gemcitabine. Transfer of exosomal content from macrophages to tumor cells was shown to induce cisplatin resistance in gastric cancer cells [84]. Zheng et al. [84] showed that miR-21-containing exosomes derived from tumor-associated M2 macrophages induce resistance to cisplatin. M2 macrophage–derived miR-21-containing exosomes were shown to be directly transferred to gastric cancer cells where they prevented cell apoptosis via activation of PI3K/AKT signaling pathway, Bcl-2 gene overexpression and downregulation of PTEN. Drug resistant cancer cells of breast and ovarian origin have also been shown to transfer miRNA based cargoes, TGF-β in their exosomes to the sensitive cancer cells, making them resistant to the chemotherapeutic drugs and therapeutic antibodies [85–87]. Pink et al. [86] showed that cisplatin-resistant ovarian cancer cells upregulate a number of microRNAs, of which miR-31 and miR-21-3p induced increased resistance to cisplatin in sensitive ovarian cancer cell line A270. Exosomes derived from cisplatin-resistant CP70 ovarian cancer cells transfer miR-21-3p to cisplatin sensitive A270 cell line, thereby transferring their resistance to the sensitive counterpart. A 30% increase in the levels of miR-21-3p were seen in A270 cells after treatment with exosomes derived from the cisplatin-resistant CP70 ovarian cancer cell line. miR-21-3p was shown to directly reduce the levels of NAV3 mRNA and protein (a tumor suppressor protein) leading to resistance against cisplatin.

Removal of chemotherapeutic drug from cancer cells via exosomes can lead to the reduction of the drug in the cancer cells and subsequent resistance and improved survival. Safaei et al. [88] [89] showed that exosomes released from cisplatin-resistant ovarian carcinoma cells contained high levels of cisplatin compared to exosomes from cisplatin sensitive ovarian carcinoma cells (2.6-fold higher), suggesting a use of endocytic pathway to remove drugs from the cancer cells. The resistant ovarian cells also contained higher exosomal concentrations of cisplatin export transporters such as MRP2, ATP7A, and ATP7B likely responsible for the exosomal removal of the drug. Similar mechanisms of cancer drug expulsion via exosomes and microvesicles have been shown in other cancer cells and chemotherapeutic drugs [90]. Other mechanisms of drug resistance involved exosomal transfer of multidrug resistance proteins (MDR-1) and P-glycoprotein (P-gp) from drug-resistant cancer cells to sensitive cancer cells. Drug-resistant breast and prostate cancer cells were shown to utilize these pathways to gain resistance against the microtubule-stabilizing drug docetaxel [91, 92].

Another mechanism of drug resistance exploited by cancer cells utilizes exosomes to neutralize therapeutic antibodies. For example, Ciravolo et al. [93] showed that HER2-overexpressing breast cancer cells released exosomes with the HER2 protein on their surface which could bind the anti-HER2 antibody trastuzumab. Further, this sequestration of anti-HER2 antibody trastuzumab by the HER2 decorated exosomes reduced the antibody remaining to bind the cancer cells and diminished the overall effect of trastuzumab. These findings suggest a crucial role of exosomes in cancer drug resistance, which could be exploited for novel therapeutic biomarkers to assess drug response and novel drug discovery targets to enhance anti-cancer drug efficacy.

7.5 Exosomes as Cancer Biomarkers

Currently available biomarker detection methods make use of molecules which are highly expressed but may lack the specificity to detect very low levels of biomolecules that may be indicative of disease. Exosomes carry cargo containing various proteins from their cell or origin, which, in turn, can be indicative of disease states [94–96]. Tanaka et al. [97] were able to observe an increased expression of exosomal microRNA-21 in human esophageal squamous cell carcinoma (ESCC). They also found that the expression of this microRNA is consistently upregulated when compared to samples from patients suffering benign diseases. Their results also showed a positive correlation between the expression of miRNA-21 and patients exhibiting resistance to docetaxel chemotherapy which could likely be a useful hallmark for doctors to determine if patient needs alternative therapy.

Melo et al. [98] identified exosomes expressing a cell surface proteoglycan Glypican-1 which was enriched on pancreatic cancer cells. It has been previously shown that Glypican-1 is over expressed in both breast and pancreatic cancers [99, 100]. The authors observed that circulating exosomes (crExos) concentrations were significantly higher in cancer patients in comparison to healthy controls. Surprisingly, the authors also observed that the size of the pancreatic cancer patient–derived crExos was significantly smaller than those isolated from healthy patients, but this was not the case with breast cancer–derived crExos. Glypican-1 expression in healthy patient–derived crExos was found to be at a baseline level whereas nearly 75% of the pancreatic cancer–derived crExos expressed Gylpican-1 at higher levels than healthy patients. The authors also showed that crExos expressing Glypican-1 performed consistently better than the current gold standard biomarker for pancreatic cancer, Carbohydrate antigen 19-9 (CA19-9) [101, 102].

Currently, screening techniques for prostate cancer include detection of PSA or prostate specific antigen. Unfortunately, multiple factors, including pathogen-mediated infections, inflammation, etc., can affect PSA levels [103–105] and therefore interfere with the assay. Researchers have shown links between the expression of the survivin protein and the prognosis as well recurrence of prostate cancer [106–108]. Khan et al. [109] were able to observe the increased expression of survivin in exosomes isolated from the plasma of pancreatic cancer patients (Fig. 7.5). A significantly increased expression of survivin in exosomes derived from patients with low-grade and high-grade prostate cancer compared to exosomes derived from normal samples was observed. The levels of survivin were also found to be higher in patients exhibiting tumor recurrence following chemotherapy. The sensitivity of this biomarker was also high; all patient-derived samples (low to high grade) exhibited survivin levels greater than 100 pg/ml compared to exosomes derived from disease-free individuals or those with BPH (Fig. 7.6). In comparison, only 60% of the low-grade tumor sample and 85% of high-grade tumor samples expressed increased PSA in comparison to normal human– and BPH-derived samples. The advantage of this exosomal biomarker system is the increased reliability in diagnosis of prostate cancer in comparison to PSA, which is the current gold standard biomarker. Although levels of survivin do not change between the different grades of tumors, they may be useful in early detection of the disease due to the significant increase in levels in comparison to a disease-free individual. Nilsson et al. [24] employed urine-derived exosomes for the detection of two common biomarkers in prostate cancer, PCA-3 and TMPRSS2:ERG. It was observed that PCA-3 was overexpressed in all patient-derived samples expressing high-grade tumors whereas TMPRSS2:ERG was found in 50% of the same samples (Fig. 7.7). Both biomarkers were undetectable in patients who had undergone androgen deprivation therapy (ADT) showing a clear correlation between tumor regression and loss of biomarker expression. The loss of expression of these biomarkers was also observed in patients who had a non-functional prostate due to bone metastases or had undergone castration/prostatectomy. Exosomes derived from urine have also been used to diagnose prostate cancer using PSA and PSMA expression levels [109].

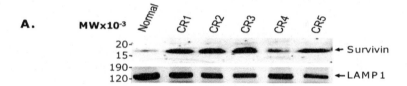

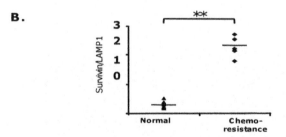

Figure 7.5 (A) Western blot for Survivin expression and (B) Densiometric analysis for Survivin/Lamp1 in normal individuals and chemoresistant prostate cancer patients. Reproduced from Khan et al. [109] with permission from PLOS one under Creative Commons Attribution License.

Patients	Survivin (pg/mL)	PSA (ng/mL)
Normal	<100 (n = 10) (**100%**)	<4 (n = 10) (**100%**)
	>100 (n = 0) (0%)	>4 (n = 0) (0%)
Gleason 6	<100 (n = 0) (0%)	<4 (n = 4) (40%)
	>100 (n = 10) (**100%**)	>4 (n = 6) (**60%**)
Gleason 9	<100 (n = 0) (0%)	<4 (n = 2) (20%)
	>100 (n = 10) (**100%**)	>4 (n = 8) (**80%**)
Recurrences	<100 (n = 0) (0%)	<4 (n = 0) (0%)
	>100 (n = 8) (**100%**)	>4 (n = 10) (**100%**)
BPH	<100 (n = 20) (**100%**)	<4 (n = 20) (**100%**)
	>100 (n = 0) (0%)	>4 (n = 0) (0%)
PCA	<100 (n = 0) (0%)	<4 (n = 3) (16%)
	>100 (n = 19) (**100%**)	>4 (n = 16) (**84%**)

Figure 7.6 Correlation of Survivin and PSA levels in plasma-derived exosomes. Reproduced from Khan et al. [109] with permission from *PLOS One* under Creative Commons Attribution License.

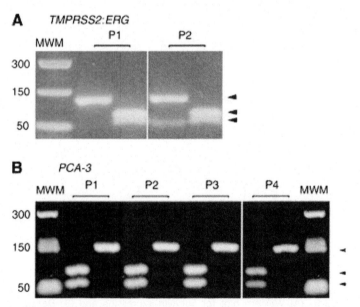

Figure 7.7 Restriction enzyme analysis to detect presence of TMPRSS2:ERG and PCA3 Left lanes show digested samples, Right lanes show undigested samples. PCA3 samples were digested with Sca1 and TMPRSS2:ERG samples were digested with HaeII. Reproduced from Nilsson et al. [24] with permission from Springer Science.

Colorectal cancers are the second leading cause of death in the United States with nearly 15 deaths per 100,000 people each year [110]. This amounts to 10% of all cancer related deaths in the US. Clinical gold standard methods of detection are the fetal occult blood test which is not a very sensitive detection method whereas sigmoidoscopy is a highly invasive method of detection. The two biomarkers currently used for detection of colorectal cancer are CA19-9 (Carbohydrate antigen 19-9) and CEA (Carcinoembroyonic Antigen) [111, 112]. Both these antigens are not sensitive in early stages of cancer which results in patients presenting late stage symptoms by the time a diagnosis has been made. This results in higher fatality rate for a relatively lower number of new cases. Hence, there is a need for efficient detecting methods for different stages of colorectal cancer.

Ogata-Kawata et al. [113] devised a microRNA (miRNA) assay based detection method using exosomes derived from serum of colorectal cancer patients. A microarray screen was used to determine the variations in miRNA expression between healthy individuals and colorectal cancer patients. It was observed that 69 miRNA out of a total of 164 were highly overexpressed in individuals with the disease compared to healthy individuals. Similarly, they also observed that 52 miRNA were overexpressed in colorectal cancer cell lines compared to control normal colon epithelial cells. A comparison of data between patient-derived samples and cell lines revealed 16 miRNAs (miR-7a, miR-1224-5p, miR-1229, miR-1246, miR-1268, miR-1290, miR-1308, miR-150, miR-181b, miR-181d, miR-1915, miR-21, miR-223, miR-23a, miR-483-5p, and miR-638) that were highly expressed in the exosomes (Fig. 7.8). Interestingly, apart from one miRNA, miR-181d, no other miRNA was found to be overexpressed in cells clearly indicating that presence of these miRNA in exosomes are not dependent on cellular expression of the miRNA. The expression levels of the 16 miRNAs did not significantly change between different stages of cancer and downregulation of 8 miRNAs (let-7a, miR-1224-5p, miR-1229, miR-1246, miR-150, miR-21, miR-223, and miR-23a) was observed after tumor resection. These miRNA may potentially be effective in determining the status of the disease. It was also observed that the true positive detection rates of miR-1246 and miR-23A were 95.5% and 92.0%, respectively, and the true positive rates for miR-21 (61.4%), miR-150 (55.7%), let-7a (50.0%) and miR-223 (46.6%) were all higher than the true positive rate of detection in the case of current gold standards CEA (30.7%) and CA19-9 (16.0%). These studies show that exosomes may be a more efficacious and potentially less cumbersome way of detecting biomarkers.

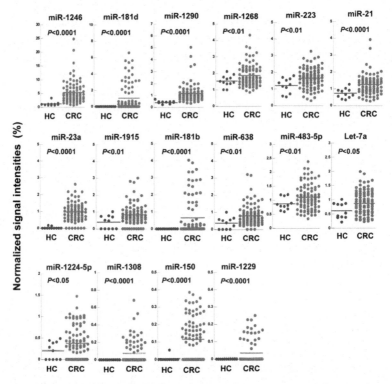

Figure 7.8 miRNA levels in serum exosomes derived from healthy controls (blue) and colorectal cancer patients (red). Reproduced from Ogata-Kawata et al. [113] with permission from *PLOS One* under Creative Commons Attribution License.

7.6 Exosomes in Cancer Immunotherapy

Cancer cells are extremely efficient at evading the human immune system through the use of diverse techniques including reducing/modifying tumor-associated antigens, secretion of immunosuppressive factors, recruitment of T-regulatory cells, etc. [114–116]. Exosomes derived from dendritic/tumor cells have been shown to function similar to their parent cells and are capable of immunomodulatory functions. These findings have led to possibilities of using exosomes for cancer immunotherapies. Andre et al. [117] used exosomes isolated from bone marrow–derived dendritic cells and pulsed them with peptides extracted from P815 mastocytoma cells. After injection of these dendritic exosomes (DEX) they observed close to 60% of the mice to be tumor free at the end of 60 days. Exosomes isolated from HLA-A2⁻ Mart1+ tumor cells were also used to pulse dendritic cells, which helped protect against subsequent tumor challenges.

Thery et al. [118] used dendritic cell-derived exosomes to indirectly activate naïve CD4+ T cells. Exosomes were isolated from dendritic cells of either male or female B6 mice and injected into a specific strain of mice named Marilyn which contain CD45.1+ T-cells, specific to a male H-Y antigen. Male mice-derived exosomes induced activation of CD4+ T-cells but female mice–derived exosomes did not. Concurrently female-derived exosomes pulsed with H-Y antigen were able to induce activation of CD4+ T-cells in female mice injected with CD45.1+ Marilyn T-cells whereas control exosomes did not. To identify if the exosomes directly activate T-cells, Thery et al. cultured Marilyn T-Cells with various doses of the H-Y antigen-presenting exosomes. It was observed that T-cells were not activated in any of the doses. Culturing of Marilyn T-cells were with dendritic cells pulsed with H-Y exosomes resulted in the activation and proliferation of T-cells, indicating the mechanism by which H-Y exosomes were activating T-cells in vivo.

Cho et al. [118] transformed two cell lines, CT26 and TA3HA, to stably express hMUC1⁵. These cell lines were then used to isolate hMUC1- expressing exosomes which were characterized for the presence of the protein by Western blotting. First, these hMUC1-expressing exosomes were used to pulse bone marrow dendritic cells, which were injected into immunized mice. These DCs were able to activate splenocytes to produce Interferon-γ. Seven-week-old mice were injected intradermally with hMUC1 exosomes and were then injected with hMUC1 or non-hMUC1 tumor cells. It was observed these exosomes had an anti-tumorigenic effect on not only hMUC1 expressing tumor cells but also on the non-hMUC1 tumor cells, which indicated that the response produced by these exosomes was not dependent on the MHC type and resulted in a cross protective effect. They also observed that the immunogenic response was not dependent on the dose as they saw a similar response to both 4 and 20 µg of exosomes.

Exosomes derived from tumor cells are capable of sensitizing T-cells resulting in an immune response against cells they were previously tolerant towards. Romagnoli et al. [119], first pulsed dendritic cells derived from monocytes with tumor exosomes isolated from SK-BR-3 cell lines. These dendritic cells were then co-cultured with CD3+ T-cells to prime them for an immune response. After a period of 14 days the CD3+ T-cells were introduced to a culture of SK-BR-3 cells which were previously treated with DC-derived exosomes (control) or tumor-derived exosomes. It was observed that IFN-γ response from CD3+ T-cells was higher in the case of tumor cells primed with either DC-derived or tumor-derived exosomes (Fig. 7.9). It was thus possible to convert immune-tolerant tumor cells into an immunogenic cell population thereby opening up new ways of fighting resistant cancers.

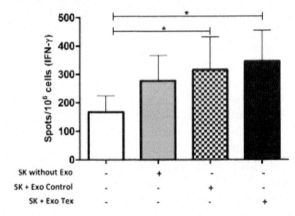

Figure 7.9 Quantification of IFN-γ in tumor-sensitized T-cells in response to exosome-treated tumor cells. Reproduced from Romagnoli et al. [119] with permission from *Frontiers in Immunology* under Creative Commons Attribution License.

Morse et al. [120] used dendritic cells isolated from peripheral blood mononuclear cells (PBMCs) of late stage non-small cell lung cancer patients to derive exosomes expressing MAGE A3 (class 1 or class 2), A4, A10, or control peptides. The patients were injected weekly twice with different treatment conditions consisting of a mixture of exosomes manufactured previously. The patients were separated into three cohorts: A (MAGE A3, A4, A10, CMV and tetanus toxoid), B (MAGE A3, A4, A10, A3 class 2, CMV and tetanus toxoid), and C (MAGE A3 class I and class II, A4 and A10). The immune response triggered by these peptides was studied and was observed to primarily mediate a NK cell response rather than a T-cell response (Fig. 7.10), likely because of T-regulatory cell activity. Survival of the patients from different cohorts was found to be in a range of 52–309 days in the case of cohort A, 280 more than 665 days in the case of cohort B and 244 to 502 days in the case of cohort C. The researchers concluded that an extended stabilization of the disease was observed in patients as compared to regular treatments in which case the disease progressed within 3 months. This small trial also showed the practicality of producing these exosomes from patient PBMCs and the efficacy of such treatments in human patients.

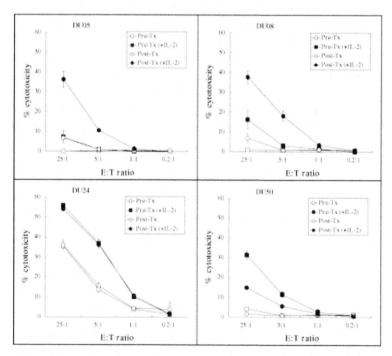

Figure 7.10 Natural killer cell-mediated cytotoxicity. Percentage lysis of tumor cells represented by effector target ratio of 0.2:1 to 25:1. Reproduced from Morse et al. [120] with permission from *Frontiers in Immunology* under Creative Commons Attribution License.

Chen et al. [121] isolated exosomes from heat-shocked (HS) A20 lymphoma cells and used the exosomes to mature bone marrow–derived dendritic cells. They observed a stronger immune response induced by DC cells previously pulsed with the HS-exosomes in comparison to those pulsed with control exosomes derived from untreated A20 cells resulting in potent anti-tumor activity. Mice treated with HS exosomes were tumor free after 90 days post injection of tumor cells subcutaneously. In comparison, 50% of the mice treated with regular exosomes presented tumors, 90 days post injection. Similarly, these exosomes also presented excellent anti-tumor activity. Mice previously injected with tumor cells and presenting tumors were treated with either saline, regular exosomes or HS exosomes. It was observed that 40% of the mice treated with HS exosomes survived until a period of 90 days or longer. It was found that both CD4$^+$ and CD8$^+$ cells were important for this anti-tumor effect as abrogation of any of these cells resulted in a drastically reduced tumor cytotoxicity.

In many gastrointestinal cancers there is an accumulation of fluid containing cancer cells and are known as malignant ascites. Most patients who present the disease with malignant ascites have a poor disease prognosis. Dai et al. [122] used exosomes derived from the ascites in order to induce an immune response against colorectal cancers. They observed that these exosomes consisted of multiple immunomodulatory biomarkers including carcinoembroynic antigen a.k.a. CEA and also other proteins including heat shock proteins (HSPs). The malignant ascites were removed from 40 patients suffering from colorectal cancers (CRCs) and then used to isolate these ascite-derived exosomes (AEX) via sucrose gradient centrifugation. The patients were injected with these AEX alone or AEX along with GM-CSF weekly once for 4 weeks. Most patients injected with these exosomes presented very mild side effects. It was observed that AEX alone produced a mild antigen specific anti-tumor response but AEX at lower dosages in combination with GM-CSF produced a much better response leading to tumor regression. The above findings suggest that exosomes are capable of producing a strong immune response against tumor cells in multiple cancer types.

7.7 Summary

Exosomes are nanoscale vesicles containing cell-specific cargo including DNA, RNA and/or proteins essential for cellular cross-communication in the body. Their ubiquitous existence, cargo diversity, target honing and target transformation capabilities make them an indispensable accomplice in diseases such as cancer where transformation of the surrounding cells and microenvironment is essential for disease progression. Advances in efficient isolation, processing and characterization of these nanovesicles have significantly increased our understanding about their cargo contents and the role they play in cancer progression, microenvironment transformation and organotypic metastases. Further, these exosomal contents provide useful information as disease biomarkers and initiators of anti-tumor immunity. As we seek to develop novel therapies against cancer, the role of exosomes in drug resistance and therapeutic antibody neutralization cannot be neglected. The next decade will seek to develop a deeper mechanistic understanding about exosomal content, their role in cancer progression, and their potential use in cancer diagnosis and immunotherapy.

References

1. Théry C, Zitvogel L, Amigorena S. Exosomes: Composition, biogenesis and function. *Nature Reviews Immunology*. 2002; 2(8): 569–579.

2. Valadi H, Ekström K, Bossios A, Sjöstrand M, Lee JJ, Lötvall JO. Exosome-mediated transfer of mRNAs and microRNAs is a novel mechanism of genetic exchange between cells. *Nature Cell Biology*. 2007; 9(6): 654.

3. Waldenström A, Gennebäck N, Hellman U, Ronquist G. Cardiomyocyte microvesicles contain DNA/RNA and convey biological messages to target cells. *PloS one*. 2012; 7(4): e34653.

4. Mathivanan S, Fahner CJ, Reid GE, Simpson RJ. ExoCarta 2012: Database of exosomal proteins, RNA and lipids. *Nucleic Acids Research*. 2011; 40(D1): D1241-D4.

5. Lässer C, Alikhani VS, Ekström K, Eldh M, Paredes PT, Bossios A, et al. Human saliva, plasma and breast milk exosomes contain RNA: uptake by macrophages. *Journal of Translational Medicine*. 2011; 9(1): 9.

6. Wahlgren J, Karlson TDL, Brisslert M, Vaziri Sani F, Telemo E, Sunnerhagen P, et al. Plasma exosomes can deliver exogenous short

interfering RNA to monocytes and lymphocytes. *Nucleic Acids Research*. 2012; 40(17): e130-e.

7. Hoshino A, Costa-Silva B, Shen T-L, Rodrigues G, Hashimoto A, Mark MT, et al. Tumour exosome integrins determine organotropic metastasis. *Nature*. 2015; 527(7578): 329–335.

8. Ludwig A-K, Giebel B. Exosomes: Small vesicles participating in intercellular communication. *The International Journal of Biochemistry & Cell Biology*. 2012; 44(1): 11–15.

9. Mathivanan S, Ji H, Simpson RJ. Exosomes: Extracellular organelles important in intercellular communication. *Journal of Proteomics*. 2010; 73(10): 1907–1920.

10. Simons M, Raposo G. Exosomes–vesicular carriers for intercellular communication. *Current Opinion in Cell Biology*. 2009; 21(4): 575–581.

11. Lee TH, D'Asti E, Magnus N, Al-Nedawi K, Meehan B, Rak J, editors Microvesicles as mediators of intercellular communication in cancer—the emerging science of cellular 'debris'. In *Seminars in immunopathology*; 2011: Springer.

12. Zhang J, Li S, Li L, Li M, Guo C, Yao J, et al. Exosome and exosomal microRNA: Trafficking, sorting, and function. *Genomics, Proteomics & Bioinformatics*. 2015; 13(1): 17–24.

13. Melo SA, Sugimoto H, O'Connell JT, Kato N, Villanueva A, Vidal A, et al. Cancer exosomes perform cell-independent microRNA biogenesis and promote tumorigenesis. *Cancer Cell*. 2014; 26(5): 707–721.

14. King HW, Michael MZ, Gleadle JM. Hypoxic enhancement of exosome release by breast cancer cells. *BMC Cancer*. 2012; 12(1): 421.

15. Costa-Silva B, Aiello NM, Ocean AJ, Singh S, Zhang H, Thakur BK, et al. Pancreatic cancer exosomes initiate pre-metastatic niche formation in the liver. *Nature Cell Biology*. 2015; 17(6): 816.

16. Peinado H, Alečković M, Lavotshkin S, Matei I, Costa-Silva B, Moreno-Bueno G, et al. Melanoma exosomes educate bone marrow progenitor cells toward a pro-metastatic phenotype through MET. *Nature Medicine*. 2012; 18(6): 883–891.

17. Liu Y, Gu Y, Han Y, Zhang Q, Jiang Z, Zhang X, et al. Tumor exosomal RNAs promote lung pre-metastatic niche formation by activating alveolar epithelial TLR3 to recruit neutrophils. *Cancer Cell*. 2016; 30(2): 243–256.

18. Kahlert C, Kalluri R. Exosomes in tumor microenvironment influence cancer progression and metastasis. *Journal of Molecular Medicine*. 2013; 91(4): 431–437.

19. Grandhi TSP, Potta T, Nitiyanandan R, Deshpande I, Rege K. Chemomechanically engineered 3D organotypic platforms of bladder cancer dormancy and reactivation. *Biomaterials.* 2017; 142: 171–185.

20. Peinado H, Zhang H, Matei IR, Costa-Silva B, Hoshino A, Rodrigues G, et al. Pre-metastatic niches: Organ-specific homes for metastases. *Nature Reviews Cancer.* 2017; 17(5): 302–317.

21. Hanahan D, Coussens LM. Accessories to the crime: Functions of cells recruited to the tumor microenvironment. *Cancer Cell.* 2012; 21(3): 309–322.

22. Taylor DD, Gercel-Taylor C. MicroRNA signatures of tumor-derived exosomes as diagnostic biomarkers of ovarian cancer. *Gynecologic Oncology.* 2008; 110(1): 13–21.

23. Kosaka N, Iguchi H, Ochiya T. Circulating microRNA in body fluid: A new potential biomarker for cancer diagnosis and prognosis. *Cancer Science.* 2010; 101(10): 2087–2092.

24. Nilsson J, Skog J, Nordstrand A, Baranov V, Mincheva-Nilsson L, Breakefield X, et al. Prostate cancer-derived urine exosomes: A novel approach to biomarkers for prostate cancer. *British Journal of Cancer.* 2009; 100(10): 1603.

25. Thakur BK, Zhang H, Becker A, Matei I, Huang Y, Costa-Silva B, et al. Double-stranded DNA in exosomes: A novel biomarker in cancer detection. *Cell Research.* 2014; 24(6): 766.

26. Li Y, Zheng Q, Bao C, Li S, Guo W, Zhao J, et al. Circular RNA is enriched and stable in exosomes: A promising biomarker for cancer diagnosis. *Cell Research.* 2015; 25(8): 981.

27. Zhang X, Yuan X, Shi H, Wu L, Qian H, Xu W. Exosomes in cancer: Small particle, big player. *Journal of Hematology & Oncology.* 2015; 8(1): 83.

28. Inamdar S, Nitiyanandan R, Rege K. Emerging applications of exosomes in cancer therapeutics and diagnostics. *Bioengineering & Translational Medicine.* 2017; 2(1): 70–80.

29. Février B, Raposo G. Exosomes: Endosomal-derived vesicles shipping extracellular messages. *Current Opinion in Cell Biology.* 2004; 16(4): 415–421.

30. Hanson PI, Shim S, Merrill SA. Cell biology of the ESCRT machinery. *Current Opinion in Cell Biology.* 2009; 21(4): 568–574.

31. Hurley JH. The ESCRT complexes. *Critical Reviews in Biochemistry and Molecular Biology.* 2010; 45(6): 463–487.

32. Hurley JH, Hanson PI. Membrane budding and scission by the ESCRT machinery: It's all in the neck. *Nature Reviews Molecular Cell Biology.* 2010; 11(8): 556–566.

33. Kleijmeer MJ, Stoorvogel W, Griffith JM, Yoshie O, Geuze HJ. Selective enrichment of tetraspan proteins on the internal vesicles of multivesicular endosomes and on exosomes secreted by human B-lymphocytes. *Journal of Biological Chemistry.* 1998; 273(32): 20121–20127.

34. Kajimoto T, Okada T, Miya S, Zhang L, Nakamura S-I. Ongoing activation of sphingosine 1-phosphate receptors mediates maturation of exosomal multivesicular endosomes. *Nature Communications.* 2013; 4: 2712.

35. Baietti MF, Zhang Z, Mortier E, Melchior A, Degeest G, Geeraerts A, et al. Syndecan-syntenin-ALIX regulates the biogenesis of exosomes. *Nature Cell Biology.* 2012; 14(7): 677–685.

36. Iavello A, Frech VS, Gai C, Deregibus MC, Quesenberry PJ, Camussi G. Role of Alix in miRNA packaging during extracellular vesicle biogenesis. *International Journal of Molecular Medicine.* 2016; 37(4): 958–966.

37. Santangelo L, Giurato G, Cicchini C, Montaldo C, Mancone C, Tarallo R, et al. The RNA-binding protein SYNCRIP is a component of the hepatocyte exosomal machinery controlling microRNA sorting. *Cell Reports.* 2016; 17(3): 799–808.

38. Perez-Hernandez D, Gutiérrez-Vázquez C, Jorge I, López-Martín S, Ursa A, Sánchez-Madrid F, et al. The intracellular interactome of tetraspanin-enriched microdomains reveals their function as sorting machineries toward exosomes. *Journal of Biological Chemistry.* 2013; 288(17): 11649–11661.

39. Trajkovic K, Hsu C, Chiantia S, Rajendran L, Wenzel D, Wieland F, et al. Ceramide triggers budding of exosome vesicles into multivesicular endosomes. *Science.* 2008; 319(5867): 1244–1247.

40. Valadi H, Ekström K, Bossios A, Sjöstrand M, Lee JJ, Lötvall JO. Exosome-mediated transfer of mRNAs and microRNAs is a novel mechanism of genetic exchange between cells. *Nature Cell Biology.* 2007; 9(6): 654–659.

41. Villarroya-Beltri C, Gutiérrez-Vázquez C, Sánchez-Cabo F, Pérez-Hernández D, Vázquez J, Martin-Cofreces N, et al. Sumoylated hnRNPA2B1 controls the sorting of miRNAs into exosomes through binding to specific motifs. *Nature Communications.* 2013; 4: 2980.

42. Hoek KS, Kidd GJ, Carson JH, Smith R. hnRNP A2 selectively binds the cytoplasmic transport sequence of myelin basic protein mRNA. *Biochemistry*. 1998; 37(19): 7021–7029.

43. Mickleburgh I, Burtle B, Hollås H, Campbell G, Chrzanowska-Lightowlers Z, Vedeler A, et al. Annexin A2 binds to the localization signal in the 3⬛ untranslated region of c-myc mRNA. *The FEBS Journal*. 2005; 272(2): 413–421.

44. Mathivanan S, Simpson RJ. ExoCarta: A compendium of exosomal proteins and RNA. *Proteomics*. 2009; 9(21): 4997–5000.

45. Stenmark H. Rab GTPases as coordinators of vesicle traffic. *Nature Reviews Molecular Cell Biology*. 2009; 10(8): 513–525.

46. Jahn R, Scheller RH. SNAREs—engines for membrane fusion. *Nature Reviews Molecular Cell Biology*. 2006; 7(9): 631–643.

47. Urbanelli L, Magini A, Buratta S, Brozzi A, Sagini K, Polchi A, et al. Signaling pathways in exosomes biogenesis, secretion and fate. *Genes*. 2013; 4(2): 152–170.

48. Ostrowski M, Carmo NB, Krumeich S, Fanget I, Raposo G, Savina A, et al. Rab27a and Rab27b control different steps of the exosome secretion pathway. *Nature Cell Biology*. 2010; 12(1): 19–30.

49. Kasai K, Ohara-Imaizumi M, Takahashi N, Mizutani S, Zhao S, Kikuta T, et al. Rab27a mediates the tight docking of insulin granules onto the plasma membrane during glucose stimulation. *Journal of Clinical Investigation*. 2005; 115(2): 388.

50. Vanlandingham PA, Ceresa BP. Rab7 regulates late endocytic trafficking downstream of multivesicular body biogenesis and cargo sequestration. *Journal of Biological Chemistry*. 2009; 284(18): 12110–12124.

51. Kalluri R, Zeisberg M. Fibroblasts in cancer. *Nature Reviews Cancer*. 2006; 6(5): 392–401.

52. Orimo A, Weinberg RA. Stromal fibroblasts in cancer: A novel tumor-promoting cell type. *Cell Cycle*. 2006; 5(15): 1597–1601.

53. Bhowmick NA, Neilson EG, Moses HL. Stromal fibroblasts in cancer initiation and progression. *Nature*. 2004; 432(7015): 332–337.

54. Pietras K, Östman A. Hallmarks of cancer: Interactions with the tumor stroma. *Experimental Cell Research*. 2010; 316(8): 1324–1331.

55. Resetkova E, Reis-Filho JS, Jain RK, Mehta R, Thorat MA, Nakshatri H, et al. Prognostic impact of ALDH1 in breast cancer: A story of stem cells and tumor microenvironment. *Breast Cancer Research and Treatment*. 2010; 123(1): 97–108.

56. Calabrese C, Poppleton H, Kocak M, Hogg TL, Fuller C, Hamner B, et al. A perivascular niche for brain tumor stem cells. *Cancer Cell.* 2007; 11(1): 69–82.

57. Condeelis J, Pollard JW. Macrophages: Obligate partners for tumor cell migration, invasion, and metastasis. *Cell.* 2006; 124(2): 263–266.

58. Hao N-B, Lü M-H, Fan Y-H, Cao Y-L, Zhang Z-R, Yang S-M. Macrophages in tumor microenvironments and the progression of tumors. *Clinical and Developmental Immunology.* 2012; 2012.

59. Cho JA, Park H, Lim EH, Lee KW. Exosomes from breast cancer cells can convert adipose tissue-derived mesenchymal stem cells into myofibroblast-like cells. *International Journal of Oncology.* 2012; 40(1): 130–138.

60. Chowdhury R, Webber JP, Gurney M, Mason MD, Tabi Z, Clayton A. Cancer exosomes trigger mesenchymal stem cell differentiation into pro-angiogenic and pro-invasive myofibroblasts. *Oncotarget.* 2015; 6(2): 715.

61. Paggetti J, Haderk F, Seiffert M, Janji B, Distler U, Ammerlaan W, et al. Exosomes released by chronic lymphocytic leukemia cells induce the transition of stromal cells into cancer-associated fibroblasts. *Blood.* 2015; 126(9): 1106–1117.

62. Webber J, Steadman R, Mason MD, Tabi Z, Clayton A. Cancer exosomes trigger fibroblast to myofibroblast differentiation. *Cancer Research.* 2010; 70(23): 9621–9630.

63. Zhang L, Wu X, Luo C, Chen X, Yang L, Tao J, et al. The 786-0 renal cancer cell-derived exosomes promote angiogenesis by downregulating the expression of hepatocyte cell adhesion molecule. *Molecular Medicine Reports.* 2013; 8(1): 272–276.

64. Schillaci O, Fontana S, Monteleone F, Taverna S, Di Bella MA, Di Vizio D, et al. Exosomes from metastatic cancer cells transfer amoeboid phenotype to non-metastatic cells and increase endothelial permeability: Their emerging role in tumor heterogeneity. *Scientific Reports.* 2017; 7(1): 4711.

65. Sun X, Ma X, Wang J, Zhao Y, Wang Y, Bihl JC, et al. Glioma stem cells-derived exosomes promote the angiogenic ability of endothelial cells through miR-21/VEGF signal. *Oncotarget.* 2017; 8(22): 36137.

66. Salimu J, Webber J, Gurney M, Al-Taei S, Clayton A, Tabi Z. Dominant immunosuppression of dendritic cell function by prostate-cancer-derived exosomes. *Journal of Extracellular Vesicles.* 2017; 6(1): 1368823.

67. Xiang X, Poliakov A, Liu C, Liu Y, Deng Zb, Wang J, et al. Induction of myeloid-derived suppressor cells by tumor exosomes. *International Journal of Cancer.* 2009; 124(11): 2621–2633.

68. Schlecker E, Stojanovic A, Eisen C, Quack C, Falk CS, Umansky V, et al. Tumor-infiltrating monocytic myeloid-derived suppressor cells mediate CCR5-dependent recruitment of regulatory T cells favoring tumor growth. *The Journal of Immunology.* 2012; 189(12): 5602–5611.

69. Wieckowski EU, Visus C, Szajnik M, Szczepanski MJ, Storkus WJ, Whiteside TL. Tumor-derived microvesicles promote regulatory T cell expansion and induce apoptosis in tumor-reactive activated CD8+ T lymphocytes. *The Journal of Immunology.* 2009; 183(6): 3720–3730.

70. Chow A, Zhou W, Liu L, Fong MY, Champer J, Van Haute D, et al. Macrophage immunomodulation by breast cancer-derived exosomes requires Toll-like receptor 2-mediated activation of NF-κB. *Scientific Reports.* 2014; 4: 5750.

71. Tao S-C, Guo S-C, Zhang C-Q. Platelet-derived extracellular vesicles: An emerging therapeutic approach. *International Journal of Biological Sciences.* 2017; 13(7): 828.

72. Hood JL, San RS, Wickline SA. Exosomes released by melanoma cells prepare sentinel lymph nodes for tumor metastasis. *Cancer Research.* 2011; 71(11): 3792–3801.

73. Zhang L, Zhang S, Yao J, Lowery FJ, Zhang Q, Huang W-C, et al. Microenvironment-induced PTEN loss by exosomal microRNA primes brain metastasis outgrowth. *Nature.* 2015; 527(7576): 100.

74. Grange C, Tapparo M, Collino F, Vitillo L, Damasco C, Deregibus MC, et al. Microvesicles released from human renal cancer stem cells stimulate angiogenesis and formation of lung premetastatic niche. *Cancer Research.* 2011; 71(15): 5346–5356.

75. Richards KE, Zeleniak AE, Fishel ML, Wu J, Littlepage LE, Hill R. Cancer-associated fibroblast exosomes regulate survival and proliferation of pancreatic cancer cells. *Oncogene.* 2017; 36(13): 1770.

76. Zhao H, Yang L, Baddour J, Achreja A, Bernard V, Moss T, et al. Tumor microenvironment derived exosomes pleiotropically modulate cancer cell metabolism. *Elife.* 2016; 5: e10250.

77. Nouraee N, Khazaei S, Vasei M, Razavipour SF, Sadeghizadeh M, Mowla SJ. MicroRNAs contribution in tumor microenvironment of esophageal cancer. *Cancer Biomarkers.* 2016; 16(3): 367–376.

78. Solaini G, Sgarbi G, Baracca A. Oxidative phosphorylation in cancer cells. *Biochimica et Biophysica Acta (BBA)-Bioenergetics*. 2011; 1807(6): 534–542.

79. Takasugi M, Okada R, Takahashi A, Chen DV, Watanabe S, Hara E. Small extracellular vesicles secreted from senescent cells promote cancer cell proliferation through EphA2. *Nature Communications*. 2017; 8: 15729.

80. Goubran H, Sabry W, Kotb R, Seghatchian J, Burnouf T. Platelet microparticles and cancer: An intimate cross-talk. *Transfusion and Apheresis Science*. 2015; 53(2): 168–172.

81. Janowska-Wieczorek A, Wysoczynski M, Kijowski J, Marquez-Curtis L, Machalinski B, Ratajczak J, et al. Microvesicles derived from activated platelets induce metastasis and angiogenesis in lung cancer. *International Journal of Cancer*. 2005; 113(5): 752–760.

82. Herrera-Abreu MT, Palafox M, Asghar U, Rivas MA, Cutts RJ, Garcia-Murillas I, et al. Early adaptation and acquired resistance to CDK4/6 inhibition in estrogen receptor–positive breast cancer. *Cancer Research*. 2016; 76(8): 2301–2313.

83. Knudsen ES, Witkiewicz AK. The Strange Case of CDK4/6 Inhibitors: Mechanisms, Resistance, and Combination Strategies. *Trends in Cancer*. 2017; 3(1): 39–55.

84. Zheng P, Chen L, Yuan X, Luo Q, Liu Y, Xie G, et al. Exosomal transfer of tumor-associated macrophage-derived miR-21 confers cisplatin resistance in gastric cancer cells. *Journal of Experimental & Clinical Cancer Research*. 2017; 36(1): 53.

85. Chen W-X, Cai Y-Q, Lv M-M, Chen L, Zhong S-L, Ma T-F, et al. Exosomes from docetaxel-resistant breast cancer cells alter chemosensitivity by delivering microRNAs. *Tumor Biology*. 2014; 35(10): 9649–9659.

86. Pink RC, Samuel P, Massa D, Caley DP, Brooks SA, Carter DRF. The passenger strand, miR-21-3p, plays a role in mediating cisplatin resistance in ovarian cancer cells. *Gynecologic Oncology*. 2015; 137(1): 143–151.

87. Martinez VG, O'Neill S, Salimu J, Breslin S, Clayton A, Crown J, et al. Resistance to HER2-targeted anti-cancer drugs is associated with immune evasion in cancer cells and their derived extracellular vesicles. *Oncoimmunology*. 2017; 6(12): e1362530.

88. Safaei R, Larson BJ, Cheng TC, Gibson MA, Otani S, Naerdemann W, et al. Abnormal lysosomal trafficking and enhanced exosomal export of cisplatin in drug-resistant human ovarian carcinoma cells. *Molecular Cancer Therapeutics*. 2005; 4(10): 1595–1604.

89. Safaei R, Katano K, Larson BJ, Samimi G, Holzer AK, Naerdemann W, et al. Intracellular localization and trafficking of fluorescein-labeled cisplatin in human ovarian carcinoma cells. *Clinical Cancer Research.* 2005; 11(2): 756–767.

90. Shedden K, Xie XT, Chandaroy P, Chang YT, Rosania GR. Expulsion of small molecules in vesicles shed by cancer cells: Association with gene expression and chemosensitivity profiles. *Cancer Research.* 2003; 63(15): 4331–4337.

91. Corcoran C, Rani S, O'Brien K, O'Neill A, Prencipe M, Sheikh R, et al. Docetaxel-resistance in prostate cancer: Evaluating associated phenotypic changes and potential for resistance transfer via exosomes. *PloS One.* 2012; 7(12): e50999.

92. Lv M-M, Zhu X-Y, Chen W-X, Zhong S-L, Hu Q, Ma T-F, et al. Exosomes mediate drug resistance transfer in MCF-7 breast cancer cells and a probable mechanism is delivery of P-glycoprotein. *Tumor Biology.* 2014; 35(11): 10773–10779.

93. Ciravolo V, Huber V, Ghedini GC, Venturelli E, Bianchi F, Campiglio M, et al. Potential role of HER2-overexpressing exosomes in countering trastuzumab-based therapy. *Journal of Cellular Physiology.* 2012; 227(2): 658–667.

94. Runz S, Keller S, Rupp C, Stoeck A, Issa Y, Koensgen D, et al. Malignant ascites-derived exosomes of ovarian carcinoma patients contain CD24 and EpCAM. *Gynecologic Oncology.* 2007; 107(3): 563–571.

95. Choi D-S, Lee J-M, Park GW, Lim H-W, Bang JY, Kim Y-K, et al. Proteomic analysis of microvesicles derived from human colorectal cancer cells. *Journal of Proteome Research.* 2007; 6(12): 4646–4655.

96. Mathivanan S, Lim JW, Tauro BJ, Ji H, Moritz RL, Simpson RJ. Proteomics analysis of A33 immunoaffinity-purified exosomes released from the human colon tumor cell line LIM1215 reveals a tissue-specific protein signature. *Molecular & Cellular Proteomics.* 2010; 9(2): 197–208.

97. Tanaka Y, Kamohara H, Kinoshita K, Kurashige J, Ishimoto T, Iwatsuki M, et al. Clinical impact of serum exosomal microRNA-21 as a clinical biomarker in human esophageal squamous cell carcinoma. *Cancer.* 2013; 119(6): 1159–1167.

98. Melo SA, Luecke LB, Kahlert C, Fernandez AF, Gammon ST, Kaye J, et al. Glypican-1 identifies cancer exosomes and detects early pancreatic cancer. *Nature.* 2015; 523(7559): 177.

99. Kleeff J, Ishiwata T, Kumbasar A, Friess H, Büchler MW, Lander AD, et al. The cell-surface heparan sulfate proteoglycan glypican-1 regulates growth factor action in pancreatic carcinoma cells and is overexpressed

in human pancreatic cancer. *The Journal of Clinical Investigation*. 1998; 102(9): 1662–1673.

100. Matsuda K, Maruyama H, Guo F, Kleeff J, Itakura J, Matsumoto Y, et al. Glypican-1 is overexpressed in human breast cancer and modulates the mitogenic effects of multiple heparin-binding growth factors in breast cancer cells. *Cancer Research*. 2001; 61(14): 5562–5569.

101. Ballehaninna UK, Chamberlain RS. Serum CA 19-9 as a biomarker for pancreatic cancer—a comprehensive review. *Indian Journal of Surgical Oncology*. 2011; 2(2): 88–100.

102. Pavai S, Yap S. The clinical significance of elevated levels of serum CA 19-9. *Medical Journal of Malaysia*. 2003; 58(5): 667–672.

103. Bjartell A, Montironi R, Berney DM, Egevad L. Tumour markers in prostate cancer II: Diagnostic and prognostic cellular biomarkers. *Acta Oncologica*. 2011; 50(sup1): 76–84.

104. Chang R, Kirby R, Challacombe B. Is there a link between BPH and prostate cancer? *The Practitioner*. 2012; 256(1750): 13–17.

105. Freedland SJ, Humphreys EB, Mangold LA, Eisenberger M, Dorey FJ, Walsh PC, et al. Risk of prostate cancer–specific mortality following biochemical recurrence after radical prostatectomy. *Jama*. 2005; 294(4): 433–439.

106. Kishi H, Igawa M, Kikuno N, Yoshino T, Urakami S, Shiina H. Expression of the survivin gene in prostate cancer: Correlation with clinicopathological characteristics, proliferative activity and apoptosis. *The Journal of Urology*. 2004; 171(5): 1855–1860.

107. Shariat SF, Lotan Y, Saboorian H, Khoddami SM, Roehrborn CG, Slawin KM, et al. Survivin expression is associated with features of biologically aggressive prostate carcinoma. *Cancer*. 2004; 100(4): 751–757.

108. Shen J, Liu J, Long Y, Miao Y, Su M, Zhang Q, et al. Knockdown of survivin expression by siRNAs enhances chemosensitivity of prostate cancer cells and attenuates its tumorigenicity. *Acta Biochimica et Biophysica Sinica*. 2009; 41(3): 223–230.

109. Khan S, Jutzy JM, Valenzuela MMA, Turay D, Aspe JR, Ashok A, et al. Plasma-derived exosomal survivin, a plausible biomarker for early detection of prostate cancer. *PLoS One*. 2012; 7(10): e46737.

110. Ries LA, Harkins D, Krapcho M, Mariotto A, Miller BA, Feuer EJ, et al. *SEER Cancer Statistics Review*. 2006; 1975–2003.

111. Stiksma J, Grootendorst DC, van der Linden PWG. CA 19-9 as a marker in addition to CEA to monitor colorectal cancer. *Clinical Colorectal Cancer*. 2014; 13(4): 239–244.

112. Vukobrat-Bijedic Z, Husic-Selimovic A, Sofic A, Bijedic N, Bjelogrlic I, Gogov B, et al. Cancer antigens (CEA and CA 19-9) as markers of advanced stage of colorectal carcinoma. *Medical Archives.* 2013; 67(6): 397.

113. Ogata-Kawata H, Izumiya M, Kurioka D, Honma Y, Yamada Y, Furuta K, et al. Circulating exosomal microRNAs as biomarkers of colon cancer. *PloS One.* 2014; 9(4): e92921.

114. Beatty GL, Gladney WL. Immune escape mechanisms as a guide for cancer immunotherapy. *Clinical Cancer Research.* 2015; 21(4): 687–692.

115. Messerschmidt JL, Prendergast GC, Messerschmidt GL. How cancers escape immune destruction and mechanisms of action for the new significantly active immune therapies: Helping nonimmunologists decipher recent advances. *The Oncologist.* 2016; 21(2): 233–243.

116. Vinay DS, Ryan EP, Pawelec G, Talib WH, Stagg J, Elkord E, et al. Immune evasion in cancer: Mechanistic basis and therapeutic strategies. *Seminars in Cancer Biology;* 2015; 35 Suppl: S185–S198..

117. Andre F, Andersen M, Wolfers J, Lozier A, Raposo G, Serra V, et al. Exosomes in cancer immunotherapy: Preclinical data. In *Progress in Basic and Clinical Immunology*: Springer; 2001. pp. 349–354.

118. Cho JA, Yeo Dj, Son HY, Kim HW, Jung DS, Ko JK, et al. Exosomes: A new delivery system for tumor antigens in cancer immunotherapy. *International Journal of Cancer.* 2005; 114(4): 613–622.

119. Romagnoli GG, Zelante BB, Toniolo PA, Migliori IK, Barbuto JAM. Dendritic cell-derived exosomes may be a tool for cancer immunotherapy by converting tumor cells into immunogenic targets. *Frontiers in Immunology.* 2015; 5: 692.

120. Morse MA, Garst J, Osada T, Khan S, Hobeika A, Clay TM, et al. A phase I study of dexosome immunotherapy in patients with advanced non-small cell lung cancer. *Journal of Translational Medicine.* 2005; 3(1): 9.

121. Chen W, Wang J, Shao C, Liu S, Yu Y, Wang Q, et al. Efficient induction of antitumor T cell immunity by exosomes derived from heat-shocked lymphoma cells. *European Journal of Immunology.* 2006; 36(6): 1598–1607.

122. Dai S, Wei D, Wu Z, Zhou X, Wei X, Huang H, et al. Phase I clinical trial of autologous ascites-derived exosomes combined with GM-CSF for colorectal cancer. *Molecular Therapy.* 2008; 16(4): 782–790.

123. McGough IJ, Vincent J-P. Exosomes in developmental signalling. *Development.* 2016; 143(14): 2482–2493.

Chapter 8

Porous Inorganic Nanomaterials for Drug Delivery

Elshaimaa Sayed,[a,b] **Yasmine Alyassin,**[a] **Aliyah Zaman,**[a]
Ketan Ruparelia,[a] **Neenu Singh,**[a] **Ming-Wei Chang,**[c,d] **and**
Zeeshan Ahmad[a]

[a]*Leicester School of Pharmacy, De Montfort University, Leicester LE1 9BH, UK*
[b]*Department of Pharmaceutics, Faculty of Pharmacy, Minia University,
Minia, Egypt*
[c]*College of Biomedical Engineering and Instrument Science,
Zhejiang University, Hangzhou 310027, China*
[d]*Zhejiang Provincial Key Laboratory of Cardio-Cerebral Vascular
Detection Technology and Medicinal Effectiveness Appraisal,
Zhejiang University, Hangzhou 310027, China*
p15249031@my365.dmu.ac.uk, zahmad@dmu.ac.uk

8.1 Introduction

Drug delivery is the method of transporting pharmaceutical drug
(in various dosage forms) to a target site (cells, tissues or organs)
to achieve a desired therapeutic action [1, 2]. One of the major
problems that confronts drug delivery scientists is how to provide

Handbook of Materials for Nanomedicine: Lipid-Based and Inorganic Nanomaterials
Edited by Vladimir Torchilin
Copyright © 2020 Jenny Stanford Publishing Pte. Ltd.
ISBN 978-981-4800-91-4 (Hardcover), 978-1-003-04507-6 (eBook)
www.jennystanford.com

the required level of therapeutic agent to a desired site of action at an ideal time. Over the last few decades, drug delivery strategies have been advanced and progressed to improve therapeutic efficacy, pharmacokinetic properties (including drug bioavailability), patient compliance and safety [3, 4]. Solubility and permeability are key factors in determining efficacy of different formulations, such as oral and parenteral dosage forms. Both factors are considered major requirements to attain the required concentration of pharmaceutical drug in blood, in order to achieve desired therapeutic response [5]. However, poor aqueous solubility and/or permeability of many pharmaceutical drugs pose major obstacles for achieving the desired clinical outcome [6, 7]. Therefore, innovative strategies have been proposed and utilized to overcome such problems. Nanosizing is one of the most employed strategies to improve drug dissolution rate, and hence, to enhance pharmacokinetic aspects (absorption and distribution) [2, 8, 9]. Consequently, various unorthodox nanotechnological tools have been developed to achieve intracellular drug targeting, reduction of undesired adverse effects and providing controlled release behaviour [10]. These include nanoparticles [11, 12], microspheres [13, 14], micelles [15] dendrimers [16, 17], liposomes [18, 19], nanoemulsions [20, 21], nanofibers [22, 23], niosomes [24, 25], proniosomes [26, 27] and loading drugs into nanosized pores of porous materials [10].

This chapter reviews specific advances in inorganic porous drug delivery systems. A brief overview detailing different types of inorganic porous materials is provided before several features of inorganic mesoporous materials (including history, structure, synthesis, biodegradation and toxicity) are covered. Recent advances in the applications of mesoporous silica-based nanomaterials as a delivery system for different therapeutic actives including (anticancer, anti-inflammatory, antimicrobial and genes) are also highlighted with selected findings illustrated in a detailed table.

8.2 Inorganic Porous Materials

Amongst different drug delivery systems, porous materials are emerging as state-of-the-art carriers with ability to entrap various types of actives within their porous structure. These unique matrices have been extensively exploited in the field of drug delivery showing a great potential in several medical and pharmaceutical applications such as cancer therapy, gene delivery and tissue engineering.

Porous materials have been utilized widely as controlled drug delivery systems owing to their optimistic features including stable structure, enhanced surface area attributes and uniform tunable pores [28–30]. Moreover, these porous carriers are characterized by good chemical and mechanical stability in different physiological conditions [30]. These promising carriers can be used successfully to deliver diverse types of drugs, proteins and genes. Porous and hydrophilic properties of these materials are beneficial in controlling the diffusion rate of encapsulated active from their nanopores. Hence, they are able to deliver targeted, sustained or controlled release features in different applications of drug delivery [30].

Porous materials are categorized according to their pore size: Porous materials with pore sizes <2 nm are called microporous (such as zeolites molecular sieves), mesoporous materials are those with pores in the range 2 to 50 nm (such as mesoporous silica (MS) and mesoporous metal oxides), and those with pore sizes >50 nm are microporous materials (such as macroporous aluminophosphate) [30, 31].

8.3 Inorganic Mesoporous Drug Carriers

The International Union of Pure and Applied Chemistry (IUPAC) has used the term meso (in between) to characterize materials which have pores between the macropore and the micropore range (2–50 nm). Additionally, based on the nature of their porous network, IUPAC identifies mesoporous materials into ordered or disordered materials [32]. Mesoporous materials are emerging as attractive platforms for designing diverse types of drug delivery systems. For years, great efforts have been witnessed to develop mesoporous materials with varying chemical structures, porous architectures and functionalities [33, 34]. Various types of mesoporous materials have been employed as drug delivery matrices such as mesoporous silica [35–41], mesoporous carbon [42, 43], alumina [44, 45], zirconia [46], titanium oxide [47] and various composites [48, 49].

The ability to design their mesostructure with different pore geometries, sizes and arrangements allows the entrapment of several types of drugs\biological molecules and provides rational, predicted and on-demand release profiles [33, 34, 50–52]. Pore size can be customized from 1.5 nm to several tens of nanometres providing the ability to encapsulate actives of various sizes [33]. Amongst different mesoporous materials, ordered mesoporous silica has displayed great promise in recent years due to their emerging applications in drug delivery and other biomedical domains [41, 53, 54]. Mesoporous silica is mainly composed of SiO_4 tetrahedra that possesses different ordered alignments of pores (voids and channels) with various morphologies. Historically, a group of ordered mesoporous silicate and alumina silicate–based materials were synthesized in 1992 using liquid crystal templating technique and they were named the M41S family [55, 56]. M41S family comprises various members with different pore morphologies and arrangements. For instance, MCM 41 (Mobil Composition of Matter 41) comprises hexagonally ordered two-dimensional cylindrical pores (2–5 nm), MCM-48 presents a unique three-dimensional gyroid structure that comprises bi-continuous cubic porous system (2–5 nm) and MCM-50 possesses a laminar arrangement (2–5 nm) [50, 57]. Zhao and co-workers synthesized another family of ordered mesoporous silica in 1998 using non-ionic triblock copolymer synthesis technique. The resulting ordered mesoporous silica possessed a pore size of 20–300 Å and was called SBA (Santa Barbara Amorphous). Two of the most well-known members of this family are cubic structured SBA-16 and hexagonally ordered SBA-15 [58]. When compared with M41S silica, SBA-15 and SBA-16 form thicker walls and wider pore sizes [59]. The most common and investigated types of ordered mesoporous silica materials are MCM-41 and SBA-15 [41]. Alternative kinds of mesoporous silica such as, TUD (Technische Universiteit Delft) [60], FSM (Folded-Sheet Mesoporous Material) [61], MSU (Michigan State University) and MPS (synthesized mesoporous silica) [62] have been synthesized using several synthetic methods. Figure 8.1 shows TEM images of selected types of mesoporous matrices.

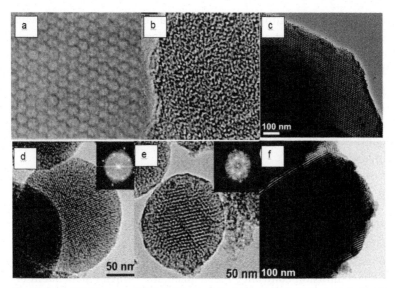

Figure 8.1   TEM images of (a) MCM-41 [56], (b) TUD [35], (c) SBA-15 [48], (d) MCM-48 [43] (e) mesoporous carbon [43] and (f) composites [48].

In 2001, Vallet-Regi et al. adopted MCM-41 mesoporous silica for the first time as a controlled drug delivery carrier [63]. Since then, several studies have investigated employing different types of mesoporous silica as targeted [54, 64], controlled [51, 65, 66], sustained (prolonged) [67, 68] and responsive [69, 70] drug delivery systems. These materials have been an area of interest in the field of drug delivery due to their chemical and thermal stability, wide surface area, biocompatibility, tunable large pores, narrow pore size distribution, high compatibility with different excipients and free reactive silanol groups [50]. These distinctive features allow the entrapment of diverse types of molecules (e.g. drugs, genes, vaccines and proteins) within their mesopores protecting them against the surrounding environment. Also, the possibility of functionalizing their surfaces with various moieties through (electrostatic or ionic) interactions with silica silanol groups enables prospective targeted therapy and co-delivery strategies [71]. In addition, these carriers have improved utilization of poorly water-soluble drugs in different applications [72, 73]. This is afforded through interactions between surface silanol groups in pore walls with organic functional groups of poorly soluble drugs. This physiochemical interaction results in drug entrapment into mesopores in an amorphous state improving dissolution properties [74]. Water molecules entrapment inside silica mesopores and on its surfaces is another attractive feature that has recently been explored and highlighted [75]. Water molecules participate in the rapid chemical exchange which occurs between H^+ (present on functional group of entrapped drug) and H^+ from adsorbed water layer. This exchange induces the mobile state of guest molecules resulting in fast solubilization and release pattern of loaded drug [75].

8.3.1 Mesoporous Silica as a Controlled Delivery System

Mesoporous silica-based platforms are able to accommodate both large and small sized actives within their mesopores [50]. The release of entrapped drugs from the matrix of mesoporous silica is found to be a diffusion-controlled mechanism and is highly dependent on several factors such as pore characteristics (e.g. size, morphology and connectivity), surface area, chemical composition of their surfaces, drug loading methodology, physiochemical properties of the entrapped drug and hydrophilicity of the substrate. Kinetics of drug release from their mesostructure can be controlled by adjusting different parameters, including pore geometry [76], pore size [65], drug loading technique [63] and surface functionalization (through introducing different organic functional groups to their surface) [51].

Mesoporous (silica) pores can be customized to several sizes and geometries via synthetic parameters (detailed later). Pore size usually exerts an important impact on drug entrapment efficiency and drug delivery rate. Vallet-Regi et al. synthesized two types of MCM-41 with dissimilar pore sizes and used ibuprofen as a model drug with a drug loading of 30% w/w. It was found that ibuprofen release profiles, under stirring conditions, were remarkably affected by MCM-41 pore size. Here, the release rate of ibuprofen was reported to decrease proportionally with reduction in pore size. Hence, controlled drug delivery can be achieved by selecting suitable pore size of mesoporous material [65]. Another study by the same group revealed that under static conditions, the release of loaded ibuprofen exhibited various patterns arising from drug loading method and was not affected by pore size [63]. This indicates that the delivery rate of drugs is also significantly affected by drug loading method into mesoporous silica.

Mesoporous materials possess variable mesostructured geometries (such as those shown by MCM-41 and MCM-48) that have an influence on the route and rate of drug diffusion. For example, a unidirectional two-dimensional hexagonal pore (e.g. MCM-41 pores) is characterized by a short and linear diffusion path that results in rapid and direct diffusion of loaded drug into the release medium. On the contrary, drug molecules loaded in bi-continuous three-dimensional cubic pores (e.g. MCM-48 gyroid system) will have to travel a longer distance through a zigzagged route which may result in a delayed or sustained release profile [77].

Moreover, the surface of mesoporous silica can be functionalized with different chemical functional groups to control release kinetics of loaded drug molecules [50, 78]. Amine functionalized SBA-15, for example, was studied as a controlled drug delivery system for bovine serum albumin (BSA) and ibuprofen. As a result, loading efficiency and release profiles for both drugs were efficiently controlled by the amine functionalized SBA-15 when compared to pristine SBA-15. Moreover, it has been found that the loading capacity and release patterns of both drugs (ibuprofen and BSA) were influenced by the time that the functionalization of SBA-15 was performed (post-synthesis or during "one-pot" synthesis). In this regard, ionic interaction between the chemical functional groups (e.g. carboxylic acid) of ibuprofen and amine groups in the surface of mesopores was favourable through post-synthesis. However, it was found that electrostatic interactions between BSA and amine functionalized matrix was favoured through one-pot synthesis technique [51].

8.3.2 Delivery Routes

Mesoporous silica–based materials have been prepared as mesoporous silica nanoparticles (MSNs) [54], hollow mesoporous silica nanoparticles (HMSNs) [79, 80], microspheres [81], beads [82] and fibres [83] or through direct adsorption/inclusion of therapeutic drug into the silica matrix [74]. Such carriers have represented effective tools for several applications involving oral, topical or injectable therapies.

Regarding oral drug delivery systems, it has been reported that ordered mesoporous silica-based materials (e.g. SBA-15 and MCM-41) can be utilized to produce fast acting formulations [74], to enhance solubility and permeability of poorly water-soluble drugs [6, 84] and consequently improve drug bioavailability [41, 85]. In this regard, drug dissolution enhancement which correlates with drug loading into mesoporous matrix is achieved by changing the phase of loaded drug from crystalline to amorphous. Improved solubility and permeability of drug assist in intestinal drug absorption and this in turn results in improved oral bioavailability [6, 76]. For example, charging ordered mesoporous silica with Itraconazole (hydrophobic drug) increases its dissolution rate and consequently improves its systemic bioavailability [41, 86]. Mesoporous silica materials (SBA-15, MCM-41, and TUD-1) have been investigated as oral delivery systems for ibuprofen and have demonstrated to reduce dependency of ibuprofen dissolution profile on the pH of in vitro release medium. As a result, in acidic medium, the release of ibuprofen has been enhanced by loading into mesoporous silica and thus improves its bioavailability [76]. Moreover, oral drug delivery system of telmisartan-loaded MSN has shown to improve the drug's permeability and to reduce its efflux, consequently telmisartan showed a greater oral absorption and high bioavailability [6].

In the field of topical drug delivery, silica nanoparticles are extensively employed in cosmetics and topical therapeutics as a carrier for different actives such as antifungals, antioxidants and UV ray filters [87]. Many flavonoidal compounds such as rutin [88] and quercetin [89] are utilized in topical cosmetic formulations as antioxidants and radical scavengers to provide protective action against photo-degradation. However, their poor physiochemical stability has limited their topical use [87]. Encapsulation of rutin (a quercetin derivative) in aminopropyl-functionalized MCM-41, for example, has demonstrated enhanced metal chelating activity without diminishing its antioxidant properties [88]. In another study, quercetin was loaded into octyl-functionalized MCM-41 and an improvement in its stability and activity was seen whilst still maintaining its chemical antioxidant property [89]. Trolox (an antioxidant compound) has previously been encapsulated into MCM-41-type mesoporous silica and showed enhanced photostability whilst conserving its action as a radical scavenging agent. This was achieved through the slow release of the active from MCM-41 mesopores [90]. Other research demonstrated that mesoporous silica can be utilized as a carrier for UV ray filters such as octal methoxycinnamate [91] and benzophenone-3 [92]. It was reported that methoxycinnamate encapsulated within the MS matrix was around 65% w/w and showed a higher SPF value by 57% when compared to the unloaded methoxycinnamate. This indicates a significant improvement in UV protection activity of methoxycinnamate by loading within mesoporous silica matrix [91].

The dermal therapeutic applications of mesoporous silica can be demonstrated using MCM-41 as a carrier for econazole nitrate. This carrier system has been indicated for use as a topical powder for the treatment of fungal infections. This inclusion complex has improved the dissolution properties of econazole nitrate and consequently enhanced its antifungal therapeutic action when compared to the commercial formulation. Moreover, MCM-41 moisture adsorption capacity plays a role in the treatment of fungal infections. MCM-41 has the ability to reduce moisture existing in skin folds, considered an ideal medium for fungal growth [93].

In terms of intravenous drug delivery, scientists have investigated MSNs as favourable delivery systems to enhance the administration of chemotherapeutic drugs [11, 94, 95] and genes [96]. Most anticancer drugs are characterized by poor aqueous solubility thus limiting their intravenous administration. Therefore, MSNs have been utilized as carriers to enhance their aqueous solubility [11] as well as showing an improvement in their permeability. As a result, a preferential accumulation of anticancer agents in tumour tissues was achieved, resulting in an improvement in their chemotherapeutic efficacy [94]. Another intravenous application of MSNs is the delivery of drugs/genes intracellularly exploiting their gate-keeping capability (discussed later) [53].

8.4 Synthesis of Mesoporous Silica

Synthesis of mesoporous materials is based on the condensation of an inorganic silica precursor around different shaped amphiphilic molecules. These amphiphilic molecules form different geometric structures in the reaction mixture that serve as a template for the production of silica walls with similar geometries. Aging, filtration, drying and calcination are then applied to remove the template and to obtain solid mesoporous silica particles [97]. Among different synthetic techniques, liquid crystal templating method is the most utilized technique for manufacturing ordered mesoporous silica [56, 58, 98, 99]. In this technique, the liquid crystal structures (i.e. ordered arrays with various geometries, e.g. hexagonal, lamellar or cubic) which form in surfactant micellar solutions act as a template for the production of ordered mesoporous silica. Here, silica molecules in the reaction mixture assemble around these structures reflecting their unique shapes and arrangements. Later, calcination process is applied to remove the surfactant template, leaving an ordered porous silicate structure [56, 98] as shown in Fig. 8.2.

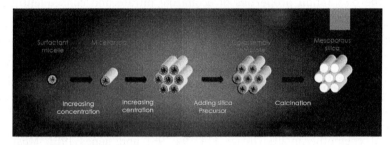

Figure 8.2 Diagram for mesoporous silica synthesis using surfactant templating method.

The characteristics of the employed surfactant (e.g. its carbon chain length, surfactant to silicon molar ratio, aqueous concentration, etc.) as well as the applied temperature are key factors in defining the nature of the final mesoporous product [56]. For instance, the pore size can be controlled by altering the applied temperature during synthesis, where increased temperature results in larger pore capacity [97]. In addition, pore size and wall thickness are critically varied by the surfactant's carbon chain length. An increased pore size is observed with an increase in the chain length of the surfactant [56].

Generally, several categories of templates can be used to produce various types of mesoporous materials. For example, ionic surfactants such as quaternary ammonium compounds (e.g. cetyltrimethylammonium bromide) are utilized to produce MCM-X type [36, 56]. Non-ionic block copolymers such as Pluronic can be employed to synthesize SBA silicate structures and finally organic non-surfactant templates (e.g. triethanolamine) provide TUD mesoporous silica type [60].

Studies conducted by Monica Cicuendez et al. showed the calcination process results in variations of the silica porous network and the silanol group population. For example, calcined mesoporous silica demonstrated a greatly opened silica network with a lesser quantity of silanol groups when compared to their non-calcined counterparts. Therefore, this study reveals the crucial role of the calcination process in determining the biocompatibility of mesoporous silica and their potential use as drug delivery systems [100].

8.5 Biodegradation and Elimination of Mesoporous Silica

Various in vitro biodegradation studies have revealed mesoporous silica and its functionalized derivatives biodegrade in human cells and different simulated body fluids (SBF) [101–104], demonstrating mesoporous silica is a potential candidate for numerous biomedical applications.

For instance, hollow mesoporous silica nanoparticles (HMSNs) were found to biodegrade inside human cells. For the first time, Zhai et al. studied the biodegradation process of HMSNs inside human umbilical endothelial cells. The biodegradation process of HMSNs occurred in both lysosomes and cytoplasm through two degradation steps. Though, the first biodegradation step occurs in both cytoplasm and lysosome, the second step is limited only to the lysosome. A fast biodegradation rate was observed in the first 48 h followed by slower kinetics in the subsequent two days [104]. Research by Chen et al. explored the ability of Stöber MSN (uniformly sized spherical colloidal silica particles; synthesized by sol-gel hydrolysis process) to completely degrade in SBF and human embryo kidney cells [105]. It was found that their biodegradation rate was greatly influenced by surface area, functionalization and initial concentration of mesoporous silica. For example, Polyethylene glycol (PEG) functionalization decreased the biodegradation rate of mesoporous silica, while phenyl functionalization resulted in faster kinetics [102].

When looking at the elimination of mesoporous silica particles, it was found that almost 100% of injected MSNs were effectively removed through hepatobiliary clearance after in vivo administration, without exerting any hepatotoxic effects [106]. Another study revealed that silica degradation products were largely excreted in urine, and their excreted amounts were significantly affected by both particle size and functionalization (PEGylation). It was observed that MSNs biodistributed mainly in the spleen and liver while smaller amounts were evident in the lungs, kidneys and heart. Also, it was found that both PEGylation and particle size reduction allowed an easier escape from capture by spleen, liver and other tissues. As a result, the small PEGylated MSNs were characterized by a slower biodegradation rate and longer blood circulation time, consequently biodegradation products of a lower amount were found in urine when compared to pristine MSNs. Overall, MSNs and its PEGylated derivative did not induce any toxicity after in vivo monitoring for one month [107].

8.6 Methods for Loading Drugs into Mesoporous Materials

The method of charging drugs into the mesoporous matrix greatly influences the properties of the resultant drug-loaded products such as the drug solid state, encapsulation efficiency and release profiles [108, 109]. Several strategies have been adopted to load different actives into mesoporous inorganic matrices such as physical mixing [93, 110], solvent impregnation–based techniques [51, 74, 86], melting [62, 93], supercritical fluids–assisted method [84, 108], spray drying [111, 112], microwave-based irradiation [113] and electrospraying [109]. Amongst these, the solvent impregnation method is the most utilized loading technique as it is considered a simple, fast and easy technique that is suitable for most drugs. Solvent impregnation involves suspension of the mesoporous carrier into a concentrated solution of drug in organic solvent, followed by stirring of the suspension for a number of hours. Solvent evaporation or centrifugation is then carried out to obtain a drug-loaded mesoporous carrier in a solid form [51, 74, 86, 109]. The incipient wetness technique [75, 114] is another solvent-based method which is conducted by reiterating the impregnation of mesoporous carriers in drug solution multiple times. Thus, complete pore accommodation is fulfilled through the successive impregnation steps (involved by the incipient wetness technique) resulting in improved drug loading efficiency [114]. Physical mixing is carried out by blending the required amounts of mesoporous inorganic materials and the drug until a uniform dispersion of drug into mesoporous carrier is obtained [93, 108, 110]. The melting technique is another simple and solvent- free method which consists of melting the mesoporous carrier and the drug at high temperatures (above the drug's melting point) to yield a homogamous dispersion [62, 93, 115].

The previously mentioned conventional methods have inherent drawbacks such as the usage of organic solvents that will involve a consequent solvent elimination step which is a costly and time consuming process [84]. In addition, a procedure such melting technique may affect both drug physicochemical stability and thermal properties making the applicability of this technique virtually limited [116]. In addition, a heterogeneous distribution of drug within the silica matrix is highly expected when using physical mixing and melting as drug loading techniques [108]. Moreover, these manual methods are applicable on a small scale however are problematic on a larger scale when large quantities of drug-loaded products are required [116]. Consequently, alternative innovative strategies have been adopted to load actives into mesoporous matrices while maintaining its physicochemical properties such as co-spray drying [111, 112], supercritical fluids method [84, 108], microwave irradiation [113] and electrospraying [109].

The supercritical fluid–assisted loading method (e.g. supercritical carbon dioxide) results in drug-loaded mesoporous products with improved delivery properties compared to those obtained by conventional solvent-based loading methods [84]. The unique properties of the supercritical fluid offer many advantages when compared to organic solvents; for example, its liquid like density enables dissolution of excess amounts of drug while its gas like diffusivity grants an ultimate access to the mesopores of the silica matrix [84, 108]. Additionally, the complexity is reduced as the process does not involve solvent elimination steps associated with techniques which use organic solvents. It is reported that supercritical fluid process parameters, such as temperature, pressure and duration time highly influence drug encapsulation into mesoporous matrix, the resultant solid state and the drug release behaviour. For instance, it was found that reduced drug encapsulation efficiency was demonstrated when applying a higher processing pressure. The supercritical fluid is an exciting strategy to be employed for loading poorly soluble actives into mesoporous carriers with the aim of enhancing their dissolution and bioavailability [84].

Microwave-based drug loading methods have been investigated to charge drugs into mesoporous matrix, which involves heating of mesoporous carriers and drugs via microwave irradiation in an optimal temperature. This technique avoids thermal degradation of the drug as it guarantees a fixed temperature throughout the process which does not exceed a specified value. This method allows the drug to be melted inside the mesopores and the resulting drug-loaded products have demonstrated an amorphous state with enhanced dissolution properties [113].

Another innovative drug loading strategy is spray drying of the mesoporous carrier and drug together. This approach induces amorphization of the drug within the silica mesopores resulting in a product with optimal physicochemical stability, improved dissolution and enhanced supersaturation when compared to solvent impregnation method [111, 112, 117].

Electro hydrodynamic atomization (EHDA) or electrospraying is a new approach that involves electrospraying of a suspension of mesoporous carrier in a drug solution. It was reported that electrospraying has the ability to achieve an ultimate amorphization of the drug entrapped within silica mesopores. EHDA demonstrated improvement in encapsulation efficiency, enhanced drug release profiles and permeability rates. For example, loading a poorly water-soluble chalcone (Kaz3) into mesoporous matrix (SBA-15 and MCM-41) using EHDA resulted in high encapsulation efficiency (~100%), 30 folds improved drug dissolution rate and high permeability rate across rats intestine when compared to pure crystalline drug or its solvent impregnation-loaded counterparts [109].

One-pot synthesis (also called surfactant-assisted loading method) is an interesting approach to load drugs into mesoporous carriers. This technique employs the presence of surfactant micellar structures as an adjuvant for encapsulating drugs into mesopores. In this technique, drug is loaded throughout the mesoporous silica synthesis process, thus enabling the drug to be efficiently captured inside the surfactant template structures. The resultant products exhibit high encapsulation efficiency, improved release behaviour and enhanced stability. Loading levofloxacin using one-pot synthesis, for example, resulted in encapsulation efficiency of 100% [100].

8.7 Mesoporous Silica-Based Drug Delivery Systems

Mesoporous silica-based materials have revolutionized the area of drug delivery showing a great potential to host several categories of drugs (such as anticancer, anti-inflammatory and antibiotics). They have been studied as exciting matrices for various emerging therapeutic and biomedical applications such as gene delivery, tissue engineering and co-delivery potential.

8.7.1 Cancer Therapy

Conventional chemotherapeutic treatments have several constrains, for example, inability to maintain adequate drug level at the tumour site, non-selective cytotoxicity and inducing multidrug resistance (MDR). Moreover, the majority of anticancer drugs possess poor bioavailability as a consequent of their low aqueous solubility, low permeability across biological membranes, first pass metabolism and rapid clearance [52, 53]. Research work in the last decade has been carried out to develop different targeting tools and smart nanocarriers to overcome such medical constrains which reduce the efficiency of chemotherapeutic drugs. [11, 53, 94, 118, 119]. Among different carriers, mesoporous silica-based delivery systems are evolving as a cutting-edge nanotechnology in cancer therapy. This novel technology is reinforced by interdisciplinary research in the application of such complex material with unique physical properties, targeting and co-delivery potentials [52, 54, 71]. In addition to the intrinsic features of mesoporous silica which include biocompatibility, rapid clearance and their ability to entrap great amounts of differently sized drugs simultaneously with the ability to deliver them to different types of cancer [118, 120]. Mesoporous silica-based nanocarriers have successfully been utilized to entrap and deliver different anticancer agents such as methotrexate [121], doxorubicin [122], paclitaxel [119] and camptothecin [11].

Mesoporous silica–based nanocarriers have demonstrated delivery of anticancer drugs selectively to tumours producing a great tumour suppressing effect whilst reducing their unselective cytotoxic adverse effects on normal tissues [94, 120]. This targeting effect is achieved via the application of three different strategies using MSNs, namely, controlled on-demand release, passive targeting and active (cell-specific) targeting [120, 123].

With regard to the on-demand release strategy, MSNs are employed as a smart tool to preserve their loaded anticancer drugs from external degradation, prevent cargo premature release and provide intracellular on-demand responsive release [53, 70, 94]. For instance, loading of doxorubicin into polyacrylic acid functionalized SBA-15 (785.7 mg/g) has controlled the diffusion of drug molecules from and into the pores. This novel oral delivery system demonstrated excellent biocompatibility and pH sensitivity. Polyacrylic acid was grafted on the outlets of SBA-15 mesopores to act as a gatekeeper to prevent the premature release of doxorubicin in the acidic gastric environment. Whilst, this gatekeeper was intended to be removed in the neutral pH environment of the colon allowing a fast doxorubicin release in the site of the tumour [124]. Balmiki Kumar et al. [69] developed another colon specific system based on guar gum–capped MSNs which has been adopted as an oral carrier for 5-flurouracil. Here, the anticancer agent release was triggered only by enzymatic degradation of guar gum capping layer through the enzymes present in the simulated colon environment. It has been also reported that this system did not display any undesired 5-flurouracil premature release in different simulated pH environments of gastrointestinal tract [69]. An intravenous delivery system of doxorubicin was developed using MSNs functionalized with Polyethylene glycol (PEG) and iminodiacetic acid (IDA). This system improved doxorubicin loading, presented longer circulation lifetime and decreased plasma elimination rate. The PEG and iminodiacetic acid grafted MSNs have the potential to prevent the undesired premature drug leakage in the systemic circulation and are able to achieve on-demand release of their payload under the acidic condition of tumour cells [125].

For passive targeting, mesoporous silica nanocarriers (100 nm or less) are exploited to prevent anticancer drug uptake by reticuloendothelial systems and to improve enhanced permeability and retention effect (EPR) [120,126]. For example, polyethyleneimine-polyethylene glycol functionalized MSNs with a reduced particle size (50 nm) demonstrated enhanced passive delivery of doxorubicin to squamous cancer cells. These nanocomplex particles have improved the EPR effect and enhanced doxorubicin cancer cell cellular uptake, consequently it was able to successfully induce cancer cell apoptosis and shrinking of tumour size without exerting significant systemic adverse effects [126]. Additional advantages of using MSNs as a form of passive delivery is to favour the therapeutic profile of poorly water-soluble anticancer drugs by improving their dissolution and permeability, consequently enhancing their cellular uptake, bioavailability and therapeutic effectiveness [11, 119] It is worth to mention not all types of tumours exhibit the EPR effect, as a result the use of MSNs as a passive targeting system is not applicable in all cases [120].

Active cancer targeting is an effective approach to overcome passive targeting limitations and is possible by decorating mesoporous silica external surfaces with targeting ligands [118, 120]. Grafting specific affinity ligands such as folic acid, aptamers, antibodies or peptides on the surface of MSNs (via covalent binding, physical adsorption or electrostatic interaction) allows targeting of certain receptors or antigens that exist specifically on the membrane of cancerous cells. These affinity ligands grant MSNs the ability to recognize and interact selectively with receptors/antigens that are overexpressed on the surface of cancer cells, and thus effectively transfer their anticancer payloads to the targeted cells [54, 94, 120, 123]. This targeting technique results in preferential accumulation of the anticancer agent into tumour tissues over the various normal organs, thus circumventing the undesired off-target toxicity [123]. According to the kind of expressed receptor/antigen on the cancer cells, a diversity of targeting ligands was employed to develop multifunctional MSN-based drug carriers [120]. For instance, folic acid was widely employed as a targeting ligand to actively bind to folate receptors that are extensively upregulated in diverse types of cancer (e.g. renal, ovarian, lung, endometrial, colorectal, and breast cancer). Functionalization of MSNs with folic acid enhances their selective internalization to these cancer cells and consequently the cellular uptake of the loaded anticancer agents is improved [54, 121]. Another interesting ligand is transferring which was grafted on the surface of MSNs to give a dual action of producing on-demand release and active targeting of hepatocyte cellular carcinoma. This unique system was used as a carrier for doxorubicin and was able to bind to the overexpressed transferrin receptors existing on the surface of cancer cells resulting in enhanced cellular uptake. Moreover, because of the gate-keeping ability of transferrin, this system inhibited the undesired premature leakage of doxorubicin [64].

Another advantage of using MSNs in cancer therapy is its ability to circumvent multi-drug resistance (MDR) displayed by several types of cancers through its potential of co-delivery of anticancer agents and siRNA to silence the genes responsible for MDR thus improving patient's response to chemotherapy [127–129].

8.7.2 Anti-Inflammatory Drugs

Mesoporous silica has been effectively employed as a delivery system for various anti-inflammatory actives (e.g., naproxen [130], sulfasalazine [80, 131, 132], piroxicam [74], resveratrol [133] and ibuprofen [51, 63, 65, 67, 76]). Mesoporous silica platforms were explored to improve the aqueous solubility and dissolution profiles of anti-inflammatory agents [74, 76], to produce regio-specific targeting action [131, 132] and to control or prolong their release patterns [63, 65, 67]. Encapsulation of piroxicam (~%14 w/w) into MCM-41-type mesoporous silica, for example, demonstrated a significant improvement of piroxicam's solubility and dissolution rate when compared to pristine crystalline drug. It was found that this improvement was not only due to entrapping piroxicam molecules within the mesopores as amorphous non-crystalline state but also attributed to the high surface area of silica substrate. As a result, this inclusion complex achieved a fast analgesic onset that was equivalent to that of analgesic effect of Brexin® (marketed product of piroxicam) [74].

On the other hand, SBA-15 was functionalized with long crosslinked hydrophobic amine bridges to produce a sustained release delivery system for ibuprofen. This unique system was able to attain high drug loading content (21% w/w) and extend the release of ibuprofen to ~75.5 h [67]. Another ibuprofen sustained release system was obtained via entrapment of ibuprofen into 3-aminopropyltriethoxysilane functionalized monodispersed MCM-41-type microspheres, which displayed a slow ibuprofen release rate as compared to its irregular shaped counterparts [40]. Such unique delivery systems can be used to prolong the therapeutic response of drugs with short half-life times (e.g. ibuprofen), and consequently increase their efficiency, decrease dosage frequency and enhance patient compliance [67]. Also, unfunctionalized mesoporous silica-based carriers (3D TUD-1, 2D SBA-15 and 2D MCM-41) for ibuprofen demonstrated an enhancement of dissolution rate at acidic pH conditions (pH 5.5). It was found that the release rates of ibuprofen from all mesoporous carriers were considerably faster than pure ibuprofen, with the fastest rate obtained from the three-dimensional mesoporous TUD-1 [76]. Notably, the release rate of ibuprofen from different silica matrices can be controlled by adjusting the pore size [65] or through surface modification of the mesoporous carrier [40, 51].

Sulfasalazine is an anti-inflammatory prodrug that is mainly used to treat inflammatory colon disease. Mesoporous silica was employed to design a unique oral delivery system for such drugs to achieve a local targeted delivery to the colon and decrease its premature release in the stomach and accordingly protects it from stomach degradation, reduces undesirable adverse effects and perfectly ensures a longer residence of sulfasalazine in the GI tract [131, 132]. For example, polymer-coated and amine-functionalized mesoporous silica (SBA-15 and MCM-41) carriers were employed as an oral delivery system for sulfasalazine. The designed nanotechnological device achieved minimal premature release of the drug in an acidic environment (pH = 1.2), while providing a sustained sulfasalazine release profile at pH = 6.8 consequently a regio-specific targeting of sulfasalazine can be obtained [132]. A similar colon targeting delivery system for sulfasalazine was designed using trimethylammonium functionalized MSNs [131].

MCM-48-type MSNs were explored to improve solubility and permeability of resveratrol (a hydrophobic nutraceutical) across cellular membrane tight junctions. The reported results demonstrated that resveratrol saturated solubility was dependant on both pore size and particle size of the mesoporous carrier. Loading of resveratrol in MSNs demonstrated an improvement of the drug's bioavailability and anti-inflammatory therapeutic response when compared with either the suspension or solution of resveratrol [133].

8.7.3   Antimicrobial Agents

There have been extensive interests to explore the prospects of using mesoporous silica-based carriers as antimicrobial adjuvants due to their potential as platforms to provide safe antibacterial delivery with minimized off-target toxicity [134, 135]. A range of antimicrobial drugs have previously been encapsulated into different mesoporous silica nanomaterials, such as amoxicillin [136], levofloxacin [100], Parmetol S15 [137] vancomycin [138], erythromycin [139, 140] gentamicin [39] and histidine kinase autophosphorylation inhibitors (HKAIs) [134]. Not only is mesoporous silica employed as simple controlled delivery systems for antimicrobial drugs to control or prolong their release [136, 139, 141], but more recently as a targeting platform to broaden their antibacterial spectrum, enhance their efficacy, improve their selectivity, target their release into the bacterial cells and reduce their undesired toxicity [134, 135, 140].

Regarding controlled delivery prospects of antimicrobials, mesoporous silica has been employed as a sustained release platform for entrapping levofloxacin via solvent impregnation or the one-pot synthesis method (surfactant-assisted). The antimicrobial efficiency of levofloxacin in this delivery system was investigated against *Escherichia coli* bacteria. The levofloxacin-loaded matrices using both loading methods demonstrated comparable in vitro release profiles with an initial burst release for 10 h followed by a sustained release of drug for 350 h. The sustained antibacterial effect of loaded mesoporous carriers revealed the suitability of these systems for the treatment of osteomyelitis and bone infections through local administration [100]. Another sustained release delivery system for treatment of osteomyelitis was designed using vancomycin-loaded sol-gel silica microspheres. Different controlled release patterns of vancomycin were achieved via optimizing different process parameters to obtain a desired therapeutic profile [138].

Controlled release kinetics of gentamicin [39] and amoxicillin [136] were also obtained through loading into ordered mesoporous SBA-15. Two kinds of SBA-15 were used as either disk conformation or calcined powder. A controlled release pattern of amoxicillin was attained using SBA-15 as disk conformations while using SBA-15 calcined powder as a matrix resulted in faster release kinetics [136]. Regarding gentamicin, no considerable difference was observed between in vitro release profiles obtained from disks or calcined powder platforms. In vitro release profiles of both formulations demonstrated fast-initial release (~60%) followed by a prolonged release pattern [39]. Controlled delivery of gentamicin and amoxicillin using SBA-15 matrix exhibited an advantageous sustained release profile when compared to instant and fast dissolution of conventional dosage forms. These antibiotic carriers result in a reduced frequency of administration as the whole required dose of an antibiotic can be administered once and released from the silica matrix in a controlled pattern [39, 136]. Erythromycin controlled delivery system was designed using three-dimensional cubic ordered mesoporous silica (large pore Ia3d silica matrix and MCM-48) as carriers. The diffusion of erythromycin from mesopores was controlled by applying mesoporous silica surface modification or changing the pore size. A reduction in the release rate was noticed as the silica pore size was decreased. It has also been observed that the hydrocarbon surface functionalization of silica reduced the delivery rate of erythromycin by six times [139].

For targeting potential, mesoporous silica has been investigated to design a delivery platform which works in such a way that its antimicrobial cargo is released only at the site of infection and only targets specific bacteria. For example, a pathogen specific drug carrier system was designed by capping of the mesopores of MSNs with FB11 antibody. The antibody served as a gate-keeper that featured the antimicrobial cargo release only in response to the interaction between the capping antibody and the O-antigen present in the *Francisella tularensis* bacteria, thereby preventing the premature release of the loaded agent and favouring its release only at the site of infection [135]. Also, encapsulation of vancomycin [140] and HKAIs [134] into ε-poly-L-lysine functionalized MSNs overcame the permeability barrier of Gram negative bacteria and thereby broadened their antibacterial spectrum. An improved delivery and targeted internalization of the antibiotic molecules (vancomycin and HKAIs) to the Gram negative bacterial cell has been achieved using these gated nanocarriers. The positively charged ε-poly-L-lysine capping layer interacts with the bacterial cell inducing damage in the bacterial wall, which grants the loaded antibiotic access to the bacterial cells [134, 140]. Recently, levofloxacin-loaded MSNs functionalized with polycationic dendrimer have proven not only to improve the antibacterial penetrability across Gram negative bacterial cell membranes but also to trigger an effective antibacterial effect on the bacterial biofilm. As a result, this unique nanoantibiotic system has demonstrated an efficient targeted antimicrobial action against Gram negative bacteria providing great potential to reduce bacterial antibiotic multi resistance [142]. Another nanosystem was designed to deliver antimicrobial proteins to target the bacterial biofilm (which is responsible for antibiotic resistance) using large cone shaped HMSNs [143]. Table 8.1 shows selected delivery systems that involve mesoporous silica materials.

8.7.4 Tissue Engineering

The pivotal pillar of tissue engineering is based on using bioactive materials that are capable of forming apatite layers on the surface of biological tissues under physiological conditions [144]. In fact, mesoporous silica-based matrices are emerging in biomedical tissue engineering research due to their exciting bioactivity, biodegradability and strong mechanical properties [145–147]. Mesoporous silica-based materials have demonstrated bioactivity when they contact the simulated biological fluids [145, 148]. Furthermore, they show capability to biologically bind an artificial tissue (e.g. bone) to a living one [146]. The highly reactive silanol moieties are able to form molecular bonds with the surfaces of biological tissues via developing a carbonated apatite layer [145, 148]. The fundamental requirements for the apatite layer formation are the surface negative charge and the porosity of the bioactive material [144]. Therefore, silanol groups [149], large surface area [146] and pores [144] of mesoporous silica are necessary to promote the apatite nucleation required for cell adhesion. However, silanol groups have a crucial role in the bioactivity of materials as they form a hydrated silica gel that serves as nucleation sites for inducing

Table 8.1 Different therapeutic drugs entrapped into various types of mesoporous silica

Mesoporous matrix	Pore size	Drug/ pharmacological activity	Route of delivery	Loading method	Results	Ref.
			A			
MCM-41	3.6 nm And 2.5 nm	Ibuprofen/anti-inflammatory	Oral	Solvent imp.	MCM-41 pore size affected the delivery rate of ibuprofen in SBF. Sustained release rate resulted from smaller pore size.	[65]
Amine - SBA-15	7886 Å	Ibuprofen and BSA/anti-inflammatory and model protein, respectively	Oral	Solvent imp.	SBA-15 amine functionalization was achieved via post-synthesis or one-pot synthesis. A significant impact of SBA-15 resulted in controlled release profiles of ibuprofen and bovine serum albumin. This was due to SBA-15 surface properties.	[51]
3 aminopropyl-triethoxy-silane-MCM41	2.1 nm and 3.7 nm	Ibuprofen/anti-inflammatory	Oral	Solvent imp.	Loading of ibuprofen into functionalized micro-sized MCM-41 spheres generated a slow release of the drug, irregular shaped particles.	[40]
MCM-41	3.21 nm	Piroxicam/anti-inflammatory	Oral	Solvent imp.	Improved solubility of drug at pH 1.2 and a faster analgesic effect were achieved by loading piroxicam into MCM-41 compared to marketed product	[74]

(Continued)

Table 8.1 *(Continued)*

Mesoporous matrix	Pore size	Drug/ pharmacological activity	Route of delivery	Loading method	Results	Ref.
			B			
Trimethyl-ammonium MCM41- type MSN	2.7 nm	sulfasalazine/ anti-inflammatory prodrug	Oral	Solvent imp.	An innovative pH trigged release approach obtained minimal release rate of sulfasalazine under stomach acidic conditions and maximum release rate in colon neutral pH.	[131]
Calcined and non-calcined mesoporous silica	8.5 nm	Levofloxacin/ antibiotic	Local bone administration	Solvent imp. and one-pot	In vitro studies showed that both loading methods of levofloxacin led to the same release profiles which illustrated an initial fast release followed by sustained release pattern. Both methods indicated effective extended antibacterial activity.	[100]
SBA-15- powder and SBA-15 disk	8 nm for calcined samples	Amoxicillin/ antibiotic	Oral	Solvent imp.	Controlled release of amoxicillin was performed using SBA-15 in the form of discs. In vitro studies showed an immediate release rate attained by using SBA-15 as powder. The administration of amoxicillin using SBA-15 discs scheme was compared to different forms including capsules, tablets and suspensions.	[136]

C

Mesoporous matrix	Pore size	Drug/pharmacological activity	Route of delivery	Loading method	Results	Ref.
SBA-15- powder and SBA-15 disk	5.5 nm	Gentamicin/antibiotic	Oral, intravenous and topical	Solvent imp.	In vitro study release of gentamicin showed no remarkable difference between calcined powder and disc. An initial immediate release (~60%) followed by an extended release period was exhibited in both types. The dissolution of gentamicin: SBA-15 was instantaneous.	[39]
MCM-48 And large pore LP-Ia3d	3.6 and 5.7 nm, respectively	Erythromycin/antibiotic	Oral	Solvent imp.	Functionalized surface of MSN achieved controlled release of erythromycin effectively. A reduction of the release rate was achieved after surface functionalization by a factor of ~6	[139]
MCM-41 Powder formulation	3.25 nm	Econazole/Antifungal	Topical	Melting	Loading of econazole into MCM-41 nanoparticles improved drug dissolution rate and increased antifungal activity. While MCM-41 concentration was increased, the antifungal activity also increased	[93]

(Continued)

Table 8.1 (*Continued*)

Mesoporous matrix	Pore size	Drug/ pharmacological activity	Route of delivery	Loading method	Results	Ref.
			D			
Functionalized MSNs with PEG and IDA	3 nm	Doxorubicin (DOX)/ Anticancer	Intravenous	Solvent imp.	The release of DOX was developed as a pH-responsive system with prolonged circulating. The avoidance of premature leakage of drug was achieved through Functionalization of MSNs with PEG and IDA (~2.39%). Loading efficiency was enhanced 1.9 times	[125]
ε-poly-L-lysine-MCM-41 type MSNs	2.65 nm	HKAIs/ antibacterial	Intravenous	Not reported	Loading antibacterials into functionalized MSNs to improve bactericidal outcomes towards Gr- bacteria of HKAIs. Thus broadening of the antibacterial spectrum.	[134]
ε-poly-L-lysine - MSNs	2.3 nm	Vancomycin/ antibacterial	Intravenous	Not reported	Novel loading of vancomycin onto mesoporous nanoparticles using ε-PLL was developed. The mesoporous capability was explored for the first time to improve the antibacterial efficiency for vancomycin.	[140]

Mesoporous matrix	Pore size	Drug/ pharmacological activity	Route of delivery	Loading method	Results	Ref.
			E			
cationic dendrimer-MSNs	2.4 nm	Levofloxacin/ antibacterial	Oral	Solvent imp.	Loading of levofloxacin into MSNs developed extended release. Bonding of dendrimers and MSNs improved the internalization in GR- bacteria. A significant antimicrobial effect on the bacterial biofilm was reported.	[142]
Poly acrylic acid– capped SBA-15	7.1 nm	DOX/anticancer	Oral	Solvent imp.	A pH-triggered, colon targeted, oral anticancer delivery based on capped acrylic acid with mesoporous silica to control drug release in and out of the pore channels. The DOX solubility enhanced through its excellent dispersion in the pore channels of SBA-15. A great loading efficiency of DOX, high biocompatibility and proper pH sensitivity were achieved.	[124]
Guar gum MSN-41-type MSN	3 nm	5 Fluorouracil/ anticancer	Oral	Not reported	Colonic enzyme responsive drug delivery system. Showed no drug leakage by passing through stomach pH environment. GG-MSN delivery systems indicated a successful oral targeting release.	[69]

(Continued)

Table 8.1 (*Continued*)

Mesoporous matrix	Pore size	Drug/ pharmacological activity	Route of delivery	Loading method	Results	Ref.
		F				
Transferrin-capped mesoporous nanoparticles system type MSN-41	2.1 nm	DOX/anticancer	Intravenous	Not reported	Important findings of this approach were to serve both gatekeeper and targeting ligands simultaneously by attaching transferrin to the surface of mesoporous silica. Slow release of DOX was observed with limited premature leakage. Using Transferrin resulted in improved cellular uptake of the drug.	[64]
Amino modified SBA-15	7.1 nm	Ibuprofen/anti-inflammatory	Oral	Not reported	Modified silica with crosslinked amine linkage improved drug loading efficiency. Sustained release of Ibuprofen was observed up to ~ 75.5 h.	[67]

Mesoporous matrix	Pore size	Drug/pharmacological activity	Route of delivery	Loading method	Results	Ref.
Eudragit-coated MCM-41 and SBA-15 type	2.3 nm to 5 nm respectively	Sulfasalazine / anti-inflammatory	Oral	Incipient wetness method	Different release rates of sulfasalazine are achieved based on pH values. Thus a regiospecific delivery of sulfasalazine in GIT.	[132]
G						
MCM-41, SBA-15	4 nm and 8 nm	Chalcone (Kaz3)/ anticancer	Oral	Electrospraying (EHDA)	First study to report use of EHDA as loading method, the results showed improved solubility and permeability and complete amorphization of the insoluble chalcone.	[109]
MCM-48	3.5 nm	Resveratrol/anti-inflammatory	Oral	Solvent Imp.	Reservatol loading into MSNs resulted in improved solubility, permeability and anti-inflammatory.	[133]

biomineralization [144, 146]. Additionally, the large pore volume of silica platform (more than 2 nm) not only induces the nucleation of apatite, but also improves the oxygenation required for tissue growth [145]. In addition, these mesopores serve as nutritious depot for cells as they are able to adsorb biomolecular ions and growth factors thereby, promote cells proliferation [150].

Bioactivity studies performed on different mesoporous structures (SBA-15, MCM-41 and MCM-48) have proven that both MCM-48 and SBA-15 demonstrated bioactivity while MCM-41 did not show any sign of bioactivity. This might be attributed to the high silanol concentration presents on MCM-48 and SBA-15 surfaces as compared to MCM-41 [144, 145]. Although, pristine MCM-41 did not demonstrate any bioactivity, mixing some elements with MCM-41 such as (10% w/w bioactive glass) has promoted its bioactive behaviour and induced the carbonated apatite layer nucleation [144].

Based on the exciting properties of mesoporous silica–based materials including its textural properties, bioactivity, biocompatibility and biodegradability, they have been designed as scaffolds for the reconstruction of hard tissues such as tooth and bone regeneration [146, 149]. Different morphologies of mesoporous silica particles have been utilized to design scaffolds suitable for tissue engineering applications such as nanoparticles [151], microparticles [147], spheres [152] and nanofibres [153].

Interestingly, coupling of both drug delivery potential and bioactivity can be achieved by loading mesoporous silica in scaffolds synthesis with therapeutic drugs (e.g. antibiotics). The drug-loaded mesoporous silica-based scaffolds provide an important advancement in the clinical applications of tissue engineering [145, 147, 154]. For example, a scaffold made of vancomycin-loaded mesoporous silica particles, gelatine and hydroxyapaptite is a novel combination that not only proved to accelerate the bone reconstruction process but also provided prolonged vancomycin release [147].

8.7.5 Gene Delivery

Gene delivery is evolving as a vital treatment for genetically caused diseases, such as hepatitis, cancer and sickle cell anaemia [155,

156]. Gene vectors should be nanosized to be suitable for cellular uptake and hence, provide efficient intracellular gene transfection [157]. Nanosized non-viral vectors such as polyplexes, lipoplexes and inorganic nanoparticles are considered safe gene carriers that are able to achieve intracellular delivery of diverse types of organic nucleic acids [155]. Amongst these, mesoporous silica–based nanomaterials have engrossed great attention for designing efficient gene vectors to be used for different gene transfection applications [71, 155, 158]. This is owing to their intrinsic features such as small size allowing easy endocytosis inside cells [96] as well as wide surface area and large pore volume that offer a space or binding position to accommodate different types of nucleic acids such as DNA, siRNA and shRNA [71]. Moreover, mesoporous silica–based nanocarriers are able to protect their entrapped nucleic agents from biodegradation such as enzymatic (Rnase or Dnase) degradation [155, 159]. Thus, allowing safe and stable transportation of DNA/siRNA directly into the cellular cytoplasm and to the nucleus in the case of DNA [96]. Successful loading of genes into MSNs is accomplished through two routes: loading inside mesopores via utilizing large pores sized MSNs [156, 160, 161] or surface attachment via cationic surface modification [71, 162, 163]. Encapsulating large sized biomolecular nucleic acids inside mesopores of MSNs is not accessible due to the small intrinsic pore volume of MSNs that are obtained by conventional synthesis methods [159–161]. Therefore, MSNs with large pore dimensions have been utilized to allow gene encapsulation inside the pores of MSNs [155, 156, 161]. The synthesis of large pore MSNs have been demonstrated via use of a swelling additive or pore expansion agent [156, 160, 161]. MSNs of a 30 nm pore size, for example, have been synthesized through the addition of excess quantity of ethyl acetate as a pore expanding agent during the synthesis process [160]. Larger pore MSNs (pore size 23 nm) have been prepared by post synthesis incubation with trimethylbenzene (a swelling additive) for 4 days. This large pore mesoporous silica platform has proven to successfully encapsulate siRNA and achieve an efficient siRNA transfection while preserving the gene integrity despite the presence of serum nucleases [161].

The surface of pristine MSNs is negative due to silanol groups thus attachment of negatively charged gene molecules on the surface is not straightforward. Therefore, a cationic surface modification of

MSNs is required to allow the electrostatic interaction between the MSNs surface and nucleic acids [164]. Although cationic modification is chiefly used to impart a positive charge to MSNs, it is also found to act as proton sponges that enable siRNA endosomal escape to intracellular fluid (cytosol) of the cells [163].

Multifunctional MSNs were used to serve other functions aside from optimizing nucleic agent surface adsorption; for example, controlling the entrapped nucleic acid release and targeting potential have also been investigated [158, 159]. Functionalizing MSNs with biodegradable poly (2-dimethylaminoethyl acrylate), for instance, has achieved siRNA-controlled release after the biodegradation of the binding polymer inside the cellular compartments [158].

8.8 Toxicity of Mesoporous Silica

Although, MSNs are considered biocompatible when used in adequate concentrations to accomplish the desired pharmaceutical applications [165, 166], the current evidence that is provided on the safety of MSNs is insufficient and controversial [165]. The free silanol groups on the surface of silica-based materials are highly reactive with a great liability to interact and impair biomolecular structures such as proteins and lipids of cellular membrane [167], they were also found to induce red blood cells (RBCs) haemolysis [168]. Hence, functionalizing surface silanol groups has proven to reduce the reactivity of particles and improve the overall biocompatibility [167].

Unlike nonporous amorphous silica nanoparticles, only a slight ratio of silanol groups of MSNs are accessible to the surrounding environment (as the majority of silanol groups exist in the internal surface of MSNs) providing enhanced biocompatibility properties [168]. For example, MSNs show an improved biocompatibility towards RBCs compared to nonporous silica particles. No RBCs haemolysis was reported upon intravenous injection of MSNs suspensions with concentrations lower than 100 µg/ml [168]. Another study found intravenous injecting of silica nanoparticles at a dose of 7 mg/kg did not induce any alternations in serum biochemical markers or haematological parameters [165].

Various controversial studies have been carried out on the cytotoxicity of silica-based particles. It has been reported that different sized and charged silica particles did not induce any cytotoxic activity in concentrations lower than 2.5 nM [169]. However, other studies have shown that cytotoxicity of mesoporous silica is greatly dependent on the concentration, particle size [103, 170] and presence of remaining surfactant molecules [103]. For instance, mesoporous particles in the nanosize range have exhibited a slight cytotoxic effect when used in concentrations lower than 25 µg/ml. On the other hand, micro-sized mesoporous silica particles have demonstrated minor cytotoxicity over a wider range of silica concentrations (up to 480 µg/ml). These results indicate that the nanoparticles are more easily endocytosed than microparticles. Also, the same study reported that biodegradation products of mesoporous silica particles showed no cytotoxicity to cells in concentrations up to 480 µg/ml [103]. However, another research study performed on the cytotoxicity of MCM-41 and its functionalized analogues found that a cytotoxic effect was observed against human neuroblastoma cells when these materials were used in concentrations of 40–800 µg/ml [171]. Moreover, Zhimin Tao et al. conducted research investigating the effect of different types of MSNs on cellular respiration. A time and concentration dependant cellular respiration inhibition was detected upon using SBA-15-type MSNs, whereas MCM-41 particles did not induce any noticeable cellular respiration inhibition [172].

Studies exploring the toxicity of MSNs towards different tissues found no histopathological abnormalities were detected in lungs, spleen, heart, liver and kidneys upon intravenous injection of silica particles into mice. This result suggests that mesoporous silica particles do not induce any tissue toxicity or inflammation. This excellent tissue compatibility is correlated to biodegradability, biocompatibility and physicochemical stability of MSNs and their biodegradation products [107].

8.9 Future Perspectives

Several physiological and anatomical barriers are confronted by drug carriers preventing them from targeting selective sites, such as distribution from the blood to the extracellular compartment of a tumour, binding to target-cell membranes before endocytosis, delivering their payload intracellularly and finally reaching subcellular target sites. To avoid such obstacles, the pharmaceutical research community has devoted many efforts to design multifunctional platforms with the aim of improving the targeting prospect of different drug delivery systems [173]. Among various platforms, mesoporous silica-based carriers are showing a great potential in achieving selective drug targeting [174] and co-delivery of different actives together such as therapeutic drugs with genes [158] peptides [173] or chemosensitizers [175].

Modifying pore size, geometry and surface of mesoporous silica can be used to design valuable controlled delivery systems. Nevertheless, these routes are incapable of achieving extraordinary on-demand release profiles which are required clinically to accomplish desired therapeutic outcomes [176]. Consequently, it is crucial to design environmentally responsive drug delivery systems enabling the release of their payload specifically to the target sites and only under the effect of certain physiological stimuli [159]. Mesoporous silica–based nanocarriers are a promising platform to achieve an on-demand release profile. Environmentally responsive MSNs which have the ability to change chemically or physically in response to external or intercellular stimuli have recently received extensive research attention to design smart targeting systems for future applications. However, their clinical consideration is currently challenging and under exploration [64, 79, 124, 125].

Environmentally responsive drug delivery systems based on MSNs can be designed by three main routes: via attaching the drug to silica matrix with the aid of environmentally cleavable linkers [177], via modifying the surface of MSNs through surface coating with responsive polymers [132] or finally through capping their mesopores by attaching certain functional groups that act as gate-keepers [64]. In these particular cases, these innovative modifications act as a barrier that prevents the premature release of drug whilst achieving on-demand release inside specific intracellular compartments under the effect of specific exogenous or endogenous stimuli. These stimuli are capable of removing (cleaving) such barriers thus releasing the encapsulated drugs [174, 178]. They are either chemical stimuli such as ionic strength, reactive oxygen species, enzymes, redox potential and pH or physical stimuli such as magnetism, ultrasound, light, temperature, and electricity. However, cell microenvironment responsive delivery systems that release their cargo under the effect of a natural process (e.g. enzymes, redox and acidic pH stimuli) serve more logical and feasible routes to achieve biomedical and clinical applications [69, 174, 179]. For instance, the pH-triggered systems based on MSNs are designed as to take advantage of the presence of a pH difference between diseased tissues (e.g. inflammatory and cancer tissues) and normal ones. Here, MSNs carrying certain therapeutic drugs (such as anticancer and anti-inflammatory drugs) tend to keep their content within their mesopores while circulating in blood but release their payloads only under the acidic environment of tumours and inflamed tissues [80, 122, 180].

For example, tannin as a gate-keeper was grafted on the surface of MSNs to design a pH-responsive system which is promising for stomach and cancer treatment. Tannin-capped mesopores have proven to entrap the model cargos efficiently thus preventing any premature leakage at pH 7.4. However, under acidic pH conditions tannin caps were cleaved, as a result a rapid release of the model cargo was achieved [180]. Enzyme responsive MSNs are another type of cell microenvironment responsive systems in which biodegradation of the capped layer through certain enzymes existing in specific regions induces the release of the entrapped drug [69, 70]. Guar gum is a biodegradable polysaccharide that undergoes hydrolysis by specific enzymes that are only present in the colon. For colon specific responsive system, guar gum was anchored on the surface of MSNs to prevent the premature cargo release and to achieve responsive release only in the colon under the response of colonic enzymatic biodegradation [69].

Cancerous cells have a tendency to develop different mechanisms to survive against the administered cytotoxic drugs. The phenomenon where cancerous cells develop resistance against several types of drugs at the same time is called MDR. Consequently, developing chemotherapy with the ability to tackle MDR is the most challenging task for scientists [175, 181]. Multifunctional MSNs were employed to achieve simultaneous delivery of different therapeutic actives together with the potential to improve the cytotoxic efficiency of anticancer drugs and to diminish the MDR [52, 173, 175, 182]. Two principal mechanisms responsible for MDR are pump and non-pump processes. Pump resistance is induced by certain membrane-bound ATP-dependent proteins (e.g. P-glycoprotein pgp) that serve as efflux pumps which expel anticancer drugs independently outside the cells, while the non-pump mechanism is activated by overexpression of anti-apoptotic proteins in cancer cells thus inhibiting cancer cell apoptosis [181]. Therefore, it is crucial to simultaneously prevent these resistance mechanisms to achieve an efficient chemotherapeutic treatment [127]. Co-delivery of anticancer drugs with siRNA is a smart approach which is utilized to circumvent MDR mechanisms and also to achieve synergistic effects, here siRNA is transfected in such a way to knockdown genes which are accountable for MDR. Several studies reported the use of multifunctional MSNs for co-delivery of anticancer drugs and silencing siRNA has demonstrated to inhibit MDR resistance mechanisms, reduce the premature extracellular leakage of anticancer drugs, enhance anticancer cellular uptake, induce synergistic cytotoxic action and finally has proven to decrease tumour growth [52, 127, 128, 182]. For example, utilizing MSNs for co-delivery of the anticancer drug sorafenib and siVEGF "siRNA" into hepatocellular carcinoma improved the cytotoxicity of sorafenib. This smart nanosystem was found to significantly induce cell cycle death, enhance the tumour targeting of sorafenib and siVEGF and achieve efficient siRNA transfection [182]. Grafting targeting groups or stimuli responsive moieties on the surface of MSNs can be further used to achieve a selective delivery of anticancer agents and siRNA to cancerous tissues and decrease the system's undesired toxic effects on normal tissues. For instance, a pH-responsive system based on MSNs was designed to achieve simultaneous delivery of survivin shRNA–expressing plasmid (shP) and doxorubicin to hepatoma cells. This unique nanocomplex system has demonstrated high encapsulation efficiency, produce on-demand drug release profiles, enhance shP endosomal escape and silencing efficacy, improve actives nucleic accumulation and hence decrease hepatoma cells proliferations (in vitro) and tumour progression (in vivo) [52].

Such molecular devices (responsive system and/or co-delivery systems) could be valuable in the future to open clinical pathways for the treatment of different pathological conditions (e.g. cancer and inflammatory diseases) through achieving maximum response of therapeutic drugs while preventing their severe adverse effects.

8.10 Concluding Remarks

Multifunctional mesoporous silica-based delivery systems, characterized by desired features such as large surface area, robustness, uniform pore size, free accessible silanol moieties, bioactivity, biocompatibility and biodegradability are demonstrating significant prospects in pharmaceutical and biomedical applications to achieve various desired clinical and therapeutic outcomes. Such versatile inorganic carriers represent a useful nanotechnological platform to customize drug delivery systems that are able to achieve selective targeting of different categories of therapeutic drugs, broadening of their therapeutic action, addressing the acquired drug resistance, achieving controlled dosing and reducing their unfavourable adverse effect. Tailoring different designs of such delivery systems can be achieved via optimizing their synthesis, functionalization and drug loading methods to provide interesting strategies for improving existing and evolving therapeutic actives.

Acknowledgements

The authors would like to acknowledge Minia University represented by Egyptian Culture Centre and Educational Bureau in London. The authors also would like to thank the EPSRC (EPSRC EHDA Network) for their continuous support and the Council for At-Risk Academics (CARA).

References

1. Jonathan, G., and Karim, A. (2016). 3D Printing in Pharmaceutics: A New Tool for Designing Customized Drug Delivery Systems, *Int. J. Pharm.*, **499**, pp. 376–394.

2. Tiwari, G., Tiwari, R., Sriwastawa, B., Bhati, L., Pandey, S., Pandey, P., and Bannerjee, S. (2012). Drug Delivery Systems: An Updated Review, *Int. J. Pharm. Invest.*, **2**, pp. 2–11.

3. Muheem, A., Shakeel, F., Jahangir, M. A., Anwar, M., Mallick, N., Jain, G. K., et al. (2014). A Review on the Strategies for Oral Delivery of Proteins and Peptides and their Clinical Perspectives, *Saudi Pharm J.*, **24**, pp. 413–428.

4. Mitragotri, S., Lammers, T., Bae, Y. H., Schwendeman, S., De Smedt, S., Leroux, J., Peer, D., Kwon, I. C. (2017). Drug Delivery Research for the Future: Expanding the Nano Horizons and Beyond, *J. Control. Release*, **246**, pp. 183–184.

5. Savjani, K. T., Gajjar, A. K., and Savjani, J. K. (2012). Drug Solubility: Importance and Enhancement Techniques, *ISRN Pharmaceutics*, **2012**, http://dx.doi.org/10.5402/2012/195727.

6. Zhang, Y., Wang, J., Bai, X., Jiang, T., Zhang, Q., and Wang, S. (2012). Mesoporous Silica Nanoparticles for Increasing the Oral Bioavailability and Permeation of Poorly Water Soluble Drugs, *Mol. Pharm.*, **9**, pp. 505–513.

7. Leuner, C., and Dressman, J. (2000). Improving Drug Solubility for Oral Delivery Using Solid Dispersions, *Eur. J. Pharm. Biopharm.*, **50**, pp. 47–60.

8. Koo, O. M., Rubinstein, I., and Onyuksel, H. (2005). Role of Nanotechnology in Targeted Drug Delivery and Imaging: A Concise Review, *Nanomed. Nanotech. Biol. Med.*, **1**, pp. 193–212.

9. Peltonen, L., Valo, H., Kolakovic, R., Laaksonen, T., and Hirvonen, J. (2010). Electrospraying, Spray Drying and Related Techniques for Production and Formulation of Drug Nanoparticles, *Expert Opin. Drug Deliv.*, **7**, pp. 705–719.

10. Sayed, E., Haj-Ahmad, R., Ruparelia, K., Arshad, M., Chang, M., and Ahmad, Z. (2017). Porous Inorganic Drug Delivery Systems—a Review, *AAPS PharmSciTech*, **18**, pp. 1507–1525.

11. Lu, J., Liong, M., Zink, J. I., and Tamanoi, F. (2007). Mesoporous Silica Nanoparticles as a delivery system for hydrophobic anticancer drugs, *Small*, **3**, pp. 1341–1346.

12. Enayati, M., Ahmad, Z., Stride, E., and Edirisinghe, M. (2010). One-step Electrohydrodynamic Production of Drug-Loaded Micro- and Nanoparticles, *J. R. Soc. Interface*, **7**, pp. 667–675.

13. Miyake, Y., Ishida, H., Tanaka, S., and Kolev, S. D. (2013). Theoretical Analysis of the Pseudo-Second Order Kinetic Model of Adsorption. Application to the Adsorption of Ag(I) to Mesoporous Silica Microspheres Functionalized with Thiol Groups, *Chem. Eng. J.*, **218**, pp. 350–357.

14. Sayed, E. G., Hussein, A. K., Khaled, K. A., and Ahmed, O. A. (2015). Improved Corneal Bioavailability of Ofloxacin: Biodegradable Microsphere-Loaded Ion-Activated in situ Gel Delivery System, *Drug Des. Devel. Ther.,* **9**, pp. 1427–1435.

15. Kazunori, K., Glenn S., K., Masayuki, Y., Teruo, O., and Yasuhisa, S. (1993). Block Copolymer Micelles as Vehicles for Drug Delivery, *J. Control. Release,* **24**, pp. 119–132.

16. Dutta, T., Agashe, H. B., Garg, M., Balasubramanium, P., Kabra, M., and Jain, N. K. (2007). Poly (Propyleneimine) Dendrimer Based Nanocontainers for Targeting of Efavirenz to Human Monocytes/ Macrophages in Vitro: Research Paper, *J. Drug Target.,* **15**, pp. 89–98.

17. Wiwattanapatapee, R., Carreño-Gómez, B., Malik, N., and Duncan, R. (2000). Anionic PAMAM Dendrimers Rapidly Cross Adult Rat Intestine in Vitro: A Potential Oral Delivery System? *Pharm. Res.,* **17**, pp. 991–998.

18. Oberoi, H. S., Yorgensen, Y. M., Morasse, A., Evans, J. T., and Burkhart, D. J. (2016). PEG Modified Liposomes Containing CRX-601 Adjuvant in Combination with Methylglycol Chitosan Enhance the Murine Sublingual Immune Response to Influenza Vaccination, *J. Controlled Release,* **223**, pp. 64–74.

19. Kaminski, G. A. T., Sierakowski, M. R., Pontarolo, R., Santos, L. A. D., and Freitas, R. A. d. (2016). Layer-by-Layer Polysaccharide-Coated Liposomes for Sustained Delivery of Epidermal Growth Factor, *Carbohydr. Polym.,* **140**, pp. 129–135.

20. Shakeel, F., Baboota, S., Ahuja, A., Ali, J., and Shafiq, S. (2008). Skin Permeation Mechanism and Bioavailability Enhancement of Celecoxib from Transdermally Applied Nanoemulsion, *J. Nanobiotechnol.,* **6**, pp. 11.

21. Shafiq, S., Shakeel, F., Talegaonkar, S., Ahmad, F. J., Khar, R. K., and Ali, M. (2007). Development and Bioavailability Assessment of Ramipril Nanoemulsion Formulation, *Eur. J. Pharm. Biopharm.,* **66**, pp. 227–243.

22. Li, X., Zhang, Q., Ahmad, Z., Huang, J., Ren, Z., Weng, W., et al. (2015). Near-Infrared Luminescent $CaTiO_3:Nd^{3+}$ Nanofibers with Tunable and Trackable Drug Release Kinetics, *J. Mater. Chem. B,* **3**, pp. 7449–7456.

23. Hartgerink, J. D., Beniash, E., and Stupp, S. I. (2001). Self-Assembly and Mineralization of Peptide-Amphiphile Nanofibers, *Science,* **294**, pp. 1684–1688.

24. Haj-Ahmad, R. R., Elkordy, A. A., and Chaw, C. S. (2015). In Vitro Characterisation of Span 65 Niosomal Formulations Containing Proteins, *Curr. Drug Deliv.,* **12**, pp. 628–639.

25. Guinedi, A. S., Mortada, N. D., Mansour, S., and Hathout, R. M. (2005). Preparation and Evaluation of Reverse-Phase Evaporation and Multilamellar Niosomes as Ophthalmic Carriers of Acetazolamide, *Int. J. Pharm.*, **306**, pp. 71–82.

26. El Maghraby, G. M., Ahmed, A. A., and Osman, M. A. (2015). Penetration Enhancers in Proniosomes as a New Strategy for Enhanced Transdermal Drug Delivery, *Saudi Pharm. J.*, **23**, pp. 67–74.

27. Yuksel, N., Bayindir, Z. S., Aksakal, E., and Ozcelikay, A. T. (2016). In situ Niosome Forming Maltodextrin Proniosomes of Candesartan Cilexetil: In vitro and in vivo Evaluations, *Int. J. Biol. Macromol.*, **82**, pp. 453–463.

28. Sher, P., Ingavle, G., Ponrathnam, S., and Pawar, A. P. (2007). Low Density Porous Carrier: Drug Adsorption and Release Study by Response Surface Methodology using Different Solvents, *Int. J. Pharm.*, **331**, pp. 72–83.

29. Ahuja, G., and Pathak, K. (2009). Porous Carriers for Controlled/Modulated Drug Delivery, *Indian J. Pharm. Sci.*, **71**, pp. 599–607.

30. Arruebo, M. (2012). Drug Delivery from Structured Porous Inorganic Materials, *Wiley Interdiscip Rev. Nanomed. Nanobiotechnol.*, **4**, pp. 16–30.

31. Brinker, C. J. (1996). Porous Inorganic Materials, *Curr. Opin. Solid State Mater. Sci.*, **1**, pp. 798–805.

32. Pal, N., and Bhaumik, A. (2013). Soft Templating Strategies for the Synthesis of Mesoporous Materials: Inorganic, Organic–inorganic Hybrid and Purely Organic Solids, *Adv. Colloid Interface Sci.*, **189–190**, pp. 21–41.

33. Vallet-Regí, M., Balas, F., and Arcos, D. (2007). Mesoporous Materials for Drug Delivery, *Angew. Chem. Int. Ed.*, **46**, pp. 7548–7558.

34. Vallet-Regí, M. (2006). Ordered Mesoporous Materials in the Context of Drug Delivery Systems and Bone Tissue Engineering, *Chem. Eur. J.*, **12**, pp. 5934–5943.

35. Heikkilä, T., Salonen, J., Tuura, J., Hamdy, M. S., Mul, G., Kumar, N., et al. (2007). Mesoporous Silica Material TUD-1 as a Drug Delivery System, *Int. J. Pharm.*, **331**, pp. 133–138.

36. Nishiwaki, A., Watanabe, A., Higashi, K., Tozuka, Y., Moribe, K., and Yamamoto, K. (2009). Molecular States of Prednisolone Dispersed in Folded Sheet Mesoporous Silica (FSM-16), *Int. J. Pharm.*, **378**, pp. 17–22.

37. Zhang, Y., Jiang, T., Zhang, Q., and Wang, S. (2010). Inclusion of Telmisartan in Mesocellular Foam Nanoparticles: Drug Loading and Release Property, *Eur. J. Pharm. Biopharm.,* **76**, pp. 17–23.

38. Popovici, R. F., Seftel, E. M., Mihai, G. D., Popovici, E., and Voicu, V. A. (2011). Controlled Drug Delivery System Based on Ordered Mesoporous Silica Matrices of Captopril as Angiotensin-Converting Enzyme Inhibitor Drug, *J. Pharm. Sci.,* **100**, pp. 704–714.

39. Doadrio, A. L., Sousa, E. M. B., Doadrio, J. C., Pérez Pariente, J., Izquierdo-Barba, I., and Vallet-Regí, M. (2004). Mesoporous SBA-15 HPLC Evaluation for Controlled Gentamicin Drug Delivery, *J. Control. Release,* **97**, pp. 125–132.

40. Manzano, M., Aina, V., Areán, C. O., Balas, F., Cauda, V., Colilla, M., et al. (2008). Studies on MCM-41 Mesoporous Silica for Drug Delivery: Effect of Particle Morphology and Amine Functionalization, *Chem. Eng. J.,* **137**, pp. 30–37.

41. Mellaerts, R., Mols, R., Jammaer, J. A. G., Aerts, C. A., Annaert, P., Van Humbeeck, J., et al. (2008). Increasing the Oral Bioavailability of the Poorly Water Soluble Drug Itraconazole with Ordered Mesoporous Silica, *Eur. J. Pharm. Biopharm.,* **69**, pp. 223–230.

42. Zhao, P., Wang, L., Sun, C., Jiang, T., Zhang, J., Zhang, Q., et al. (2012). Uniform Mesoporous Carbon as a Carrier for Poorly Water Soluble Drug and its Cytotoxicity Study, *Eur. J. Pharm. Biopharm.,* **80**, pp. 535–543.

43. Kim, T. (11). Structurally Ordered Mesoporous Carbon Nanoparticles as Transmembrane Delivery Vehicle in Human Cancer Cells, *Nano Lett.,* **8**, pp. 3724–3727.

44. Kapoor, S., Hegde, R., and Bhattacharyya, A. J. (2009). Influence of Surface Chemistry of Mesoporous Alumina with Wide Pore Distribution on Controlled Drug Release, *J. Control. Release,* **140**, pp. 34–39.

45. Borbane, S., Pande, V., Vibhute, S., Kendre, P., and Dange, V. (2015). Design and Fabrication of Ordered Mesoporous Alumina Scaffold for Drug Delivery of Poorly Water Soluble Drug, *Austin Ther.,* **2**, p. 1015.

46. Tang, S., Huang, X., Chen, X., and Zheng, N. (2010). Hollow Mesoporous Zirconia Nanocapsules for Drug Delivery, *Adv. Funct. Mater,* **20**, pp. 2442–2447.

47. Gedda, G., Pandey, S. S., Khan, S., Talib, A., and Wu, H. F. (2015). Synthesis of Mesoporous Titanium Oxide for Control Release and High Efficiency Drug Delivery of Vinorelbin Bitartrate, *RSC Adv.,* **6**, pp. 13145–13151.

48. Huang, S., Li, C., Cheng, Z., Fan, Y., Yang, P., Zhang, C., et al. (2012). Magnetic Fe_3O_4@mesoporous Silica Composites for Drug Delivery and Bioadsorption, *J. Colloid Interface Sci.,* **376**, pp. 312–321.

49. Reddy, M. N., Cheralathan, K., and Sasikumar, S. (2015). In vitro Bioactivity and Drug Release Kinetics Studies of Mesoporous Silica-Biopolymer Composites, *J. Porous Mater.,* **22**, pp. 1465–1472.

50. Wang, S. (2009). Ordered Mesoporous Materials for Drug Delivery, *Microporous Mesoporous Mater.,* **117**, pp. 1–9.

51. Song, S., Hidajat, K., and Kawi, S. (2005). Functionalized SBA-15 Materials as Carriers for Controlled Drug Delivery: Influence of Surface Properties on Matrix-Drug Interactions, *Langmuir,* **21**, pp. 9568–9575.

52. Li, Z., Zhang, L., Tang, C., and Yin, C. (2017). Co-Delivery of Doxorubicin and Survivin shRNA-Expressing Plasmid Via Microenvironment-Responsive Dendritic Mesoporous Silica Nanoparticles for Synergistic Cancer Therapy, *Pharm. Res.,* **34**, pp. 2829–2841.

53. Shahbazi, M., Herranz, B., and Santos, H. A. (2012). Nanostructured Porous Si-Based Nanoparticles for Targeted Drug Delivery, *Biomatter,* **2**, pp. 296–312.

54. Bharti, C., Nagaich, U., Pal, A. K., and Gulati, N. (2015). Mesoporous Silica Nanoparticles in Target Drug Delivery System: A Review, *Int. J. Pharm. Invest.,* **5**, pp. 124–133.

55. Zeng, W., Qian, X., Zhang, Y., Yin, J., and Zhu, Z. (2005). Organic Modified Mesoporous MCM-41 through Solvothermal Process as Drug Delivery System, *Mater. Res. Bull.,* **40**, pp. 766–772.

56. Beck, J., Vartuli, J., Roth, W. J., Leonowicz, M., Kresge, C., Schmitt, K., et al. (1992). A New Family of Mesoporous Molecular Sieves Prepared with Liquid Crystal Templates, *J. Am. Chem. Soc.,* **114**, pp. 10834–10843.

57. Hoffmann, F., Cornelius, M., Morell, J., and Fröba, M. (2006). Silica-based Mesoporous Organic–Inorganic Hybrid Materials, *Angew. Chem. Int. Ed.,* **45**, pp. 3216–3251.

58. Zhao, D., Huo, Q., Feng, J., Chmelka, B. F., and Stucky, G. D. (1998). Nonionic Triblock and Star Diblock Copolymer and Oligomeric Surfactant Syntheses of Highly Ordered, Hydrothermally Stable, Mesoporous Silica Structures, *J. Am. Chem. Soc.,* **120**, pp. 6024–6036.

59. Giraldo, L., López, B., Pérez, L., Urrego, S., Sierra, L., and Mesa, M. (2007). Mesoporous silica applications. *Macromolecular Symposia,* 258(1) pp. 129–141.

60. Jansen, J., Shan, Z., Marchese, L., Zhou, W., vd Puil, N., and Maschmeyer, T. (2001). A New Templating Method for Three-Dimensional Mesopore Networks, *Chem. Commun.,* pp. 713–714.

61. Inagaki, S., Koiwai, A., Suzuki, N., Fukushima, Y., and Kuroda, K. (1996). Syntheses of Highly Ordered Mesoporous Materials, FSM-16, Derived from Kanemite, *Bull. Chem. Soc. Jpn.,* **69**, pp. 1449–1457.

62. Uejo, F., Limwikrant, W., Moribe, K., and Yamamoto, K. (2013). Dissolution Improvement of Fenofibrate by Melting Inclusion in Mesoporous Silica, *Asian J. Pharmacol.,* **8**, pp. 329–335.

63. Vallet-Regi, M., Ramila, A., Del Real, R., and Pérez-Pariente, J. (2001). A New Property of MCM-41: Drug Delivery System, *Chem. Mater,* **13**, pp. 308–311.

64. Chen, X., Sun, H., Hu, J., Han, X., Liu, H., and Hu, Y. (2017). Transferrin gated mesoporous silica nanoparticles for redox-responsive and targeted drug delivery, *Colloids Surf. B: Biointerfaces*, **152**, pp. 77–84.

65. Horcajada, P., Ramila, A., Perez-Pariente, J., and Vallet-Regı, M. (2004). Influence of Pore Size of MCM-41 Matrices on Drug Delivery Rate, *Microporous Mesoporous Mater.,* **68**, pp. 105–109.

66. Deodhar, G. V., Adams, M. L., and Trewyn, B. G. (2017). Controlled Release and Intracellular Protein Delivery from Mesoporous Silica Nanoparticles, *Biotechnol. J.,* **12, doi: 10.1002/biot.201600408.**

67. Rehman, F., Ahmed, K., Airoldi, C., Gaisford, S., Buanz, A., Rahim, A., et al. (2017). Amine Bridges Grafted Mesoporous Silica, as a Prolonged/ Controlled Drug Release System for the Enhanced Therapeutic Effect of Short Life Drugs, *Mater. Sci. Eng. C,* **72**, pp. 34–41.

68. Li-hong, W., Xin, C., Hui, X., Li-li, Z., Jing, H., Mei-juan, Z., et al. (2013). A Novel Strategy to Design Sustained-Release Poorly Water-Soluble Drug Mesoporous Silica Microparticles Based on Supercritical Fluid Technique, *Int. J. Pharm.,* **454**, pp. 135–142.

69. Kumar, B., Kulanthaivel, S., Mondal, A., Mishra, S., Banerjee, B., Bhaumik, A., et al. (2017). Mesoporous Silica Nanoparticle Based Enzyme Responsive System for Colon Specific Drug Delivery through Guar Gum Capping, *Colloids Surf. B Biointerfaces,* **150**, pp. 352–361.

70. Song, Y., Li, Y., Xu, Q., and Liu, Z. (2016). Mesoporous Silica Nanoparticles for Stimuli-Responsive Controlled Drug Delivery: Advances, Challenges, and Outlook, *Int. J. Nanomed.,* **12**, pp. 87–110.

71. Dilnawaz, F., and Sahoo, S. K. (2018). Augmented Anticancer Efficacy by Si-RNA Complexed Drug Loaded Mesoporous Silica Nanoparticles in Lung Cancer Therapy, *ACS Appl. Nano Mater.,* 1(2), pp. 730–740.

72. O'shea, J. P., Nagarsekar, K., Wieber, A., Witt, V., Herbert, E., O'driscoll, C. M., et al. (2017). Mesoporous Silica-based Dosage Forms Improve Bioavailability of Poorly Soluble Drugs in Pigs: Case Example Fenofibrate, *J. Pharm. Pharmacol.*, **69**, pp. 1284–1292.

73. Thomas, M. J. K., Slipper, I., Walunj, A., Jain, A., Favretto, M. E., Kallinteri, P., et al. (2010). Inclusion of Poorly Soluble Drugs in Highly Ordered Mesoporous Silica Nanoparticles, *Int. J. Pharm.*, **387**, pp. 272–277.

74. Ambrogi, V., Perioli, L., Marmottini, F., Giovagnoli, S., Esposito, M., and Rossi, C. (2007). Improvement of Dissolution Rate of Piroxicam by Inclusion into MCM-41 Mesoporous Silicate, *Eur. J. Pharm. Sci.*, **32**, pp. 216–222.

75. Azaïs, T., Laurent, G., Panesar, K., Nossov, A., Guenneau, F., Sanfeliu Cano, C., et al. (2017). Implication of Water Molecules at the Silica-Ibuprofen Interface in Silica-Based Drug Delivery Systems Obtained through Incipient Wetness Impregnation, *J. Phys. Chem. C*, **121**, pp. 26833–26839.

76. Heikkilä, T., Salonen, J., Tuura, J., Kumar, N., Salmi, T., Murzin, D. Y., et al. (2007). Evaluation of Mesoporous TCPSi, MCM-41, SBA-15, and TUD-1 Materials as API Carriers for Oral Drug Delivery, *Drug Deliv.*, **14**, pp. 337–347.

77. He, Q., and Shi, J. (2011). Mesoporous Silica Nanoparticle Based Nano Drug Delivery Systems: Synthesis, Controlled Drug Release and Delivery, Pharmacokinetics and Biocompatibility, *J. Mater. Chem.*, **21**, pp. 5845–5855.

78. Van Speybroeck, M., Mellaerts, R., Martens, J. A., Annaert, P., Van den Mooter, G., and Augustijns, P. (2011). *Ordered Mesoporous Silica for the Delivery of Poorly Soluble Drugs. Controlled Release in Oral Drug Delivery*, pp. 203–219, Springer.

79. Zou, Z., Li, S., He, D., He, X., Wang, K., Li, L., et al. (2017). A Versatile Stimulus-Responsive Metal–organic Framework for Size/Morphology Tunable Hollow Mesoporous Silica and pH-Triggered Drug Delivery, *J. Mater. Chem. B*, **5**, pp. 2126–2132.

80. Ghasemi, S., Farsangi, Z. J., Beitollahi, A., Mirkazemi, M., Rezayat, S. M., and Sarkar, S. (2017). Synthesis of hollow mesoporous silica (HMS) nanoparticles as a candidate for sulfasalazine drug loading, *Ceramics Int.*, **43**, pp. 11225–11232.

81. Zhang, C., Hou, T., Chen, J., and Wen, L. (2010). Preparation of Mesoporous Silica Microspheres with Multi-Hollow Cores and their Application in Sustained Drug Release, *Particuology*, **8**, pp. 447–452.

82. Sathe, T. R., Agrawal, A., and Nie, S. (2006). Mesoporous Silica Beads Embedded with Semiconductor Quantum Dots and Iron Oxide Nanocrystals: Dual-Function Microcarriers for Optical Encoding and Magnetic Separation, *Anal. Chem.*, **78**, pp. 5627–5632.

83. Mao, C., Wang, F., and Cao, B. (2012). Controlling Nanostructures of Mesoporous Silica Fibers by Supramolecular Assembly of Genetically Modifiable Bacteriophages, *Angew. Chem.*, **124**, pp. 6517–6521.

84. Ahern, R. J., Crean, A. M., and Ryan, K. B. (2012). The Influence of Supercritical Carbon Dioxide ($SC-CO_2$) Processing Conditions on Drug Loading and Physicochemical Properties, *Int. J. Pharm.*, **439**, pp. 92–99.

85. McCarthy, C. A., Faisal, W., O'Shea, J. P., Murphy, C., Ahern, R. J., Ryan, K. B., Griffin, B. T., and Crean, A. M. (2017). In vitro dissolution models for the prediction of in vivo performance of an oral mesoporous silica formulation, *J. Control. Release*, **250**, pp. 86–95.

86. Mellaerts, R., Aerts, C. A., Van Humbeeck, J., Augustijns, P., Van den Mooter, G., and Martens, J. A. (2007). Enhanced Release of Itraconazole from Ordered Mesoporous SBA-15 Silica Materials, *Chem. Commun*, **13**, pp. 1375–1377.

87. Nafisi, S., Schäfer-Korting, M., and Maibach, H. I. (2015). Perspectives on Percutaneous Penetration: Silica Nanoparticles, *Nanotoxicology*, **9**, pp. 643–657.

88. Berlier, G., Gastaldi, L., Sapino, S., Miletto, I., Bottinelli, E., Chirio, D., et al. (2013). MCM-41 as a Useful Vector for Rutin Topical Formulations: Synthesis, Characterization and Testing, *Int. J. Pharm.*, **457**, pp. 177–186.

89. Berlier, G., Gastaldi, L., Ugazio, E., Miletto, I., Iliade, P., and Sapino, S. (2013). Stabilization of Quercetin Flavonoid in MCM-41 Mesoporous Silica: Positive Effect of Surface Functionalization, *J. Colloid Interface Sci.*, **393**, pp. 109–118.

90. Gastaldi, L., Ugazio, E., Sapino, S., Iliade, P., Miletto, I., and Berlier, G. (2012). Mesoporous Silica as a Carrier for Topical Application: The Trolox Case Study, *Phys. Chem. Chem. Phys.*, **14**, pp. 11318–11326.

91. Chen-Yang, Y. W., Chen, Y. T., Li, C. C., Yu, H. C., Chuang, Y. C., Su, J. H., et al. (2011). Preparation of UV-Filter Encapsulated Mesoporous Silica with High Sunscreen Ability, *Mater. Lett.*, **65**, pp. 1060–1062.

92. Ambrogi, V., Perioli, L., Marmottini, F., Latterini, L., Rossi, C., and Costantino, U. (2007). Mesoporous Silicate MCM-41 Containing Organic Ultraviolet Ray Absorbents: Preparation, Photostability and in Vitro Release, *J. Phys. Chem. Solids*, **68**, pp. 1173–1177.

93. Ambrogi, V., Perioli, L., Pagano, C., Marmottini, F., Moretti, M., Mizzi, F., et al. (2010). Econazole Nitrate-loaded MCM-41 for an Antifungal Topical Powder Formulation, *J. Pharm. Sci.*, **99**, pp. 4738–4745.

94. Lu, J., Liong, M., Li, Z., Zink, J. I., and Tamanoi, F. (2010). Biocompatibility, Biodistribution, and Drug-delivery Efficiency of Mesoporous Silica Nanoparticles for Cancer Therapy in Animals, *Small*, **6**, pp. 1794–1805.

95. Lee, C., Cheng, S., Huang, I., Souris, J. S., Yang, C., Mou, C., et al. (2010). Intracellular pH-responsive Mesoporous Silica Nanoparticles for the Controlled Release of Anticancer Chemotherapeutics, *Angew. Chem.*, **122**, pp. 8390–8395.

96. Xue, Z., Liang, D., Li, Y., Long, Z., Pan, Q., Liu, X., et al. (2005). Silica Nanoparticle is a Possible Safe Carrier for Gene Therapy, *Chin. Sci. Bull*, **50**, pp. 2323–2327.

97. Barrabino, A. (2011). Synthesis of Mesoporous Silica Particles with Control of Both Pore Diameter and Particle Size. Master Thesis. *Chalmers University of Technology.*

98. Bagshaw, S. A., Prouzet, E., and Pinnavaia, T. J. (1995). Templating of Mesoporous Molecular Sieves by Nonionic Polyethylene Oxide Surfactants, *Science*, **269**, pp. 1242–1244.

99. Kresge, C., Leonowicz, M., Roth, W., Vartuli, J., and Beck, J. (1992). Ordered Mesoporous Molecular Sieves Synthesized by a Liquid-Crystal Template Mechanism, *Nature*, **359**, pp. 710–712.

100. Cicuéndez, M., Izquierdo-Barba, I., Portolés, M. T., and Vallet-Regí, M. (2013). Biocompatibility and Levofloxacin Delivery of Mesoporous Materials, *Eur. J. Pharm. Biopharm.*, **84**, pp. 115–124.

101. He, Q., Shi, J., Zhu, M., Chen, Y., and Chen, F. (2010). The Three-Stage in vitro Degradation Behavior of Mesoporous Silica in Simulated Body Fluid, *Microporous Mesoporous Mater.*, **131**, pp. 314–320.

102. Cauda, V., Schlossbauer, A., and Bein, T. (2010). Bio-Degradation Study of Colloidal Mesoporous Silica Nanoparticles: Effect of Surface Functionalization with Organo-Silanes and Poly(Ethylene Glycol), *Microporous Mesoporous Mater.*, **132**, pp. 60–71.

103. He, Q., Zhang, Z., Gao, Y., Shi, J., and Li, Y. (2009). Intracellular Localization and Cytotoxicity of Spherical Mesoporous Silica Nano- and Microparticles, *Small*, **5**, pp. 2722–2729.

104. Zhai, W., He, C., Wu, L., Zhou, Y., Chen, H., Chang, J., et al. (2012). Degradation of Hollow Mesoporous Silica Nanoparticles in Human Umbilical Vein Endothelial Cells, *J. Biomed. Mater. Res. Part B Appl. Biomater.*, **100**, pp. 1397–1403.

105. Chen, G., Teng, Z., Su, X., Liu, Y., and Lu, G. (2015). Unique Biological Degradation Behavior of Stöber Mesoporous Silica Nanoparticles from their Interiors to their Exteriors, *J. Biomed. Nanotechnol.*, **11**, pp. 722–729.

106. Kumar, R., Roy, I., Ohulchanskky, T. Y., Vathy, L. A., Bergey, E. J., Sajjad, M., et al. (2010). In vivo Biodistribution and Clearance Studies using Multimodal Organically Modified Silica Nanoparticles, *ACS Nano*, **4**, pp. 699–708.

107. He, Q., Zhang, Z., Gao, F., Li, Y., and Shi, J. (2011). In Vivo Biodistribution and Urinary Excretion of Mesoporous Silica Nanoparticles: Effects of Particle Size and PEGylation, *Small*, **7**, pp. 271–280.

108. Ahern, R. J., Hanrahan, J. P., Tobin, J. M., Ryan, K. B., and Crean, A. M. (2013). Comparison of Fenofibrate–Mesoporous Silica Drug-Loading Processes for Enhanced Drug Delivery, *Eur. J. Pharm. Sci.*, **50**, pp. 400–409.

109. Sayed, E., Karavasili, C., Ruparelia, K., Haj-Ahmad, R., Charalambopoulou, G., Steriotis, T., et al. (2018). Electrosprayed Mesoporous Particles for Improved Aqueous Solubility of a Poorly Water Soluble Anticancer Agent: In Vitro and Ex Vivo Evaluation, *J. Control. Release*, **278**, pp. 142–155.

110. Qian, K. K., Suib, S. L., and Bogner, R. H. (2011). Spontaneous Crystalline-to-amorphous Phase Transformation of Organic Or Medicinal Compounds in the Presence of Porous Media, Part 2: Amorphization Capacity and Mechanisms of Interaction, *J. Pharm. Sci.*, **100**, pp. 4674–4686.

111. Hong, S., Shen, S., Tan, D. C. T., Ng, W. K., Liu, X., Chia, L. S., et al. (2016). High Drug Load, Stable, Manufacturable and Bioavailable Fenofibrate Formulations in Mesoporous Silica: A Comparison of Spray Drying Versus Solvent Impregnation Methods, *Drug Deliv.*, **23**, pp. 316–327.

112. Shen, S., Ng, W. K., Chia, L., Dong, Y., and Tan, R. B. (2010). Stabilized Amorphous State of Ibuprofen by Co-Spray Drying with Mesoporous SBA-15 to Enhance Dissolution Properties, *J. Pharm. Sci.*, **99**, pp. 1997–2007.

113. Waters, L. J., Hussain, T., Parkes, G., Hanrahan, J. P., and Tobin, J. M. (2013). Inclusion of Fenofibrate in a Series of Mesoporous Silicas using Microwave Irradiation, *Eur. J. Pharm. Biopharm.*, **85**, pp. 936–941.

114. Charnay, C., Bégu, S., Tourné-Péteilh, C., Nicole, L., Lerner, D. A., and Devoisselle, J. M. (2004). Inclusion of Ibuprofen in Mesoporous Templated Silica: Drug Loading and Release Property, *Eur. J. Pharm. Biopharm.*, **57**, pp. 533–540.

115. Shen, S., Ng, W. K., Hu, J., Letchmanan, K., Ng, J., and Tan, R. B. H. (2017). Solvent-Free Direct Formulation of Poorly-Soluble Drugs to Amorphous Solid Dispersion via Melt-Absorption, *Adv. Powder Technol.*, **28**, pp. 1316–1324.

116. Limnell, T., Santos, H. A., Mäkilä, E., Heikkilä, T., Salonen, J., Murzin, D. Y., et al. (2011). Drug Delivery Formulations of Ordered and Nonordered Mesoporous Silica: Comparison of Three Drug Loading Methods, *J. Pharm. Sci.*, **100**, pp. 3294–3306.

117. Letchmanan, K., Shen, S., Ng, W. K., and Tan, R. B. H. (2017). Dissolution and Physicochemical Stability Enhancement of Artemisinin and Mefloquine co-Formulation via Nano-Confinement with Mesoporous SBA-15, *Colloids Surf. B: Biointerfaces*, **155**, pp. 560–568.

118. Lebold, T. (2009). Nanostructured Silica Materials as Drug-Delivery Systems for Doxorubicin: Single Molecule and Cellular Studies, *Nano Lett.*, **9**, pp. 2877–2883.

119. Lu, J., Liong, M., Sherman, S., Xia, T., Kovochich, M., Nel, A. E., et al. (2007). Mesoporous Silica Nanoparticles for Cancer Therapy: Energy-Dependent Cellular Uptake and Delivery of Paclitaxel to Cancer Cells, *Nanobiotechnology*, **3**, pp. 89–95.

120. Feng, Y., Panwar, N., Tng, D. J. H., Tjin, S. C., Wang, K., and Yong, K. (2016). The Application of Mesoporous Silica Nanoparticle Family in Cancer Theranostics, *Coordination Chem. Rev.*, **319**, pp. 86–109.

121. Rosenholm, J. M., Peuhu, E., Bate-Eya, L. T., Eriksson, J. E., Sahlgren, C., and Lindén, M. (2010). Cancer-Cell-Specific Induction of Apoptosis using Mesoporous Silica Nanoparticles as Drug-Delivery Vectors, *Small*, **6**, pp. 1234–1241.

122. He, Y., Luo, L., Liang, S., Long, M., and Xu, H. (2017). Amino-Functionalized Mesoporous Silica Nanoparticles as Efficient Carriers for Anticancer Drug Delivery, *J. Biomater. Appl.*, **32**, pp. 524–532.

123. Poonia, N., Lather, V., and Pandita, D. (2018). Mesoporous Silica Nanoparticles: A Smart Nanosystem for Management of Breast Cancer, *Drug Discov. Today*, **23**, pp. 315–332.

124. Tian, B., Liu, S., Wu, S., Lu, W., Wang, D., Jin, L., et al. (2017). pH-Responsive Poly (Acrylic Acid)-Gated Mesoporous Silica and its Application in Oral Colon Targeted Drug Delivery for Doxorubicin, *Colloids Surf. B: Biointerfaces*, **154**, pp. 287–296.

125. Zhang, Q., Zhao, H., Li, D., Liu, L., and Du, S. (2017). A Surface-Grafted Ligand Functionalization Strategy for Coordinate Binding of Doxorubicin at Surface of PEGylated Mesoporous Silica Nanoparticles:

Toward pH-Responsive Drug Delivery, *Colloids Surf. B: Biointerfaces*, **149**, pp. 138–145.

126. Meng, H., Xue, M., Xia, T., Ji, Z., Tarn, D. Y., Zink, J. I., et al. (2011). Use of Size and a Copolymer Design Feature to Improve the Biodistribution and the Enhanced Permeability and Retention Effect of Doxorubicin-Loaded Mesoporous Silica Nanoparticles in a Murine Xenograft Tumor Model, *ACS Nano*, **5**, pp. 4131–4144.

127. Chen, A. M., Zhang, M., Wei, D., Stueber, D., Taratula, O., Minko, T., et al. (2009). Co-delivery of Doxorubicin and Bcl-2 siRNA by Mesoporous Silica Nanoparticles Enhances the Efficacy of Chemotherapy in Multidrug-Resistant Cancer Cells, *Small*, **5**, pp. 2673–2677.

128. Meng, H., Mai, W. X., Zhang, H., Xue, M., Xia, T., Lin, S., et al. (2013). Codelivery of an Optimal Drug/siRNA Combination using Mesoporous Silica Nanoparticles to Overcome Drug Resistance in Breast Cancer in Vitro and in Vivo, *ACS Nano*, **7**, pp. 994–1005.

129. Gary-Bobo, M., Hocine, O., Brevet, D., Maynadier, M., Raehm, L., Richeter, S., et al. (2012). Cancer Therapy Improvement with Mesoporous Silica Nanoparticles Combining Targeting, Drug Delivery and PDT, *Int. J. Pharm.*, **423**, pp. 509–515.

130. Halamová, D., Badaničová, M., Zeleňák, V., Gondová, T., and Vainio, U. (2010). Naproxen Drug Delivery using Periodic Mesoporous Silica SBA-15, *Appl. Surf. Sci.*, **256**, pp. 6489–6494.

131. Lee, C., Lo, L., Mou, C., and Yang, C. (2008). Synthesis and Characterization of Positive-Charge Functionalized Mesoporous Silica Nanoparticles for Oral Drug Delivery of an Anti-Inflammatory Drug, *Adv. Funct. Mater*, **18**, pp. 3283–3292.

132. Popova, M., Trendafilova, I., Zgureva, D., Kalvachev, Y., Boycheva, S., Novak Tušar, N., and Szegedi, A. (2018). Polymer-Coated Mesoporous Silica Nanoparticles for Controlled Release of the Prodrug Sulfasalazine, *J. Drug Deliv. Sci. Technol.*, **44**, pp. 415–420.

133. Juère, E., Florek, J., Bouchoucha, M., Jambhrunkar, S., Wong, K. Y., Popat, A., et al. (2017). In Vitro Dissolution, Cellular Membrane Permeability, and Anti-Inflammatory Response of Resveratrol-Encapsulated Mesoporous Silica Nanoparticles, *Mol. Pharm.*, **14**, pp. 4431–4441.

134. Velikova, N., Mas, N., Miguel-Romero, L., Polo, L., Stolte, E., Zaccaria, E., et al. (2017). Broadening the Antibacterial Spectrum of Histidine Kinase Autophosphorylation Inhibitors Via the use of Epsilon-Poly-L-Lysine Capped Mesoporous Silica-Based Nanoparticles, *Nanomedicine*, **13**, pp. 569–581.

135. Ruehle, B., Clemens, D. L., Lee, B., Horwitz, M. A., and Zink, J. I. (2017). A Pathogen-Specific Cargo Delivery Platform Based on Mesoporous Silica Nanoparticles, *J. Am. Chem. Soc.,* **139**, pp. 6663–6668.

136. Vallet-Regí, M., Doadrio, J. C., Doadrio, A. L., Izquierdo-Barba, I., and Pérez-Pariente, J. (2004). Hexagonal Ordered Mesoporous Material as a Matrix for the Controlled Release of Amoxicillin, *Solid State Ion,* **172**, pp. 435–439.

137. Michailidis, M., Sorzabal-Bellido, I., Adamidou, E. A., Diaz-Fernandez, Y. A., Aveyard, J., Wengier, R., et al. (2017). Modified Mesoporous Silica Nanoparticles with a Dual Synergetic Antibacterial Effect, *ACS Appl. Mater. Interf.,* **9**, pp. 38364–38372.

138. Radin, S., Chen, T., and Ducheyne, P. (2009). The Controlled Release of Drugs from Emulsified, Sol Gel Processed Silica Microspheres, *Biomaterials,* **30**, pp. 850–858.

139. Izquierdo-Barba, I., Martinez, Á, Doadrio, A. L., Pérez-Pariente, J., and Vallet-Regí, M. (2005). Release Evaluation of Drugs from Ordered Three-Dimensional Silica Structures, *Eur. J. Pharm. Sci.,* **26**, pp. 365–373.

140. Mas, N., Galiana, I., Mondragón, L., Aznar, E., Climent, E., Cabedo, N., et al. (2013). Enhanced Efficacy and Broadening of Antibacterial Action of Drugs Via the use of Capped Mesoporous Nanoparticles, *Chem. Eur. J.,* **19**, pp. 11167–11171.

141. Doadrio, A. L., Sousa, E. M. B., Doadrio, J. C., Pérez Pariente, J., Izquierdo-Barba, I., and Vallet-Regí, M. (2004). Mesoporous SBA-15 HPLC Evaluation for Controlled Gentamicin Drug Delivery, *J. Control. Release,* **97**, pp. 125–132.

142. González, B., Colilla, M., Díez, J., Pedraza, D., Guembe, M., Izquierdo-Barba, I., et al. (2018). Mesoporous Silica Nanoparticles Decorated with Polycationic Dendrimers for Infection Treatment, *Acta Biomater.,* **68**, 261–271.

143. Xu, C., He, Y., Li, Z., Yusilawati, A. N., and Ye, Q. (2018). Nanoengineered Hollow Mesoporous Silica Nanoparticles for the Delivery of Antimicrobial Protein into Biofilm, *J Mater Chem B,* **6**, 1899–1902.

144. Horcajada, P., Rámila, A., Boulahya, K., González-Calbet, J., and Vallet-Regí, M. (2004). Bioactivity in Ordered Mesoporous Materials, *Solid State Sci.,* **6**, pp. 1295–1300.

145. Izquierdo-Barba, I., Ruiz-González, L., Doadrio, J. C., González-Calbet, J. M., and Vallet-Regí, M. (2005). Tissue Regeneration: A New Property of Mesoporous Materials, *Solid State Sci.,* **7**, pp. 983–989.

146. Xu, Y., Gao, D., Feng, P., Gao, C., Peng, S., Ma, H., et al. (2017). A Mesoporous Silica Composite Scaffold: Cell Behaviors, Biomineralization and Mechanical Properties, *Appl. Surf. Sci.*, **423**, pp. 314–321.

147. Ezazi, N. Z., Shahbazi, M., Shatalin, Y. V., Nadal, E., Mäkilä, E., Salonen, J., et al. (2018). Conductive Vancomycin-Loaded Mesoporous Silica Polypyrrole-Based Scaffolds for Bone Regeneration, *Int. J. Pharm.*, **536**, pp. 241–250.

148. Vallet-Regí, M. (2006). Ordered Mesoporous Materials in the Context of Drug Delivery Systems and Bone Tissue Engineering, *Chem. Eur. J.*, **12**, pp. 5934–5943.

149. Vallet-Regí, M., Ruiz-González, L., Izquierdo-Barba, I., and González-Calbet, J. M. (2006). Revisiting Silica Based Ordered Mesoporous Materials: Medical Applications, *J. Mater. Chem.*, **16**, pp. 26–31.

150. Shadjou, N., and Hasanzadeh, M. (2015). Bone Tissue Engineering using Silica-Based Mesoporous Nanobiomaterials:Recent Progress, *Mater. Sci. Eng., C*, **55**, pp. 401–409.

151. Luo, Z., Deng, Y., Zhang, R., Wang, M., Bai, Y., Zhao, Q., et al. (2015). Peptide-Laden Mesoporous Silica Nanoparticles with Promoted Bioactivity and Osteo-Differentiation Ability for Bone Tissue Engineering, *Colloids Surf. B*, **131**, pp. 73–82.

152. Mortera, R., Onida, B., Fiorilli, S., Cauda, V., Brovarone, C. V., Baino, F., et al. (2008). Synthesis and Characterization of MCM-41 Spheres Inside Bioactive Glass–ceramic Scaffold, *Chem. Eng. J.*, **137**, pp. 54–61.

153. Mi, H., Jing, X., Napiwocki, B. N., Li, Z., Turng, L., and Huang, H. (2018). Fabrication of Fibrous Silica Sponges by Self-Assembly Electrospinning and their Application in Tissue Engineering for Three-Dimensional Tissue Regeneration, *Chem. Eng. J.*, **331**, pp. 652–662.

154. Vallet-Regí, M., Izquierdo-Barba, I., Rámila, A., Pérez-Pariente, J., Babonneau, F., and González-Calbet, J. M. (2005). Phosphorous-Doped MCM-41 as Bioactive Material, *Solid State Sci.*, **7**, pp. 233–237.

155. Hartono, S. B., Yu, M., Gu, W., Yang, J., Strounina, E., Wang, X., et al. (2014). Synthesis of Multi-Functional Large Pore Mesoporous Silica Nanoparticles as Gene Carriers, *Nanotechnology*, **25**, p. 055701.

156. Kim, M., Na, H., Kim, Y., Ryoo, S., Cho, H. S., Lee, K. E., et al. (2011). Facile Synthesis of Monodispersed Mesoporous Silica Nanoparticles with Ultralarge Pores and their Application in Gene Delivery, *ACS Nano*, **5**, pp. 3568–3576.

157. Slowing, I. I., Vivero-Escoto, J. L., Wu, C., and Lin, V. S. (2008). Mesoporous Silica Nanoparticles as Controlled Release Drug Delivery

and Gene Transfection Carriers, *Adv. Drug Deliv. Rev.,* **60**, pp. 1278–1288.

158. Hartono, S. B., Phuoc, N. T., Yu, M., Jia, Z., Monteiro, M. J., Qiao, S., et al. (2014). Functionalized Large Pore Mesoporous Silica Nanoparticles for Gene Delivery Featuring Controlled Release and Co-Delivery, *J. Mater. Chem. B,* **2**, pp. 718–726.

159. Mamaeva, V., Sahlgren, C., and Lindén, M. (2013). Mesoporous Silica Nanoparticles in medicine—Recent Advances, *Adv. Drug Deliv. Rev.,* **65**, pp. 689–702.

160. Kwon, D., Cha, B. G., Cho, Y., Min, J., Park, E., Kang, S., et al. (2017). Extra-Large Pore Mesoporous Silica Nanoparticles for Directing in vivo M2 Macrophage Polarization by Delivering IL-4, *Nano Lett.,* **17**, pp. 2747–2756.

161. Na, H., Kim, M., Park, K., Ryoo, S., Lee, K. E., Jeon, H., et al. (2012). Efficient Functional Delivery of siRNA Using Mesoporous Silica Nanoparticles with Ultralarge Pores, *Small,* **8**, pp. 1752–1761.

162. Li, Y., Hei, M., Xu, Y., Qian, X., and Zhu, W. (2017). Ammonium Salt Modified Mesoporous Silica Nanoparticles for Dual Intracellular Stimuli-Responsive Gene Delivery, *J. Control. Release,* **259**, pp. e139–e140.

163. Chang, J., Tsai, P., Chen, W., Chiou, S., and Mou, C. (2017). Dual Delivery of siRNA and Plasmid DNA using Mesoporous Silica Nanoparticles to Differentiate Induced Pluripotent Stem Cells into Dopaminergic Neurons, *J. Mater. Chem. B,* **5**, pp. 3012–3023.

164. Zhou, Y., Quan, G., Wu, Q., Zhang, X., Niu, B., Wu, B., Huang, Y., Pan, X. (2018). Mesoporous Silica Nanoparticles for Drug and Gene Delivery, *Acta Pharm. Sinica B,* **8**, pp. 165–177.

165. Ivanov, S., Zhuravsky, S., Yukina, G., Tomson, V., Korolev, D., and Galagudza, M. (2012). In Vivo Toxity of Intravenously Administered Silica and Silicon Nanoparticles, *Materials,* **5**, pp. 1873–1889.

166. Petushkov, A., Ndiege, N., Salem, A. K., and Larsen, S. C. (2010). Toxicity of Silica Nanomaterials: Zeolites, Mesoporous Silica, and Amorphous Silica Nanoparticles, *Adv. Mol. Tox.,* **4**, pp. 223–266.

167. Tang, F., Li, L., and Chen, D. (2012). Mesoporous Silica Nanoparticles: Synthesis, Biocompatibility and Drug Delivery, *Adv. Mater.,* **24**, pp. 1504–1534.

168. Slowing, I. I., Wu, C., Vivero-Escoto, J. L., and Lin, V. S. (2009). Mesoporous Silica Nanoparticles for Reducing Hemolytic Activity Towards Mammalian Red Blood Cells, *Small,* **5**, pp. 57–62.

169. Malvindi, M. A., Brunetti, V., Vecchio, G., Galeone, A., Cingolani, R., and Pompa, P. P. (2012). SiO 2 Nanoparticles Biocompatibility and their Potential for Gene Delivery and Silencing, *Nanoscale,* **4**, pp. 486–495.

170. Napierska, D., Thomassen, L. C., Rabolli, V., Lison, D., Gonzalez, L., Kirsch-Volders, M., et al. (2009). Size-Dependent Cytotoxicity of Monodisperse Silica Nanoparticles in Human Endothelial Cells, *Small,* **5**, pp. 846–853.

171. Di Pasqua, A. J., Sharma, K. K., Shi, Y., Toms, B. B., Ouellette, W., Dabrowiak, J. C., et al. (2008). Cytotoxicity of Mesoporous Silica Nanomaterials, *J. Inorg. Biochem.,* **102**, pp. 1416–1423.

172. Tao, Z., Morrow, M. P., Asefa, T., Sharma, K. K., Duncan, C., Anan, A., et al. (2008). Mesoporous Silica Nanoparticles Inhibit Cellular Respiration, *Nano Lett.,* **8**, pp. 1517–1526.

173. Luo, G., Chen, W., Liu, Y., Lei, Q., Zhuo, R., and Zhang, X. (2014). Multifunctional Enveloped Mesoporous Silica Nanoparticles for Subcellular Co-Delivery of Drug and Therapeutic Peptide, *Sci. Rep.,* **4**, pp. 1–10.

174. Zhu, C., Wang, X., Lin, Z., Xie, Z., and Wang, X. (2014). Cell Microenvironment Stimuli-Responsive Controlled-Release Delivery Systems Based on Mesoporous Silica Nanoparticles, *J. Food Drug Anal.,* **22**, pp. 18–28.

175. Jia, L., Li, Z., Shen, J., Zheng, D., Tian, X., Guo, H., et al. (2015). Multifunctional Mesoporous Silica Nanoparticles Mediated Co-Delivery of Paclitaxel and Tetrandrine for Overcoming Multidrug Resistance, *Int. J. Pharm.,* **489**, pp. 318–330.

176. Ma, J., Lin, H., Xing, R., Li, X., Bian, C., Xiang, D., et al. (2014). Synthesis of pH-Responsive Mesoporous Silica Nanotubes for Controlled Release, *J. Sol Gel Sci. Technol.,* **69**, pp. 364–369.

177. Ng, D. K., Wong, R., Fong, W., and Lo, P. (2017). Encapsulating pH-Responsive Doxorubicin-Phthalocyanine Conjugates in Mesoporous Silica Nanoparticles for Combined Photodynamic Therapy and Controlled Chemotherapy, *Chem. Eur. J.,* **23**, pp. 16505–16515.

178. Sun, R., Wang, W., Wen, Y., and Zhang, X. (2015). Recent Advance on Mesoporous Silica Nanoparticles-Based Controlled Release System: Intelligent Switches Open Up New Horizon, *Nanomaterials,* **5**, pp. 2019–2053.

179. Bernardos, A., Aznar, E., Marcos, M. D., Martínez-Máñez, R., Sancenón, F., Soto, J., et al. (2009). Enzyme-Responsive Controlled Release using Mesoporous Silica Supports Capped with Lactose, *Angew. Chem.,* **121**, pp. 5998–6001.

180. Hu, C., Yu, L., Zheng, Z., Wang, J., Liu, Y., Jiang, Y., et al. (2015). Tannin as a Gatekeeper of pH-Responsive Mesoporous Silica Nanoparticles for Drug Delivery, *RSC Adv.*, **5**, pp. 85436–85441.

181. Castillo, R. R., Colilla, M., and Vallet-Regí, M. (2017). Advances in Mesoporous Silica-Based Nanocarriers for Co-Delivery and Combination Therapy Against Cancer, *Expert Opin. Drug Deliv.*, **14**, pp. 229–243.

182. Zheng, G., Zhao, R., Xu, A., Shen, Z., Chen, X., and Shao, J. (2018). Co-Delivery of Sorafenib and siVEGF Based on Mesoporous Silica Nanoparticles for ASGPR Mediated Targeted HCC Therapy, *Eur. J. Pharm. Sci.*, **111**, pp. 492–502.

Chapter 9

Silica Nanoparticles for Diagnosis, Imaging and Theranostics

Jessica Rosenholm[a] and Tuomas Näreoja[b]
[a]*Pharmaceutical Sciences Laboratory, Faculty of Science and Engineering,*
Åbo Akademi University, BioCity, Tykistökatu 6A, FI - 20521 Turku, Finland
[b]*Department of Laboratory Medicine, Division of Pathology, Karolinska Institutet,*
Stockholm, 171 77, Sweden
jerosenh@abo.fi, tuomas.nareoja@ki.se

9.1 Introduction

Silicon (Si) is the second most abundant element in the earth's crust preceded only by oxygen, making Si the most abundant metallic element. Their combination (SiO_2), i.e., silica, makes up about 60% of the crust in the form of, e.g., quartzite and sandstone [1]. These are crystalline forms of silica, whereas the amorphous form is well known to, e.g., the pharmaceutical, cosmetic and food industries since decades, usually in the form of colloidal (or fumed) silica. As a pharmaceutical excipient, colloidal silica is used as an adsorbent, anticaking agent, emulsion stabilizing agent, glidant, suspending agent, tablet and capsule disintegrant, and viscosity-increasing agent [2]. Silica is also the main constituent of bioglass, where its

Handbook of Materials for Nanomedicine: Lipid-Based and Inorganic Nanomaterials
Edited by Vladimir Torchilin
Copyright © 2020 Jenny Stanford Publishing Pte. Ltd.
ISBN 978-981-4800-91-4 (Hardcover), 978-1-003-04507-6 (eBook)
www.jennystanford.com

relative amount dictates the bioactive properties [3]. In the SiO_2–CaO–Na_2O–P_2O_5 system that constitutes bioglass, bonding to both bone and soft tissue is possible at 52 wt% SiO_2, whereas 52–60% SiO_2 bonds only to bone.

While the conventional colloidal silica is produced via flame hydrolysis of chlorosilanes on industrial scale, more sophisticated methods have been developed to produce silica materials with controlled properties on the nanoscale for research purposes. A colloidal approach for preparing ceramic amorphous metal oxides with high reactivity is the sol-gel process, which not only allows different morphologies (powders, fibers, monoliths, coatings and thin films) to be prepared, but also introduces porosity into the matrix—both properties of which cannot be obtained with conventional ceramics. The porosity of these matrixes paved the way for them to be loaded with active substances, and incorporation of drugs into sol-gel derived silica matrices was introduced as early as 1983 [4] after which sol-gel processed porous silica materials have proven to possess considerable potential as drug delivery carriers. The simplicity and yet simultaneous versatility of the sol-gel derived materials lies in it being a low-temperature process, where "green state" ceramics can be prepared at room temperature and ambient pressure. Consequently, the method allows for encapsulation of active and even highly fragile or sensitive molecules, such as biomolecules, into the material already at the synthesis step. This feature thus also allows molecular imaging agents to be incorporated into these materials, and benefit from the stabilizing ceramic matrix in the same manner as incorporated drugs or biomolecules.

Since amorphous silica in itself is optically transparent, the utilization of silica materials usually requires the incorporation of imaging agents into the silica matrix to render the material detectable. This is in contrast to most inorganic materials, that do possess inherent visibility, e.g., by some optical or magnetic imaging modality [5]. Conversely, the optical signal of incorporated imaging agents (molecular/nanoparticular) should not suffer considerably by encapsulation into a silica matrix, but instead gain benefit from the shelter of the ceramic matrix (as mentioned above). This approach has been exploited for many years to construct non-porous fluorescent silica nanoparticles incorporated with a range of different molecular fluorescent dyes, utilized for a variety of bio/

med/tech-related applications [6]. The most known silica NPs of this type would perhaps be the "C-dots" (Cornell Dots) that entered clinical trials in 2010 as an intravenously administered cancer diagnosis imaging agent. In all of these NP designs, the dye-rich core is protected by a solid silica outer shell.

With the advent of porous silica materials, and especially so the class of ordered mesoporous silicas in the beginning of the 1990s, another strategy for incorporation of active molecules into silica materials emerged. Namely, in these cases, where the synthesis strategy bears origins with zeolite synthesis but instead of using molecular templates for creating the pores, mesoporous materials are synthesized using supramolecular surfactant aggregates as pore templates leading to much larger pore sizes than possible via the zeolite approach. This synthesis approach in principle allows for a molecule of any size to be fit inside the pores into the ready material, thus constituting a very versatile carrier for active compounds. In the beginning of the new millennium, these materials were introduced into the biomedical field as drug delivery carriers, and a few years later when mesoporous silica in nanoparticulate form (i.e., mesoporous silica nanoparticles, MSNs) took over this research field, the necessity for labeling the MSNs emerged, as these were studied in cellular systems for intracellular delivery. Nevertheless, since most commercially available fluorophores are designed for bioconjugation with proteins, these could equally well be applied to MSNs bearing surface functional groups—mostly amines, as the fluorophores are also usually amine-reactive dyes. This has become a standard procedure within the last decade, but applying such materials as imaging agents presents itself with some further complexities that will be discussed in the following sections.

9.2 Synthesis of Silica Nanoparticles

The prospects of using silica as a construct in nanoscopic imaging agents are multifold. First, for non-porous silica nanoparticles, molecular dyes can be incorporated into the silica matrix itself, or in the interior forming a dye-rich core that finally is encapsulated into a solid silica shell; or the dyes can be introduced via a layer-by-layer technique where dye-rich layers are sandwiched between layers of solid silica. The outermost silica layer serves to protect the incorporated dyes from the surroundings. Second, for mesoporous silica, the imaging agents can be conjugated to the pore walls via surface functional groups, or the molecular agents can be loaded into the pores without chemical bonding, or ionic species can even be doped into the silica matrix itself. Third, both porous and non-porous silica shells can be coated onto inherently detectable nanoparticles, where the shells can further incorporate imaging agents in accordance with the above-listed approaches to create multimodal imaging agents [5]. In the following, we shall outline in brief the synthesis approaches used for the fabrication of these different types of silica-based imaging probes.

9.2.1 Non-Porous Silica

Non-porous silica nanoparticles are essentially synthesized via two different approaches:

- the so-called Stöber process
- the reverse microemulsion technique

Both of these synthesis methods rely on sol-gel chemistry. On the most fundamental level, the sol-gel process involves the preparation of nanoscale particles (colloids), forming a dispersion, i.e., the sol, using inorganic or more commonly, organic alkoxide precursors. The sol reacts further to form another, reversed, dispersion, i.e., the gel. If the precursor is a silicon alkoxide, $Si(OR)_4$, molecular polymerization through a series of hydrolysis and condensation reactions occur via the following reactions:

$$\text{hydrolysis}$$
$$\equiv Si - OR + H_2O \leftrightarrow \equiv Si - OH + ROH$$
$$\text{(esterification)}$$

$$\text{alcohol condensation (alcoxolation)}$$
$$\equiv Si - OR + HO - Si \equiv \leftrightarrow \equiv Si - O - Si \equiv + ROH$$
$$\text{(alcoholysis)}$$

$$H_2O \text{ condensation (oxolation)}$$
$$\equiv Si - OH + HO - Si \equiv \leftrightarrow \equiv Si - O - Si \equiv + H_2O$$
$$\text{(hydrolysis)}$$

Commonly ammonia (in basic preparations) or a mineral acid (in acidic syntheses) are employed as catalysts. Iler proposed that polymerization of silica occurs through a mechanism where silicic acid monomers first polymerize into colloidal nanoparticles forming a new population of monomers that, in turn, increase in size and aggregate into chains and networks, finally forming a gel [7]. Both reactions proceed in parallel but in different extent depending on the conditions employed. The three-dimensional particles serve as nuclei and further growth proceeds by an Ostwald ripening mechanism. In this process, particles grow in size and decrease in number as highly soluble small particles dissolve and reprecipitate on larger, less soluble nuclei. Growth ends when the difference in solubility between the smallest and the largest particles become insignificant (dissolution-reprecipitation). The rate of reactions depends on factors such as temperature, pH and component concentrations.

Employing this chemistry in a controlled fashion, the Stöber method is still one of the most used chemical approaches employed to prepare non-porous silica materials of uniform size, and was discovered in 1968 by Werner Stöber et al. [8], but built on earlier work of by G. Kolbe in 1956 [9]. Kolbe developed the synthesis of monodispersed silica particles based on the hydrolysis and subsequent condensation of silicon alkoxides in ethanol, whereas Stöber et al. systematically worked on the experimental conditions of this reaction and investigated the controlled growth of spherical silica particles, which is now the well-known Stöber process. The use of ammonia as morphological catalyst and the resulting pH was concluded to be responsible for particle monodispersity. Via this synthesis approach, silica nanoparticles with controlled sizes from 50 nm to 2 μm can be prepared. Despite appearing like solid silica particles after synthesis, the macroscopic particle actually consist of granular silica nanoparticles of a few nanometers in size. Consequently, the Stöber silica particles are in general microporous, and this microporosity may even be tuned by slightly modifying the synthesis conditions [10].

Given that the a silicon alkoxide $Si(OR)_4$ used as silica source in sol-gel processing can be partly substituted by organosilanes, $Si(OR)_3$-X, where X is an organic linker terminated with a functional group (or the functional group directly) provides considerable versatility in introducing organic groups into the silica matrix. Since the functional groups also serve as reactive sites for molecular imaging agents such as fluorophores, pre-reaction of organosilanes with dyes serves as the basis for the synthesis of fluorescent silica nanoparticles. There are, in essence, three strategies for incorporating dye molecules into non-porous silica nanoparticles (Fig. 9.1). Type 1 is the physical incorporation of the dye molecules during the synthesis, whereby the dye molecules are encapsulated into the nanoparticles during formation via the sol-gel process by simply adding dye molecules into the synthesis mixture. Literature studies have revealed that entrapped dye molecules via this route exhibit a higher quantum yield and enhanced photostability than corresponding free dye molecules [11]. This procedure is also often referred to as "doping." Type 2 is the above-described, commonly employed strategy involving chemical conjugation of the dye molecules within the silica matrix, which is realized via co-condensation between the main silica source and a dye-conjugated organosilane during the synthesis. Here, the ratio between the silanes need to be carefully adjusted so as not to distort the formation of the nanoparticles, but compared to the Type 1 approach the incorporation yield of the dye into the silica matrix under non-covalent bonding is poor and dependent on the adsorption force between the dye itself and the silica precursor. In type 3, the dye molecules are introduced via post-synthesis procedures. Either the nanoparticle is of co-condensed type, whereby dye molecules can be reacted to the functional group on the particle surface; or the pristine silica surface of the nanoparticle is reacted with pre-reacted organosilane-dye conjugates. The organosilane is typically an aminosilane, most often 3-aminopropyltriethoxysilane (APTES). Although surface conjugation is relatively more straightforward than direct incorporation into the silica matrix, the long-term stability under physiological environments is considered as the limitation during application since all the dye molecules are exposed on the particle surface.

The reverse microemulsion method is the other common method used for the production of non-porous silica nanoparticles, whereby the resultant silica particles usually yield within a diameter range of tens to a few hundred nanometers. This method is based on the formation of silica nanoparticles in inverse micelles compartmentalized by a suitable surfactant in a nonpolar organic solvent, usually cyclohexane (Fig. 9.2). This method was reported in 1999 by Arriagada et al., who synthesized ultrafine silica nanoparticles within an optimized nonionic water-in-oil microemulsion system [13]. The size of the particles can be altered by changing the kinetics of hydrolysis and condensation processes and the ratios among the contents of the microemulsion (i.e., continuous phases, co-solvent amounts and surfactant) but is in general more difficult to control than NPs synthesized via the Stöber synthesis.

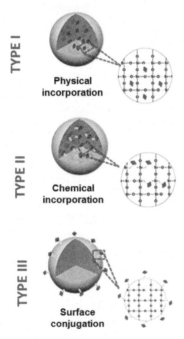

TYPE I — Physical incorporation

TYPE II — Chemical incorporation

TYPE III — Surface conjugation

Figure 9.1 Incorporation of fluorophores into non-porous silica NPs of Stöber type. Type 1: physical incorporation; type 2: chemical incorporation by organosilane chemistry; type 3: Surface conjugation. Adapted from [12].

The dye incorporation is essentially realized via either doping (addition of dye into the water droplets, in which the silica NPs form) or pre-reaction with an organosilane that is subsequently used in the synthesis [14]. Typical for NPs synthesized via the reverse microemulsion method is that a second silica layer is deposited around the dye-rich core, which prevents the dye molecules from leaching out from the core (in the case of doped dye molecules) and protects the incorporated dyes from the environment (in both cases) from affecting the dye properties, which may vary significantly depending on the surrounding conditions. To this layer may then be added, e.g., APTES-organosilanes to facilitate bioconjugation, i.e., coupling of biomolecules to the particle surface. Furthermore, a second set of fluorophores could also be incorporated into this second silica layer and conjugated to the surface, thus enhancing the fluorescence; denoted fluorescent double-layered silica nanoparticles [15]. This outer silica layer is typically grown using the Stöber approach, and is thus susceptible to hydrolytic stability issues if the main aim is to protect the dyes in the core from leaching [16]. This is due to the mode of silica growth resulting from the Stöber synthesis discussed above, i.e., very small silica nanoparticles aggregating together to form larger structures. As mentioned above, the resultant materials may appear as macroscopically solid but actually are microporous. This microporosity renders these materials susceptible to hydrolytic degradation in aqueous environments. More recent methods have thus been developed to create more dense silica shells, such as for instance biosilicification related methods in water (in which amino acid residues are used to control the reaction) that have shown to generate silica shells that are resistant against dissolution around Stöber silica cores in which the dyes are incorporated [16].

In addition to the reverse microemulsion method, also the oil-in-water microemulsion system (instead of water-in-oil) can be used to prepare silica NPs that uses "normal" micelles instead of reversed ones (Fig. 9.2). In this method, the non-polar core is used for hydrolysis and condensation of an organosilane as silica precursor, e.g., vinyltriethoxysilane (VTES) [17]. Here, the surfactant properties dictates the size of the final silica NPs. Owing to the "nanoreactor" conditions in both microemulsion based methods (where the micelles serve as the nanoreactors for the formation of the silica NPs) the reverse microemulsion method (polar core) is especially suitable for encapsulating hydrophilic dyes or nanoparticles, while the opposite is true for the approach using regular micelles (non-polar core).

Many names can be discerned that have been devoted to fluorescent silica NPs. Besides the already-mentioned C-dots or Cornell Dots [18, 19], developed at Cornell University, likewise, the so-called FloDots derived their name from being developed at the University of Florida [20]. While the C-dots are prepared via a modified Stöber process (also involving deposition of a second, solid silica layer around the dye-rich core NP) the FloDots can be prepared both via the Stöber and reverse microemulsion methods. The oil-in-water microemulsion method has been mostly used to synthesize so-called ORMOSIL (organically modified silica) nanoparticles [21, 22].

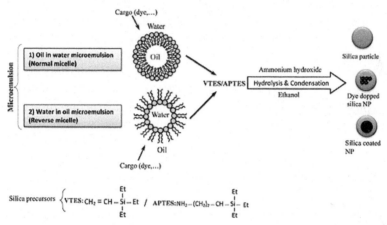

Figure 9.2 Overview of microemulsion methods to synthesize silica NPs. (1) Oil-in-water microemulsion method, in which the hydrolysis and condensation of an organosilanes occurs within normal micelles and then APTES is used for presenting amine groups on the surface of the nanoparticles. (2) Water-in-oil microemulsion (reverse micelle) method. Various cargos such as organic fluorescent dyes or other nanoparticles can be doped or encapsulated within silica particles through both of these methods. From [17].

9.2.2 Mesoporous Silica Nanoparticles (MSNs)

The synthesis of mesoporous silica materials dates back to the beginning of the 1990s, when the groups of Kato et al. and Mobil Oil researchers independently from each other reported on the synthesis of a silicate material with hexagonally ordered pore structure, but via different synthesis approaches [23, 24]. The Mobil Oil approach included synthesis of ordered mesoporous materials by using isotropic supramolecular surfactant aggregates as templates, around which inorganic material deposits via the sol-gel process forming a mesoscopically ordered hybrid organic-inorganic composite material. The organic template was subsequently removed via calcination, leaving the porous silicate network with ordered pore structure. This provided a novel route to make highly ordered nanocomposites that are difficult to prepare by traditional routes, and allows a precise control and design of architecture of the ceramic materials on the nanometer-scale. The original syntheses were based on self-assembly principles in surfactant solutions by the crystallization of aluminosilicate or silicate gels in a concentrated solution of alkyltrimethylammonium ions [25]. These materials are characterized by their well-ordered structure, tunable pore size from 1.5 to 10 nm, and simple preparation methods. The main components in the synthesis are essentially the same as for Stöber silica NPs, i.e., a source of silica, a solvent, and a catalyst, typically an acid or a base—with the addition of structure-directing surfactants as pore templates. In principle, the self-assembled liquid crystals from the surfactant act as a template to support the growth of the ceramic materials (Fig. 9.3) [26]. The amount of reported syntheses and accompanied synthesis mechanisms in the literature are extensive, but in 2009 a review was published that compiled a large number of verified syntheses for mesoporous materials [27].

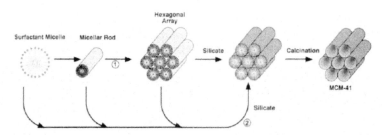

Figure 9.3 Possible mechanistic pathways for the formation of hexagonally ordered MCM-41 as originally proposed by the Mobil Oil Research group: (1) Liquid crystal initiated and (2) silicate anion initiated. From [25].

In the 2000s, when mesoporous silica entered the biomedical research field, the particle size also became important to control on the nanoscale. Essentially the particle size can be controlled by employing the principles from the Stöber synthesis; but given the dilute conditions used, the yield is typically quite low. In order to improve the reaction yield of nanoparticles, the addition of growth quenchers such as other surfactants, triethanolamine or ethylene glycol was introduced instead [28]. Considering the reaction conditions: pH, reaction temperature, type of silica precursor and even stirring rate can have a profound impact on particle size [29]. Numerous syntheses on MSNs of different sizes, typically within the range of 30–300 nm have been reported, but it seems inconclusive that a general rule for size control over the full nanoscale would exist to date. Regardless of syntheses yielding monodisperse MSNs in the synthesis solution, it is usually the post-synthesis handing and treatment that has proven to be crucial for maintaining the dispersability of the MSNs. To avoid self-aggregation during synthesis, the template removal is usually carried out via solvent extraction methods for MSNs. Most often, the next step involves surface functionalization, to prevent aggregation during further processing (labeling/drug loading/biofunctionalization/introduction of stimuli-responsive functions/etc.) or drying steps, and in later stages to ensure proper interactions at the nano-bio interface [30]. Surface functionalization is generally carried out for MSNs via two different approaches: co-condensation or post-synthesis grafting of organosilanes (i.e., the same as for the non-porous silica NPs). In the co-condensation approach, the functional organosilanes are added already in the synthesis step together with the main silica source (most often TEOS) and thus the functional groups are homogeneously distributed throughout the silica matrix (Fig. 9.4B). In the post-synthesis grafting approach, the organosilane is reacted with surface silanols of the already formed MSNs (Fig. 9.4A). This approach usually leads to a higher amount of accessible surface groups than the co-condensation approach, but typically the groups are congregated close to the pore entrances.

Similarly as for non-porous silica NPs, the incorporation of fluorescent dyes is most often realized by pre-reacting with the organosilane (most often APTES) [31] that is subsequently used for either co-condensation or post-synthesis grafting reactions. Depending on the dye, care must be taken that it is not subject to any inactivating conditions during these reactions, e.g., high temperature, incompatible solvents and so forth [32]. To avoid such incompatibilities, the dye can also be conjugated to an already surface functionalized MSN. Most commercial dyes are amine-reactive, so APTES is suitable also in this case. If a higher dye loading is desired, the dye can also be conjugated at later stages if amine-rich polymers (such as polyethylene imine, PEI) is used for further surface functionalization [33–35]. Dye loading degrees achieved via covalent conjugation are typically in the range of a couple of weight per cent at maximum. In the case of fluorescent dyes, however, the maximal amount loaded does not at all necessarily correlate with optimal fluorescence intensity signal; which is an issue we will return to in Section 9.3 below.

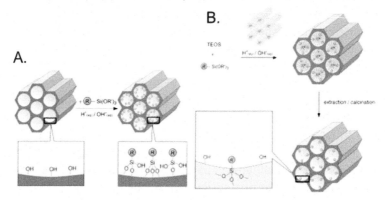

Figure 9.4 Surface functionalization with functional organosilanes via the co-condensation process (B) versus the post-synthesis grafting approach (A) [31]. "R" is the functional group.

9.2.3 Core@Shell Particles

9.2.3.1 Non-porous shells

Among all core@shell constructs, inorganic@inorganic nanoparticles have been claimed to be the most important class of core@shell nanoparticles and amongst these, silica is the most common [36]. The use of silica as a coating material for other inorganic nanoparticles is thus widely applied, and the rationale is multifold. Coating with non-porous silica is commonly applied to enhance the dispersability of nanoparticles, especially in aqueous solvent, due to the inherent hydrophobicity of most other inorganic nanoparticles. Deposition of a silica coating further provides enhanced colloidal/chemical/photo/thermal stability to the core material, controlled porosity and surface chemistry, facile processing, and optical transparency. A typical example is coating of iron oxides with silica, which introduces hydrophilicity and biocompatibility and simultaneously allows for easy further functionalization, while protecting the core material against pH changes in the environment, which otherwise can lead to, e.g., oxidation or even full dissolution of the core material. Protection from an aqueous environment is also one reason for frequently coating UCNPs (upconverting nanophosphors) with silica, as contact with water is prone to quench the optical signal for many UCNPs [37, 38]. In the case of quantum dots (QD) silica coatings are often applied to enhance the biocompatibility and aqueous dispersability of the QDs. Silica coatings have also been used, e.g., in photoacoustic imaging for increasing the photothermal stability of the Au [39] or superparamagnetic iron oxide (SPION) core materials [40], or in magnetic resonance imaging (MRI) to modulate the relaxivity of maghemite (γ-Fe_2O_3) nanoparticles [41].

Coating of inorganic nanoparticles with non-porous silica shells can be conducted via both the Stöber method as well as the reverse microemulsion method. Prior to coating via the Stöber method, an organosilane is usually first used to "activate" the surface of the NP, after which the hydrolysis and condensation reactions of added TEOS can proceed from the NP surface. By adjusting the amount of silica precursor, it has been reported that this approach can be used to control the resulting shell thickness from a couple of nm to several hundred nm. Also the reverse microemulsion technique can reportedly be utilized to control the thickness of the silica coating on inorganic NPs within a few tens of nanometers. If the core NPs are hydrophilic, the NPs can be solubilized in the water-filled micelles directly; otherwise a silane coupling agent (organosilane) like in the case of coating via the Stöber method, or polymer stabilizer are needed to render the NP water dispersible prior to coating. Gold (Au), silver (Ag), nanodiamonds (ND) [42], different types of iron oxides and QDs of type CdSe, CdSe@ZnS and CdSe@ZnSe@ZnS have been coated using these methods [17].

9.2.3.2 Mesoporous shells

The prospect of coating inorganic NPs with mesoporous shells was manifested from the attempt to create imaging probes with in-built drug delivery capacity (at the time referred to as "theranostic probes") and was reported for the first time in 2006 for hydrophobic iron oxide and quantum dot nanocrystals [43]. In these syntheses, the structure-directing agent (SDA), most often CTAB, serves a double purpose. In the first step, it functions as a phase-transfer agent to transfer the hydrophobic cores from non-polar organic phase to aqueous phase. In the following step, the SDA should also serve its general purpose as a pore template. The formation of the mesoporous shell subsequently proceeds as a liquid-phase seeded growth approach [44], where the pre-existing cores serve as nucleation seeds for the propagation of the mesoporous shell. The final size and uniformity of the resulting core-shell particles are greatly affected by the self-assembly of CTAB/silica mesostructured assemblies onto the CTAB-stabilized cores [45]. The self-assembly process follows the same principles as the synthesis of pristine MSNs, with the distinct difference that no seeded growth is taking place [46]. Similarly as for pristine MSNs, pore swelling can be attempted also for porous shells; albeit the synthesis conditions need to be more tightly controlled in the case of shell synthesis, and the nature of the pore swelling agent determines the prevailing mechanism (see Fig. 9.5). For hydrophilic core materials, no phase transfer needs to take place but the cores are dispersed directly into the synthesis medium. Here, it is crucial that CTAB is used in excess to form a double layer around the core material, in order for it not to precipitate the core from the aqueous phase. Further, successful formation of a porous shell instead of all-silica MSNs is highly dependent on the ratio of the used solvents. For smaller cores, a mesoporous silica coating is often directly deposited while for larger cores a middle layer of non-porous silica is often deposited first to assure proper interactions for further mesoporous silica deposition. For iron oxide beads, the non-porous silica coating may also needed for protection of the core material from further mesoporous silica coating, which generally takes place under either acidic or basic conditions whereby acidic conditions could dissolve the core material. This approach can also, conversely, be exploited for the synthesis of hollow mesoporous silica particles (H-MSNs), where the hollow interior space is templated by nanoscaled cores of a material that can be easily leached out after the synthesis of a porous shell; so-called "hard templating." When one or several core materials are incorporated into the hollow void of H-MSNs they are usually referred to as rattle-like structures. Such hollow or rattle structures have been constructed out of, e.g., UCNPs to provide for luminescence imaging [38] or manganese oxide (MnO) for magnetic resonance imaging (MRI) [47]. Core@shell nanostructures based on hydrophilic core nanoparticles coated with mesoporous shells have to date been synthesized from core materials including, e.g., metals such as platinum, gold, and silver or other inorganic NPs such as quantum dots, nanodiamonds (NDs) and upconverting nanophosphors (UCNPs) that are interesting for optical imaging applications. For the construction of Au@MSN NPs, even a one-pot synthesis have been developed [48]. It may also be interesting to note that, if the nanomaterial to be incorporated is too small for core-shell composite formation, other types of nanocomposites can be constructed besides core@shell structures. For instance, in the case of luminescent Carbon Dots (CDs) with typical sizes well below 10 nm, these can instead be incorporated either into the pores of MSNs via in situ synthesis using the mesopores as nanoreactors [49], or outside the pores as gatekeepers [50].

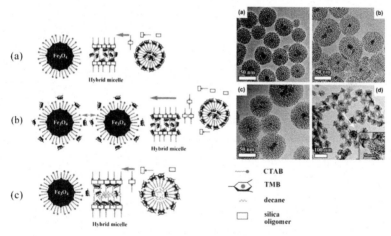

Figure 9.5 Schematic illustration of the proposed mechanism for the growth process of silica/CTAB mesophases on magnetic cores in the presence of different pore swelling agents: (a) TMB only, at low TMB/CTAB molar ratio; (b) TMB only, at high TMB/CTAB molar ratio; (c) Joint incorporation of TMB and decane, at high pore swelling agent/CTAB molar ratio. The left-pointing red arrow represents the growth process of mesostructure on the nanocrystal core, and the bidirectional red arrow represents the interparticle interaction. From [46].

9.3 Silica Nanoparticles for Imaging

Given that silica in itself does not possess any imaging activity, the rationale for using all-silica nanoparticles as imaging probes relates to their use as carriers for molecular imaging agents or other, smaller nanoparticles. There are several advantages that can be achieved by incorporating imaging agents into nanoscaled carrier particles:

- Sensitive agents can be provided *long-term physical, chemical and photostability*.
- Protection of molecular agents against potential *enzymatic or hydrolytic degradation* in the body is possible.
- Particle carrier can be designed to *provide access to sites unreachable by the molecule itself* (crossing of physiological barriers).
- *Cellular targeting and uptake* → enhanced accumulation of imaging agent at the target site → lowering of administered dose can be achieved.
- Particles can be designed to be *long-circulating* (stealth properties).

9.3.1 Silica Nanoparticles as Carriers for Optical Imaging Agents

Many advantages of silica-based nanoparticulate fluorescent probes have been identified early on, including the following [20]:

- *High emission intensity*: a large number of fluorescent molecules can be encapsulated into a single NP, giving rise to strong emission signals under the right conditions.
- *Excellent photostability*: photobleaching is one of the major problems for traditional fluorescent dyes, but owing to the protection and shielding effect of the silica matrix, the incorporated dye molecules are well protected from the surroundings including environmental oxygen, leading to constant fluorescence enabling accurate measurements.
- *Water dispersability and efficient conjugation*: silica is a hydrophilic material that is biocompatible and readily dispersible in aqueous media, both of which are required for biomedical applications. In addition, the silica surface is

flexibly surface functionalized via different methods enabling facile conjugation of biomolecules or other active moieties to the silica particle surface.

However, the properties of the incorporated fluorescent dyes are not additive, but the resulting fluorescent properties largely depend on the architecture of the silica NP construct as well as the vicinity to neighboring dyes and polarity of the most immediate surroundings, amongst other things. In NP structures encompassing a dye-rich core, fluorescence quenching is prone to occur via either intraparticle energy transfer or some other nonradiative pathway (e.g., molecule–molecule interaction, electron transfer, isomerization) within the solid matrix [19]. Webb and co-workers aimed to control the photophysical properties of C-dots through nanoparticle architecture, and studied three different structures to investigate how the particle structure could either ameliorate quenching or even lead to fluorescence enhancement (Fig. 9.6).

Figure 9.6 Schematic representation of three different silica nanoparticle architectures, denoted from left to right as (1) a compact core–shell particle containing dye surrounded by a silica shell, (2) a slightly expanded core/shell particle, and (3) homogeneous particle with dye molecules sparsely embedded within the matrix. Blue designates silica without dye molecules; orange designates a composite silica dye matrix. From [19].

They observed a greater than threefold increase in the quantum efficiency of fluorescence of tetramethylrhodamine by optimizing the NP structure, and each architecture gave a twofold enhanced radiative rate for the constituent dye. The reduction in nonradiative rate varied by a factor of 3 between the architectures, with the lowest nonradiative rate observed for the homogeneous particle. These changes in nonradiative rates correlate with the restricted rotational mobility of the within the particle covalently bound dyes: the more restricted the mobility, the smaller the nonradiative rate. No direct evidence for intraparticle energy transfer between dye molecules was observed. As a conclusion, the authors suggested that dyes with large absorption cross-section and low quantum efficiency might benefit from being encapsulated within NPs. This property is dependent on the spectral characteristics of added dye, e.g., the overlap of excitation and emission spectra, i.e., Stokes' shift.

In the case of fluorescent metal–antenna-chelates that employ Förster resonance energy transfer (FRET) from the chelate to, e.g., a lanthanide ion, doping within silica in the reverse microemulsion strategy helps to serve three goals: A simple chelate structure could suffice as other ligands could not freely compete from the coordination sites of the metal ion, dynamic quenching of the long-lived fluorescence could be excluded, and very hydrophobic constructs could be used as the surrounding silica shell would counteract any problems in solubility [51]. Lanthanide chelate-labels have extremely long Stokes' shift due to the FRET between chelate-antenna and the emitting metal-ion. This property allows high local concentrations of the label to be packed inside the NPs, because self-quenching observed with many organic fluorophores is nearly eliminated by the low-energy (higher wavelength) emission [52].

For mesoporous silica the situation may become even more complicated, due to the additional confinement effects created by the constraints of the nanosized mesopores, including overlapping surface potentials, differing conditions inside the pores compared to the outside, and so forth [53, 54]. Also in the case of mesoporous silica, the dyes can either be loaded into the pores without any chemical bond, or conjugated to the pore surfaces by in situ (co-condensation) or post-synthesis functionalization methods. Before doing so, the dye properties should be taken into account to avoid inactivation of the dye during incorporation. Such contributing factors may include solvent incompatibilities, inappropriate pH or temperature exposure, creating of high local abundance of acidic or basic groups that may affect the fluorescence properties of the incorporated dye and so on (Fig. 9.7).

In the case represented in Fig. 9.7, the most common fluorophore in MSN context, fluorescein isothiocyanate (FITC), was used. Fluorescein is well known to be pH-sensitive, and thus, not only the surrounding pH will have a marked effect on the signal readout but also the local surroundings of the dye in the mesopores will have an impact. In Fig. 9.7, while the MSNs are not surface functionalized (except for the APTES used to conjugate FITC to the silica matrix) the fluorescence behavior follows what could be expected from the surrounding pH, while after functionalization with the polybase PEI, the local pH effect of PEI becomes dominant. Further, a pronounced absorbance peak shift is observed as a color change in the figure, indicating a change in polarity in the immediate surroundings of the dye.

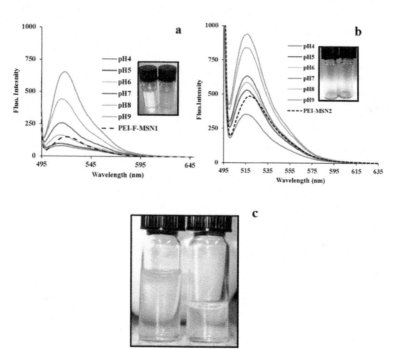

Figure 9.7 Dependence of FITC-conjugated MSN (F-MSN) fluorescence intensity as a function of pH. (a) F-MSN fluorescence intensity variability with solvent pH. Coating with 25k PEI (10 wt%) via electrostatic adsorption results in the same emission spectra regardless of solvent pH. Inset: F-MSNs (left) vs. PEI-F-MSNs (right) in HEPES buffer. (b) Same as previous repeated for a second set of MSNs with surface-grafted PEI, resulting in an enhanced local pH drop due to PEI residing also inside the mesopores. Inset: dry FITC-conjugated MSNs vs. PEI-MSNs. Non-porous (Stöber) control particle. On the left PEI-SiO$_2$ (+39 mV surface charge, $\lambda_{max,Abs}$ = 501.5 nm, $I_{max,fluo}$ = 174) and on the right the same particle but further succinylated (–42 mV surface charge, $\lambda_{max,Abs}$ = 488 nm, $I_{max,fluo}$ = 547) as measured in HEPES buffer at pH 7.2. From [35].

Further, the dye concentration inside the pores as well as the particle concentration will have a profound impact on the fluorescence intensity. Upon investigation of the dependence of dye loading degree of fluorophores in MSNs on the fluorescence intensity, the loading degree of both hydrophilic fluorescein (incorporated by conjugation) [55] and hydrophobic carbocyanine dyes (incorporated by adsorption) [32] was found to be around 1 wt-%. Beyond this dye loading degree, rapid self-quenching occurs most likely due to a FRET (fluorescence, or Förster, resonance energy transfer) effect (see Fig. 9.8a).

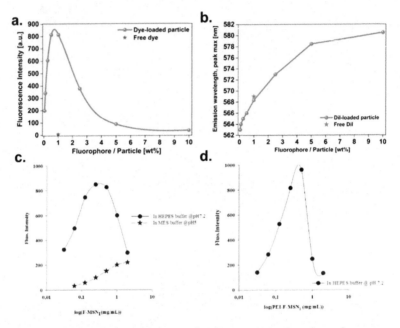

Figure 9.8 Dye loading degree and MSN concentration dependence on fluorescence emission properties. (a) Fluorescence intensity as a function of hydrophobic DiI fluorophore loading into MSNs; (b) Shift in maximum emission wavelength as a function of hydrophobic fluorophore loading into MSNs. Fluorescence intensity as a function of MSN concentration for (c) F-MSNs; (d) PEI$_{ads}$-F-MSNs. From [32].

A maximum emission wavelength peak shift of up to almost 20 nm can also be observed with gradually increasing loading degrees from 0,1 to 10 wt% for hydrophobic fluorophore loading into MSNs (Fig. 9.8b). Interestingly, the same intensity dependence is observed as a function of MSN concentration (Fig. 9.8c,d) and again, for non-surface functionalized MSNs the intensity variation is further dependent on the surrounding pH (Fig. 9.8c). These observations suggest that great care should be taken when attempting to use fluorescence intensity as a quantitative measure, especially for pH-sensitive fluorophores. Depending on intracellular location, the pH tends to vary between 5 and 7 and further, NPs tend to accumulate in intracellular compartments, which, according to above, could lead to quenching effects.

An even greater concern upon using fluorophore labeling is the imminent risk of dye leaching, especially for mesoporous NPs with an open pore structure with efficient water access. As the detection of MSNs is completely relying on the detection of attached labels, is must be assured the label is indeed still associated with the NP carrier. As mentioned above, significant dye leaching has been observed even in the case when the dyes are incorporated within the core of a non-porous particle, surrounded by a siliceous shell (Fig. 9.9) [16]. This is due to the hydrolysis of the silica shell, the kinetics of which is promoted by the microporosity of Stöber derived silica coatings.

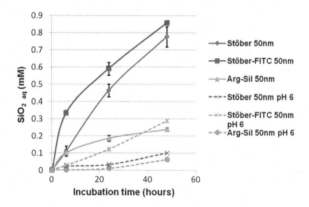

Figure 9.9 Dissolution of SiO_2 NPs monitored by the molybdenum blue assay incubated in buffer (100 mg/ml, HEPES pH 7.4, NaCl 148 mM, $CaCl_2$ 1 mM at 37°C and HEPES pH 6, NaCl 148 mM, $CaCl_2$ 1 mM at 37°C). From [16].

MSNs with typical wall thicknesses of 1 nm are consequently even more sensitive to hydrolysis, which inevitably in parallel leads to leaching of attached dye molecules even if the chemical bond itself would be stable against hydrolysis. In Fig. 9.10, FITC- and TRITC (tetrarhodamine isothiocyanate)-conjugated MSNs with and without PEI-coating were investigated for their dye leaching kinetics. Very different leaching behavior was observed that could be ascribed to the different characteristics of the dye molecules and their interactions with the silica surface (modified or not). Due to the static conditions used, re-adsorption of dye could also be observed under favorable conditions. Overall, a combined dissolution-reprecipitation (of silica) and adsorption-desorption (of fluorescent labels) equilibria is what is most likely being observed [35]. Under application (in vivo) conditions, re-precipitation and re-adsorption effects are more likely to be absent due to much higher liquid volumes available combined with continuous solvent exchange. Under intracellular conditions, however, the situation may be very different.

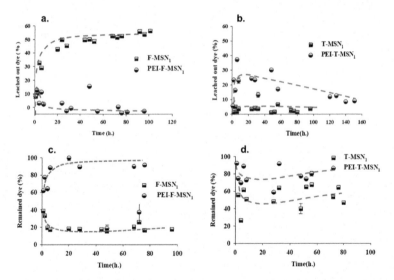

Figure 9.10 Relative fluorophore leaching percentage based on measurement from the supernatant for F-MSNs (a) and T-MSNs (b). Remaining fluorophore content in the leached MSNs, obtained by dissolution of the remaining MSN cake after separation from the leaching solution for F-MSNs (c) and T-MSNs (d) From [35].

It stands clear from the above that the application conditions should not only be kept in mind when designing appropriate imaging probes. A relatively unexplored area is using the MSNs as carriers for loaded fluorophores, with the intent of delivering the dye inside cells with subsequent intracellular release the dye to stain the cell. As hydrophobic drug molecules are very suitable for this purpose [29], one would expect similar advantages to be gained for hydrophobic dye molecules. On this note, we conducted a study where we incorporated a range of different dye molecules into differently surface modified MSNs to yield an imaging probe for cellular labeling (Fig. 9.11).

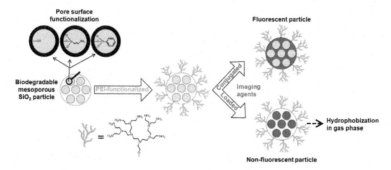

Figure 9.11   Schematic representation of the particle designs studied in ref. [32].

The rationale behind this study was to find the most appropriate design in terms of surface chemistry (hydrophilic vs. hydrophobic), mode of dye incorporation (conjugation vs. loading) and dye characteristics (hydrophilic vs. hydrophobic) for the most efficient cellular labeling and long-term cellular tracking in vitro and in vivo. The results revealed that the most durable (long-term) imaging probe design out of the almost 40 combinations studied, was an MSN loaded with a hydrophobic dye that could release the dye cargo inside the cells in a sustained manner. The intracellular release ability was further confirmed by FRAP (fluorescence recovery after photobleaching) measurements [56]. The labeling efficiency was significantly improved as compared to that of quantum dots of similar emission wavelength, highlighting the potential to utilize the carrier capability of MSNs as a self-generating cellular label instead of inherently detectable NPs (see also Section 9.1.4.2).

9.3.2 Silica Nanoparticles as Carriers for MRI, PET, SPECT Labels

Introduction of metal atoms/ions into silica nanoparticles can generally follow three routes: (1) direct substitution, (2) doping, and (3) chelation. In the third case, radiolabeling and attachment of paramagnetic complexes to MSNs essentially follow the same procedure, which in the initial stages is not very different to dye incorporation. Both radionuclides (for PET and SPECT imaging) and paramagnetic ions (for MR-imaging) are usually complexed into organic chelates, which can be attached as labels to NP systems. For radiolabeling, the subsequent complexation of radionuclides requires specialized facilities and follows established radiochemical methods, and will thus not be discussed in detail here. We will concentrate on the incorporation of the chelates to silica NPs, for which the complexation of paramagnetic ions can take place either before or after immobilization on or into a NP matrix. Since silica is an inorganic material, direct doping of ions into the silica matrix can also be used as an approach to incorporate different ions [57]. This approach has been studied to some extent for the incorporation of paramagnetic ions such as Gd^{3+} or Mn^{2+} for MRI activity [58], both of which mainly provide T_1-weighted or positive (white) contrast MR-imaging. Especially for elemental silicon (Si) based materials, radiolabeling via direct incorporation of, e.g., fluorine-18 ions [^{18}F] to the surface of the material can also be utilized, where the mechanism of incorporation depends on the material characteristics (Fig. 9.12) [59]. Especially porous silicon (PSi) materials have been utilized successfully for these purposes [5, 60].

In this case, for thermally hydrocarbonized porous silicon (THCPSi) the radiolabeling reaction has been suggested to proceed via a direct substitution of ^{18}F$^-$ to a silyl hydrogen; while for thermally carbonized porous silicon (TCPSi) the incorporation of ^{18}F$^-$ is more likely to occur by a nucleophilic attack to the Si–O–Si bridges on the thin silica layer on the Si_x–C_y surface. In addition to the attack to both a silyl hydrogen or to a Si–O–Si bridge, the isotopic exchange of ^{18}F for the residual silyl hydrogen is possible in thermally oxidized porous silicon (TOPSi) with a silica surface. Nevertheless, it is feasible that in all the materials, a Si–^{18}F bond is created [59].

THCPSi

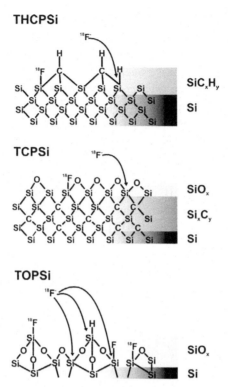

TCPSi

TOPSi

Figure 9.12 Plausible mechanisms for [^{18}F] incorporation to PSi surfaces [60].

The more conventional method, i.e., conjugation of organic chelating groups such as tetraazacyclododecane-1,4,7,10-tetraacetic acid (DOTA), 1,4,7-triazacyclononane-triacetic acid (NOTA) or diethylenetriamine-pentaacetic acid (DTPA), to NP systems follows essentially the same rationale as dye molecules. The chelates can be entrapped into the NP matrix during synhesis via co-condensation, or conjugated to the surface post-synthesis. These chelates correspond to the Gd-based contrast agents that are in use in the clinic under trade names such as Dotarem® (Gd-DOTA) and Magnevist® (Gd-DTPA). The same chelating agents are commercially available with a reactive linker group, again, analogous to most commercially available fluorescent dyes. For the most part, the metal centre (Gd^{3+} ion) is complexed after incorporation to the MSNs, whereas it should be noted that the accessibility of the chelate will be largely dependent on the mode of incorporation [61]. In other cases, the

Gd^{3+} ion can be complexed before conjugation to the NP, as the complexation is quite strong. In some cases, the complexation is even carried out prior to being added into the synthesis mixture. In this case, the chelates have been pre-reacted with the aminosilane APTES, whereafter the Gd^{3+} complexation takes place before the silyl-derived Gd^{3+} complexes are used in a co-condensation reaction in the synthesis of MSNs (Fig. 9.13). In this specific study, MSNs were synthesized using 10 wt% Gd-1 and 10-40 wt% Gd-2 complexes to investigate the loading efficiency of Gd(III) chelates using this approach. The co-condensation procedure affords MSNs with much higher loadings of Gd(III) chelates into MSNs, but the r_1 relaxivities still appear to be smaller on a per Gd basis, presumably owing to the reduced accessibility of the Gd(III) chelates to the water molecules [62].

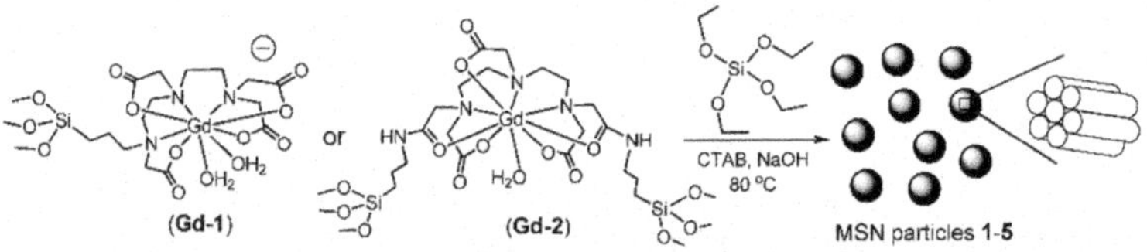

Figure 9.13 Formation of (1) 3-aminopropyl(trimethoxysilyl)-diethylenetriamine tetraacetic acid (Si-DTTA) and bis(3-aminopropyl triethoxysilyl)-diethylenetriamine pentaacetic acid (Si$_2$-DTPA) followed by (2) Gd^{3+} complexation to yield derivatives Gd-1 and Gd-2, and (3) co-condensation of these in the synthesis of MSNs.

Consequently, the accessibility of the chelates can be a challenge both in terms of (1) complexation of Gd^{3+} ions, if the chelating groups are conjugated prior to complexation or (2) water accessibility upon MR-imaging, whereas interaction between the magnetic centers and water is essential for the MRI activity (i.e., contrast enhancement). Davis et al. compared the co-condensation approach to post-grafting of Gd-DOTA complexes, and found significant differences in relaxometric properties depending on the location of the Gd-DOTA complex [63]. This provided them with a means to control the relaxivity by location-tuning of the complex by "slow-delay" vs. "long-delay" co-condensation as compared to post-grafting. Deliberately restricting the conjugation of Gd(III) chelates to the outer surface of MSNs, thus, should maximize the relaxivity of the attached complexes [61], since they will all be rendered active in terms of MR-imaging (water accessibility). The advantage of using a porous matrix in this case may of course be questionable, since a non-porous NP would serve the same purpose. Notably, while the formed bonds between the incorporated chelates and MSN matrix has been deemed to be stable [64], it is worth keeping in mind the hydrolytic stability of material itself (c.f. dye leaching) when considering the stability of the system in vivo.

When other rare-earth transition metals, with photoluminescent energy transitions, are complexed into such chelates, the same system can be utilized for optical imaging if the metal ion is, e.g., Eu^{3+}, Dy^{3+}, Nb^{3+}, or Tb^{3+} [65]. Contrary to the case of Gd-complexes and MRI, these complexes need to be protected from water, as water exposure will lead to luminescence quenching or suboptimal coordination of the antenna chelates [66]. Here, the metal complexes can preferably be incorporated inside the mesopores, after which the MSN may be coated with a water-impermeable (polymeric) coating to maximize the luminescence under aqueous conditions. The same but opposite strategy is frequently applied in the case of non-porous silica coating of chelate carrying NPs, whereby, e.g., Eu-chelates are entrapped into a polymeric NP matrix that is subsequently coated with silica.

Given the size of the chelates as well as the restriction from water if the bulky chelates are located inside mesopores, pore-expanded MSNs are perhaps to prefer in order to render the portion of Gd-complexes incorporated inside mesopores active as well. Another aspect to consider is the theranostic potential of these carriers (see Section 9.1.5.2) whereby the mesopore space may be visioned to be saved for drug loading. In this case, doping of metal ions into the silica matrix can be attempted instead, as these will be incorporated into the matrix itself and not occupy pore space. Guillet-Nicolas et al. [67] prepared MSNs with 3D and 2D pore network connectivity by introducing the Gd^{3+} ions into MSNs by the incipient wetness technique, thus creating $GdSi_xO_y$ MSN hybrid systems and compared their relaxiometric properties. The 3D pore connectivity provided a significant increase in r_1 relaxivity compared to the 2D pore geometry (and 4.6 times higher relaxivity than free Gd-DTPA) most likely due to the more open structure and hence, increased water accessibility and diffusivity within the 3D pore structure. These results were corroborated by Şen Karaman et al. [68], who studied a range of different incorporation methods and MSN structures to find the optimal preparation route for maximizing the r_1 relaxivity (Fig. 9.14). Parameters under study included the structure of the MSN matrix, post-synthesis treatment protocols, as well as the source and incorporation routes of paramagnetic Gd^{3+} centers. Also here, the best results in terms of relaxivity were obtained with a hollow MSN structure with expanded pore size, most likely due to the enhanced water accessibility compared to regular MSN structures. Such modulation of the MSN structures to maximize relaxivity thus allows for minimization of the Gd dose, which is of utmost importance from a clinical perspective since Gd is quite a toxic element. The enhanced relaxivity frequently observed for NP-immobilized MRI contrast agents is not only owing to water accessibility (which may be more relevant for porous systems) but also the resulting slow global tumbling rates [69]. Additionally, all the benefits of having imaging agents incorporated in a NP matrix is highly valid also in the case of MR-imaging; besides the lowered dose also, e.g., prolonged imaging time-frame, crossing of biological barriers, cellular uptake and retention, cellular and tissue targeting potential with long-circulating properties, leading to enhanced accumulation at the target site and so forth.

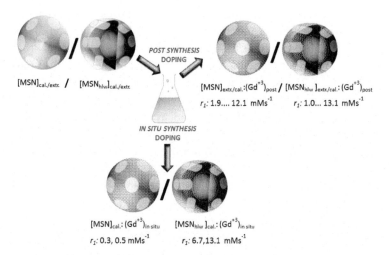

Figure 9.14 Incorporation of Gd-ions into MSN matrices by varying in situ synthesis and post-synthesis methods, along with resulting relaxivities. From [68].

9.4   Applications in Imaging and Diagnostics

9.4.1   In vitro Diagnostics with Silica Nanoparticles

Nanoparticles have been widely used as labels in diagnostic and imaging applications as they possess high specific activity, and even single binding events can be observed due to the extremely intense, e.g., photoluminescence signal of the particles as compared to molecular labels. The higher the signal the easier and more sensitive the detection will be. This is mainly due to a high concentration of label units within a nanoparticle, novel particle materials or high absorption cross section. In solid silica particles, the label content may exceed 20% (w/w) amplifying the signal from one binding reaction when compared with organic fluorophores. Furthermore, the nanoparticle shell provides protection for the fluorophore, reducing both dynamic quenching and probability of reactions leading to irreversible photobleaching [11]. In many cases, this outcome is achieved simply by excluding oxygen and water molecules from the vicinity of the fluorophore [66].

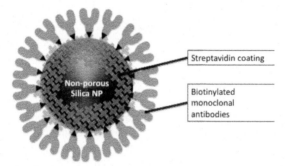

Figure 9.15 Schematic representation of bioconjugation where a non-porous silica NP is first coated with streptavidin and immediately prior to assay biotinylated antibodies are added to the particles to facilitate specific detection of antigens.

These particles may subsequently be coated with bioactive molecules such as antibodies and used in immunoassays or imaging. On the labeled NPs there is a large surface area available for accommodating the bioactive molecules to facilitate binding or selective uptake. To achieve efficient coating, there are various technologies ranging from nonspecific physical adsorption, elaborate bioconjugate chemistry to site-specific bioconjugation of antibody fragments and over to use of adaptor proteins, e.g., avidins, reviewed in [70]. The amount of molecules depends on the particle surface area and biomolecule size, e.g., in the case of antibodies approximately one hundred active binding sites may be achieved on a (non-porous) 60 nm NP. This, in turn, will make use of avidity of the binders and improve the observed overall affinity of the label higher than that of a labeled single binder. Furthermore, the large surface-to-volume ratio allows a part of the surface to be exchanged to another bioactive molecule to facilitate, e.g., electrochemical detection. Printed electronics offer cost-effective means of detection, but would often benefit from amplified signals. By utilizing HRP-antibody double coated particles, a 30-fold amplification of electrochemical signal has been obtained [71].

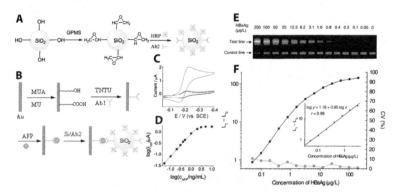

Figure 9.16 In vitro diagnostics applications using non-porous silica NPs. (A) The NPs are double functionalized with HRP and antibodies. (B) The work electrode is coated with another antibody and subsequently sample is added and then the NPs. (C) Finally, cyclic voltammogram is recorded in the presence of thioene and H_2O_2. (D) Concentration of the analyte is plotted against electrocatalytic current. (F) Eu (III) nanoparticle-based LFIA of HBsAg, calibrator images of test strips (membrane area) and a negative control in assay buffer. (E), Calibration curves (•) and imprecision profiles (□) of panel A. The inset illustrates the linear range region of the plot. Mean value of test line luminosity (Lt) and background luminosity (Lb). Adapted from [52, 71].

However, hydrolysis of silica will limit the coating strategies to some extent as high pH may not be used, and additionally, the coated bioactive nanoparticles cannot be stored for long periods of time, because the optimal storage conditions for proteins, e.g., antibodies and silica NPs are different. For these reasons, the more stable non-porous silica NPs are predominantly used in diagnostic applications along with a coating on the NPs with a stable adaptor protein, e.g., streptavidin. This will facilitate storage times of several weeks for the bioactive silica NPs [72].

9.4.2 Silica Nanoparticles in Live Cell Imaging

Positively charged silica NPs are readily taken up by various cell types. This property has been utilized in many studies to detect and stain cells with inherently labeled particles or with particles loaded with fluorescent molecules. The most common method is, as discussed above, to functionalize silica NPs with APTES and subsequently conjugate the formed primary amino groups with FITC; thus facilitating convenient tracking of the stained cells [33]. The dissolution of silica NPs is considerably slowed down by the confined (and acidic) intracellular environment. In dividing cells, the particles are diluted evenly amongst the daughter cells and decay of the signal depends on the degradation or exocytosis of the particles, amount divisions that dilute the particles and amount of dye attached to the particles. For MSNs, typical dye loading degrees achieved by conjugation (attachment) amount to a maximum of a couple of wt%, while adsorption strategies can yield loading degrees up to 50 wt%. The loss of signal due to dilution between daughter cells may be circumvented by this strategy, i.e., loading the particles with a dye that is released from the particles in a sustained manner over time [32]. In addition to labeling and tracking of entire cells, silica NPs allow more elaborate mechanistical studies on intracellular membrane trafficking that has, e.g. been able to demonstrate cargo dependent motility of endosomes [73].

9.5 Core@Shell Nanoparticles for Multimodal Imaging and Theranostics

The most prominent advantage of silica as a construct in the design of any nanostructure is, as already established, its synthetic flexibility. Silica can be constructed as part of nanocomposites (inorganic-inorganic) or hybrid (inorganic-organic) materials, in porous or non-porous form, and in varying morphologies (most often nanoparticles, or coatings in the case of nanostructures). Since silica is not inherently detectable, its function as construct in nanostructured imaging agents is either to add functionality to the system or act as a barrier to separate two other material constructs, or protecting another construct from the surroundings. The former can be exemplified by introducing pores to a nanosystem, that can subsequently be loaded with active molecules (drugs, molecular imaging agents...) or simply maximizing the available surface area for further surface functionalization [74]. A typical example of the latter would be separating a magnetic core from a surface layer of fluorescent dyes with a silica layer, as direct contact between the dye and the magnetic NP would lead to quenching of the dye fluorescence. Other barrier functions may include protection of the core material from water (e.g., UCNPs) which would lead to luminescence quenching, or from the surrounding pH conditions that could lead to corrosion or dissolution of the core material (e.g., SPIONs). In all cases, a silica surface (porous or non-porous) provides the nanosystem with good water dispersability (hydrophilicity), enhanced colloidal stability, easy further functionalization via all techniques developed for silica surfaces, chemical inertness and biocompatibility. In this section, we will thus concentrate on such structures where silica is used as coating material, i.e., core@shell structures. This is perhaps the most flexible approach for constructing either multimodal or theranostic agents, as the two or more imaging modalities and/ or imaging and therapeutic functions can be distinctly separated, if the core is responsible for one function and the shell of another. There are surely multiple examples also of constructs where both active entities are molecular agents, e.g., where one is loaded into the core material, which can be porous or non-porous silica; and a second, e.g., imaging agent is attached to the particle surface. From a functionality point of view, also these constructs can be regarded core@shell materials, but here we will concentrate on the physical core@shell constructs where the core and the shell are of different materials, and thus represent different functionalities.

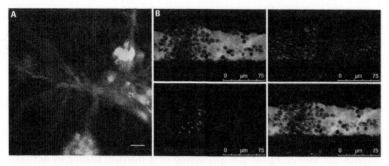

Figure 9.17 Tracking of cells labeled with fluorescent MSNs. (A) A coculture of fibroblasts labeled with MSNs (magenta) and LNCaP cells (green), collagen is shown in white. Scale bar 30 um. (B) Two-photon imaging of injected MSN stained cancer cells (magenta) circulating in CAM-model, blood pool (green) stained with dextran conjugated dye and nuclei stained with Hoechst.

9.5.1 Multimodal Imaging

The rationale for multimodal imaging is gaining complementary information in "one shot" via more than one imaging technique, or to approximately localize stained cells with a non-invasive method prior to a closer more detailed examination. Every imaging modality has advantages and disadvantages, and they are thus highly complementary. For instance, whereas optical imaging reveals pathologies at the cellular or sub-cellular level, MRI generally display physiological differences at the level of tissues and organs [69]. This combination, optical imaging (OI) with MRI is consequently also the perhaps most common combination for the construction of bimodal imaging agents. The most typical example of a core@shell construct of this type is a magnetic core surrounded by a shell carrying fluorescent molecular dyes. As mentioned above, a solid silica shell can be used here as barrier layer to separate these, while a porous silica shell can be coated to maximize the amount of dyes that can be carried (Fig. 9.18) [74].

As outlined in Section 9.3.1, maximizing the amount of dyes in the shell does not necessarily correspond to the highest imaging signal (in this case, intensity). The same holds true for the core material;

an imaging signal originating from the activity of the core NP will most likely be altered upon coating of a shell. Non-porous coatings are commonly used to protect core NPs from surrounding water if water contact may lead to luminescence quenching, as is the case for UCNPs [37]. However, when porous coatings are considered, the pores are most often radially aligned within the shell, which means that water still has direct access to the core material. In the case of magnetic cores, such as iron oxides intended for use as contrast agents for MR-imaging, the presence of a porous shell has actually shown to quite drastically enhance the relaxometric properties of the core. Primarily, the properties of the magnetic core material itself, including magnetic parameters (magnetic susceptibility (χ), saturation magnetization (m_s), anisotropy (K), Néel relaxation time (τ_N), Brownian relaxation time (τ_B)) and surface functionalities of the magnetic NPs can be tuned via the magnetic NP size, composition, and surface chemistry [75]. Additionally, not only the presence of a coating but further, the properties of the coating material will further alter the relaxivity of the resulting NP system. In the case of porous silica coatings, it is mainly the pore size of the shell that will have a profound impact on the resultant relaxivity (Fig. 9.19).

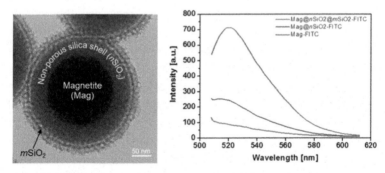

Figure 9.18 Core@shell@shell (Mag@nSiO$_2$@mSiO$_2$) constructs consisting of a superparamagnetic iron oxide core (Mag) and a non-porous silica inner shell (nSiO$_2$) separating the core from the mesoporous silica outer shell (mSiO$_2$). The TEM image shows the morphology and structure of the construct while the fluorescence intensity measurements show the impact of the barrier layer on the fluorescence of attached dyes (FITC). The porous outer layer allows maximization of the amount dye molecules that can be attached to the nanosystem.

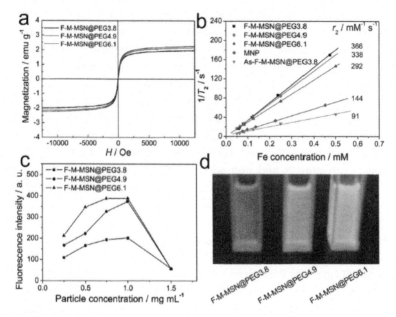

Figure 9.19 Pore size dependence on relaxivity. (a) Field-dependent magnetization curves at 300 K. An enlargement of the central part of the curves is shown in the inset. (b) T_2 relaxivity (r_2) plots for the aqueous suspensions of F-M-MSN@PEG with different pore sizes, the slopes indicate r_2. (c) Peak fluorescence emission as a function of the particle concentration. d) The photograph of F-M-MSN@PEG suspensions (0.5 mg mL–1) irradiated with UV light. The numbers after the sample names indicate the pore sizes in nm.

As seen in Fig. 9.18, the highest relaxivity was obtained with the smallest pore size of 3.8 nm, while the relaxivity markedly decreased with increasing pore size. This can be understood, as the diffusion of water in mesopores slows down with a decrease in the pore size [76]. Consequently, a longer confinement effect and thus, residence time can be achieved for the protons in smaller mesopores, prolonging their interaction within the magnetic NP-generated local magnetic fields, leading to enhanced relaxivity [77]. Notably, for the non-extracted sample, i.e., pores not liberated from the template yet (thus resembling that of a non-porous coating) the relaxivity is less than that for the pure iron oxide core. Further observable in the figure, is that the fluorescence of the fluorescent dyes conjugated inside the mesopores follow the opposite trend. Here, an enlarged pore size provides for enhanced fluorescence due to less self-quenching processes taking place in the confined space of the mesopores (as discussed in Section 9.1.3.1). From this demonstration follows that the design should be judiciously chosen depending on the foreseen application. Given the flexibility of the silica coating procedures, virtually any core material can be chosen for a core@shell design encompassing a porous layer. This opens up for the possibilities of constructing core@shell material from optically active cores (QD, UCNP, ND) with porous shells, while also the optical properties of the cores may be affected by the coating [78]. Both porous and non-porous silica coatings can further be utilized for the attachment of paramagnetic complexes, doping of ions f or MR-imaging or radiolabeling via chelator-based or chelator-free methods for PET imaging [79]. There are also a few examples of using silica NPs as agents for ultrasound (US) imaging, e.g., superhydrophobic MSNs exhibited a significant and strong US contrast intensity compared with other nanoparticles by enhanced stabilization of microbubbles [80]. Based on these modularities, a multitude of complex architectures can be found in the literature, and there are several comprehensive reviews that have covered these lately that the interested reader is kindly referred to, e.g., the following ones: [5, 17, 81].

9.5.2 Intracellular Sensing

Silica NP based optical pH sensors coupled with photoluminescent indicator dyes facilitate measurements without a direct contact or perturbing the observed system. Here as well, the use of bright fluorescent silica NPs enables sensitive detection and precise localization [85]. The probes can be miniaturized to the size of intracellular organelles, readily internalized in living cells and even be used for in vivo imaging. The luminescent pH indicators have a dynamic range of 3–4 pH units that usually covers the range of pH changes in living systems. A crucial aspect that may be achieved with silica NPs is referencing of the sensor signals to ascertain that changes in the intensity are not caused by local changes in probe concentration. Moreover, the core@shell strategy may be employed to detect multiple parameters from intracellular milieu like in simultaneous detection of dissolved oxygen and pH [86].

9.5.3 Theranostic Prospects

The most exploited designs for core@shell structures involving a porous shell are, nevertheless, not multimodal imaging probes, but prospective theranostic agents. In this context, "theranostic agents" refer to nanosystems capable of both imaging (diagnostics) and drug delivery (therapy). The obvious division of labor is the core material catering for the imaging modality, while the porous shell functions as drug carrier. Taken into account that the porous shell in itself influences the imaging signal, it is palpable that the loading of drug molecules into the porous layer surrounding the core material will have an impact on the same. This has, however, not been well investigated in the literature but rather a multitude of different design combinations can be found, which are too vast to cover here. Nevertheless, all of the above-discussed design aspects can be utilized and combined in different ways to create a multifunctional nanosystem encompassing both diagnostic and therapeutic capabilities. Furthermore, other types of advantages can be exploited based on the properties of the individual constructs, such as magnetic targeting and/or magnetically enhanced cell uptake in the case of magnetic nanocomposites (Fig. 9.20). Certain materials inherently encompass properties rendering them useful for therapy in themselves, such as photodynamic and photothermal therapy, or hyperthermia (magnetic NPs) that could be further combined with the delivery of drugs [82].

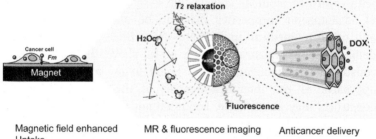

Magnetic field enhanced Uptake   MR & fluorescence imaging   Anticancer delivery

Figure 9.20 Schematic representation of triple-functional M-MSNs. The nanocomposite has a Fe_3O_4 nanocrystal core coated with a mesoporous silica shell doped with FITC, and can enhance the proton T_2 relaxation in water. The cellular uptake of M-MSNs was enhanced by the magnetic field. With large pore volumes, the nanoparticles could be used for magnetic delivery of anticancer drugs [83].

The main advantage obtained by the integration of a diagnostic functionality into a drug delivery system is still most likely for promoting the development of nanotherapeutics in enabling monitoring of the drug delivery process, as well as following the individual's therapeutic response. This not only enables patient pre-selection of the individuals that are most likely to respond well to the therapy, but also aids in putting forward the nanotherapeutics with the highest clinical translatability [84]. Despite the high hopes devoted to nanomedicines, especially in cancer therapy, and the substantial research efforts that have been directed toward the development of methods to improve site-specific delivery of chemotherapeutics, this largely still remains an unattainable goal [5]. Here, imaging can play a crucial role in the discovery of more successful methods for studying the targeting efficiency in order to improve the specificity of nanotherapeutic delivery. Further, imaging promotes the understanding the interactions between nanomaterials and biological systems, i.e., the "nano-bio interface." These aspects constitute where "nanotheranostics" is anticipated to be of critical importance toward the development of personalized medicines, and improved efficacy and safety of nanomedicine-based therapies.

9.6 Conclusions

Silica has indeed been proven to be a highly flexible material with regard to modular and flexible design options, and a range of advantageous properties that can be utilized in bioimaging and other biomedical applications. Nevertheless, the conceptual demonstrations illustrated here is to show that each novel construction of imaging agents should follow a rational design approach, as the interdependencies of the properties of the constructs may be multiple. For multimodal imaging probes, the situation becomes even more complex, when the dependency of several imaging modalities need to be taken into account and the dependencies are seldom linear (such as concentration dependent) as in the case of both luminescence and relaxivity, as have been demonstrated with several examples throughout this chapter. A multitude of designs exist in the literature, with recent emphasis on multifunctional and theranostic systems. The greatest benefit of these nanoscopic imaging, diagnostic or theranostic agents will most likely be the information that can be gained through imaging for promoting the development of new nanotherapeutic systems. Eventually, multifunctional and/or multimodal imagining agents could aid in generating a new generation of nanotherapeutics with improved specificity, efficacy, and safety.

References

1. Kuhlmann, A. M. (1963) The second most abundant element in the earth's crust. *JOM* **15**, 502–505.

2. Rowe, R. C., Sheskey, P. J., Cook, W. G., Fenton, M. E. (eds.) (2012) *Handbook of Pharmaceutical Excipients - 7th edition*, Pharmaceutical Press and American Pharmacists Association.

3. Hench, L. L., and Jones, J. R. (2015) Bioactive glasses: Frontiers and challenges. *Front. Bioeng. Biotechnol.*, **3**, 194.

4. Unger, K., Rupprecht, H., Valentin, B., and Kircher, W. (1983) The use of porous and surface modified silicas as drug delivery and stabilizing agents. *Drug Dev. Ind. Pharm.*, **9**, 69–91.

5. Karaman, D. Ş., Sarparanta, M. P., Rosenholm, J. M., and Airaksinen, A. J. Multimodality Imaging of Silica and Silicon Materials in vivo. *Adv. Mater.*, **0**, 1703651.

6. Burns, A., Ow, H., and Wiesner, U. (2006) Fluorescent core–shell silica nanoparticles: towards "Lab on a Particle" architectures for nanobiotechnology. *Chem. Soc. Rev.*, **35**, 1028–1042.

7. Iler, R. K. (1979) *The Chemistry of Silica: Solubility, Polymerization, Colloid and Surface Properties and Biochemistry of Silica*, Wiley.

8. Stöber, W., Fink, A., and Bohn, E. (1968) Controlled growth of monodisperse silica spheres in the micron size range. *J. Colloid Interface Sci.*, **26**, 62–69.

9. Kolbe, G. (1956) Das komplexchemische Verhalten der Kieselsäure.

10. Bazuła, P. A., Arnal, P. M., Galeano, C., Zibrowius, B., Schmidt, W., and Schüth, F. (2014) Highly microporous monodisperse silica spheres synthesized by the Stöber process. *Microporous Mesoporous Mater.*, **200**, 317–325.

11. Auger, A., Samuel, J., Poncelet, O., and Raccurt, O. (2011) A comparative study of non-covalent encapsulation methods for organic dyes into silica nanoparticles. *Nanoscale Res. Lett.*, **6**, 328.

12. Xu, Z., Ma, X., Gao, Y.-E., Hou, M., Xue, P., Li, C. M., and Kang, Y. (2017) Multifunctional silica nanoparticles as a promising theranostic platform for biomedical applications. *Mater. Chem. Front.*, **1**, 1257–1272.

13. Osseo-Asare, K., and Arriagada, F. J. (1999) Growth kinetics of nanosize silica in a nonionic water-in-oil microemulsion: A reverse micellar pseudophase reaction model. *J. Colloid Interface Sci.*, **218**, 68–76.

14. Tavernaro, I., Cavelius, C., Peuschel, H., and Kraegeloh, A. (2017) Bright fluorescent silica-nanoparticle probes for high-resolution STED and confocal microscopy. *Beilstein J. Nanotechnol.*, **8**, 1283–1296.

15. Yoo, H., and Pak, J. (2013) Synthesis of highly fluorescent silica nanoparticles in a reverse microemulsion through double-layered doping of organic fluorophores. *J. Nanoparticle Res.*, **15**, 1609.

16. Mahon, E., Hristov, D. R., and Dawson, K. A. (2012) Stabilising fluorescent silica nanoparticles against dissolution effects for biological studies. *Chem. Commun.*, **48**, 7970–7972.

17. Shirshahi, V., and Soltani, M. (2015) Solid silica nanoparticles: Applications in molecular imaging. *Contrast Media Mol. Imaging*, **10**, 1–17.

18. Ow, H., Larson, D. R., Srivastava, M., Baird, B. A., Webb, W. W., and Wiesner, U. (2005) Bright and stable core–shell fluorescent silica nanoparticles. *Nano Lett.*, **5**, 113–117.

19. Larson, D. R., Ow, H., Vishwasrao, H. D., Heikal, A. A., Wiesner, U., and Webb, W. W. (2008) Silica nanoparticle architecture determines radiative properties of encapsulated fluorophores. *Chem. Mater.*, **20**, 2677–2684.

20. Yao, G., Wang, L., Wu, Y., Smith, J., Xu, J., Zhao, W., Lee, E., and Tan, W. (2006) FloDots: Luminescent nanoparticles. *Anal. Bioanal. Chem.*, **385**, 518–524.

21. Das, S., Jain, T. K., and Maitra, A. (2002) Inorganic-organic hybrid nanoparticles from n-octyl triethoxy silane. *J. Colloid Interface Sci.*, **252**, 82–88.

22. Sharma, R. K., Das, S., and Maitra, A. (2004) Surface modified ormosil nanoparticles. *J. Colloid Interface Sci.*, **277**, 342–346.

23. Yanagisawa, T., Shimizu, T., Kuroda, K., and Kato, C. (1990) The preparation of alkyltriinethylaininonium–kaneinite complexes and their conversion to microporous materials. *Bull. Chem. Soc. Jpn.*, **63**, 988–992.

24. Beck, J. S., Vartuli, J. C., Roth, W. J., Leonowicz, M. E., Kresge, C. T., Schmitt, K. D., Chu, C. T. W., Olson, D. H., Sheppard, E. W., McCullen, S. B., et al. (1992) A new family of mesoporous molecular sieves prepared with liquid crystal templates. *J. Am. Chem. Soc.*, **114**, 10834–10843.

25. Kresge, C. T., Leonowicz, M. E., Roth, W. J., Vartuli, J. C., and Beck, J. S. (1992) Ordered mesoporous molecular sieves synthesized by a liquid-crystal template mechanism. *Nature*, **359**, 710–712.

26. Liu, J., Kim, A. Y., Wang, L. Q., Palmer, B. J., Chen, Y. L., Bruinsma, P., Bunker, B. C., Exarhos, G. J., Graff, G. L., Rieke, P. C., et al. (1996) Self-assembly in the synthesis of ceramic materials and composites. *Adv. Colloid Interface Sci.*, **69**, 131–180.

27. Meynen, V., Cool, P., and Vansant, E. F. (2009) Verified syntheses of mesoporous materials. *Microporous Mesoporous Mater.*, **125**, 170–223.

28. M. Rosenholm, J., Sahlgren, C., and Lindén, M. (2010) Towards multifunctional, targeted drug delivery systems using mesoporous silica nanoparticles – opportunities & challenges. *Nanoscale*, **2**, 1870–1883.

29. Maleki, A., Kettiger, H., Schoubben, A., Rosenholm, J. M., Ambrogi, V., and Hamidi, M. (2017) Mesoporous silica materials: From physico-chemical properties to enhanced dissolution of poorly water-soluble drugs. *J. Control. Release*, **262**, 329–347.

30. Gomes, M. C., Cunha, Â., Trindade, T., and Tomé, J. P. C. (2016) The role of surface functionalization of silica nanoparticles for bioimaging. *J. Innov. Opt. Health Sci.*, **09**, 1630005.

31. Hoffmann, F., Cornelius M., Morell J., and Fröba M. (2006) Silica-based mesoporous organic–inorganic hybrid materials. *Angew. Chem. Int. Ed.*, **45**, 3216–3251.

32. Rosenholm, J. M., Gulin-Sarfraz, T., Mamaeva, V., Niemi, R., Özliseli, E., Desai, D., Antfolk, D., von Haartman, E., Lindberg, D., Prabhakar, N., et al. (2016) Prolonged dye release from mesoporous silica-based imaging probes facilitates long-term optical tracking of cell populations in vivo. *Small Weinh. Bergstr. Ger.*, **12**, 1578–1592.

33. Karaman, D. S., Desai, D., Senthilkumar, R., Johansson, E. M., Råtts, N., Odén, M., Eriksson, J. E., Sahlgren, C., Toivola, D. M., and Rosenholm, J. M. (2012) Shape engineering vs organic modification of inorganic nanoparticles as a tool for enhancing cellular internalization. *Nanoscale Res. Lett.*, **7**, 358.

34. Prabhakar, N., Näreoja, T., von Haartman, E., Karaman, D. Ş., Jiang, H., Koho, S., Dolenko, T. A., Hänninen, P. E., Vlasov, D. I., Ralchenko, V. G., et al. (2013) Core-shell designs of photoluminescent nanodiamonds with

porous silica coatings for bioimaging and drug delivery II: Application. *Nanoscale*, **5**, 3713–3722.

35. Desai, D., Karaman, D. S., Prabhakar, N., Tadayon, S., Duchanoy, A., Toivola, D. M., Rajput, S., Näreoja, T., and Rosenholm, J. M. (2014) Design considerations for mesoporous silica nanoparticulate systems in facilitating biomedical applications. *Mesoporous Biomater.*, **1**, 16–43.

36. Ghosh Chaudhuri, R., and Paria, S. (2012) Core/shell nanoparticles: classes, properties, synthesis mechanisms, characterization, and applications. *Chem. Rev.*, **112**, 2373–2433.

37. Liu, D., Xu, X., Wang, F., Zhou, J., Mi, C., Zhang, L., Lu, Y., Ma, C., Goldys, E., Lin, J., et al. (2016) Emission stability and reversibility of upconversion nanocrystals. *J. Mater. Chem. C*, **4**, 9227–9234.

38. Duan, C., Liang, L., Li, L., Zhang, R., and Xu, Z. P. (2018) Recent progress in upconversion luminescence nanomaterials for biomedical applications. *J. Mater. Chem. B*, **6**, 192–209.

39. Chen, Y.-S., Frey, W., Kim, S., Homan, K., Kruizinga, P., Sokolov, K., and Emelianov, S. (2010) Enhanced thermal stability of silica-coated gold nanorods for photoacoustic imaging and image-guided therapy. *Opt. Express*, **18**, 8867–8878.

40. Alwi, R., Telenkov, S., Mandelis, A., Leshuk, T., Gu, F., Oladepo, S., and Michaelian, K. (2012) Silica-coated super paramagnetic iron oxide nanoparticles (SPION) as biocompatible contrast agent in biomedical photoacoustics. *Biomed. Opt. Express*, **3**, 2500–2509.

41. Wu, W., Wu, Z., Yu, T., Jiang, C., and Kim, W.-S. (2015) Recent progress on magnetic iron oxide nanoparticles: Synthesis, surface functional strategies and biomedical applications. *Sci. Technol. Adv. Mater.* **16**, 023501.

42. Bumb, A., Sarkar, S. K., Billington, N., Brechbiel, M. W., and Neuman, K. C. (2013) Silica encapsulation of fluorescent nanodiamonds for colloidal stability and facile surface functionalization. *J. Am. Chem. Soc.*, **135**, 7815–7818.

43. Kim, J., Lee, J. E., Lee, J., Yu, J. H., Kim, B. C., An, K., Hwang, Y., Shin, C.-H., Park, J.-G., Kim, J., et al. (2006) Magnetic fluorescent delivery vehicle using uniform mesoporous silica spheres embedded with monodisperse magnetic and semiconductor nanocrystals. *J. Am. Chem. Soc.*, **128**, 688–689.

44. Nooney, R. I., Thirunavukkarasu, D., Chen, Y., Josephs, R., and Ostafin, A. E. (2003) Self-assembly of mesoporous nanoscale silica/gold composites. *Langmuir*, **19**, 7628–7637.

45. Cai, Q., Luo, Z.-S., Pang, W.-Q., Fan, Y.-W., Chen, X.-H., and Cui, F.-Z. (2001) Dilute solution routes to various controllable morphologies of MCM-41 silica with a basic medium. *Chem. Mater.*, **13**, 258–263.

46. Zhang, J., Li, X., Rosenholm, J. M., and Gu, H. (2011) Synthesis and characterization of pore size-tunable magnetic mesoporous silica nanoparticles. *J. Colloid Interface Sci.*, **361**, 16–24.

47. Kim, T., Momin, E., Choi, J., Yuan, K., Zaidi, H., Kim, J., Park, M., Lee, N., McMahon, M. T., Quinones-Hinojosa, A., et al. (2011) Mesoporous silica-coated hollow manganese oxide nanoparticles as positive t_1 contrast agents for labeling and mri tracking of adipose-derived mesenchymal stem cells. *J. Am. Chem. Soc.*, **133**, 2955–2961.

48. Chen, J., Zhang, R., Han, L., Tu, B., and Zhao, D. (2013) One-pot synthesis of thermally stable gold@mesoporous silica core-shell nanospheres with catalytic activity. *Nano Res.*, **6**, 871–879.

49. Nelson, D. K., Razbirin, B. S., Starukhin, A. N., Eurov, D. A., Kurdyukov, D. A., Stovpiaga, E. Y., and Golubev, V. G. (2016) Photoluminescence of carbon dots from mesoporous silica. *Opt. Mater.*, **59**, 28–33.

50. Jiao, J., Liu, C., Li, X., Liu, J., Di, D., Zhang, Y., Zhao, Q., and Wang, S. (2016) Fluorescent carbon dot modified mesoporous silica nanocarriers for redox-responsive controlled drug delivery and bioimaging. *J. Colloid Interface Sci.*, **483**, 343–352.

51. Hai, X., Tan, M., Wang, G., Ye, Z., Yuan, J., and Matsumoto, K. (2004) Preparation and a time-resolved fluoroimmunoassay application of New Europium fluorescent nanoparticles. *Anal. Sci.*, **20**, 245–246.

52. Xia, X., Xu, Y., Zhao, X., and Li, Q. (2009) Lateral flow immunoassay using europium chelate-loaded silica nanoparticles as labels. *Clin. Chem.*, **55**, 179–182.

53. Rosenholm, J. M., Czuryszkiewicz, T., Kleitz, F., Rosenholm, J. B., and Lindén, M. (2007) On the Nature of the Brønsted Acidic Groups on Native and Functionalized Mesoporous Siliceous SBA-15 as Studied by Benzylamine Adsorption from Solution. *Langmuir*, **23**, 4315–4323.

54. Valetti, S., Feiler, A., and Trulsson, M. (2017) Bare and effective charge of mesoporous silica particles. *Langmuir ACS J. Surf. Colloids*, **33**, 7343–7351.

55. Gulin-Sarfraz, T., Sarfraz, J., Karaman, D. Ş., Zhang, J., Oetken-Lindholm, C., Duchanoy, A., Rosenholm, J. M., and Abankwa, D. (2014) FRET-reporter nanoparticles to monitor redox-induced intracellular delivery of active compounds. *RSC Adv.*, **4**, 16429–16437.

56. von Haartman, E., Lindberg, D., Prabhakar, N., and Rosenholm, J. M. (2016) On the intracellular release mechanism of hydrophobic cargo

and its relation to the biodegradation behavior of mesoporous silica nanocarriers. *Eur. J. Pharm. Sci.*, **95**, 17–27.

57. Bérubé, F., Khadraoui, A., Florek, J., Kaliaguine, S., and Kleitz, F. (2015) A generalized method toward high dispersion of transition metals in large pore mesoporous metal oxide/silica hybrids. *J. Colloid Interface Sci.*, **449**, 102–114.

58. Kim, S. M., Im, G. H., Lee, D.-G., Lee, J. H., Lee, W. J., and Lee, I. S. (2013) Mn^{2+}-doped silica nanoparticles for hepatocyte-targeted detection of liver cancer in T_1-weighted MRI. *Biomaterials*, **34**, 8941–8948.

59. Sarparanta, M. (2013) ^{18}F-Radiolabeled porous silicon particles for drug delivery : Tracer development and evaluation in rats. Doctoral dissertation, Helsinki: Unigrafia.

60. Sarparanta, M., Mäkilä, E., Heikkilä, T., Salonen, J., Kukk, E., Lehto, V.-P., Santos, H. A., Hirvonen, J., and Airaksinen, A. J. (2011) 18F-labeled modified porous silicon particles for investigation of drug delivery carrier distribution in vivo with positron emission tomography. *Mol. Pharm.*, **8**, 1799–1806.

61. Carniato, F., Tei, L., Arrais, A., Marchese, L., and Botta, M. (2013) Selective anchoring of gdiii chelates on the external surface of organo-modified mesoporous silica nanoparticles: A new chemical strategy to enhance relaxivity. *Chem. Eur. J.*, **19**, 1421–1428.

62. Taylor-Pashow, K. M. L., Rocca, J. D., and Lin, W. (2011) Mesoporous silica nanoparticles with co-condensed gadolinium chelates for multimodal imaging. *Nanomaterials*, **2**, 1–14.

63. Davis, J. J., Huang, W.-Y., and Davies, G.-L. (2012) Location-tuned relaxivity in Gd-doped mesoporous silica nanoparticles. *J. Mater. Chem.*, **22**, 22848–22850.

64. Laprise-Pelletier, M., Bouchoucha, M., Lagueux, J., Chevallier, P., Lecomte, R., Gossuin, Y., Kleitz, F., and Fortin, M.-A. (2015) Metal chelate grafting at the surface of mesoporous silica nanoparticles (MSNs): Physico-chemical and biomedical imaging assessment. *J. Mater. Chem. B*, **3**, 748–758.

65. Li, Y.-J., and Yan, B. (2009) Lanthanide (Eu^{3+}, Tb^{3+})/β-Diketone modified mesoporous sba-15/organic polymer hybrids: chemically bonded construction, physical characterization, and photophysical properties. *Inorg. Chem.*, **48**, 8276–8285.

66. Zhang, J., Prabhakar, N., Näreoja, T., and Rosenholm, J. M. (2014) Semiconducting polymer encapsulated mesoporous silica particles with conjugated europium complexes: toward enhanced luminescence

under aqueous conditions. *ACS Appl. Mater. Interfaces*, **6**, 19064–19074.

67. Guillet-Nicolas, R., Bridot, J.-L., Seo, Y., Fortin, M.-A., and Kleitz, F. (2011) Enhanced relaxometric properties of MRI "positive" contrast agents confined in three-dimensional cubic mesoporous silica nanoparticles. *Adv. Funct. Mater.*, **21**, 4653–4662.

68. Karaman, D. Ş., Desai, D., Zhang, J., Tadayon, S., Unal, G., Teuho, J., Sarfraz, J., Smått, J.-H., Gu, H., Näreoja, T., et al. (2016) Modulation of the structural properties of mesoporous silica nanoparticles to enhance the T_1-weighted MR imaging capability. *J. Mater. Chem. B*, **4**, 1720–1732.

69. Verwilst, P., Park, S., Yoon, B., and Kim, J. S. (2015) Recent advances in Gd-chelate based bimodal optical/MRI contrast agents. *Chem. Soc. Rev.*, **44**, 1791–1806.

70. Sapsford, K. E., Algar, W. R., Berti, L., Gemmill, K. B., Casey, B. J., Oh, E., Stewart, M. H., and Medintz, I. L. (2013) Functionalizing nanoparticles with biological molecules: Developing chemistries that facilitate nanotechnology. *Chem. Rev.*, **113**, 1904–2074.

71. Wu, Y., Chen, C., and Liu, S. (2009) Enzyme-functionalized silica nanoparticles as sensitive labels in biosensing. *Anal. Chem.*, **81**, 1600–1607.

72. Moore, C. J., Montón, H., O'Kennedy, R., Williams, D. E., Nogués, C., Crean, C. (née Lynam), and Gubala, V. (2015) Controlling colloidal stability of silica nanoparticles during bioconjugation reactions with proteins and improving their longer-term stability, handling and storage. *J. Mater. Chem. B*, **3**, 2043–2055.

73. Aoyama, M., Yoshioka, Y., Arai, Y., Hirai, H., Ishimoto, R., Nagano, K., Higashisaka, K., Nagai, T., and Tsutsumi, Y. (2017) Intracellular trafficking of particles inside endosomal vesicles is regulated by particle size. *J. Control. Release*, **260**, 183–193.

74. Gulin-Sarfraz, T., Zhang, J., Desai, D., Teuho, J., Sarfraz, J., Jiang, H., Zhang, C., Sahlgren, C., Lindén, M., Gu, H., et al. (2014) Combination of magnetic field and surface functionalization for reaching synergistic effects in cellular labeling by magnetic core–shell nanospheres. *Biomater. Sci.*, **2**, 1750–1760.

75. Shin, T.-H., Choi, Y., Kim, S., and Cheon, J. (2015) Recent advances in magnetic nanoparticle-based multi-modal imaging. *Chem. Soc. Rev.*, **44**, 4501–4516.

76. Walther Hansen, E., Schmidt, R., Stöcker, M., and Akporiaye, D. (1995) Self-diffusion coefficient of water confined in mesoporous MCM-41 materials determined by 1H nuclear magnetic resonance spin-echo measurements. *Microporous Mater.*, **5**, 143–150.

77. Zhang, J., Rosenholm, J. M., and Gu, H. (2012) Molecular confinement in fluorescent magnetic mesoporous silica nanoparticles: Effect of pore size on multifunctionality. *ChemPhysChem*, **13**, 2016–2019.

78. Liu, N., and Yang, P. (2013) Highly luminescent hybrid SiO_2-coated CdTe quantum dots: Synthesis and properties. *Luminescence*, **28**, 542–550.

79. Ni, D., Jiang, D., Ehlerding, E. B., Huang, P., and Cai, W. (2018) Radiolabeling silica-based nanoparticles via coordination chemistry: Basic principles, strategies, and applications. *Acc. Chem. Res.*, **51**, 778–788.

80. Jin, Q., Lin, C.-Y., Kang, S.-T., Chang, Y.-C., Zheng, H., Yang, C.-M., and Yeh, C.-K. (2017) Superhydrophobic silica nanoparticles as ultrasound contrast agents. *Ultrason. Sonochem.*, **36**, 262–269.

81. Caltagirone, C., Bettoschi, A., Garau, A., and Montis, R. (2015) Silica-based nanoparticles: A versatile tool for the development of efficient imaging agents. *Chem. Soc. Rev.*, **44**, 4645–4671.

82. Huang, H., and Lovell, J. F. (2017) Advanced functional nanomaterials for theranostics. *Adv. Funct. Mater.*, **27**, 1603524.

83. Liu, Q., Zhang, J., Xia, W., and Gu, H. (2012) Towards magnetic-enhanced cellular uptake, MRI and chemotherapeutics delivery by magnetic mesoporous silica nanoparticles. *J. Nanosci. Nanotechnol.*, **12**, 7709–7715.

84. Arranja, A. G., Pathak, V., Lammers, T., and Shi, Y. (2017) Tumor-targeted nanomedicines for cancer theranostics. *Pharmacol. Res.*, **115**, 87–95.

85. Burns, A., Sengupta, P., Zedayko, T., Baird, B., and Wiesner, U. (2006) Core/shell fluorescent silica nanoparticles for chemical sensing: Towards single-particle laboratories. *Small*, **2**, 723–726.

86. Wang, X., Stolwijk, J. A., Lang, T., Sperber, M., Meier, R. J., Wegener, J., and Wolfbeis, O. S. (2012) Ultra-small, highly stable, and sensitive dual nanosensors for imaging intracellular oxygen and pH in cytosol. *J. Am. Chem. Soc.*, **134**, 17011–17014.

Chapter 10

Silica Nanoparticles for Drug Delivery

Marisa Adams[a] and Brian G. Trewyn[b]
[a]*Department of Chemistry, Colorado School of Mines,*
1500 Illinois St., Golden, Colorado 80401, USA
[b]*Department of Chemistry, Colorado School of Mines,*
1500 Illinois St., Golden, Colorado 80401, USA
madams2@mymail.mines.edu, btrewyn@mines.edu

10.1 History, Biocompatibility, and Endocytosis

The bioavailability of drug molecules is a crucial component to the success of any treatment and is dependent on the overall capacity of the free drug to make it intact to the target site. In order to enhance bioavailability, nanoscale drug delivery systems, including polymeric nanoparticles, dendrimers, liposomes, solid lipid nanoparticles, self-emulsifying drug delivery systems, nanostructured lipid carriers, hydrogels, and carbon nanotubes, have been investigated. Silica nanoparticles are particularly appealing candidates due to their biocompatibility, typically high drug loading, and facile surface functionalization [1–5]. As an example, targeting moieties and stimuli responsive organic groups (which act to prevent premature

Handbook of Materials for Nanomedicine: Lipid-Based and Inorganic Nanomaterials
Edited by Vladimir Torchilin
Copyright © 2020 Jenny Stanford Publishing Pte. Ltd.
ISBN 978-981-4800-91-4 (Hardcover), 978-1-003-04507-6 (eBook)
www.jennystanford.com

drug release) can be attached and ensure cytotoxic drugs are delivered specifically to cancer cells.

The successful synthesis of monodisperse nonporous silica nanoparticles was first demonstrated in 1968 and named after the author: Stöber [6]. Although these particles had substantial potential and showed diverse applications almost immediately, drug delivery was not among them. In the early 1990s mesoporous (pores between 2 and 50 nm in diameter) silica nanoparticles were codiscovered by the Mobil Oil Company and Waseda University, Japan [7, 8]. The earliest and most commonly reported of these was termed MCM-41 and the 2D hexagonally ordered mesoporous structure significantly increased the particle surface area over nonporous silica particles and in turn advanced potential applications into drug delivery. Another mesoporous silica demonstrating drug delivery is SBA-15, which also has a 2D pore structure but a less uniform and larger morphology, a wider tunable pore size range, and greater hydrothermal stability [9]. This capacity for drug delivery was first explored in 2001 by Vallet-Regí and co-workers, who showed MCM-41 could successfully load and release ibuprofen. Although novel, the proposed system lacked any sort of control over when and where the encapsulated drug was released [10]. Soon thereafter and owing to the facile functionalization of silica through silane chemistry, this was quickly rectified and systems that released drug molecules in response to specific stimuli began to emerge [11–14]. The field expanded quickly and continues to be the subject of significant research [15–17].

Importance of biocompatibility

Once inside the body, an efficient drug delivery vehicle must transport cargo into cells, where the drug exerts its therapeutic effects. Silica nanoparticles are successful drug carriers because they are capable of undergoing endocytosis, though the mechanism by which this occurs is dictated by the particle shape [18]. Once inside the cell, negatively charged particles undergo endosomal escape via the proton sponge effect [19]. This ability to escape from endosomes and deliver cargo directly into the cytoplasm is an advantageous property since cargo is not degraded by the harsh conditions inside the endosome. Drug release is governed by the ability of the medium to access the surface area and wet the silica surface, after which physisorbed drug molecules can simply diffuse into solution [20].

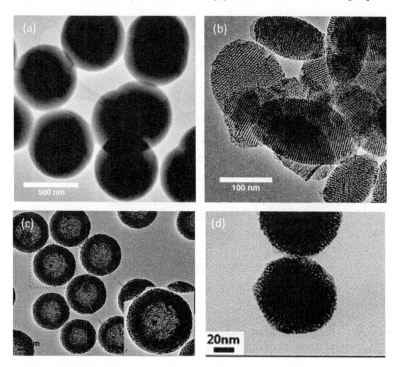

Figure 10.1 Transmission electron micrographs of (a) nonporous Stöber silica, (b) mesoporous silica nanoparticles, (c) core-shell mesoporous silica, Reprinted with permission from Luo, Z., Ding, X., Hu, Y., Wu, S., Xiang, Y., Zeng, Y., Zhang, B., Yan, H., Zhang, H., Zhu, L., Liu, J., Li, J., Cai, K. and Zhao, Y. (2013). Engineering a hollow nanocontainer platform with multifunctional molecular machines for tumor-targeted therapy in vitro and in vivo, *ACS Nano,* **7,** pp. 10271–10284. Copyright 2013 American Chemical Society, and (d) magnetic nanoparticle incorporated mesoporous silica. Reprinted with permission from Lee, J., Kim, H., Kim, S., Lee, H., Kim, J., Kim, N., Park, H. J., Choi, E. K., Lee, J. S. and Kim, C. (2012). A multifunctional mesoporous nanocontainer with an iron oxide core and a cyclodextrin gatekeeper for an efficient theranostic. platform, *J. Mater. Chem.,* **22,** pp. 14061-14067. Copyright 2012 Royal Society of Chemistry.

Aside from successfully delivering drug to a target site, it is also imperative that the carrier not cause damage while doing so. That is to say, the carrier must not be harmful to living tissue.

Nanoparticles with sizes above 10 nm are known to accumulate in the reticuloendothelial system, which has raised concerns about the biocompatibility. However, silica is hydrolytically unstable and dissolves into water-soluble silicic acid, which is then excreted renally. The main factor in determining the rate of degradation is the porosity of the silica material, with increased porosity leading to quicker degradation. Another factor contributing to degradation rate is the silica surface chemistry where more silanols result in faster hydrolysis rates [21]. In turn, those particles which degraded more quickly were shown to be less toxic [22]. A recent study demonstrated the degradation of silica nanoparticles occurs faster under intracellular conditions as opposed to in buffer or culture medium, though occurred through a similar, hydrolytic mechanism [23].

In the remainder of this chapter, we discuss a variety of silica nanoparticles. Specifically, nonporous, mesoporous, hollow, rattle-type, and magnetic silica nanoparticles and their applications to drug delivery will be reviewed. For reference, some representative TEM images can be seen in Fig. 10.1.

10.2 Nonporous Silica

Synthesis of silica nanoparticles relies on the hydrolysis and condensation of alkoxysilanes in a mixture of water and alcohol with a base catalyst, ammonium in the case of Stöber silica. Size and shape of the nanoparticles are highly dependent on the silica precursor concentration, as well as the rate of precursor addition to the solvent [24]. Nonporous silica continues to be investigated as a drug delivery vehicle due to its hydrophilic surface, biocompatibility, silane surface chemistry, and ease and low cost of large-scale synthesis. Indeed, in 2011 ultrasmall (7 nm) nonporous silica nanoparticles were approved for the first in-human clinical trial and used for cancer imaging [25]. Cargo can either be encapsulated within the silica and released when the silica matrix degrades or conjugated to the particle surface and released in response to a given stimulus, as illustrated in Scheme 10.1 [26].

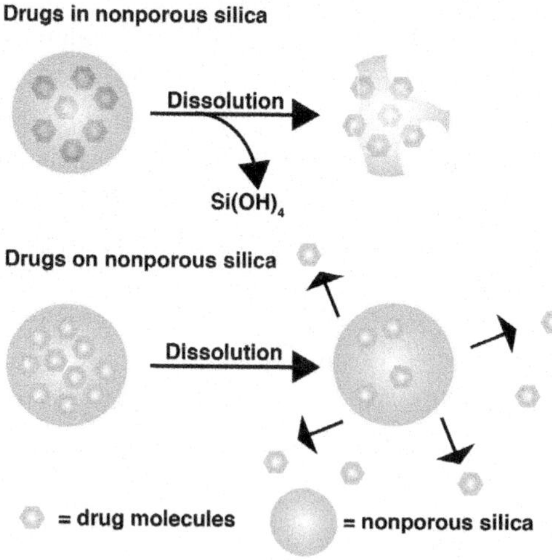

Drugs in nonporous silica

Dissolution

Si(OH)₄

Drugs on nonporous silica

Dissolution

○ = drug molecules   ● = nonporous silica

Scheme 10.1 Release mechanisms of drug molecules from nonporous silica nanoparticles through (a) degradation of the silica and (b) release from the silica surface through cleavage of a covalent bond or desorption of physisorbed drug molecules.

A nonporous silica shell was used as a transient core-stabilizing layer by Li et al. in a small interfering ribonucleic acid (siRNA) delivery vehicle. The siRNA was incorporated into a polycation core which was encased in dissolvable nonporous silica and subsequently coated with another polycation layer through electrostatic interactions with the negatively charged silica surface. This polycation layer was further modified with polyethylene glycol (PEG), which enhanced the biocompatibility during blood circulation but could be detached under reducing conditions. Without PEG, the polycation layer could disrupt the endosomal membrane, allowing the nanoparticles to escape the endosome and enter the cell. Here, the silica layer, which functions to protect the siRNA from nonspecific binding and dissociation, could dissolve, allowing for siRNA release. Altogether, the system showed effective gene silencing in a luciferase assay against HuH7-Luc, A549-Luc, and SKOV3-Luc cells without substantial toxicity. Additionally, in vitro and in vivo studies using OS-RC-2 human renal cancer cells with vascular endothelial growth factor as the target gene further demonstrated the gene silencing efficacy and tumor reducing potential of these nanocomposites [27].

Li et al. synthesized silica nanoparticles in the presence of DOX and an *N*-isopropylacrylamide/acrylamide copolymer, leading to incorporation of the organic materials within the silica matrix. The lower critical solution temperature (LCST) of the polymer was tuned to 39°C, so at physiological temperatures the nanoparticles remained intact. When the temperature was raised above the LCST, the polymer contracted into an insoluble globular form, which caused the particle to break apart. From here both the silica and polymer could be biodegraded, allowing for drug release. Acidic media was also shown to accelerate drug release from the nanoparticles due to hastened degradation of the silica. These self-decomposing particles showed no cytotoxicity in HeLa, HELF, or MCF-7 cells up to 160 μg mL^{-1} over 12 h without any thermal treatment, but could deliver doxorubicin (DOX) into HeLa cells when exposed to a thermal treatment [28].

Schoenfisch and co-workers varied the hydrophobicity of amine functionalized silica nanoparticles to generate a nitric oxide releasing system with controllable release rates. The amines were converted into *N*-diazeniumdiolate, which spontaneously releases nitric oxide in aqueous conditions. Hydrocarbons and fluorocarbons were variably grafted onto the surface of these particles and used to regulate the rate at which water could access the nitric oxide donors, and thus how quickly nitric oxide was released. These particles were electrospun with a polymer to form fibers, where the hydrocarbon groups (ethyl and butyl) improved the particle stability in the fibers, decreasing particle leaching from the fibers in phosphate buffer saline (PBS) at 37°C for 7 days. Within these fibers, nitric oxide flux increased with decreasing particle hydrophobicity, and the most stable fibers showed the longest nitric oxide release duration [29].

Within cancer cells, the pH is generally reduced from 7.4 to 6.0–7.0 and the glutathione (GSH) levels are increased to 2–8 mM; thus, both are suitable triggers for selectively releasing drug molecules in cancer cells. In a proof of concept study, Xu et al. covalently encapsulated two anticancer drugs, camptothecin and DOX, into silica matrices through disulfide and hydrazone bonds, respectively. Drug containing silanes were synthesized and then co-condensed into nonporous silica nanoparticles through the Stöber process, which should prevent drug leaching. Without GSH only 6% of the camptothecin was released, as opposed to 65.5% in the presence of 10 mM GSH. For the DOX, 71.5% was released at pH 5.0 while only 10% was released at pH 7.4. This is due to reduction of the disulfide bonds (by GSH) or cleavage of the hydrazone bonds (by low pH) within the silica particles, which caused them to degrade and release the drugs. The silica served as a physical barrier which slows diffusion into the particles, leading to an extended release profile. The particles were effective against HeLa cells while the non-drug loaded particles showed no cytotoxicity [30].

Davidson et al. synthesized ibuprofen loaded, bioinspired silica nanoparticles under benign conditions. Inspired by the biomineralization found in diatoms and other organisms, amine analogues of the biomolecules involved were used to rapidly condense silica. Ibuprofen was added during the silica synthesis and the drug was both encased in and physisorbed onto the resulting particles. The synthetic conditions were optimized and it was shown that drug loading increased with amine concentration, drug release with silicate concentration, and synthesis under acidic conditions improved drug loading and thus the total mass of drug released. Using rat gut that had been removed from the rat, the authors reported ~22% of administered silica passed through the gut wall through passive diffusion. In hemolytic experiments, only ~2% of the red blood cells were lysed at 500 µg mL^{-1}, and at the concentration of nanoparticles which passed through the gut wall (~250 µg mL^{-1}) ~0.6% of the red blood cells were lysed. This demonstrated the bioinspired silica nanoparticles were also biocompatible [31].

10.3   Mesoporous Silica

Mesoporous silicas result from the condensation of silica precursors directed by a liquid crystal array of surfactant micelles. Depending on the experimental conditions, such as pH, temperature, molar ratios, and templates, diverse mesophases can be synthesized [32, 33]. Methods for tuning the pore sizes from less than 2 nm up to 30 nm include the use of pore swelling agents like mesitylene, adjusting the length of the carbon chain in the surfactant, or hydrothermal treatment [8, 34, 35]. Two mechanisms for the supramolecular surfactant aggregates and subsequent porous silica nanoparticle formation have been proposed. The liquid crystal templating mechanism states that surfactant above the critical micelle concentration first forms a liquid crystal structure, which exists before the silica precursors are added and serves as the template [36]. In the second mechanism, the addition of the silica precursor and subsequent condensation triggers the formation of this liquid crystal array, resulting in the final mesopore ordering [37]. According to this proposition, porous silica nanoparticles could be prepared at surfactant concentrations below the critical micelle concentration [38].

A typical synthesis involves the aqueous hydrolysis and condensation of tetramethylorthosilicate or tetraethylorthosilicate with a cetyltrimethylammonium bromide surfactant under basic conditions. The Stucky research group further explored these reaction conditions and instead made use of a series of block copolymers as structural directing agents in acidic environments [39–42].

Organic groups can be added to the silica surfaces via an in situ co-condensation method in which the organoalkoxysilane of choice is added with the silica precursor during the synthesis at controlled concentrations or through post-synthetic grafting. The former results in functionality on the interior pore and exterior particle surfaces, while the latter yields organic groups concentrated on the exterior surface. Precursor concentration and type greatly influence the particle morphology in a co-condensation process, resulting in spherical or rod-shaped MCM-41-type mesoporous silica nanoparticles (MSN) [12].

The advantage of porous silica nanoparticles over nonporous is the increased surface area that originates with the pores. Cargo is loaded into these pores, sheltered therein, and then released. Since functionalization of the silica surface is well understood, this release can be selectively triggered. A great deal of work has been done in this regard and been extensively reviewed [14, 43, 44]. For clarity, several of these stimuli responsive drug release mechanisms are generally represented in Scheme 10.2.

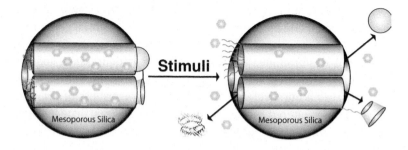

Scheme 10.2 Examples of stimuli-responsive release mechanisms of drug molecules from mesoporous silica nanoparticles.

pH Responsive

The enhanced growth of cancer cells generates a distinctive extracellular microenvironment. Anaerobic metabolism leads to increased intracellular lactic acid production and thus an acidic pH relative to healthy tissues. Although toxic to normal cells, cancer cells have adapted to this decreased pH. Generally speaking, the pH around tumor cells is ~6.4 as opposed to 7.4 found in most noncancer cells [45, 46]. This difference can be exploited for drug delivery by using either pH responsive polymers or molecular caps anchored to the MSN surface through pH sensitive bonds.

Perhaps the simplest strategy was employed by Bierbach and co-workers. Using carboxylate functionalized large pore MSN and the model drug [PtCl(en)(*N*-[acridin-9-ylaminoethyl]-*N* methylpropionamidine)] dinitrate salt, a platinum-acridine anticancer agent, the authors achieved a system in which virtually no premature leakage was observed at neutral pH owing to the electrostatic interactions between the particles and drug. At acidic pH these interactions were mitigated, leading to drug release. A phospholipid bilayer was employed to increase colloidal stability and the carriers were noncytotoxic towards PANC-1 and BxP3 pancreatic cancer cells. The drug-loaded carriers proved to be as effective as the free drug in arresting the S-phase of cell growth and, interestingly, did not undergo endosomal escape. Rather, these nanoparticles released drug directly into the nucleus [47].

In the literature, many pH responsive drug delivery systems rely on polymers. Using polydopamine, a biomimetic material generated from the aqueous self-polymerization of dopamine, Chang et al. coated desipramine, a drug used to treat depression, loaded MSN with a pH degradable organic layer. Without this coating, MSN showed a burst release of desipramine regardless of pH. The polydopamine altered this behavior, showing minimal drug leakage at pH 7.4 while lower pH environments resulted in higher release percentages and a more sustained profile. Effective endocytosis was demonstrated in HeLa cells using DOX as a fluorescent model drug and biocompatibility of the MSN carrier (up to 500 µg mL^{-1} for empty MSN) confirmed. Cell viability experiments showed the polydopamine-coated drug-loaded MSN was in fact more cytotoxic than free DOX. Similarly, the inhibitory effects of desipramine-loaded MSN against acid sphingomyelinase, an enzyme which catalyzes the breakdown of one of the sphingolipids in the cell membranes in the myelin sheath, sphingomyelin, showed the nanocomposites were more effective than free drug [48].

Ross and co-workers used MCM-48, which has a cubic pore structure, coated with succinylated ε-polylysine for the colon specific release of prednisolone. At the relatively acidic pH values in the stomach and small intestine (1.9 and 5.0, respectively), the polymer covers the pore entrances. Ionization of the polylysine at pH 7.4, similar to what is observed in the colon, leads to expansion of the polymer matrix and thus drug release. Carrier toxicity was evaluated in RAW 264.7 macrophages as well as LS 174T and Caco-2 adenocarcinoma intestinal epithelial cells and the particles were found to be biocompatible up to 100 µg mL^{-1}. Using the cell membrane impermeable dye sulforhodamine B, it was found that the polymer-coated particles were successfully internalized into all three cell lines through active transport [49].

Zhang et al. coated MSN with a polymer-lipid layer using Pluronic P123 grafted 1,2-dioleoyl-sn-glycero-3-phosphoethanolamine. The system combined the pH sensitive phospholipid end group with the drug efflux transport inhibiting behavior of the block copolymer. Irinotecan, a drug used for colon cancer treatment, was encapsulated within the MSN and it was shown the particles entered MCF-7 and MCF-7/BCRP multidrug resistant cells through endocytosis. Once at endosomal pH, the polymer-lipid disassembles, leading to a burst release of drug. Pluronic copolymer has been shown to translocate into multidrug resistant cells, localize in the mitochondria, and inhibit complex I and IV in the respiratory chain. This leads to decreased oxygen consumption and therefore less ATP. The authors observed this phenomenon; the Pluronic P123 went on to deplete ATP levels and inhibit the drug efflux protein BCRP selectively in the multidrug resistant cells. Thus, drug could be effectively delivered and retained within cancer cells. In a BALB/c nude mice MCF-7/BCRP drug resistance tumor orthotopic xenograft model, the nanocomposites showed tumor reduction comparable to high doses of free drug (at a relatively lower dose) without the associated drop in survivability. Images of these tumors, the changes in tumor weight, tumor volume, mice body weights, survival curves, and histological analysis of the tumor tissue across all tested groups can be observed in the original publication [50].

Han et al. proposed a triple-stage targeted delivery system that used DOX-loaded MSN modified with TAT peptide and acid-cleavable PEG groups. Additionally, an anionic shell consisting of galactose-modified poly(allylamine hydrochloride)-citraconic anhydride, a polymer with hepato-carcinoma-targeting and charge-reversal properties, was adhered to the positively charged MSN surface through electrostatic interactions. From the outside in, the PEG groups allowed for passive accumulation of the particles in tumor tissue where the slightly lower pH hydrolyzed the acid-cleavable linkage to the MSN surface. The freshly exposed galactose ligands then facilitated internalization into hepato-carcinoma cells and, in the more acidic lysosomal environment, led to charge reversal and thus shedding of the polymer. Finally, the TAT peptide mediated delivery of DOX specifically into the nucleus, leading to cell death. In vitro apoptosis and cytotoxicity studies were conducted using QGY-7703 human hepato-carcinoma cells and showed that, though the nanocomposite itself were nontoxic up to 1 mg mL^{-1}, the drug-loaded nanoparticles were more effective than free DOX. The novelty of this triple-stage delivery system is to address the rapid clearance from the circulating blood by opsonization and re-entry from tumors into the circulatory system resulting from poor cellular uptake and massive exocytosis of the intracellular cargos. In hepato-carcinoma xenograft bearing mice, the drug-loaded nanoparticles showed both 10–40× higher accumulation in the tumor sites and markedly enhanced tumor reduction when compared to free drug [51].

In addition to polymer coatings on mesoporous silica materials, molecular caps have shown to have properties that can be exploited for controlled release applications. Chen and co-workers anchored adamantane to DOX MSN through a pH sensitive benzoic-imine bond, then capped the pores using β-CD. At pH 7.4, the supramolecular complex successfully prevented drug release, though when the pH decreased the benzoic-imine bond was hydrolyzed and the pores uncapped. In the absence of this acid labile bond DOX release was not significantly affected by pH. In vitro cytotoxicity was evaluated in HepG2 and HeLa cancer cells and it was determined that these nanocomposites did not induce cell death at concentrations up to 10 mg mL^{-1}. In both cell lines, cytotoxicity of the drug-loaded nanoparticles was comparable to free DOX. The necessity of the pH sensitive bond was demonstrated again; without the benzoic-imine anchored, the supramolecular adamantane-β-CD assembly remained over the pores and no cytotoxicity was observed [52].

Khatoon et al. synthesized quaternary amine and carboxylic acid bearing MSN as a zwitterionic gatekeeper. The carboxylic acid group was anchored through an acid labile maleic amide linkage such that at physiological pH the particles were slightly negative, but upon entering the tumor microenvironment this bond cleaved. Here, the particles became positively charged and were able to undergo endocytosis. These zwitterionic particles by themselves were shown to be non-cytotoxic up to 200 μg mL^{-1} in SCC7 cells. When the MSN was charged with DOX cytotoxicity increased dramatically at pH 6.5 for 200 μg mL^{-1} particle concentration. In SCC7 tumor-bearing Sprague Dawley rats, the MSN were shown to accumulate in the tumor tissue owing to the enhanced permeability and retention (EPR) effect; however, though less, accumulation in the liver and kidneys was also observed. Doxorubicin-loaded zwitterionic MSN proved to be more effective at reducing tumor volume in SCC7 tumor-bearing nude mice than free DOX and bare DOX-loaded MSN [53].

Lei et al. encapsulated DOX into MSN and then capped the pores with tLyP-1-modified tungsten disulfide quantum dots through benzoic-imine bonds. The tLyP-1 peptide served as both a tumor targeting and a penetrating moiety, recognizing the upregulated

NRP-1 on the cell membranes of tumor cells in nutrient-deficient conditions and activate endocytosis through the CendR pathway. This allowed for drug delivery to be restricted to cancer cells. In the lower pH of the tumor microenvironment, the benzoic-imine bonds became labile, releasing the quantum dots from the surface, resulting in DOX delivery. Additionally, these tungsten disulfide quantum dots could also be used for deep penetrating near infrared light (NIR) triggered photothermal therapy. Effective endocytosis, biocompatibility, and acid-dependent drug release were determined in 4T1 cancer cells. These nanocomposites were more cytotoxic at pH 6.8 than at 7.4, as opposed to free DOX, which was similarly cytotoxic regardless of pH. At higher DOX concentrations, the DOX-loaded MSN was more toxic than free DOX. Near infrared light irradiation significantly enhanced this toxicity owing to the photothermal therapy contributed by the tungsten disulfide quantum dots. Studies using 4T1-cell-based multicellular tumor spheroids confirmed the deep-penetration tendency of the peptide-modified quantum dots released from the DOX-loaded MSN. These particles were shown to accumulate in the tumor tissue in a 4T1-tumor-bearing mice model, but also in the liver. Upon NIR irradiation temperatures of ~47°C was reached in the tumor, showing the efficacy of the system for photothermal therapy. All together, these nanocomposites displayed superior tumor inhibition, up to 93%, than free drug or photothermal therapy alone [54].

Redox Responsive

Owing to the difference in the reducing or oxidizing environment between healthy and diseased cells, a variety of redox responsive drug release systems have been developed. These systems take advantage of the presence of reactive oxygen species or GSH levels. For instance, extracellular GSH tends to be ~10 μM and the intracellular GSH concentration ranges from 1–10 mM in normal cells, but in cancer cells these levels tend to be elevated [55, 56]. Specifically, breast, head and neck, ovarian, and lung cancers display increased GSH concentrations, though brain and liver cancers exhibit lower GSH levels when compared to healthy cells [57].

Cleavage of a disulfide bond by intracellular GSH is a very popular method for redox responsive drug delivery. Kim and co-workers used a peptide with an induced turn structure as a gatekeeper for calcein-loaded MSN. The authors demonstrated the turn structure efficacy with a CPGC peptide sequence, which showed very little dye leakage in PBS. To generate a stimuli-responsive system, a CGGC sequence was used instead. This peptide possessed a disulfide bond between the two cysteine residues which forced the chain into a loop over the MSN pores and prevented dye leakage in PBS. Upon the addition of 1 mM GSH, the disulfide bond was cleaved, the peptide adopted a random structure, and a burst release of dye was observed. In experiments without this disulfide bond, no apparent capping of the pores took place [58].

Transferrin was grafted to the MSN surface through a GSH cleavable disulfide bond by Chen et al. The protein could function as both an obstruction to drug leaching from the MSN pores and a targeting moiety owing to the upregulated transferrin receptors on cancer cell surfaces. These particles were loaded with DOX and showed a burst release with 10 mM GSH regardless of the pH. More than 90% cell viability was reported at MSN concentrations from 1–500 µg mL^{-1} in Huh-7 hepatoma cells, while the DOX-loaded particles showed similar cytotoxicity to the free drug [59].

Mirkhani and co-workers attached a bioactive polyoxometalate $(TBA)_4H_3[GeW_9V_3O_{40}]$ antitumor drug, a low cost and novel inorganic molecule, which has demonstrated efficacy against U87 human glioblastoma brain cancer cells, to DOX-loaded MSN through a disulfide bond. The MSN served as a vehicle into cells and provided a platform for the intracellular GSH responsive dual inorganic and organic drug release. An attached dye allowed for the tracking of the polyoxometalate in the cell environment and was used to confirm successful delivery into U87 glioblastoma cancer cells. Doxorubicin release was shown to be GSH concentration dependent with minimal leakage occurring in the absence of the reducing agent. In normal cells very little cytotoxicity was observed after treatment with the DOX-loaded nanocomposites; however, in the U87 cancer cells significant cell death occurred [60].

Frequently the cargo of choice for these studies is fluorescent in order to track successful delivery into cells. Lai et al. developed a system where this was not the case by incorporating a fluorescence resonance energy transfer (FRET) system into MSN. The FRET donor-acceptor pair, coumarin and fluorescein isothiocyanate, were held in close proximity through the host–guest interactions of an adamantane and β-cyclodextrin (β-CD). For the purposes of this study, DOX was used as a model drug. Coumarin was anchored to the MSN surface using a cysteine and the thiol was reacted with 1-adamantanethiol to form a disulfide bond. Fluorescein isothiocyanate-β-CD was then added and the resulting host–guest interactions led to effective capping of the MSN pores. Under normal conditions, the proximity between the coumarin and fluorescein isothiocyanate resulted in FRET from coumarin to fluorescein isothiocyanate; however, in the intracellular reducing environment, cleavage of the disulfide bond by GSH resulted in a loss of FRET as the drug was released. These particles were shown to be non-cytotoxic in HeLa cells up to 20 μg mL^{-1} and cell viability was inversely proportional to FRET signal across GSH concentrations [61].

Chen et al. capped MSN with a gadolinium-based bovine serum albumin complex that could act as both a drug delivery system and contrast agent for magnetic resonance imaging. Hyaluronic acid was also anchored to the surface and helped direct the particles towards cancer cells owing to the overexpressed CD44 receptors on the 4T1 cell surface. At elevated GSH concentrations, the cap is released owing to the disulfide bond cleavage. Biocompatibility up to 500 μg mL^{-1} and effective endocytosis was demonstrated in 4T1 cancer cells, and the nanoparticles showed favorable hemocompatibility. When loaded with DOX, the nanocomposites were more effective against 4T1 cells in vitro and in tumor-bearing mice when compared to free DOX. Indeed, mice that underwent treatment had no significant weight loss and histological studies showed no signs of inflammation or tissue lesions in the heart, liver, spleen, lung, or kidney of Balb/c mice after 1 week postinjection [62].

Another strategy is to utilize elevated levels of peroxide species to trigger drug release from MSN. Hu et al. co-loaded MSN with DOX and a reactive oxygen species generator α-tocopheryl succinate and blocked the pores using a β-CD anchored through a thioketal. PEG was conjugated with adamantane and attached to the nanocomposites via host–guest interactions with the β-CD. Upon exposure to the reactive oxygen species within cancer cells, some of these thioketals were cleaved, leading to DOX and α-tocopheryl succinate release, which caused the generation of additional reactive oxygen species. In this way, the system demonstrated self-accelerating drug release. This behavior was confirmed in

MCF-7 human breast cancer cells, which are rich in reactive oxygen species, yet in 293T human embryonic normal cells very little DOX release or reactive oxygen species generation was observed. The carriers proved to be biocompatible up to 100 µg mL^{-1} in both cell lines and, without the α-tocopheryl succinate, a 69% cell viability was still observed in MCF-7 cells for DOX-loaded particles. In vivo experiments using MCF-7 tumor-bearing nude mice established that the system was more effective at tumor volume reduction than free DOX without pathological abnormalities in the heart, liver, spleen, lungs, and kidney or cardiotoxicity [63].

Metal chelators are a possible treatment for Alzheimer's disease owing to the elevated levels of trace metals that attribute to the disease state, but the resultant subacute myelo-optic neuropathy limits their long-term use. To address this problem and take advantage of the increased levels of hydrogen peroxide associated with Alzheimer's disease, Geng et al. used human IgG to cap clioquinol-loaded MSN as a hydrogen peroxide responsive system. The MSN was first functionalized with phenylboronic acid groups, which then formed cyclic esters with the saccharide diols on the glycoprotein. Upon exposure to hydrogen peroxide, these esters oxidized to phenols, releasing the protein from the MSN surface and uncapping the pores. Cargo release was shown to be directly dependent on the hydrogen peroxide concentration. In rat pheochromocytoma PC12 cells, the presence of the drug-loaded MSN showed zero premature release and, upon addition of hydrogen peroxide, increased cell survivability over the control [64].

Thermally Responsive

The advantage of a thermally responsive system is that heat stimuli can be targeted to a specific location, reducing drug release in healthy tissue, and the procedure is noninvasive. Generally speaking, these systems are advantageous if thermally responsive polymers are either grafted to or grafted from the MSN surface. A common selection in the literature is poly(*N*-isopropylacrylamide) (PNIPAM), which has found extensive use in biomedical applications, though other methods are also prevalent [65].

Two types of MSN with differently sized pores (3.5 and 5.0 nm) were investigated by Ugazio et al. for the topical delivery of quercetin, an antioxidant. Thermally responsive PNIPAM-*co*-

MPS [3-(methacryloxypropyl)trimethoxysilane] was radically polymerized inside the MSN mesopores such that at cutaneous temperatures the polymer chains collapsed and allowed for drug release. Biocompatibility was established in human keratinocyte HaCaT cells and the accumulation/permeation of the functional MSN was investigated through porcine skin, showing modest accumulation. The functional activity of released drug was evaluated relative to the antiradical and metal chelating activities, and it was determined that quercetin was still active upon release from MSN, indicating the potential of the silica carrier to shelter the antioxidant from degradation during storage. Although both the small and large pore MSN showed drug release, the MSN with 5.0 nm pores exhibited more thermoresponsive behavior [66].

Ribeiro et al. combined MSN with imbedded fluorescent molecules (a perylenediimide derivative) with thermally responsive PEG-acrylate random copolymers to generate a temperature controlled drug release system. At temperatures below the LCST (37°C in water), the polymer shell is hydrophilic and exists in an expanded matrix into which the loaded model molecule (sulforhodamine B) diffuses. Then, upon an increase in temperature to 40–50°C, the polymer becomes hydrophobic and extrudes the molecule in a "pumping" mechanism which increased release rate in a controlled manner. Release kinetics were dependent on the polymer shell type: linear or cross-linked. While diffusion across the extended polymer did not depend on the shell thickness or structure, the "squeezing-out" of the cargo was faster in the case of the cross-linked polymer than either of the linear polymer systems studied [67].

Using the melting of a single helical peptide, de la Torre et al. were able to control the release of safranin O. The 17-mer peptide was specifically designed to preserve the tendency to fold into an α-helix and minimize alternative secondary structures, resulting in a final H-SAAEAYAKRIAEALAKG-OH sequence. Zero dye release was observed at temperatures between 4 and 40°C, when the peptide was in a helical conformation. With an increase in temperature above 40°C, the peptide transitioned into a random coil and safranin O diffused out of the MSN. This transition from α-helix to random coil was shown to be reversible; thus, the nanocomposites were reloaded with additional dye and the release profile was similar to that observed initially, indicating the reusable potential of this system [68].

Similarly, Kros and co-workers used a coiled-coil peptide motif as a thermally responsive valve for the controlled release of fluorescein. Simple MSN was modified with thiol groups via a post synthetic grafting method and the first peptide [C(EIAALEK)$_3$] was anchored to the surface through a disulfide bond. To minimize premature leakage, it was important to attach the peptide close to the silica surface. After fluorescein loading, the complimentary peptide [(KIAALKE)$_3$], or this peptide with an added PEG group, was added and formed a coiled-coil complex with the anchored peptide. Near zero fluorescein release was observed at 20°C, even after 4 h, but when the coiled-coil complex melted at 80°C, the model compound diffused very quickly from the pores [69].

Light Responsive

Light is an attractive stimulus owing to its noninvasive nature, deep tissue penetration, and the ease with which a specific location can be targeted. Upon exposure to a designated wavelength, certain bonds can isomerize or break, leading to drug release in systems where capping agents are attached through such bonds. Alternatively, exposure to specific wavelengths can induce a change in the local environment, leading to drug release through a secondary mechanism.

Guardado-Alvarez et al. functionalized MSN with a hydrophobic coumarin-based molecule, which could non-covalently associate with β-cyclodextrin (β-CD), resulting in MSN pore blockage. One or two photon excitation could cleave the bond through which the coumarin was anchored, resulting in uncapping and cargo release. Rhodamine B was used as a model compound to study the release kinetics. No rhodamine B or free coumarin derivative were observed without laser irradiation, even when the system was heated, indicating removal of the snap-top is indeed caused by the laser light [70].

Using a singlet oxygen sensitive bis(alkylthio)alkene linker, Kim and co-workers covalently anchored the model drug 5-[(2-aminoethyl)amino]naphthalene-1-sulfonic acid to the MSN surface. Zinc phthalocyanine was loaded into the pores and, upon exposure to light between 600–700 nm, generated singlet oxygen which cleaved the linker and resulted in drug release. Stability of the photosensitizer was determined by exposing zinc phthalocyanine–loaded MSN without further modification to a long-pass-filtered halogen lamp in medium with 10% serum. The residual photosensitizer was then leached out into dimethyl sulfoxide and the concentration determined using UV-Vis absorption. Nonspecific release of the zinc phthalocyanine was not observed until after 6 h of incubation, indicating MSN delivery system ensured high stability of the loaded photosensitizer. Photoirradiation was shown to dramatically increase drug release, reaching approximately 90% after 3 h of irradiation as opposed to only 10% over the same time in the dark [71].

He et al. generated a system whereby a photoinduced pH jump released graphene oxide from the MSN surface to uncap the pores. A photoacid generator was loaded into the pores with the model drug DOX and the graphene oxide sheets were anchored to the surface through an acid-labile boroester linkage. The graphene oxide itself was functionalized with amine terminated PEG, to which folic acid was attached to decrease nonspecific uptake. Very little cellular uptake was observed in L02 cells with low cytotoxicity in HeLa cells without: ultraviolet (UV) light irradiation, DOX loading, or photoacid generator loading. The complete system, however, showed a high killing efficacy against cancerous cells [72].

A calcium phosphate gatekeeper was employed by Choi et al. for the smart delivery of nitric oxide and used in a corneal wound-healing study. Diazeniumdiolates were attached to the MSN surface, a pH-jump reagent 2-nitrobenzaldehyde was loaded into the pores, and the surface was coated with calcium phosphate (MSN-CaP-NO). Upon UV irradiation, the 2-nitrobenzaldehyde undergoes an intramolecular excited-state hydrogen transfer, which then drops the local environmental pH and dissolves the calcium phosphate. The previously sheltered diazeniumdiolates are thus exposed to physiological conditions, under which they are known to decompose, resulting in the generation of nitric oxide. In vitro studies showed the nanocomposites were biocompatible; in fact, cell viability increased to 175% with UV irradiation, indicating the system's therapeutic potential. Particles that had been pre-activated with UV light were then administered topically to treat corneal wounds, showing both successful re-epithelialization and rapid clearance. A schematic representation and wound-healing results from in vivo experiments are shown in Fig. 10.2 [73].

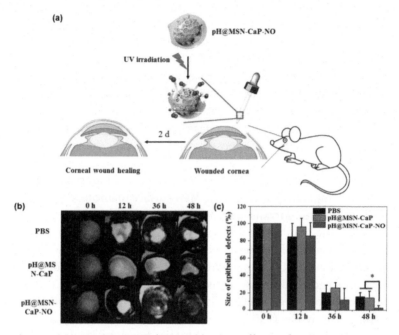

Figure 10.2 In vivo corneal wound-healing effects of pH@MSN-CaP-NO. (a) Schematic description of overall in vivo experiments. (b) Representative images of corneal wounds treated with light-exposed PBS (control), pH@MSN-CaP, and pH@MSN-CaP-NO. (c) Quantification of the wounded area after treatment with light-exposed PBS (control), pH@MSN-CaP, and pH@MSN-CaP-NO. Reprinted with permission from Choi, H. W., Kim, J., Kim, J., Kim, Y., Song, H. B., Kim, J. H., Kim, K. and Kim, W. J. (2016). Light-induced acid generation on a gatekeeper for smart nitric oxide delivery, *ACS Nano,* **10**, pp. 4199-4208. Copyright 2016 American Chemical Society.

Stoddart and co-workers developed a multifunctional system that served as both an imaging agent and a drug release vehicle. Paclitaxel was loaded into the MSN and a ruthenium(II) dipyridophenazine complex immobilized over the pores through a monodentate ligand, benzonitrile, which had been covalently attached to the MSN surface. The ruthenium complex had a dual purpose: it served as a luminescent imaging agent and a deoxyribonucleic acid (DNA) intercalator. Unlike the free complex, the MSN bound ruthenium(II) dipyridophenazine showed luminescence in water due to shielding from the MSN surface. When exposed to visible light, the complex underwent a selective ligand exchange of the benzonitrile with water, leading to uncapping of the pores and release of both anticancer molecules. Cytotoxicity against two strains of breast cancer cells was tested. In the dark, empty MSN showed no significant cytotoxicity against MDA-MB-231 cells and only some cytotoxicity against MDA-MB-468 cells; however, upon light activation cytotoxicity in both cell lines increased dramatically. It should be noted that these particles were still less effective than free paclitaxel, perhaps due to the delayed cargo release from the nanoparticles in a cellular environment [74].

Biostimuli Responsive

Some groups have generated drug delivery systems that release cargo in response to biological molecules such as enzymes. Although diverse, these mechanisms tend to be highly specific by nature.

Bhat et al. functionalized the surface of anticoagulant-loaded MSN with a peptide containing a thrombin-specific cleavage sequence to delay the blood clotting process. Acenocoumarol was loaded into the MSN pores and then a LVPRGS containing peptide was attached to the surface through "click" chemistry, effectively preventing anticoagulant leakage prior to thrombin exposure. The enzymatic concentration needed to be at least 100 nM to trigger drug release (for reference, thrombin concentrations in human blood are ~1 nM under normal conditions and >500 nM during clot formation); hence, this system could remain dormant in the blood and become active only during coagulation. In rabbit blood plasma, the peptide functionalized MSN induced no coagulation, while in the presence of human thrombin, clotting was slowed from 2.6 ± 0.3 min with just calcined MSN to 5.0 ± 0.3 min with the thrombin triggered nanocomposite [75].

Zink and co-workers used the FB11 anti-O-antigen antibody for the O-antigen of the lipopolysaccharide of *Francisella tularensis* (FT), the causative agent of tularemia and a Tier 1 Select Agent of bioterrorism, as a nanovalve for fluorescein release. Broadly, this system would release antibiotics only in response to a specific pathogen and at the specific site of infection. Modified FT lipopolysaccharide O-antigen was attached to the MSN surface and the antibody capped the pores through non-covalent antibody-antigen interactions. This prevented antibiotic leakage under normal conditions, but in the presence of natural O-antigen, a competitive displacement of the antibody cap occurred, resulting in cargo release. In live *Francisella tularensis*, the MSN nanocarriers released nearly 100% of their loaded cargo in only 1 h, while very little release was observed when the particles were incubated with live *Francisella novocida*, a closely related bacteria with similar O-antigen structures [76].

With the goal of generating a stimuli-responsive pulsatile delivery system, Villalonga and co-workers used a lactose-modified esterase to cap the pores of loaded MSN through boronic acid cyclic ester bonds with the lactose residues. Release could be triggered in two waves: displacement of the lactose with *D*-glucose leading to partial uncapping of the pores, then acid-induced cleavage of the boronic acid cyclic esters with addition of ethyl butyrate, which is converted to a butyric acid by the esterase and decreases the local pH. Both result in removal of the neoglycoesterase from the MSN and thus uncapping of the pores. Some leakage was observed in HeLa cells, which is attributed to the slightly acidic microenvironment hydrolyzing some of the ester bonds, but overall the DOX-loaded carriers were non-cytotoxic in the absence of trigger molecules. Treatment with D-glucose induced moderate cell death while ~50% of cells were either dead or in the process of cell death after the addition of ethyl butyrate [77].

Anchoring a sandwich-type DNA structure composed of two single-stranded DNA arms and an adenosine-5'-triphosphate (ATP) aptamer to the surface of $Ru(bipy)_3^{2+}$-loaded MSN, He et al. designed a potential drug release system that responded to ATP. In the absence of ATP, the pores were tightly capped by the DNA complex and showed no premature dye leakage, but when ATP was added to the system release promptly occurred. This is due to the preferential binding of the aptamer to ATP, which leaves only the short, flexible single-stranded DNA on the MSN surface, allowing for diffusion out of the pores. The system showed good selectivity for ATP over ATP analogues and stability in mouse serum [78].

In a unique example using MCM-48 with a cubic pore structure, Popat et al. used bacterial azoreductase to release a sulfasazine prodrug for inflammatory bowel disease treatment. The cubic pore structure showed superior adsorption capacity and mass transport over the two-dimensional hexagonally structured pores of the more commonly used MCM-41. Only in the presence of the enzyme is the covalently bound prodrug released, preventing the possibility of premature drug leakage. Once released, the sulfasalazine can be reduced to 5-aminosalicylic acid and sulfapyridine selectively inside the colon. In simulated intestinal fluid, the system showed no pH dependence with zero release at both pH 1.2 and 7.4. The presence of azoreductase resulted in a release of 32 mg of 5-aminosalicylic acid per gram MSN [79].

Willner and co-workers developed a biomarker responsive drug delivery device based on hairpin DNA structures. The biomarkers could be either the complimentary nucleic acid strand or an aptamer substrate and, upon drug release, the biomarkers were released as well. These could then go on to trigger additional pore uncapping (Scheme 10.3). Some leakage was observed in all models. The system efficacy was tested in both MDA-231 breast cancer cells and MCF-10a normal breast cells using camptothecin as a model drug and significantly higher cell death was observed in the cancer cells owing to their increased metabolic activity and thus biomarker concentration. Approximately 25% of normal cells, however, also died [80].

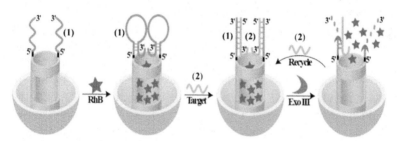

Scheme 10.3 Unlocking of hairpin-capped mesoporous SiO_2 NPs and the release of rhodamine B, RhB, using an analyte–DNA biomarker (**2**) as activator for opening the hairpins and implementing Exo III as biocatalyst for the regeneration of the DNA–biomarker. Reprinted with permission from Zhang, Z., Balogh, D., Wang, F., Sung, S. Y., Nechushtai, R. and Willner, I. (2013). Biocatalytic release of an anticancer drug from nucleic-acids-capped mesoporous SiO_2 using DNA or molecular biomarkers as triggering stimuli, *ACS Nano*, **7**, pp. 8455–8468. Copyright 2013 American Chemical Society.

Cai and co-workers took advantage of overexpressed matrix metalloprotease 2 in the tumor microenvironment to generate a multilayered drug delivery vehicle. Doxorubicin-loaded MSN were functionalized with folic acid and then coated in crosslinked gelatin. To this biopolymer PEG was conjugated, which increases the particle biocompatibility but decreases endocytosis. Thus, the particles were stable in the absence of the enzyme and very little DOX release was observed in human L02 human normal liver cells or in HT-29 human colon cancer cells with a matrix metalloprotease inhibitor. No hemolysis was observed after 3 h even with particle concentrations of 3,200 μg mL^{-1} owing to this PEG layer. However, when the enzyme digested the gelatin, the PEG was shed and the underlying folic acid exposed, leading to endocytosis into cancer cells and significant cell death in HT-29 cells. In vivo studies in a HT-29 mouse xenograft tumor model showed the folic acid functionalized, gelatin coated, PEG-conjugated MSN in its entirety was more effective than even free DOX in tumor reduction with no statistically significant body weight loss [81].

Other Controlled Release

Aside from pH, redox, thermal, light, and biostimuli responsive systems, a variety of other controlled release mechanisms have been investigated.

Paik and co-workers used α-synuclein-coated gold nanoparticles to prepare protein mediated "raspberry-type" MSN nanocomposites for Ca^{2+} ion controlled drug release. α-Synuclein is an amyloidogenic protein that is responsible for the Lewy body formation in Parkinson's disease. In this system, it also tightly capped the gold nanoparticles over the MSN pores in an acid-stable manner. In the presence of divalent cations the protein's conformation changed, such that the hydrodynamic diameter of the gold–protein nanocomposites decreased by 2 nm, opening the MSN pores while remaining attached to the MSN surface. Calcium was chosen because there is a relatively high intracellular concentration of free ions within the cytoplasm and surges in calcium concentration have been implicated in cancer, neurodegenerative diseases, and cardiovascular diseases. No release was observed without Ca^{2+} or in the presence of ethylenediaminetetraacetic acid, Na$^+$, or K$^+$ ions. In vitro studies were done using DOX as a model drug in HeLa cells. The carriers were determined to be biocompatible; however, DOX-loaded nanoparticles were cytotoxic, especially in the presence of an endoplasmic reticulum Ca^{2+} pump inhibitor thapsigargin, which increases the intercellular Ca^{2+} concentration. When a membrane permeable Ca^{2+} chelator was added, very little cytotoxicity was observed [82].

Huang et al. used carboxylate-substituted pillar[6]arene-valved MSN functionalized with dimethylbenzimidazolium or bipyridinium stalks for the pH, metal ion, and, most interestingly, competitive binding mediated release of DOX. The supramolecular nanovalve dethreaded when exposed to 1,1'-dimethyl-4-4'-bipyridinium, which took the place of the stalks in both systems as the guest inside the pilararene. Alone, the pilararene functionalized MSN was shown to be biocompatible in A549 cells [83].

Cytosine-rich DNA was grafted onto the surface of MSN as a smart molecule-gated system for drug release by He et al. The cytosine groups hybridized into metal-dependent pairs, mediated by Ag^+ ions, which could be displaced with thiol-containing molecules. This resulted in the loss of the Ag^+-induced duplex structure and thus guest molecule release. It was found that this process was entirely reversible, so the nano-containers could be reloaded and used again. The authors demonstrated successful endocytosis, biocompatibility, and release of a model compound, $Ru(bipy)_3^{2+}$, into HeLa cells [84].

Li and co-workers synthesized hierarchically structured MSN by controlling the feeding ratio of pre-prepared MSN and the second addition of the silica precursor. The resulting particles possessed two layers with different pore sizes, with some of the pores in the inner layer of silica becoming blocked by deposition of the outer layer, and variable thicknesses of the outer layer. Particles with thinner outer layers displayed a burst release of the model drug DOX before tapering off, while thicker silica outer layers resulted in sustained release with near zero-order pharmacokinetics. Successful delivery of DOX into HeLa cells, as well as biocompatibility of the DOX-free MSN, was demonstrated [85].

In work done by Vallet-Regi and co-workers, ultrasound was used to release DOX molecules from MSN coated with the random copolymer of 2-(2-methoxyethoxy)ethyl methacrylate and 2-tetrahydropyranyl methacrylate. The thermally responsive 2-(2-methoxyethoxy)ethyl groups allowed the MSN to be loaded at 4°C, when the copolymer is in an extended, random coiled state, but collapse over the pores at physiological pH, blocking the pore entrances. The polymer was first synthesized, then attached to a silane, which was post synthetically grafted onto the MSN through silanol chemistry. Upon ultrasound irradiation, the hydrophobic tetrahydropyranyl groups were cleaved, which led to an increase in the LCST above 37°C and thus the transition of the polymer gates from compact (hydrophobic) to extended (hydrophilic). Without drug loading, the polymer-coated MSN showed biocompatibility up to 500 μg mL^{-1} and the ultrasound irradiation did not harm LNCaP human prostate cancer cells. Without ultrasound exposure, cell viability remained high with DOX-loaded MSN, but dropped significantly with ultrasound, indicating successful, on demand release of DOX into cells [86].

Multistimuli Responsive Systems

In many instances, a combination of stimuli has been employed to increase the specificity of drug release.

Taking advantage of several traits of the tumor microenvironment, Zhang and co-workers engineered a targeted drug delivery system. At the MSN surface a β-CD was anchored through a disulfide bond, and then a peptide sequence containing an RGD motif and matrix metalloproteinase substrate PLGVR was attached using the host–guest interactions between the β-CD and an adamantane. To this, a polyanion, polyaspartic acid, was attached. The polymer served as a protective layer that prevented nonspecific cell uptake and was shed in the presence of matrix metalloproteinase, which hydrolyzes the PLGVR peptide sequence. This exposed the RGD peptide, which could bind to the overexpressed integrin receptors on the cancer cells. Once inside the cell, high levels of GSH cleaved the disulfide bond, removing the β-CD from the MSN pores and allowing for drug release. These particles were effectively internalized by SCC-7 squamous cell carcinoma cells and HT-29 human colon cancer cells and the loaded DOX was successfully released, resulting in high cytotoxicity. Human embryonic kidney 293 transformed cells, on the other hand, did not experience cytotoxicity when treated with DOX-loaded particles. Indeed, no uptake into the noncancerous cells as observed at all [87].

By capping DOX-loaded MSN with hyaluronic acid through a disulfide bond, Yang et al. generated a system that could effectively deliver DOX into DOX-resistant MCF-7/ADR human breast cancer cells. The particles could accumulate at the tumor site through the EPR effect, enter cells through active transport, and effectively escape from the endosomes due to a histidine moiety on the hyaluronic acid coating. Once in the cytoplasm, dual responsive release occurred owing to hyaluronidase digestion of the biopolymer and cleavage of the disulfide bond by intracellular GSH. The carriers themselves were shown to be biocompatible, while DOX-loaded MSN was cytotoxic. In MCF-7/ADR tumor-bearing mice, the MSN nanocomposite showed more effective tumor reduction than free DOX without the accompanying weight loss [88].

Du and co-workers developed a γ-CD based simultaneous cascade drug delivery system in which the γ-CD served as both a cap for the MSN mesopores and a drug carrier itself. MSN was co-modified with GSH cleavable disulfide linked carbamoylphenylboronic acid groups and amines. The γ-CD secondary hydroxyls were able to bind to the boronic acid moieties with the assistance of the amines at physiological pH, and at acidic pH the boronate bonds hydrolyzed. In addition, the drugs inside the γ-CD cavity were protonated and expelled from the cavity. Alternatively, monosaccharides could competitively bind to the phenylboronic acid, supplanting the γ-CD and uncapping the MSN pores. Finally, in the presence of a reducing agent the disulfide bond could be cleaved, yielding the same result. Synergistically, dual drug release could be accomplished under a variety of cancer characteristic conditions. High GSH levels and a weakly acidic environment in A549 tumor cells demonstrated the efficacy of the system in vivo [89].

Combining the desirable properties of gold nanoparticles and MSN, Wang et al. used carboxylatopillar[5]arene-modified gold nanoparticles to cap the pores of quaternary ammonium salt functionalized MSN. The positively charged salt was encircled by the electron-rich carboxylatopillar[5]arene cavity, effectively holding the gold nanoparticle over the mesopores and preventing premature drug release. In the presence of a competitive binding agent, in this case ethanediamine, this host–guest interaction was disrupted, leading to rapid cargo release. External heat could also be used to disrupt this interaction. In this way, irradiation of the gold nanoparticles accomplished both local hyperthermia and drug release in in vivo drug delivery applications [90].

Using a light-responsive azobenzene derivative, β-CD, and diimine, Liu and co-workers developed a drug delivery system that displayed AND logic. That is to say, both light irradiation and decreased pH were required for substantial drug release to occur. The azobenzene was covalently attached to the MSN surface and served as a guest for the β-CD, which was also anchored to the surface through a diimine bond. Upon light irradiation, the azobenzene underwent a *trans-* to *cis-* isomerization and was ejected from the β-CD core; however, the β-CD remained over the pore entrances and thus blocked drug release. In the instance of a decrease in pH, the diimine bond was hydrolyzed, but the host–guest interactions between the azobenzene and β-CD kept the pores covered. Only with both stimuli were the pores uncapped, though this process proved to be reversible such that cargo delivery could be achieved incrementally. Dye-loaded particles were injected into zebrafish tail veins and the AND logic requirement was demonstrated in vivo [91].

Ma et al. used single-stranded DNA anchored to the MSN surface through a disulfide bond to control the release of loaded DOX. A 33-mer DNA strand hybridized with the anchored single strand 15-mer DNA to block the pore entrances. Heating the system above the DNA melting temperature or exposure to reducing agents triggered shedding of the cap and subsequent drug unloading. The carriers themselves were shown to be effectively endocytosed into HeLa cells and biocompatible. Doxorubicin delivery was also demonstrated through the enhanced cytotoxicity of DOX-loaded MSN (10% cell viability after 48 h) [92].

Hei et al. developed a series of PEG-conjugated poly(acrylamide-*co*-acrylonitrile) polymers with an upper critical solution temperature of 42°C and anchored them to the MSN surface through a reductively cleavable disulfide bond. At physiological temperatures, the polymer existed in a globular, hydrophobic state, which collapsed over the mesopores and prevented premature drug release. With local heating or the addition of a reducing agent, the polymer either became hydrophilic and extended away from the pores or was cleaved from the surface entirely. Either stimulus resulted in drug release. These particles were shown to be noncytotoxic to SK-BR-3 breast cancer cells at both 37°C and 42°C. Successful endocytosis into this cell line was also observed. When loaded with DOX, the system proved highly cytotoxic when triggered either by an increase in temperature above the upper critical solution temperature or the addition of GSH [93].

Using a crosslinked polymer that was thermally, redox, and pH sensitive, Chang et al. developed a triply responsive drug delivery system for targeting the tumor microenvironment. The polymer shell used was a copolymer of the thermally responsive *N*-vinylcaprolactam and pH responsive methacrylic acid crosslinked through disulfide bonds. These two components affect each other such that when the pH becomes more acidic the volume phase transition temperature drops. The result is that when the particles enter the more acidic tumor microenvironment of lysosomes, the transition between hydrophilic and hydrophobic occurs at a temperature below 37°C, leading to drug release. Additionally, in the presence of 10 mM GSH (intracellular conditions), the polymer shell is shed entirely. At standard physiological conditions only 5% of the loaded DOX was released, while at pH 6.5, 37°C, and 10 mM GSH nearly 100% release was achieved. PEG groups were attached to the polymer shell and the nanocomposites were shown to be biocompatible in MCF-7 cancer cells and 293 normal cells up to 500 μg mL^{-1}. The DOX-loaded particles displayed high cytotoxicity against MCF-7 cancer cells, but were noncytotoxic in 293 cells [94].

Delivery of Macromolecules

Macromolecular therapeutics possess many advantages, such as increased specificity in their action and decreased interference with normal biological processes. Nevertheless, delivery of macromolecules can prove difficult due to their fragility and large size. Thus, modifications to the MSN are typical. The delivery of proteins using MSN has been recently reviewed [95].

To achieve fast and painless transdermal administration of insulin for diabetes treatment, Xu et al. used insulin and glucose oxidase-loaded MSN integrated with microneedles. Upon insertion into the skin, the microneedles released the MSN into the tissue. The MSN surface was modified with 4-(imidazoyl carbamate) phenylboronic acid pinacol ester, which served as a host for α-CD (the pore capping agent). The glucose oxidase generated hydrogen peroxide during the conversion of glucose to gluconic acid, which oxidized the phenylboronic ester and disrupted the host–guest interaction between the ester and α-CD, leading to pore uncapping.

A rapid release of insulin was observed at a hydrogen peroxide concentration of 5 mM and at a glucose concentration of 20 mM, with minimal release observed in the absence of both. The hypoglycemic effect was investigated in diabetic rats and indicated that the MSN delivered transdermally from the microneedles dropped the blood glucose levels into the normal glycemic range. These levels were sustained for 4.5 h, much longer than the duration obtained from the subcutaneous injection. Glucose oxidase was essential for successful insulin release [96].

Stroeve and co-workers used pore-expanded MCM-41 type MSN for the controlled release of bovine hemoglobin. Trimethylbenzene was added during the MSN synthesis as a pore expanding agent, resulting in worm-like mesopores approximately 15 nm in diameter, which is sufficiently large to encapsulate 5 nm bovine hemoglobin. The interior of the pores was selectively functionalized with PEG groups, which decreased the amount of protein directly adsorbed to the silica surface and improved protein release. Since proteins can denature when adsorbed onto glass surfaces, surface PEGylation assisted in maintaining the protein's secondary structure. In order to better control protein release from these particles PNIPAM was grafted on the surface and partially blocked the pores below the 37°C, the LCST. Above this temperature, proteins diffused from the pores more quickly and a higher percentage was released due to a lack of physical restriction [97].

Using a GSH cleavable disulfide bond, Griebenow and co-workers covalently immobilized sulfosuccinimidyl 6-[3'(2-pyridyldithio)-propionamido]-hexanoate-modified carbonic anhydrase inside the thiol functionalized MSN pores through a thiol-disulfide interchange. Near zero enzyme release was observed in PBS alone or in the presence of 1 µM GSH, simulating extracellular plasma conditions. However, at 10 mM GSH (intracellular conditions), nearly 100% of the carbonic anhydrase was released over 500 h, and the released enzyme retained most of its activity. Biocompatibility of the enzyme-MSN particles was demonstrated in HeLa cells [98].

Deodhar et al. took a different approach to large protein delivery and loaded the dissociated subunits of a model protein, *Concanavalin A*, into pore-expanded MCM-41 type MSN. Similar to the pore-expanding method above, mesitylene was used during the synthesis, resulting in hexagonally ordered pores with a bimodal pore distribution (5.8 nm and 15.9 nm). It was demonstrated that these dissociated protein subunits were effectively sheltered from proteases within the MSN pores and, upon release, reassociated into active protein [99].

Chen et al. used MSN as a carrier for transactivator protein-conjugated superoxide dismutase. Transactivator protein served to internalize the nanocomposites within cells via a transactivator protein-mediated nonendocytosis mechanism, thereby avoiding any degradation inside endosomes. The MSN was labeled with fluorescein isothiocyanate and a nitrilotriacetic acid, which reacted with nickel metal ions and allowed the histidine tagged protein complex to bind tightly to the MSN surface. In this way, the protein complex could be purified and conjugated to the MSN in a single step. Superoxide dismutase was denatured with concentrated urea such that no extracellular activity was observed. Once inside the cell, intracellular chaperones such as Hsp90 refolded the enzyme and functionality was restored. This effective delivery and restoration of activity was demonstrated in HeLa cells, where the ability of the MSN–protein complex to protect cells from oxidative damage was observed [100].

Instead of conventional MSN, Kwon et al. used iron oxide nanoparticles as seeds for the ethyl acetate mediated growth of extra-large pore (30 nm) MSN. The large pores allowed for significantly higher protein loading than conventional MSN. Although the extra-large pore MSN induced some reactive oxygen species, no pro-inflammatory cytokines were generated in bone marrow-derived macrophages. These particles were loaded with the cytokine IL-4, which was delivered into macrophages for the purpose of M2 macrophage polarization. On its own, IL-4 has a very short half-life, but sheltering inside the MSN allowed for extended protein release. It was found that IL-4-loaded MSN injected intraperitoneally showed no toxicity and effective M2 macrophage polarization in mice [101].

In an interesting example, Hartono et al. used poly-L-lysine modified large pore MSN with a cubic mesostructure for the delivery of siRNA against minibrain-related kinase and polo-like kinase 1 in osteosarcoma KHOS cancer cells. A cubic mesostructure was selected because the interconnected pores are less likely to become blocked upon modification with a polymer. Using mesitylene as a swelling agent, the authors were able to obtain particles with

27.9 nm cavities and 13.4 nm pore entrances. The polycation was grafted to the MSN surface and showed high RNA loading due to electrostatic interactions as well as specific binding between the polymer and the adenine-thymine sequences. Unfortunately, these same interactions decreased siRNA release efficiency in this system, though effective inhibition of KHOS cell growth was still observed (cell viability reduced by 30%). The particles alone showed effective cell uptake and high biocompatibility up to 100 μg mL^{-1} [102].

10.4 Hollow Silica

In order to increase the loadable volume, MSN can be synthesized around a preexisting nanoparticle, which is subsequently removed to leave a hollow chamber accessible by mesopores. Typically, these sacrificial templates are removed through calcination, as is the case with polystyrene, or chemical etching [103]. The resulting hollow MSN maintains the benefits of MSN, such as reactive surface silanol groups and increased drug loading capabilities.

Lu and co-workers took octadecyl-functionalized hollow MSN and applied a coating of the amphiphilic block copolymer of [7-(didodecylamino(coumarin-4-yl] methyl methacrylate with hydroxyethylacrylate and N-(3-aminopropyl) methacrylamide hydrochloride. The polymer was functionalized with folic acid, which targeted folate receptor expressing cancer cells, and served a dual purpose: tracking the particles via the coumarin moiety fluorescence and removable capping of the hollow MSN mesopores which prevented premature drug leakage. Upon excitation by a femtosecond NIR light laser, the polymer degrades via a two-photon absorption process also arising from the coumarin moiety. Doxorubicin was loaded into the hollow core and negligible release was observed in the dark. Upon exposure to NIR light, ~60% of the loaded DOX was released into PBS at 37°C and, in experiments with KB cells, cell viability decreased dramatically when incubated with DOX-loaded hollow MSN and exposed to NIR light. This is more notable when the high cell viability of non-DOX-loaded hollow MSN irradiated with NIR light and DOX-loaded hollow MSN in the dark. It was also observed that the folic acid functionalized particles were internalized into folic acid receptor expressing KB cells vs. almost no endocytosis into A549 cells [104].

In a recent study, Geng et al. reported methods for controlling the shape of hollow MSN with multifunctional capping. Sacrificial Fe_2O_3 templates with varying shapes (capsules, ellipsoids, rhomboids and cubes) were used to synthesize hollow MSN which replicated the shape of the template. The silica was then functionalized with PEG, to improve biocompatibility, and folic acid, for targeting tumor cells that overexpress folate receptors. Doxorubicin was used as a model drug and demonstrated a pH-dependent release, owing to the disruption of the electrostatic interactions between the drug and silica at lower pH, with cubic hollow MSN showing the fastest release rate. It was shown that the nonspherical shapes enhanced cellular uptake into HeLa cells over more traditional spherical particles, with rhombus-like particles showing the highest intercellular efficacy. It is proposed that the sharp corners might facilitate membrane deformation and thus initiate endocytosis more efficiently. All four shapes showed negligible cytotoxicity in A549 and HeLa cells without DOX [105].

Guo et al. produced DOX-loaded carboxylic acid functionalized hollow MSN and used insoluble calcium salts (calcium carbonate and hydroxyapatite) to cap the pores. These materials are both biocompatible and pH sensitive, becoming soluble in the weakly acidic cancer cell microenvironment and dissolving, leading to sustained drug release over 5 days. Both the calcium carbonate and hydroxyapatite-coated DOX-loaded MSN showed low cytotoxicity in V79-4 normal cells; however, in Hep G2 cancer cells substantial cell death was observed. The specificity of DOX towards cancer cells was also improved by loading the drug onto hollow MSN, and, though not specifically functionalized to target cancer cells, the EPR effect led to an accumulation of particles in cancer cells over normal cells [106].

Cai and co-workers generated a cascade pH-responsive system in which DOX-loaded hollow MSN was coated in two layers. First, a β-CD was attached to the surface through a boronic acid-catechol ester to act as a gatekeeper. Second, an adamantane-conjugated PEG group was attached through the hydrophobic interactions between the adamantane and β-CD core and served to increase biocompatibility. The PEG and adamantane were linked through a benzoic imine bond, which is stable under normal biological conditions but cleaves in the slightly acidic tumor microenvironment; thus, the PEG group can be shed upon reaching the tumor site, facilitating cellular uptake. This process is illustrated in Scheme 10.4a. Once inside the cancer cell lysosomes, the even lower pH hydrolyzes the ester bond and uncaps the pores, releasing DOX (Scheme 10.4b). The in vitro and in vivo results showed the carrier had minimal toxic side effects while the DOX-loaded particles successfully inhibited tumor growth in nude mice with less weight loss and an increased survival time over free DOX [107].

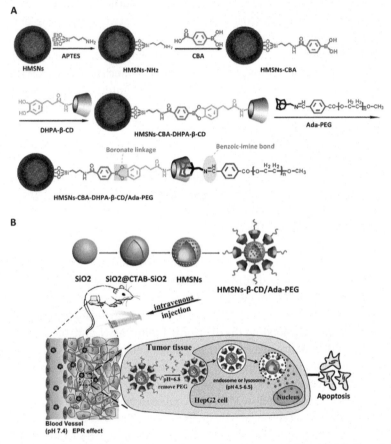

Scheme 10.4 (A) Fabrication of cascade pH-responsive HMSNs-based drug delivery system; and (B) Schematic illustration of drug delivery process of HMSNs-b-CD/Ada-PEG system under tumor microenvironment in vivo. Reprinted with permission from Liu, J., Luo, Z., Zhang, J., Luo, T., Zhou, J., Zhao, X. and Cai, K. (2016). Hollow mesoporous silica nanoparticles facilitated drug delivery via cascade pH stimuli in tumor microenvironment for tumor therapy, *Biomaterials*, **83**, pp. 51–65. Copyright 2016 Elsevier.

A triple responsive drug delivery system was synthesized by Zhang et al. and tested in HeLa cells. pH responsive poly(2-(diethylamino)ethyl methacrylate) was grafted on the silica surface through surface initiated atomic transfer radical polymerization. The initiator was anchored to the silica surface through two essential bonds: a reductively cleavable disulfide bond and a UV-cleavable *o*-nitrobenzyl ester. Using DOX as a model drug, the authors demonstrated that drug molecules were released under slightly acidic (pH 5.0) conditions, in the presence of a reducing agent (10 mM dithiothreitol), upon UV irradiation, and low pH with a reducing agent. Drug was effectively released into HeLa cells, which contain both a low pH environment and the reducing agent GSH, after 10 min UV irradiation and showed comparable cytotoxicity to free DOX at the concentrations investigated. Without DOX, the polymer-coated hollow MSN showed negligible cytotoxicity, even with 10 min of UV irradiation [108].

When compared to spherical nanoparticles, those with sharp points are expected to have higher endocytosis efficiencies. To this end, Xu and co-workers synthesized star-like hollow MSN in varying sizes for use in synergistic gene therapy and chemotherapy. β-cyclodextrin cores with ethanolamine-functionalized poly(glycidyl methacrylate) were attached to the surface through adamantane terminated disulfide bonds, resulting in redox-triggered drug release. The cytotoxicity increased with carrier concentration in HepG2 and COS7 cells; however, these particles were still less toxic than standard branched polyethylenimine. Additionally, these star-like hollow MSN showed higher transfection efficiencies than standard branched polyethylenimine and spherical hollow MSN. The sharp points also resulted in superior cell internalization. As a model drug, 10-hydroxy-camptothecin was encapsulated into the hollow MSN and plasmid DNA was electrostatically adsorbed onto the positively charged polymer coating. The resulting complexes were only 250 nm in diameter and had a positive surface charge, with no obvious drug leakage in the absence of GSH. The 5-fluorocytosine/ *Escherichia coli* cytosine deaminase suicide gene therapy system was employed to assess the synergistic anticancer potential. Here, 5-fluorocytosine is converted through deamination to the cytotoxic drug 5-fluorouracil. Together with the 10-hydroxy-camptothecin in HepG2 cells, cell viability was less than 30% [109].

Luo et al. used hollow MSN as a base for a tumor targeting, GSH responsive drug delivery system. Tetraethylene glycol chains were anchored to the silica surface through a disulfide bond and α-CD was threaded onto the chains using the high affinity between the ethyl oxide and hydrophobic CD cavity. Folic acid was then added to the end of the tetraethylene glycol chains using click chemistry, serving as both a stopper for the α-CD and as the tumor-targeting component through the over-expressed FA receptors on many kinds of cancer cells. Doxorubicin was loaded into the hollow MSN and, at physiological conditions, a negligible amount was released, indicating the efficacy of the α-CD cap. At 10 mM GSH, ~85% of the DOX was released over 26 h. Biocompatibility, effective specific receptor mediated endocytosis of the carriers, and intracellular DOX release was demonstrated in HeLa cells. Nude mice were used to investigate the curative effects of the DOX-loaded nanocomposites. Over 21 days the experimental mice showed better tumor growth inhibition than free DOX alone or bare hollow MSN [110].

10.5 Silica-Shell/Rattle Type Silica

Similar to the hollow MSN, "rattle-type" MSN possess a mesoporous silica shell that are synthesized around another type of nanoparticle. Either the silica can be grown using this nanoparticle as a base, or the non-silica nanoparticles can be synthesized within the center cavity of a hollow MSN. These extra nanoparticles complement the drug delivery function of the mesoporous silica, either by triggering drug release, assisting in imaging, or treating the disease state through another method altogether.

Kang et al. synthesized a multifunctional drug delivery system consisting of $Gd_2O_3:Eu^{3+}$ luminescent nanoparticles within thermally responsive PNIPAM/ acrylic acid acrylamide copolymer filled hollow MSN. The $Gd_2O_3:Eu^{3+}$ could also function as a magnetic resonance contrast agent, allowing for facile visualization of the silica nanoparticles, while the polymer allowed for the thermally controlled release of indomethacin. PNIPAM hydrogels have poor mechanical properties and thus encapsulation within the silica shell protects the hydrogel from degradation. At 20°C virtually no drug release was observed while at 45°C, the drug was released quickly. This was attributed to the swollen state of the polymer at temperatures below the LCST, which led to the polymer occupancy of all residual space in the hollow core as well as the mesopore volume. When the temperature is above the LCST, the polymer contracts into a hydrophobic state, "squeezing out" the drug molecules. The authors demonstrated biocompatibility of these multifunctional nanoparticles through a methylthiazolyldiphenyl-tetrazolium bromide assay using L929 cells and successful uptake into SKOV3 ovarian cancer cells [111].

In order to generate a gold nanorod based system that avoids the common challenges associated with gold nanorods, namely low surface area and aggregation (which shifts the wavelength necessary to induce localized surface plasmon resonance into the visible region and hence prevents deep tissue use), Chen and co-workers coated gold nanorods in a mesoporous silica shell. These nanoparticles were then loaded with DOX and demonstrated thermal and pH-dependent drug release resulting from the disruption of the DOX-silica electrostatic interactions. Biocompatibility of the non-DOX-loaded particles and the cancer killing potential was shown in A549 human lung cancer cells. The gold nanorod core served as a hyperthermia agent resulting from NIR irradiation and was useful in two-photon imaging. Together with the DOX-loaded silica shell, created a system with two NIR irradiation triggered anticancer therapeutic modes [112].

Huang and co-workers developed core-shell nanoparticles for cancer-targeted photothermo-chemotherapy. Hydrophobic, mesoporous graphitic carbon cores provided an environment for hydrophobic drugs (DOX in this study) and high NIR photothermal conversion efficiency. An ordered mesoporous shell surrounded the carbon, improving the biocompatibility, and both PEG groups and SP13 peptides were conjugated to the silica surface. The PEG group increased biocompatibility while the peptide targeted the nanoparticles towards HER2 oncoprotein expressing cancer cells. The particles demonstrated a laser power intensity and concentration dependent photothermal heating effect with pH and NIR dependent sustained drug release. This resulted from decreased electrostatic interaction between DOX and silica and hydrophobic interactions between DOX and carbon. The SP13 peptide led to selective uptake into HER2-positive SK-BR-3 breast cancer cells over nontumorigenic MCF-10A breast epithelial cells through receptor mediated cellular uptake. Combined synergistic therapy resulted in significantly higher cell death over both DOX only and photothermal only strategies [113].

Ma et al. synthesized MSN encapsulated Bi_2S_3 nanoparticles which could be used for radiotherapy. These particles exhibited negligible cytotoxicity against HK-2 human renal cells and at a dosage of 7.5 mg kg^{-1} for 28 days no difference from saline was observed in mice, demonstrating biocompatibility. The efficacy of these particles in external radiotherapy using X-ray irradiation was investigated and it was found that the tumor volume in nude mice reduced more substantially in the presence of the Bi_2S_3 containing MSN than with X-ray irradiation alone. Interstitial radiotherapy using ^{32}P-enriched gelatin chromic phosphate colloid in PC3 cells was also studied and cell viability decreased more significantly with the irradiated MSN than with ^{32}P alone. With regard to drug delivery, DOX release from the MSN mesopores was pH dependent, with a faster rate at pH 5.0 than 7.4. In PC3 cells, there was no observed difference between free DOX and DOX-loaded Bi_2S_3-MSN at the same DOX concentration; however, the addition of ^{32}P tremendously added to the cytotoxicity of the DOX-MSN. In tumor baring nude mice, injection of Bi_2S_3-MSN showed the MSN alone had no apparent tumor growth inhibition, but the addition of ^{32}P reduced tumor volume by ~21% in comparison to ^{32}P alone. It was also found that Bi_2S_3-MSN with ^{32}P reduced the tumors more effectively than with X-rays [114].

10.6   Magnetic Silica

Silica nanoparticles can be synthesized to contain specifically magnetic cores, which imparts the properties of both materials to a single nanocomposite. Typically, an iron oxide nanoparticle is coated with mesoporous silica, generating multifunctional systems that have the potential to act as drug delivery vehicles and hyperthermia agents.

Jeonghun Lee et al. used an iron oxide core as a magnetic resonance imaging agent. The mesoporous silica shell was loaded with DOX and functionalized with β-CD through a GSH cleavable disulfide bond. In addition, PEG groups (M_w = 5000) were added to enhance silica nanoparticle solubility and biocompatibility, which allowed the particles to maintain dispersion in PBS for several days. Initial release studies showed that the CD cap was able to prevent DOX release for up to five days without GSH, while the addition of the reducing agent caused a burst release of drug over three hours. Glutathione responsive behavior was further confirmed through the significant apoptosis observed in A549 cancer cells and in mice bearing A549 tumors. The magnetic particle core allowed the authors to image particle accumulation and clearance. Following intravenous injection of the nanocomposites, tumor growth was slowed, even more so than injected free drug; however, when the particles began to clear after one week, tumor growth accelerated again [115].

In a study by Ming Ma et al., nanoellipsoids with a magnetic core and mesoporous shell were coated with gold nanorods and used for: chemotherapy, photo-thermotherapy, in vivo infrared thermal, optical, and MR imaging. The gold was functionalized with PEG and carboxylic acid groups and conjugated to the amine-modified silica surface through carbodiimide crosslinker chemistry (Au NRs-MMSNEs). The resulting nanocomposites were shown to be non-toxic in MCF-7 cells at concentrations up to 100 µg mL^{-1}. Due to the magnetic core, the particles could be directed towards specific sites, using an external magnetic field (Fig. 10.3), at which doxorubicin release was triggered under the slightly acidic endosomal conditions. At acidic pH, the simple electrostatic interactions between the negatively charged silica surface and the positively charged drug were disrupted, allowing the drug to diffuse out of the mesopores. This combined with the localized hyperthermia from the gold nanorods gave the system synergistic photothermal- and chemo-therapy anti-cancer potential, as demonstrated in an MCF-7 cell proliferation assay [116].

Crosslinked gelatin was used to coat mesoporous silica with an iron oxide core and regulate the release of paclitaxel. In this study, the magnetic component was also successfully used to guide the particles to the tumor site in a mouse model, decreasing particle accumulation in normal tissues. The tumor reduction study showed tumor growth in the presence of these paclitaxel-loaded particles with an external magnetic field was significantly delayed and obvious body weight loss was not observed relative to the same particles without the magnetic field and commercial Taxol® at the same dose. Gelatin was electrostatically adsorbed to the amine functionalized silica surface and then crosslinked around the particles using gluteraldehyde, creating a diffusion barrier that yielded a sustained release of paclitaxel [117].

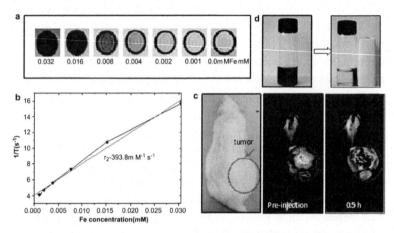

Figure 10.3 (a) T_2 phantom images of Au NRs-MMSNEs at different Fe concentrations; (b) Relaxation rate $1/T_2$ of Au NRs-MMSNEs as a function of Fe concentration; (c) In vivo MRI of a mouse before and after intratumor injection of Au NRs-MMSNEs; (d) Photographs of Au NRs-MMSNEs dispersed in water before (left) and after (right) an external magnetic attraction. Reprinted with permission from Ma, M., Chen, H., Chen, Y., Wang, X., Chen, F., Cui, X. and Shi, J. (2012). Au capped magnetic core/mesoporous silica shell nanoparticles for combined photothermo-/chemo-therapy and multimodal imaging, *Biomaterials,* **33,** pp. 989-998. Copyright 2012 Elsevier.

Yufang Zhu and coworkers reported a magnetic silica nanoparticle based system in which the magnetic core was used for hyperthermia, DOX was released from the silica mesopores in at low pH, and graphene quantum dot caps served as local photothermal generators. Additionally, the quantum dots were immobilized over the mesopore entrances through electrostatic and hydrogen bonding interactions, preventing premature DOX release. However, when the environmental pH dropped (pH 5.0), these were disrupted, allowing the DOX to diffuse out of the pores. Under an alternating magnetic field or NIR irradiation, the particles were able to generate heat and, when combined with the released DOX, were more effective at killing cancer cells than chemotherapy, magnetic hyperthermia or photothermal therapy individually, as shown in a 4T1 breast cancer cell study. These particles without DOX, an alternating magnetic field, or NIR irradiation were shown to be noncytotoxic to 4T1 cells even at a concentration of 200 μg mL^{-1} [118].

The magnetic core in these silica-based nanocomposites can also be used to stimulate on-command drug delivery. A 2012 communication demonstrated the use of an alternating magnetic field to trigger methylene blue release from MSN with a superparamegnetic iron oxide nanocrystal core. Here, the loaded particles were capped with a lipid bilayer and at physiological temperatures without an alternating magnetic field, no dye release was observed for several weeks. However, both at 50°C and with the magnetic field the vast majority of the dye was released over a short period of time owing to the increased permeability of the bilayer under these conditions. On-command release was thus attributed to a combination of the local warming and vibration of the superparamagnetic iron oxide nanocrystals in the presence of an alternating magnetic field. These particles were shown to be non-toxic up to concentrations of 0.5 mg mL^{-1} in human nervous A172, brain BE(2)-C, liver HEPG2, kidney 293T, heart HCM, colon SW480, and skin A431 cells [119].

Eduardo Guisasola et al. developed a dual temperature responsive system based on a *N*-isopropylacrylamide/*N*-hydroxymethyl acrylamide copolymer such that when the polymer is cross linked it pulls away from the MSN pores and allows for cargo release, but in its linear form collapses over the pores and prevents drug release when the temperature is increased. Conversely, in the cross-linked system the pores are capped at low temperatures while drug is able to diffuse through the linear polymer chains at the same temperature. This is based on the thermally responsive behavior of the PNIPAM component, which undergoes a change from hydrophilic to hydrophobic when the temperature is raised above its LCST for the linear chains. Similarly, in the cross-linked system this transition occurs above the volume phase transition temperature. The poly(*N*-hydroxymethyl acrylamide) (PNHMA) component increases hydrophilicity and modulates this phase transition temperature from the 32°C observed in PNIPAM homopolymers, depending on the PNHMA percentage present in the copolymer. For example, a 90:10

PNIPAM:PNHMA copolymer possesses a LCST of 42°C. Although both undergo the same transition from hydrophilic to hydrophobic at 41–43°C, the hydrophobic stacking of polymer chains in the cross-linked system results in free space around the pores, while linear polymer becomes globular and blocks the pore entrances. When the linear polymer is in a swollen, hydrophilic state the chains are less condensed, which allows the drug to diffuse. In the same swollen state, the cross-linked matrix extends across the pores, leading to diffusion barrier. The mesoporous silica was grown around an acid-coated magnetic SPIONS thermoseed, which generates heat when exposed to an alternating magnetic field, and the copolymer was grafted from the surface [120].

10.7 Conclusions and Outlook

Silica nanoparticles show great promise as drug delivery vehicles due to their biocompatibility, high drug loading, and facile functionalization which leads to cell targeting and stimuli-responsive delivery. Although well established in in vitro drug delivery, historically in vivo studies were lacking. This has recently been rectified, with many groups utilizing mouse or rat models to demonstrate the efficacy of silica nanoparticles for targeted delivery without observing negative effects associated with free drug in, specifically, cancer therapy. Clinical trials in humans are still lacking. Furthermore, much research involving silica nanoparticles has focused on cancer treatment. These materials' potential goes beyond that, as has been demonstrated above. Given the facile manner in which silica nanoparticles can be combined with other nanotherapeutics, such as polymers and metal nanoparticles, the potential to target other disease states should be explored in greater depth.

References

1. Slowing, I. I., Trewyn, B. G., Giri, S. and Lin, V. S.-Y. (2007). Mesoporous silica nanoparticles for drug delivery and biosensing applications, *Adv. Funct. Mater.*, **17,** pp. 1225–1236.

2. Tsou, Y. H., Zhang, X. Q., Zhu, H., Syed, S. and Xu, X. (2017). Drug delivery to the brain across the blood-brain barrier using nanomaterials, *Small,* **13,** 1701921.

3. Dening, T. J., Rao, S., Thomas, N. and Prestidge, C. A. (2016). Oral nanomedicine approaches for the treatment of psychiatric illnesses, *J. Control. Release,* **223,** pp. 137–156.

4. Hamidi, M., Azadi, A. and Rafiei, P. (2008). Hydrogel nanoparticles in drug delivery, *Adv. Drug Deliv. Rev.,* **60,** pp. 1638–1649.

5. Bianco, A., Kostarelos, K. and Prato, M. (2005). Applications of carbon nanotubes in drug delivery, *Curr. Opin. Chem. Biol.,* **9,** pp. 674–679.

6. Stöber, W., Fink, A. and Bohn, E. (1968). Controlled growth of monodisperse silica spheres in the micron size range, *J. Colloid. Interface. Sci.,* **26,** pp. 62–69.

7. Inagaki, S., Fukushima, Y. and Kuroda, K. (1993). Synthesis of highly ordered mesoporous materials from a layered polysilicate, *J. Chem. Soc. Chem. Commun,* **8,** pp. 680–682.

8. Beck, J. S., Vartuli, J. C., Roth, W. J., Leonowicz, M. E., Kresge, C. T., Schmitt, K. D., Chu, C. T.-W., Olson, D. H., Sheppard, E. W., McCullen, S. B., Higgins, J. B. and Schlenker, J. L. (1992). A new family of mesoporous molecular sieves prepared with liquid crystal templates, *J. Am. Chem. Soc.,* **114,** pp. 10834–10843.

9. Song, S.-W., Hidajat, K. and Kawi, S. (2005). Functionalized SBA-15 materials as carriers for controlled drug delivery: influence of surface properties on matrix-drug interactions, *Langmuir,* **21,** pp. 9568–9575.

10. Vallet-Regí, M., Rámila, A., DelReal, R. P. and Pérez-Pariente, J. (2001). A new property of MCM-41- drug delivery system, *Chem. Mater.,* **13,** pp. 308–311.

11. Lai, C.-Y., Trewyn, B. G., Jeftinija, D. M., Jeftinija, K., Xu, S., Jeftinija, S. and Lin, V. S.-Y. (2003). A mesoporous silica nanosphere-based carrier system with chemically removable CdS nanoparticle caps for stimuli-responsive controlled release of neurotransmitters and drug molecules, *J. Am. Chem. Soc.,* **125,** pp. 4451–4459.

12. Huh, S., Wiench, J. W., Yoo, J.-C., Pruski, M. and Lin, V. S.-Y. (2003). Organic functionalization and morphology control of mesoporous silicas via a co-condensation synthesis method, *Chem. Mater.,* **15,** pp. 4247–4256.

13. Liu, Y. H., Lin, H. P. and Mou, C. Y. (2004). Direct method for surface silyl functionalization of mesoporous silica, *Langmuir,* **20,** pp. 3231–3239.

14. Trewyn, B. G., Slowing, I. I., Giri, S., Chen, H.-T. and Lin, V. S.-Y. (2007). Synthesis and functionalization of a mesoporous silica nanoparticle based on the sol-gel process and applications in controlled release, *Acc. Chem. Res.,* **40,** pp. 846–853.

15. Chen, F., Hableel, G., Ruike Zhao, E. and Jokerst, J. V. (2018). Multifunctional nanomedicine with silica: role of silica in nanoparticles for theranostic, imaging, and drug monitoring, *J. Colloid Interface Sci.,* **521,** pp. 261–279.

16. Trewyn, B. G., Giri, S., Slowing, I. I. and Lin, V. S. Y. (2007). Mesoporous silica nanoparticle based controlled release, drug delivery, and biosensor systems, *Chem. Comm.,* pp. 3236–3245.

17. Vallet-Regí, M., Balas, F. and Arcos, D. (2007). Mesoporous materials for drug delivery, *Angew. Chem. Int. Ed. Engl.,* **46,** pp. 7548–7558.

18. Hao, N., Li, L. and Tang, F. (2016). Shape matters when engineering mesoporous silica-based nanomedicines, *Biomater. Sci.,* **4,** pp. 575–591.

19. Slowing, I. I., Trewyn, B. G. and Lin, V. S. (2006). Effect of surface functionalization of MCM-41-type mesoporous silica nanoparticles on the endocytosis by human cancer cells, *J. Am. Chem. Soc.,* **128,** pp. 14792–14793.

20. McCarthy, C. A., Ahern, R. J., Devine, K. J. and Crean, A. M. (2018). Role of drug adsorption onto the silica surface in drug release from mesoporous silica systems, *Mol. Pharm.,* **15,** pp. 141–149.

21. Croissant, J. G., Fatieiev, Y. and Khashab, N. M. (2017). Degradability and clearance of silicon, organosilica, silsesquioxane, silica mixed oxide, and mesoporous silica nanoparticles, *Adv. Mater.,* **29,** pp. 1604634.

22. Yu, T., Greish, K., McGill, L. D., Ray, A. and Ghandehari, H. (2012). Influence of geometry, porosity, and surface characteristics of silica nanoparticles on acute toxicity- their vasculature effect and tolerance threshold, *ACS Nano,* **6,** pp. 2289–2301.

23. Shi, Y., Helary, C., Haye, B. and Coradin, T. (2018). Extracellular versus intracellular degradation of nanostructured silica particles, *Langmuir,* **34,** pp. 406–415.

24. Nozawa, K., Gailhanou, H., Raison, L., Panizza, P., Ushiki, H., Sellier, E., Delville, J. P. and Delville, M. H. (2005). Smart control of monodisperse Stöber silica particles- effect of reactant addition rate on growth process, *Langmuir,* **21,** pp. 1516–1523.

25. Benezra, M., Penate-Medina, O., Zanzonico, P. B., Schaer, D., Ow, H., Burns, A., DeStanchina, E., Longo, V., Herz, E., Iyer, S., Wolchok, J.,

Larson, S. M., Wiesner, U. and Bradbury, M. S. (2011). Multimodal silica nanoparticles are effective cancer-targeted probes in a model of human melanoma, *J. Clin. Invest.,* **121,** pp. 2768–2780.

26. Tang, L. and Cheng, J. (2013). Nonporous silica nanoparticles for nanomedicine application, *Nano Today,* **8,** pp. 290–312.

27. Suma, T., Miyata, K., Anraku, Y., Watanabe, S., Christie, R. J., Takemoto, H., Shioyama, M., Gouda, N., Ishii, T., Nishiyama, N. and Kataoka, K. (2012). Smart multilayered assembly for biocompatible siRNA delivery featuring dissolvable silica, endosome-disrupting polycation, and detachable PEG, *ACS Nano,* **6,** pp. 6693–6705.

28. Li, A., Zhang, J., Xu, Y., Liu, J. and Feng, S. (2014). Thermoresponsive copolymer/SiO_2 nanoparticles with dual functions of thermally controlled drug release and simultaneous carrier decomposition, *Chemistry,* **20,** pp. 12945–12953.

29. Carpenter, A. W., Johnson, J. A. and Schoenfisch, M. H. (2014). Nitric oxide-releasing silica nanoparticles with varied surface hydrophobicity, *Colloids Surf. A,* **454,** pp. 144–151.

30. Xu, Z., Liu, S., Kang, Y. and Wang, M. (2015). Glutathione- and pH-responsive nonporous silica prodrug nanoparticles for controlled release and cancer therapy, *Nanoscale,* **7,** pp. 5859–5868.

31. Davidson, S., Lamprou, D. A., Urquhart, A. J., Grant, M. H. and Patwardhan, S. V. (2016). Bioinspired silica offers a novel, green, and biocompatible alternative to traditional drug delivery systems, *ACS Biomate. Sci. Eng.,* **2,** pp. 1493–1503.

32. Hoffmann, F., Cornelius, M., Morell, J. and Fröba, M. (2006). Silica-based mesoporous organic–inorganic hybrid materials, *Angew. Chem. Int. Ed.,* **45,** pp. 3216–3251.

33. Wan, Y. and Zhao (2007). On the controllable soft-templating approach to mesoporous silicates, *Chem. Rev.,* **107,** pp. 2821–2860.

34. Sayari, A., Liu, P., Kruk, M. and Jaroniec, M. (1997). Characterization of large-pore MCM-41 molecular sieves obtained via hydrothermal restructuring, *Chem. Mater.,* **9,** pp. 2499–2506.

35. Widenmeyer, M. and Anwander, R. (2002). Pore size control of highly ordered mesoporous silica MCM-48, *Chem. Mater.,* **14,** pp. 1827–1831.

36. Kresge, C. T., Leonowicz, M. E., Roth, W. J., Vartuli, J. C. and Beck, J. S. (1992). Ordered mesoporous molecular sieves synthesized by a liquid-crystal template mechanism, *Nature,* **359,** pp. 710–712.

37. Vartuli, J. C., Schmitt, K. D., Kresge, C. T., Roth, W. J., Leonowicz, M. E., McCullen, S. B., Hellring, S. D., Beck, J. S. and Schlenker, J. L. (1994).

Effect of surfactant/silica molar ratios on the formation of mesoporous molecular sieves: inorganic mimicry of surfactant liquid-crystal phases and mechanistic implications, *Chem. Mater.*, **6**, pp. 2317–2326.

38. Q, C., Lin, W. L., Xiao, F. S., Pang, W. Q., Chen, H. and Zou, B. S. (1999). The preparation of highly ordered MCM-41 with extremely low surfactant concentration, *Microporous Mesoporous Mater.*, **32**, pp. 1–15.

39. Huo, Q., Leon, R., Petroff, P. M. and Stucky, G. D. (1995). Mesostructure design with gemini surfactants: supercage formation in a three-dimensional hexagonal array, *Science*, **268**, pp. 1324.

40. Huo, Q., Margolese, D. I. and Stucky, G. D. (1996). Surfactant control of phases in the synthesis of mesoporous silica-based materials, *Chem. Mater.*, **8**, pp. 1147–1160.

41. Zhao, D., Feng, J., Huo, Q., Melosh, N., Fredrickson, G. H., Chmelka, B. F. and Stucky, G. D. (1998). Triblock copolymer syntheses of mesoporous silica with periodic 50 to 300 Angstrom pores, *Science*, **279**, pp. 548.

42. Zhao, D., Huo, Q., Feng, J., Chmelka, B. F. and Stucky, G. D. (1998). Nonionic triblock and star diblock copolymer and oligomeric surfactant syntheses of highly ordered, hydrothermally stable, mesoporous silica structures, *J. Am. Chem. Soc.*, **120**, pp. 6024–6036.

43. Song, Y., Li, Y., Xu, Q. and Liu, Z. (2017). Mesoporous silica nanoparticles for stimuli-responsive controlled drug delivery: advances, challenges, and outlook, *Int. J. Nanomed*, **12**, pp. 87–110.

44. Zhao, Y., Vivero-Escoto, J. L., Slowing, I. I., Trewyn, B. G. and Lin, V. S. (2010). Capped mesoporous silica nanoparticles as stimuli-responsive controlled release systems for intracellular drug:gene delivery, *Expert. Opin. Drug. Deliv.*, **7**, pp. 1013–1029.

45. Kato, Y., Ozawa, S., Miyamoto, C., Maehata, Y., Suzuki, A., Maeda, T. and Baba, Y. (2013). Acidic extracellular microenvironment and cancer, *Cancer Cell Int.*, **13**, pp.

46. Chen, L. Q. and Pagel, M. D. (2015). Evaluating pH in the extracellular tumor microenvironment using CEST MRI and other imaging methods, *Adv. Radiol.*, **2015**, pp.

47. Zheng, Y., Fahrenholtz, C. D., Hackett, C. L., Ding, S., Day, C. S., Dhall, R., Marrs, G. S., Gross, M. D., Singh, R. and Bierbach, U. (2017). Large-pore functionalized mesoporous silica nanoparticles as drug delivery vector for a highly cytotoxic hybrid platinum-acridine anticancer agent, *Chemistry*, **23**, pp. 3386–3397.

48. Chang, D., Gao, Y., Wang, L., Liu, G., Chen, Y., Wang, T., Tao, W., Mei, L., Huang, L. and Zeng, X. (2016). Polydopamine-based surface

modification of mesoporous silica nanoparticles as pH-sensitive drug delivery vehicles for cancer therapy, *J. Colloid. Interface. Sci.,* **463,** pp. 279–287.

49. Nguyen, C. T., Webb, R. I., Lambert, L. K., Strounina, E., Lee, E. C., Parat, M. O., McGuckin, M. A., Popat, A., Cabot, P. J. and Ross, B. P. (2017). Bifunctional succinylated ε-polylysine-coated mesoporous silica nanoparticles for pH-responsive and intracellular drug delivery targeting the colon, *ACS Appl. Mater. Interfaces,* **9,** pp. 9470–9483.

50. Zhang, X., Li, F., Guo, S., Chen, X., Wang, X., Li, J. and Gan, Y. (2014). Biofunctionalized polymer-lipid supported mesoporous silica nanoparticles for release of chemotherapeutics in multidrug resistant cancer cells, *Biomaterials,* **35,** pp. 3650–3665.

51. Han, L., Tang, C. and Yin, C. (2016). pH-Responsive core-shell structured nanoparticles for triple-stage targeted delivery of doxorubicin to tumors, *ACS Appl. Mater. Interfaces,* **8,** pp. 23498–23508.

52. Chen, L., Zhang, Z., Yao, X., Chen, X. and Chen, X. (2015). Intracellular pH-operated mechanized mesoporous silica nanoparticles as potential drug carries, *Microporous Mesoporous Mater.,* **201,** pp. 169–175.

53. Khatoon, S., Han, H. S., Lee, M., Lee, H., Jung, D. W., Thambi, T., Ikram, M., Kang, Y. M., Yi, G. R. and Park, J. H. (2016). Zwitterionic mesoporous nanoparticles with a bioresponsive gatekeeper for cancer therapy, *Acta Biomater.,* **40,** pp. 282–292.

54. Lei, Q., Wang, S. B., Hu, J. J., Lin, Y. X., Zhu, C. H., Rong, L. and Zhang, X. Z. (2017). Stimuli-responsive "cluster bomb" for programmed tumor therapy, *ACS Nano,* **11,** pp. 7201–7214.

55. Balendiran, G., Dabur, R. and Fraser, D. (2004). The role of glutathione in cancer, *Cell Biochem. Funct.,* **22,** pp. 343–352.

56. Saito, G., Swanson, J. and Lee, K.-D. (2003). Drug delivery strategy utilizing conjugation via reversible disulfide linkages- role and site of cellular reducing activities, *Adv. Drug Deliv. Rev.,* **55,** pp. 199–215.

57. Gamcsik, M. P., Kasibhatla, M. S., Teeter, S. D. and Colvin, O. M. (2012). Glutathione levels in human tumors, *Biomarkers,* **17,** pp. 671–691.

58. Lee, J., Kim, H., Han, S., Hong, E., Lee, K. H. and Kim, C. (2014). Stimuli-responsive conformational conversion of peptide gatekeepers for controlled release of guests from mesoporous silica nanocontainers, *J. Am. Chem. Soc.,* **136,** pp. 12880–12883.

59. Chen, X., Sun, H., Hu, J., Han, X., Liu, H. and Hu, Y. (2017). Transferrin gated mesoporous silica nanoparticles for redox-responsive and targeted drug delivery, *Colloids Surf. B,* **152,** pp. 77–84.

60. Karimian, D., Yadollahi, B. and Mirkhani, V. (2017). Dual functional hybrid-polyoxometalate as a new approach for multidrug delivery, *Microporous Mesoporous Mater.*, **247**, pp. 23–30.

61. Lai, J., Shah, B. P., Garfunkel, E. and Lee, K.-B. (2013). Versatile fluorescence resonance energy transfer-based mesoporous silica nanoparticles for real-time monitoring of drug release, *ACS Nano*, **7**, pp. 2741–2750.

62. Chen, L., Zhou, X., Nie, W., Zhang, Q., Wang, W., Zhang, Y. and He, C. (2016). Multifunctional redox-responsive mesoporous silica nanoparticles for efficient targeting drug delivery and magnetic resonance imaging, *ACS Appl. Mater. Interfaces*, **8**, pp. 33829–33841.

63. Hu, J.-J., Lei, Q., Peng, M.-Y., Zheng, D.-W., Chen, Y.-X. and Zhang, X.-Z. (2017). A positive feedback strategy for enhanced chemotherapy based on ROS-triggered self-accelerating drug release nanosystem, *Biomaterials*, **128**, pp. 136–146.

64. Geng, J., Li, M., Wu, L., Chen, C. and Qu, X. (2012). Mesoporous silica nanoparticle-based H_2O_2 responsive controlled-release system used for Alzheimer's disease treatment, *Adv. Healthc. Mater.*, **1**, pp. 332–336.

65. Nagase, K., Yamato, M., Kanazawa, H. and Okano, T. (2018). Poly(*N*-isopropylacrylamide)-based thermoresponsive surfaces provide new types of biomedical applications, *Biomaterials*, **153**, pp. 27–48.

66. Ugazio, E., Gastaldi, L., Brunella, V., Scalarone, D., Jadhav, S. A., Oliaro-Bosso, S., Zonari, D., Berlier, G., Miletto, I. and Sapino, S. (2016). Thermoresponsive mesoporous silica nanoparticles as a carrier for skin delivery of quercetin, *Int. J. Pharm.*, **511**, pp. 446–454.

67. Ribeiro, T., Coutinho, E., Rodrigues, A. S., Baleizao, C. and Farinha, J. P. S. (2017). Hybrid mesoporous silica nanocarriers with thermovalve-regulated controlled release, *Nanoscale*, **9**, pp. 13485–13494.

68. de la Torre, C., Agostini, A., Mondragon, L., Orzaez, M., Sancenon, F., Martinez-Manez, R., Marcos, M. D., Amoros, P. and Perez-Paya, E. (2014). Temperature-controlled release by changes in the secondary structure of peptides anchored onto mesoporous silica supports, *Chem Comm.*, **50**, pp. 3184–3186.

69. Martelli, G., Zope, H. R., Capell, M. B. and Kros, A. (2013). Coiled-coil peptide motifs as thermoresponsive valves for mesoporous silica nanoparticles, *Chem Comm.*, **49**, pp. 9932–9934.

70. Guardado-Alvarez, T. M., Sudha Devi, L., Russell, M. M., Schwartz, B. J. and Zink, J. I. (2013). Activation of snap-top capped mesoporous silica nanocontainers using two near-infrared photons, *J. Am. Chem. Soc.*, **135**, pp. 14000–14003.

71. Lee, J., Park, J., Singha, K. and Kim, W. J. (2013). Mesoporous silica nanoparticle facilitated drug release through cascade photosensitizer activation and cleavage of singlet oxygen sensitive linker, *Chem Comm.,* **49,** pp. 1545–1547.

72. He, D., He, X., Wang, K., Zou, Z., Yang, X. and Li, X. (2014). Remote-controlled drug release from graphene oxide-capped mesoporous silica to cancer cells by photoinduced pH-jump activation, *Langmuir,* **30,** pp. 7182–7189.

73. Choi, H. W., Kim, J., Kim, J., Kim, Y., Song, H. B., Kim, J. H., Kim, K. and Kim, W. J. (2016). Light-induced acid generation on a gatekeeper for smart nitric oxide delivery, *ACS Nano,* **10,** pp. 4199–4208.

74. Frasconi, M., Liu, Z., Lei, J., Wu, Y., Strekalova, E., Malin, D., Ambrogio, M. W., Chen, X., Botros, Y. Y., Cryns, V. L., Sauvage, J. P. and Stoddart, J. F. (2013). Photoexpulsion of surface-grafted ruthenium complexes and subsequent release of cytotoxic cargos to cancer cells from mesoporous silica nanoparticles, *J. Am. Chem. Soc.,* **135,** pp. 11603–11613.

75. Bhat, R., Ribes, A., Mas, N., Aznar, E., Sancenón, F., Marcos, M. D., Murguía, J. R., Venkataraman, A. and Martínez-Máñez (2016). Thrombin-responsive gated silica mesoporous nanoparticles as coagulation regulators, *Langmuir,* **32,** pp. 1195–1200.

76. Ruehle, B., Clemens, D. L., Lee, B.-Y., Horwitz, M. A. and Zink, J. I. (2017). A pathogen-specific cargo delivery platform based on mesoporous silica nanoparticles, *J. Am. Chem. Soc.,* **139,** pp. 6663–6668.

77. Díez, P., Sánchez, A., de la Torre, C., Gamella, M., Martínez-Ruíz, P., Aznar, E., Martínez-Máñez, R., Pingarrón, J. M. and Villalonga, R. (2016). Neoglycoenzyme-gated mesoporous silica nanoparticles- toward the design of nanodevices for pulsatile programmed sequential delivery, *ACS Appl. Mater. Interfaces,* **8,** pp. 7657–7665.

78. He, X., Zhao, Y., He, D., Wang, K., Xu, F. and Tang, J. (2012). ATP-responsive controlled release system using aptamer-functionalized mesoporous silica nanoparticles, *Langmuir,* **28,** pp. 12909–12915.

79. Popat, A., Ross, B. P., Liu, J., Jambhrunkar, S., Kleitz, F. and Qiao, S. Z. (2012). Enzyme-responsive controlled release of covalently bound prodrug from functional mesoporous silica nanospheres, *Angew. Chem. Int. Ed.,* **51,** pp. 12486–12489.

80. Zhang, Z., Balogh, D., Wang, F., Sung, S. Y., Nechushtai, R. and Willner, I. (2013). Biocatalytic release of an anticancer drug from nucleic-acids-capped mesoporous SiO_2 using DNA or molecular biomarkers as triggering stimuli, *ACS Nano,* **7,** pp. 8455–8468.

81. Zou, Z., He, X., He, D., Wang, K., Qing, Z., Yang, X., Wen, L., Xiong, J., Li, L. and Cai, L. (2015). Programmed packaging of mesoporous silica nanocarriers for matrix metalloprotease 2-triggered tumor targeting and release, *Biomaterials,* **58,** pp. 35–45.

82. Lee, D., Hong, J. W., Park, C., Lee, H., Lee, J. E., Hyeon, T. and Paik, S. R. (2014). Ca^{2+}-dependent intracellular drug delivery system developed with "raspberry-type" particles-on-a-particle comprising mesoporous silica core and α-synuclein-coated gold nanoparticles, *ACS Nano,* **8,** pp. 8887–8895.

83. Huang, X. and Du, X. (2014). Pillar[6]arene-valved mesoporous silica nanovehicles for multiresponsive controlled release, *ACS Appl. Mater. Interfaces,* **6,** pp. 20430–20436.

84. He, D., He, X., Wang, K., Chen, M., Cao, J. and Zhao, Y. (2012). Reversible stimuli-responsive controlled release using mesoporous silica nanoparticles functionalized with a smart DNA molecule-gated switch, *J. Mater. Chem.,* **22,** pp. 14715.

85. Wu, W., Ye, C., Xiao, H., Sun, X., Qu, W., Li, X., Chen, M. and Li, J. (2016). Hierarchical mesoporous silica nanoparticles for tailorable drug release, *Int. J. Pharm.,* **511,** pp. 65–72.

86. Paris, J. L., Cabañas, M. V., Manzano, M. and Vallet-Regí, M. (2015). Polymer-grafted mesoporous silica nanoparticles as ultrasound-responsive drug carriers, *ACS Nano,* **9,** pp. 11023–11033.

87. Zhang, J., Yuan, Z. F., Wang, Y., Chen, W. H., Luo, G. F., Cheng, S. X., Zhuo, R. X. and Zhang, X. Z. (2013). Multifunctional envelope-type mesoporous silica nanoparticles for tumor-triggered targeting drug delivery, *J. Am. Chem. Soc.,* **135,** pp. 5068–5073.

88. Yang, D., Wang, T., Su, Z., Xue, L., Mo, R. and Zhang, C. (2016). Reversing cancer multidrug resistance in xenograft models via orchestrating multiple actions of functional mesoporous silica nanoparticles, *ACS Appl. Mater. Interfaces,* pp. 22431–22441.

89. Zhou, S., Sha, H., Ke, X., Liu, B., Wang, X. and Du, X. (2015). Combination drug release of smart cyclodextrin-gated mesoporous silica nanovehicles, *Chem Comm.,* **51,** pp. 7203–7206.

90. Wang, X., Tan, L. L., Li, X., Song, N., Li, Z., Hu, J. N., Cheng, Y. M., Wang, Y. and Yang, Y. W. (2016). Smart mesoporous silica nanoparticles gated by pillararene-modified gold nanoparticles for on-demand cargo release, *Chem Comm.,* **52,** pp. 13775–13778.

91. Zhao, J., He, Z., Li, B., Cheng, T. and Liu, G. (2017). AND logic-like pH- and light-dual controlled drug delivery by surface modified

mesoporous silica nanoparticles, *Mater. Sci. Eng. C Mater. Biol. Appl.*, **73,** pp. 1–7.

92. Ma, X., Ong, O. S. and Zhao, Y. (2013). Dual-responsive drug release from oligonucleotide-capped mesoporous silica nanoparticles, *Biomater. Sci.*, **1,** pp. 912–917.

93. Hei, M., Wang, J., Wang, K., Zhu, W. and Ma, P. X. (2017). Dually responsive mesoporous silica nanoparticles regulated by upper critical solution temperature polymers for intracellular drug delivery, *J. Mater. Chem. B*, **5,** pp. 9497–9501.

94. Chang, B., Chen, D., Wang, Y., Chen, Y., Jiao, Y., Sha, X. and Yang, W. (2013). Bioresponsive controlled drug release based on mesoporous silica nanoparticles coated with reductively sheddable polymer shell, *Chem. Mater.*, **25,** pp. 574–585.

95. Deodhar, G. V., Adams, M. L. and Trewyn, B. G. (2017). Controlled release and intracellular protein delivery from mesoporous silica nanoparticles, *Biotechnol. J.*, **12,** pp.

96. Xu, B., Jiang, G., Yu, W., Liu, D., Zhang, Y., Zhou, J., Sun, S. and Liu, Y. (2017). H_2O_2-Responsive mesoporous silica nanoparticles integrated with microneedle patches for the glucose-monitored transdermal delivery of insulin, *J. Mater. Chem. B*, **5,** pp. 8200–8208.

97. Yu, E., Lo, A., Jiang, L., Petkus, B., Ileri Ercan, N. and Stroeve, P. (2017). Improved controlled release of protein from expanded-pore mesoporous silica nanoparticles modified with co-functionalized poly(*N*-isopropylacrylamide) and poly(ethylene glycol) (PNIPAM-PEG), *Colloids Surf. B*, **149,** pp. 297–300.

98. Mendez, J., Monteagudo, A. and Griebenow, K. (2012). Stimulus-responsive controlled release system by covalent immobilization of an enzyme into mesoporous silica nanoparticles, *Bioconjug. Chem.*, **23,** pp. 698–704.

99. Deodhar, G. V., Adams, M. L., Joardar, S., Joglekar, M., Davidson, M., Smith, W. C., Mettler, M., Toler, S. A., Davies, F. K., Williams, S. K. R. and Trewyn, B. G. (2018). Conserved activity of reassociated homotetrameric protein subunits released from mesoporous silica nanoparticles, *Langmuir,* **34,** pp. 228–233.

100. Chen, Y. P., Chen, C. T., Hung, Y., Chou, C. M., Liu, T. P., Liang, M. R., Chen, C. T. and Mou, C. Y. (2013). A new strategy for intracellular

delivery of enzyme using mesoporous silica nanoparticles: superoxide dismutase, *J. Am. Chem. Soc.,* **135,** pp. 1516–1523.

101. Kwon, D., Cha, B. G., Cho, Y., Min, J., Park, E. B., Kang, S. J. and Kim, J. (2017). Extra-large pore mesoporous silica nanoparticles for directing *in vivo* M2 macrophage polarization by delivering IL-4, *Nano Letters,* **17,** pp. 2747–2756.

102. Hartono, S. B., Gu, W., Kleitz, F., Liu, J., He, L., Middelberg, A. P. J., Yu, C., Lu, G. Q. and Qiao, S. Z. (2012). Poly-L-lysine functionalized large pore cubic mesostructured silica nanoparticles as biocompatible carriers for gene delivery, *ACS Nano,* **6,** pp. 2104–2117.

103. Tang, F., Li, L. and Chen, D. (2012). Mesoporous silica nanoparticles: synthesis, biocompatibility and drug delivery, *Adv. Mater.,* **24,** pp. 1504–1534.

104. Ji, W., Li, N., Chen, D., Qi, X., Sha, W., Jiao, Y., Xu, Q. and Lu, J. (2013). Coumarin-containing photo-responsive nanocomposites for NIR light-triggered controlled drug release via a two-photon process, *J. Mater. Chem. B,* **1,** pp. 5942–5949.

105. Geng, H., Chen, W., Xu, Z. P., Qian, G., An, J. and Zhang, H. (2017). Shape-controlled hollow mesoporous silica nanoparticles with multifunctional capping for *in vitro* cancer treatment, *Chem. Eur. J,* **23,** pp. 10878–10885.

106. Guo, Y., Fang, Q., Li, H., Shi, W., Zhang, J., Feng, J., Jia, W. and Yang, L. (2016). Hollow silica nanospheres coated with insoluble calcium salts for pH-responsive sustained release of anticancer drugs, *Chem Comm.,* **52,** pp. 10652–10655.

107. Liu, J., Luo, Z., Zhang, J., Luo, T., Zhou, J., Zhao, X. and Cai, K. (2016). Hollow mesoporous silica nanoparticles facilitated drug delivery via cascade pH stimuli in tumor microenvironment for tumor therapy, *Biomaterials,* **83,** pp. 51–65.

108. Zhang, Y., Ang, C. Y., Li, M., Tan, S. Y., Qu, Q., Luo, Z. and Zhao, Y. (2015). Polymer-coated hollow mesoporous silica nanoparticles for triple-responsive drug delivery, *ACS Appl. Mater. Interfaces,* **7,** pp. 18179–18187.

109. Zhao, N., Lin, X., Zhang, Q., Ji, Z. and Xu, F. J. (2015). Redox-triggered gatekeeper-enveloped starlike hollow silica

nanoparticles for intelligent delivery systems, *Small,* **11,** pp. 6467–6479.

110. Luo, Z., Ding, X., Hu, Y., Wu, S., Xiang, Y., Zeng, Y., Zhang, B., Yan, H., Zhang, H., Zhu, L., Liu, J., Li, J., Cai, K. and Zhao, Y. (2013). Engineering a hollow nanocontainer platform with multifunctional molecular machines for tumor-targeted therapy *in vitro* and *in vivo, ACS Nano,* **7,** pp. 10271–10284.

111. Kang, X., Cheng, Z., Yang, D., Ma, P. a., Shang, M., Peng, C., Dai, Y. and Lin, J. (2012). Design and synthesis of multifunctional drug carriers based on luminescent rattle-type mesoporous silica microspheres with a thermosensitive hydrogel as a controlled switch, *Adv. Funct. Mater.,* **22,** pp. 1470–1481.

112. Zhang, Z., Wang, L., Wang, J., Jiang, X., Li, X., Hu, Z., Ji, Y., Wu, X. and Chen, C. (2012). Mesoporous silica-coated gold nanorods as a light-mediated multifunctional theranostic platform for cancer treatment, *Adv. Mater.,* **24,** pp. 1418–1423.

113. Wang, Y., Wang, K., Zhang, R., Liu, X., Yan, X., Wang, J., Wanger, E. and Huang, R. (2014). Synthesis of core shell graphitic carbon@ silica nanospheres with dual-ordered mesopores for cancer-targeted photothermochemotherapy, *ACS Nano,* **8,** pp. 7870–7879.

114. Ma, M., Huang, Y., Chen, H., Jia, X., Wang, S., Wang, Z. and Shi, J. (2015). Bi_2S_3-embedded mesoporous silica nanoparticles for efficient drug delivery and interstitial radiotherapy sensitization, *Biomaterials,* **37,** pp. 447–455.

115. Lee, J., Kim, H., Kim, S., Lee, H., Kim, J., Kim, N., Park, H. J., Choi, E. K., Lee, J. S. and Kim, C. (2012). A multifunctional mesoporous nanocontainer with an iron oxide core and a cyclodextrin gatekeeper for an efficient theranostic platform, *J. Mater. Chem.,* **22,** pp. 14061–14067.

116. Ma, M., Chen, H., Chen, Y., Wang, X., Chen, F., Cui, X. and Shi, J. (2012). Au capped magnetic core/mesoporous silica shell nanoparticles for combined photothermo-/chemo-therapy and multimodal imaging, *Biomaterials,* **33,** pp. 989–998.

117. Che, E., Gao, Y., Wan, L., Zhang, Y., Han, N., Bai, J., Li, J., Sha, Z. and Wang, S. (2015). Paclitaxel/gelatin coated magnetic mesoporous silica nanoparticles: preparation and antitumor

efficacy in vivo, *Microporous Mesoporous Mater.,* **204,** pp. 226–234.

118. Yao, X., Niu, X., Ma, K., Huang, P., Grothe, J., Kaskel, S. and Zhu, Y. (2017). Graphene quantum dots-capped magnetic mesoporous silica nanoparticles as a multifunctional platform for controlled drug delivery, magnetic hyperthermia, and photothermal therapy, *Small,* **13,** pp.

119. Bringas, E., Koysuren, O., Quach, D. V., Mahmoudi, M., Aznar, E., Roehling, J. D., Marcos, M. D., Martinez-Manez, R. and Stroeve, P. (2012). Triggered release in lipid bilayer-capped mesoporous silica nanoparticles containing SPION using an alternating magnetic field, *Chem Comm.,* **48,** pp. 5647–5649.

120. Guisasola, E., Baeza, A., Talelli, M., Arcos, D. and Vallet-Regí, M. (2016). Design of thermoresponsive polymeric gates with opposite controlled release behaviors, *RSC Adv.,* **6,** pp. 42510–42516.

Chapter 11

Hyaluronan-Functionalized Inorganic Nanohybrids for Drug Delivery

Hongbin Zhang and Zhixiang Cai
Department of Polymer Science and Engineering,
Shanghai Jiao Tong University, Shanghai 200240, China
hbzhang@sjtu.edu.cn

The development of inorganic nanomaterials for biomedical applications, including diagnosis and therapy, has been undergoing a dramatic expansion in the past few years because of their unique physiochemical properties, such as controllable size and shape, dimensions, large surface area, and tunable functionalities. In particular, considerable research has focused on drug delivery using inorganic nanoparticles (i.e., magnetic nanoparticles, quantum dots, carbon dots, graphene and its derivatives, carbon nanotubes, mesoporous silica) due to their versatile properties. However, these inorganic nanomaterials exhibit normally poor water dispersibility and biodegradability, potential toxicity, and lack of cell-specific functions, which considerably limit their extensive applications in the drug delivery field. Inorganic nanomaterials have to be subjected to chemical and/or biological functionalization with various

Handbook of Materials for Nanomedicine: Lipid-Based and Inorganic Nanomaterials
Edited by Vladimir Torchilin
Copyright © 2020 Jenny Stanford Publishing Pte. Ltd.
ISBN 978-981-4800-91-4 (Hardcover), 978-1-003-04507-6 (eBook)
www.jennystanford.com

polymers or targeting ligands to meet the stringent requirements for drug delivery. In particular, extensive research has been conducted to develop hyaluronan or hyaluronic acid (HA)-functionalized inorganic nanomaterials for drug delivery because many types of tumor cells overexpress HA receptors. In this chapter, we summarize the recent progress in the development of HA-functionalized inorganic nanohybrids as nanocarriers for drug delivery and discuss the promising results of these nanohybrids for cancer therapy.

11.1　Introduction

Cellular delivery involving the transfer of drugs through the cell membrane into the cells has attracted increasing attention because of its importance in medicine and drug delivery [1]. However, until recently, the direct intracellular delivery of drugs to the target site is generally inefficient and suffers from problems, such as poor water solubility, inefficient cellular uptake, low bioavailability, undesirable side effects, and lack of cell-specific functions [1–3]. To overcome these hurdles, considerable effort has been dedicated to the development of effective drug delivery systems (DDSs) that reduce the undesirable effects of therapeutic agents, deliver a relatively large amount of drugs without any premature release before reaching the target site, and simultaneously improve the therapeutic efficacy [4, 5]. With the development of nanotechnology, nanoscale drug delivery vehicles have emerged as a promising approach to improve the efficacy of drugs [6–8]. Among available nanomaterials for drug delivery, inorganic nanomaterials (i.e., calcium phosphate, gold, carbon materials, quantum dots (QDs), mesoporous silica, iron oxide, and layered double hydroxide (LDH)) have become ideal candidates as drug delivery vehicles because of their versatile physiochemical properties, including wide availability, rich functionality, good biocompatibility, and controlled release behavior [9–15]. However, inorganic nanomaterial-based drug delivery vehicles have several drawbacks, including potential cytotoxicity and lack of cell-specific function [4, 16]. A recent attractive approach to address these limitations is to fabricate biological molecule- or polysaccharide-functionalized inorganic nanomaterials with diverse functionality that can directly deliver drugs to the target site [17, 18].

Hyaluronan or hyaluronic acid (HA) is a natural polysaccharide found in the extracellular matrix and synovial fluids of the body. It consists of repeating units of D-glucuronic acid and *N*-acetylglycosamine, linked via β-(1,4) and β-(1,3) glycosidic bonds [19]. HA has abundant functional groups, such as hydroxyl, and carboxylic acid groups, which can be utilized for various types of chemical modifications. HA has many important physiological functions in the human body, such as structural and space filling, joint lubrication, shock absorption, tissue and water absorption, and retention capabilities, all of which are closely related to its unique rheological and physiochemical properties [20]. HA plays an important role in a wide range of biological processes, including cellular signaling, wound healing, morphogenesis, tissue regeneration, and cancer prognosis [21]. HA can bind to cell surface receptors, such as CD44, RHAMM, and LYVE-1 receptors, which are overexpressed in various kinds of malignant tumor cells. Given its outstanding hydrophilicity, nontoxicity, nonimmunogenicity, biocompatibility, biodegradability, and selective targeting of focus sites, HA possesses great potential in biomedical and pharmaceutical applications [22], such as drug delivery [23], cancer diagnosis [24], tissue engineering [25], and molecular imaging [26].

Given the unique biological merits of HA and excellent physiochemical characteristic of inorganic nanomaterials, the number of studies on HA-functionalized inorganic nanomaterials used for drug delivery has increased dramatically. HA-functionalized inorganic nanomaterials can endow inorganic nanomaterial-based DDSs with specific targeting capacities via CD44 receptor-mediated endocytosis into tumor cells. In this chapter, we introduce the recent progress in the drug delivery of HA-functionalized inorganic nanomaterials (i.e., mesoporous silica, carbon nanomaterials, quantum dots, carbon dots, QDs, CDs, magnetic nanoparticles) as nanocarriers. The promising results of these HA-functionalized inorganic nanohybrids for cancer therapy will be discussed.

11.2 Multifunctional Drug Delivery Vehicles Based on HA-Functionalized Mesoporous Silica Nanoparticles (HA-MSNs)

Since the first discovery of MCM-41 in 1992 by Mobil researchers, significant research progress has been made in morphology control and surface modification of a variety of mesoporous silica nanoparticle (MSN) materials [27, 28]. MSNs have a honeycomb-like porous structure with hundreds of uniform, controllable, and empty mesopore channels that allow entrapping of relatively large amounts of drugs [29]. Compared with the variety of inorganic materials investigated for drug delivery, MSNs possess great potential as nanocarriers for targeted drug delivery and controlled release of chemotherapeutic drugs because of their attractive characteristics, such as large specific surface area and pore volume, tunable pore sizes, good biocompatibility, high chemical and thermal stabilities, and easy chemical functionalization of their inner pore and particle surface [30]. Inspired by these highly attractive features, many researchers have exerted considerable effort in recent years to create DDSs containing MSNs as host materials [31]. MSN-based DDSs have received many excellent reviews in the past years [4, 27, 28, 30–32]. However, certain issues should be overcome to improve the colloidal dispersity in physiological fluids and the targeting capability of MSNs for further clinical application [33, 34]. The functionalization of the external surface of MSNs with poly(ethylene glycol) (PEG) has been a traditional strategy to prevent the agglomeration of MSN materials and thus improve the colloidal stability in water and under physiological conditions [35, 36]. In addition, a widely used strategy to enhance the targeting capability of MSNs to cancer cells is to design MSN-based DDSs that can effectively transport drugs to the targeted cells and tissues without any premature release into the blood circulation [37]. The functionalization of the surface of MSNs by using targeting agents that selectively target receptors overexpressed in various cell types or tissues is a promising strategy to address these problems [38]. In this section, we critically discuss the recent developments related to the use of HA-functionalized MSNs as vehicles for targeted drug delivery.

To date, researchers have extensively investigated DDSs based on HA-functionalized MSNs, all of which exhibit excellent properties by synergistically combining the advantages of MSNs and HA. For instance, Ma et al. found that HA-conjugated MSNs, with excellent colloidal stability in physiological fluids, high tumor-targeting efficiency, and good biocompatibility, have great potential as nanovehicles in the drug delivery field [39]. Yu's group used HA as targeting agent of MSNs to fabricate drug delivery vehicles for the targeted delivery of doxorubicin hydrochloride (DOX) to CD44-overexpressing cells [40]. Similarly, Liu and coworkers fabricated HA-tagged MSNs as drug nanocarriers for colon cancer therapy. Their results showed that these nanohybrids could enhance the cellular uptake of anticancer drug through CD44-mediated endocytosis, thus resulting in high antitumor efficacy [41]. These drug delivery vehicles for the targeted delivery of anticancer drugs to tumors were simply fabricated by using HA-modified MSNs.

Although considerable effort has been devoted to the development of HA-MSNs for targeted drug delivery to cells or tissues, simple HA-modified MSN materials are still challenged with intracellular biological barriers, mainly involving uncontrolled drug release at the target sites and lysosomal degradation [42]. An ideal DDS should rapidly release the drug from MSNs in the target sites. Therefore, functionalized MSN-based DDSs responsive to various internal or external stimuli, such as changes in redox potential, pH, concentrations of enzymes, and temperature, have been developed rapidly [43–45]. For example, an enzyme-responsive nanoreservoir composed of HA and MSNs with loaded DOX was designed by Chen et al. [46]. The release system presents simultaneous biocompatibility, tumor recognition capability, and enzyme-responsive drug release.

In particular, redox-responsive nanomaterials are used to construct DDSs because of the significant difference in the concentration of glutathione (GSH) between extra- and intracellular compartments [47]. The concentration of GSH in the cytosol and nuclei (approximately 10 mM) is higher than that in the extracellular fluids (approximately 2–20 mM). In addition, the cytosolic GSH level in tumor cells has been found to be at least fourfold higher than that in normal cells [45]. In recent years, a few studies of the redox-responsive controlled release on HA-MSN DDSs have been conducted. For example, Zhao et al. reported an HA oligosaccharide-modified thiol-functionalized MSN drug vehicle and observed its redox-responsive release characteristics; it can deliver the mercapto-containing drug into HA-receptor-overexpressed tumor cells via CD44 receptor-mediated endocytosis without any associated drug release [48]. Similarly, a redox and enzyme dual-stimuli responsive delivery system (MSN-SS-HA) from HA-conjugated MSNs through disulfide (SS) bonds was prepared to achieve triggered drug release from MSNs, thus providing a high antitumor efficacy, as depicted in Fig. 11.1 [49]. A variety of HA-functionalized MSNs as nanovehicles are summarized Table 11.1.

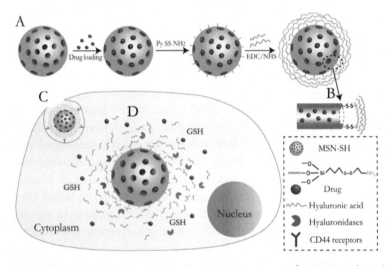

Figure 11.1   Schematic illustration of preparation process of MSN-SS-HA based dual-stimuli responsive targeted drug delivery system. (A) Synthesis of drug-loaded MSN-SS-HA, (B) magnified image of HA conjugated mesopore containing disulfide bond, (C) cell uptake via CD44 receptor-mediated endocytosis into tumor cells, and (D) drug release by GSH and enzyme in tumor cell [49].

These representative examples highlight the potential of redox-triggered DDSs. However, at present, stimuli-responsive DDSs (sDDSs) based on HA-MSNs responsive to various internal or external stimuli are still relatively sparse. Therefore, considerable effort is still needed to develop sDDSs based on HA-MSNs and intensively investigate their controlled drug release behavior at the target site. Although the recent progress in DDSs based on HA-MSNs are encouraging and show great potential for practical applications,

Table 11.1 Various DDSs using HA-MSNs nanomaterials

Systems	Drugs	Stimuli	Applications	Tumor Model	Ref.
HA-MSNs	DOX	None	Targeted delivery of drugs to CD44 over-expressing tumors	HCT-116 cells	[40]
HA-MSNs	None	None	Cancer treatment by photodynamic therapy	HCT-116 cells	[51]
HA-MSNs	Carboplatin	None	Targeted delivery of anticancer drug	Hela cells	[39]
HA-MSNs	5-fluorouracil	None	Targeted drug delivery system for colon cancer therapy	colo-205 cancer cells	[41]
HA coated C60 fullerene-silica nanoparticle	DOX	None	Targeted drug delivery system to cancer stem-like cells	MCF-7 and MDA-MB-231 human breast cancer cells	[52]
Oligosaccharide HA-MSNs	6-mercaptopurine	Redox	A stimulus-responsive targeted drug delivery system	HCT-116 cells	[48]
MSN-SS-HA	DOX	enzyme/redox	Targeted drug delivery to CD44-overexpressing cancer cells	HCT-116 cells	[49]
HA-conjugated C and Si nanocrystal-MSNs	PTX	None	Cancer theranostics	MCF-7 and MDA-MB-468 cells	[53]

several challenging tasks need to be overcome to practically use the highly promising HA-MSN-based sDDSs as drug delivery vehicles [50]. First, more attention should be focused on the long-term biocompatibility and in vivo biodistribution and degradation behavior of MSNs. Second, protocols for the reproducible synthesis and functionalization of HA-MSNs are also critical for this process. In general, a bright future may be foreseen for HA-MSN-based DDSs.

11.3 Multifunctional Drug Delivery Vehicles Based on HA-Functionalized Graphene and Its Derivatives

As a class of two-dimensional sp^2 carbon nanoscale materials, graphene and its derivatives (such as graphene oxide (GO), reduced graphene oxide (rGO), Q-graphene, and GO nanocomposites) have drawn tremendous attention in a broad range of applications, ranging from nanoelectronics, transparent conductors, energy research, catalysis, and biomaterials, because of their unique chemical, optical, electrical, and mechanical properties [54, 55]. In recent years, graphene and its derivatives have been explored as new biomaterials, such as biosensors, drug delivery nanovehicles, probes for molecular imaging, cell growth control, and cancer therapeutic agents, because of their relative biocompatibility and surface functionalizability [54, 56–58]. In particular, graphene and its derivatives have emerged as promising materials for drug delivery because the graphene surface is available for loading a variety of drugs by π–π stacking, hydrophobic interaction, and hydrogen bonding [59–61]. However, graphene-based nanomaterials generally tend to aggregate in aqueous medium with high concentrations of salts, proteins, or other ions. They also exhibit dose-dependent cytotoxicity. Thus, the poor aqueous stability of graphene-based nanomaterials under physiological conditions and their safety in vivo have hindered their applications in the biomedical field. Recently, a considerable number of studies have been conducted to improve the stability and biocompatibility of graphene and its derivatives under physiological conditions [62].

In particular, PEGylation is the most widely adopted technique to improve the stability and biocompatibility of graphene and its derivatives under physiological conditions [59, 63]. In addition to PEG, a series of studies has reported that HA endows HA-functionalized graphene nanomaterials with enhanced physiological stability and biocompatibility in blood circulation. For example, Li et al. constructed an HA-GO nanomaterial (Fig. 11.2A) with a high loading of photosensitizers and developed it as a cancer cell targeted and photoactivity switchable nanoplatform for photodynamic therapy (PDT) [64]. In this work, the functionalization significantly improved the stability of HA-GO nanomaterial in deionized water, PBS, and cell culture medium (Fig. 11.2B) and also enhanced the biocompatibility of HA-GO in physiological conditions. Miao et al. produced a tumor-targeting DDS constructed using rGO with cholesteryl HA (CHA) [65]. The CHA in the DDS increased the rGO nanosheet colloidal stability and safety in physiological conditions.

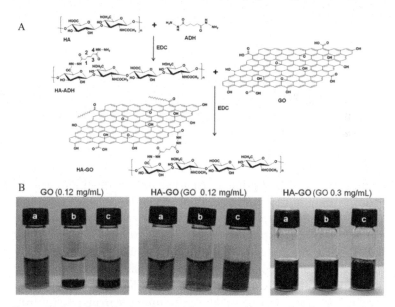

Figure 11.2 (A) Synthetic scheme of HA-GO nanomaterial, (B) Images of HA-GO conjugates and pristine GO sheets dispersed in deionized water (a), PBS (b) and cell culture medium (c) [64].

Surface modification using HA improves the physiological stability and biocompatibility of graphene-based nanovehicles in vivo and also endows DDS with active targeting capacities through selective interactions with HA-specific receptor CD44 into tumor cells. For instance, HA-conjugated GO exhibits target-specific delivery of DOX to tumors via CD44-mediated endocytosis and improves anticancer efficacy [66]. In another study, Miao et al. used CHA coated onto rGO, and the resultant nanohybrids exhibited higher drug-loading capacity, CD44-mediated delivery of DOX, and improved anticancer efficacy than rGO nanosheets [65]. The corresponding examples of HA-functionalized graphene nanomaterials for drug delivery are summarized in Table 11.2. Considerable achievements on specific drug delivery of HA-functionalized graphene nanomaterials have been attained. However, intracellular biological barriers, mainly involving uncontrolled drug release at the target sites, remain a challenge.

Recently, HA-functionalized graphene-based controlled DDSs responsive to endogenous stimuli (such as pH, redox potential, and enzyme) and external physical stimuli (such as light, magnetic field, and temperature) have been developed to overcome the poorly controlled drug delivery and thus enhance the therapeutic efficacy. Song and coworkers developed a multifunctional HA-functionalized GO nanohybrid for pH-responsive drug delivery [67]. In this work, HA was anchored onto the GO surface via H-bonding interactions. DOX was loaded onto the HA-GO surface via π–π stacking. The HA-GO-DOX nanohybrids exhibited pH-responsive drug release behavior in the slightly acidic environment of tumor cells and a high tumor inhibition rate, which have potential clinical applications for drug delivery. To our knowledge, disulfide bonds in the redox-responsive nanocarriers are susceptible to cleavage under a high intracellular concentration of GSH involving thiol–disulfide exchange reactions. Thus, the redox-responsive nanocarriers can release the drug rapidly in the cytosol (high GSH concentration), whereas these nanocarriers are quite stable in the extracellular environment (low GSH concentration) [3, 54]. In addition, the GSH level in tumor cells is four times higher than that in normal cells [68]. Therefore, the GSH level difference in cells can be used to promote drug release from DDSs. Recently, Yin and coworkers constructed a redox-triggered tumor targeting and release system by using HA-conjugated GO via disulfide linkages, as shown in Fig. 11.3A [42]. DOX in the HA-ss-GO nanohybrid (HSG-DOX) can be released in response to high GSH concentrations, thus leading to a highly efficient cancer treatment (Fig. 11.3B). A summary of the different types of endogenous stimuli-responsive DDS based on HA-functionalized graphene nanohybrids is shown in Table 11.2.

Table 11.2   Recent progress in drug delivery of HA-functionalized graphenes

Graphene structures	Drugs	Stimuli	Applications	Tumor model	Ref.
HA-GO	Cy3-labeled antisense miR-21 peptide nucleic acid	—	A cancer theranostic tool in miRNA-targeted therapy	MBA-MB231 cells	[74]
HA-modified Q-graphene	DOX	—	Tracking and monitoring targeted drug delivery for efficiently killing drug-resistant cancer cells	A549 cells and MRC-5 cells	[75]
HA-CHI(chitosan)-GO	SNX-2112	—	Treatment of lung cancer	A549 cells	[76]
HA and Arg-Gly-Asp peptide (RGD) modified GO	DOX	pH	A promising platform for targeted cancer therapeutic	SKOV-3 and HOSEpiC cells	[77]
HA-GO	DOX	pH	Targeted, and pH-responsive drug delivery	HepG2 cells	[67]
HA-ss-GO	DOX	Redox and NIR	NIR-and pH-responsive drug delivery to overcome multiple biological barriers for cancer treatment	MDA-MB-231 cells	[42]
HA-GO	chlorine 6 (Ce6)	NIR	Targeted photodynamic therapy	HeLa cells	[64]
HA-Ag/GO	—	NIR	Synergistic therapy of bacterial infection	S. aureus bacterial strains	[78]
HA and polyaspartamide functionalized GO	Irinotecan (IT)	NIR	Localized chemotherapy and photothermal therapy for solid tumors	HCT-116 cells	[73]
rGO-PDA@ MSN/HA Nanocomposite	Ce6	NIR	Enhance PDT therapeutic efficiency	HT-29 cells HCT-116 cells NIH-3T3 cells	[72]
rGO-HA	ZnO	NIR	A novel multi-synergistic platform for enhanced apoptotic cancer therapy	MDA-MB-231 and NIH3T3 cells	[79]

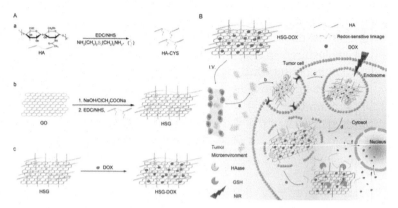

Figure 11.3 (A) Preparation of HSG-DOX nanohybrids. (B) Molecular mechanism of the HSG-DOX nanohybrid for drug delivery. NIR irradiation-controlled endo/lysosomal escape for tumor cytoplasm-selective delivery and redox-responsive rapid release of DOX in cytoplasm [42].

Beyond endogenous-responsive DDSs, a number of DDSs take advantage of external physical stimuli, such as light, temperature changes, magnetic fields, and electric fields [43, 69]. In the past decade, a variety of photoactivated DDSs have been constructed to promote drug release in response to the illumination of a specific wavelength (UV or near-infrared (NIR) regions) at the target sites because of their noninvasiveness and remote spatiotemporal control [43, 70]. Thus far, graphene-based nanomaterials have shown great promise when applied to photothermal therapy (PTT) and PDT because of their strong NIR optical absorption capabilities, large specific surface area, and abundant functional groups on the surface. In a recent study, HA-GO can be loaded with a photosensitizer (i.e., Ce6) via π–π stacking and/or hydrophobic interactions. Ce6-loaded HA-GO showed improved photodynamic therapeutic efficiency compared with free Ce6 [64]. In contrast to photosensitizer-loaded GO-based delivery systems using a photosensitizer for generating reactive oxygen species, DDSs that use graphene and its derivatives as PTT agents have demonstrated success. GO with high optical absorbance in the NIR region can effectively convert light into hyperthermia, leading to photothermal ablation of the cancer cells and subsequent cell death [71, 72]. For example, Yin and coworkers reported an HA-ss-GO nanohybrid that exhibited mild tumor inhibition in vivo under NIR irradiation [42]. In another study, Fiorica and coworkers functionalized GO with HA and polyaspartamide via covalent crosslinking (Fig. 11.4), yielding a multifunctional double-network-structured nanogel with remarkable biocompatibility, high loading of anticancer drug (irinotecan), and strong NIR optical absorbance; thus, such functionalized GO can be utilized for localized photothermal ablation of solid tumors [73]. Their study evidently showed that the combination of PTT and chemotherapy is as efficient solution for tumor killing. A series of studies conducted by various research groups for the exploration of HA-functionalized graphene-based nanomaterials in PDT and PPT is summarized in Table 11.2.

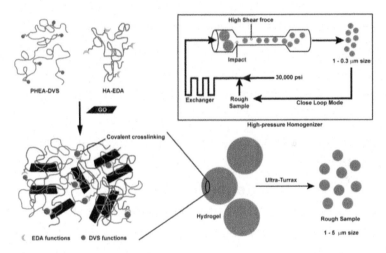

Figure 11.4 Schematic representation of HA and polyaspartamide functionalized GO nanogel synthesis processes [73].

In conclusion, DDSs based on HA-functionalized graphene and its derivatives have attracted considerable attention in the past few years mostly because of the remarkable physiochemical properties of graphene-based nanomaterials and the unique biological properties of HA. However, despite the previously mentioned effectiveness and obvious achievements on cancer therapy reported in recent years, DDSs based on HA-functionalized graphene and its derivatives are still far from clinical translation. Some remaining challenges should be addressed. First, one of the most important considerations of these nanomaterials is the profound understanding of graphene's interaction with living cells (tissues and organs) and the in vivo degradation/excretion behavior of graphene-based nanomaterials. Second, according to previous reports, HA-functionalized graphene-based nanomaterials show no evident toxicity to treated cells compared with their nonfunctionalized naked counterparts. Animal experiments conducted on rodent models may have different results from primates and humans. To address this question, numerous preclinical toxicity studies using in vivo animal models, particularly long-term toxicity studies, are required to demonstrate the biocompatibility of HA-functionalized graphene-based nanomaterials that can satisfy further clinical requirements. Third, the toxicity of graphene-based nanomaterials depends on their size distributions and surface modifications. The unpredictable physiochemical properties of these HA-functionalized graphene-based nanomaterials may severely limit their further translation to clinical applications. Hence, the surface properties and sizes of these nanomaterials should be critical considerations for clinical applications. Finally, novel and versatile materials should be explored for the construction of multifunctional HA–graphene-based DDSs that can be translated into clinical applications.

11.4 Multifunctional Drug Delivery Vehicles Based on HA-Functionalized Carbon Nanotubes

Carbon nanotubes (CNTs) can be regarded as graphene sheet rolled up and seamlessly wrapped into a cylindrical tube [80]. CNTs are classified as single-walled (SWCNTs) and multiwalled (MWCNTs). In recent years, CNTs have attracted tremendous attention as one of the most promising nanostructured materials in the biomedical field as drug delivery vehicles and diagnostic tools because of their extraordinary intrinsic physical, chemical, and physiological properties, including unique optical, mechanical, and electronic properties, ultra-large surface area, remarkable cell membrane penetrability, high drug-loading capability, and tunable surface functionalities [81–84]. In particular, given the highly hydrophobic nature of the pristine CNT sidewall surface, insoluble chemotherapy drugs can adsorb spontaneously onto the CNT sidewall [85]. CNTs as a highly efficient vehicle can transport drugs cross the cell membranes in mammalian cells [86]. In addition, the intrinsic stability and structural flexibility of CNTs may prolong the circulation time and the bioavailability of drug molecules conjugated to CNTs [87]. These unique structures and properties of CNTs have made them potential candidates for the highly efficient delivery of drugs and biomolecules [85, 88, 89]. However, a major problem of CNTs as a drug delivery platform is their complete insolubility and strong tendency to form bundles in water because of the graphitic nature of their sidewalls, which may hamper their extensive applications in nanomedicine [84]. To address this problem, many feasible approaches to the functionalization of CNT surfaces have been used to enhance the solubility and dispersion of CNTs in aqueous solution [90]. Two modification strategies, i.e., covalent and noncovalent methods, are widely employed for CNT surface functionalization. Another issue worthy of consideration is that pristine CNTs have shown toxicity to cells and animals [91]. Furthermore, the cytotoxicity of CNTs is highly dependent on their surface chemistry.

Chemical functionalization strategies of CNTs should be introduced to improve their biocompatibility and dispersion in aqueous solution. Studies have confirmed that the surface functionalization of CNTs with polysaccharides is an outstanding strategy to improve the solubility and biocompatibility of CNT conjugates [92, 93]. In recent years, many researchers have developed CNT-based DDSs functionalized with HA. Researchers have explored many strategies, including physical interaction or covalent bonding, to integrate HA onto the CNT surface successfully. For example, Dvash and coworkers developed HA-functionalized CNTs by using phospholipids as the linking arm between the HA and CNTs; the HA-functionalized CNTs exhibited excellent water solubility and low cytotoxicity [94]. Similarly, Yao and coworkers developed a DDS that is based on distearoylphosphatidylethanolamine–hyaluronic acid (DSPE-HA)-functionalized CNTs and found that the DSPE-HA-functionalized CNTs have good dispersibility, enhanced biocompatibility, and targetability to CD44-overexpressed cancer cell [95]. These studies showed strong indication that HA-functionalized CNTs can increase the biocompatibility and dispersion of CNTs in aqueous solution.

CNTs as drug delivery nanocarriers exhibit high drug-loading capacity and controlled delivery properties because of their previously mentioned unique physiochemical properties [96]. However, certain issues should be overcome to improve the colloidal dispersibility in physiological fluids and enhance the targeting capability of CNTs. Targeting molecules, such as folic acid (FA) [97], and antibodies [98] can be further incorporated into the drug-loaded CNTs via covalent or noncovalent methods to confer active receptor-mediated targeting capabilities to further augment the efficacy of CNT-based DDS. As mentioned previously, HA has been widely used for targeted drug delivery because of its unique and outstanding biological properties. To date, considerable effort has been exerted to prepare HA-functionalized CNTs as new nanomaterials for the delivery of anticancer drugs. HA-functionalized CNTs have shown great promise for synergistically combining the advantage of HA and CNTs. In this section, we will discuss the most significant progress in the field of HA-functionalized CNT nanohybrids as drug delivery vehicles.

By employing the noncovalent bonding technique, Yao et al. developed a DDS based on chitosan (CHI)-coated SWCNTs functionalized with HA (SWCNTs-CHI-HA) [99]. The obtained nanocarrier exhibited high cellular uptake and therapeutic efficiency to gastric cancer stem cells, which indicated the great potential application of this nanocarrier in overcoming the recurrence and metastasis of gastric cancer and improving gastric cancer treatment. Similarly, a drug delivery nanovehicle comprising SWCNTs-CHI-HA loaded with DOX via electrostatic interactions was reported by Mo and coworkers [100]. They observed that the obtained DOX-loaded SWCNTs-CHI-HA showed high stability in the physiologic solution, high drug-loading capacity, and targeting to cancer cells, ensuring enhanced anticancer efficacy and reduced adverse side effects. Based on these DDSs, HA was immobilized in CNT sidewalls through noncovalent methodology.

An alternative approach is the functionalization of CNTs with HA through covalent methodology to develop new DDSs for cancer therapy. For instance, Liu et al. developed a "smart" nanosystem based on HA-functionalized MWCNTs for tumor-targeted delivery of the anticancer agent DOX [101]. The functionalization was conducted through EDC/NHS reaction, as shown in Fig. 11.5. DOX was loaded onto MWCNTs coated with HA via π–π stacking interactions. The obtained biocompatible nanovehicle augmented the antitumor activity of DOX against HA receptors overexpressing cancer cells and exhibited reduced drug-associated cardiotoxicity. Cao and coworkers conducted a similar investigation by using polyethyleneimine (PEI)-modified MWCNTs conjugated with fluorescein isothiocyanate and HA as a new nanocarrier system for targeted drug delivery to cancer cells overexpressing CD44 receptors [102]. In addition to high stability and biocompatibility, the resultant nanovehicles exhibited growth inhibition effect on the cancer cells. These results confirmed the feasibility of HA-functionalized MWCNTs as a target-specific drug delivery carrier for tumor-targeted chemotherapy. A range of DDSs based on HA-functionalized CNTs to enhance the efficacy of tumor therapy have been proposed and are summarized in Table 11.3.

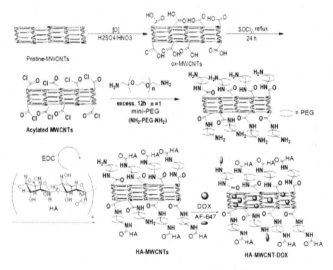

Figure 11.5 Schematic representation of functionalization of MWCNTs with HA [101].

As mentioned previously, with continuous research, an increasing number of CNT-based nanomaterials have been used in the drug delivery field. However, the increasing utilization of MWCNTs increases the risk of occupational and environmental human exposures. Several studies have shown that MWCNT exposure can lead to inflammation, fibrosis, and granuloma formation in the lungs [103]. To address this problem, researchers have conducted several

Table 11.3 Recent progress in drug delivery of HA-functionalized CNTs

CNTs structures	Drugs	Applications	Tumor Model	Ref.
DSPE-HA functionalized CNTs	DOX	Drug carrier for Multidrug resistance cancer	A549 cells	[95]
CHI and HA functionalized SWNTs	DOX	Delivery system with high therapeutic efficacy and minimal adverse side effects	Hela cells	[100]
MWCNTs/PEI–FI–HA	DOX	an efficient anticancer drug carrier for tumor-targeted chemotherapy	HeLa and L929 cells	[102]
SWNTs-CHI-HA	Salinomycin (SAL)	Overcome the recurrence and metastasis of gastric cancer and improve gastric cancer treatment	Gastric cancer stem cells	[99]
HA-MWCNTs	DOX	A "smart" platform for tumor-targeted delivery of anticancer agents	A549 cells	[101]
Indocyanine Green-HA-SWCNTs	PTT and PDT	CD44 targeted and image-guided dual PTT and PDT cancer therapy	SCC7 cells	[105]
cholanic acid (CA)-HA-SWCNTs	DOX	Deliver therapeutics and acts as a sensitizer to influence drug uptake and induce apoptosis	OVCAR8 and OVCAR8/ADR cells	[106]
HA-MWCNTs	—	Reduces pulmonary injury after MWCNTs exposure	primary human bronchial epithelial cells and primary human lung	[104]

studies to reduce the hazardous effects of MWCNTs. In a typical example, Hussain and coworkers have functionalized MWCNTs with HA; the HA-functionalized MWCNTs exhibited reduced postexposure lung inflammation, fibrosis, and mucus cell metaplasia [104].

As mentioned previously, HA-functionalized CNTs as promising candidates for drug delivery have strongly attracted the attention of researchers because of their excellent physiochemical and biological properties. Despite the encouraging results shown by various groups, researchers need to overcome several obstacles to improve the suitability of HA-functionalized CNTs for further clinical application. One general limitation of HA-functionalized CNT-based nanovehicles is their potential long-term toxicity. Although many studies have reported that HA-functionalized CNT-based nanovehicles can be safely used to treat animals at certain doses, most of the reported animal experiments are conducted on rodent models, which are different from primates and humans [107]. Thus, more research on HA-functionalized CNT-based nanovehicles should be devoted to fully elucidate their long-term toxicity on humans. Another critical consideration is that CNTs hardly degrade in biological systems, which may limit their extensive applications in the drug delivery field. Another challenging aspect of HA-functionalized CNTs is the polydispersity of CNTs in physiological fluids, which limits the reproducibility of the results and often yields inconsistent data. Although translating HA-functionalized CNT-based nanovehicles from a promising nanomaterial to effective drug carriers in clinical trials is still at the nascent stages of development, considerable effort should be devoted to overcoming the challenging aspects to transform HA-functionalized CNT-based nanovehicles into clinical reality.

11.5 Multifunctional Drug Delivery Nanoplatforms Based on HA-Functionalized QDs and Carbon Dots (CDs)

Fluorescent semiconductor nanoparticles, commonly referred to as QDs, contain approximately 200–10,000 atoms with 2–10 nm diameter [108]. QDs have shown great potential for many biomedical applications ranging from bioimaging to drug delivery because of their novel physical, chemical, and optical properties [109, 110]. In addition, QDs have been widely recognized as an alternative for traditional organic dyes in molecular, cellular, and in vivo imaging because of their high quantum yields [111, 112]. The integration of the diagnosis and chemotherapy functions into QDs allows real-time and noninvasive monitoring of therapeutic effects during a chemotherapy treatment course [15]. These QDs exhibit excellent sensitivity to fluorescent imaging, making them highly promising for optical imaging-guided drug delivery and therapy [15]. However, many obstacles need to be overcome. First, QDs generally tend to aggregate in physiological fluids, which may restrict their many possible applications in the biomedical field. Considerable concern has been raised over the toxicity of these QDs in living cells and animals [113, 114]. Reasonable surface modification of QDs can not only improve the aqueous solubility and stability of QDs but also reduce the inherent toxicity of QDs [115, 116]. Surface modification of QDs by ligands, polysaccharides, or organic compounds has been one of the most widely accepted methods to improve their biocompatibility and stability while preserving their optical properties [117–119]. HA stimulated the interest for surface modification of QDs as a major ligand because of its unique biological properties. In this section, we discuss recent developments in the synthesis of HA-functionalized QDs and their use as nanovehicles for imaging-guided targeted drug delivery.

In light of the unique physiochemical properties of QDs and biological properties of HA, scientists have been testing HA-functionalized QD nanomaterials in various bioimaging applications [120, 121]. In one notable example, Bhang and coworkers used HA-QDs as a model system to examine the real-time visualization of changes in lymphatic vessels [120]. In addition to bioimaging applications, HA-functionalized QDs have also been investigated as theranostic platform for imaging-guided drug delivery. In 2010, Kim and coworkers used HA derivatives to modify QDs for bioimaging. The bioimaging results showed that the HA derivatives are promising drug delivery carriers for the treatment of various chronic liver diseases [122]. In a recent work, multifunctional HA-functionalized QDs were developed for cancer therapy by Park and coworkers to allow monitoring of drug delivery by imaging [123]. In this work, HA was conjugated with amphiphilic-PEI derivative-coated QDs by electrostatic assemblies, which could be used for cell-specific targeted labeling. The results indicated that the QD platform can be exploited for various applications, including cellular labeling, targeting, gene delivery, and ratiometric oxygen sensing. In another work, Cai and coworkers found that ZnO QDs functionalized with dicarboxyl-terminated PEG and HA (HA-PEG-QDs) are promising pH-responsive targeted drug delivery platforms (Fig. 11.6) [124]. In this work, the HA-PEG-QDs efficiently transported the loaded DOX into cancer cells and exhibited a response to trigger drug release in the acidic environment of the tumors, resulting in enhanced antitumor activity.

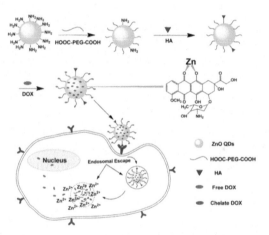

Figure 11.6 A schematic illustration of HA-PEG-QDs nanohybrid for pH-sensitive and targeted drug delivery [124].

Despite the tremendous effort, QDs are known for their inherent toxicity even at low concentrations, which inhibited their further clinical applications. Therefore, the search for benign alternatives continued. CDs, a new kind of biocompatible carbon nanomaterials, have attracted extensive investigations in biomedicine as an emerging class of fluorescent materials [125, 126]. In comparison with the traditional metal-based semiconductor QDs, CDs are highly beneficial because of their remarkable optical properties, excitation wavelength-dependent emission, high water solubility, stable photoluminescence, good biocompatibility, and low cytotoxicity, which are highly desirable for biomedical applications [126–128].

Taking advantage of their promising optical and biocompatible capabilities, CDs have been extensively utilized as imaging-guided nanocarriers for drug and gene delivery [129–131].

To date, researchers have extensively investigated imaging-guided DDSs based on HA-functionalized CDs. The conjugation of CDs with HA endows them with cancer cell targetability and can be useful in drug delivery applications. Abdullah and coworkers reported that catechol-modified HA was introduced on the surface of graphene QDs (HA-GQDs) by means of its catechol adhesive nature, integrating the optical properties of GQDs and the cancer-targeting properties of HA for the simultaneous targeted drug delivery and fluorescent tracking [132]. On the basis of in vitro and in vivo studies, HA-GQDs can be a fluorescent probe and strong drug carrier in imaging-guided targeted DDS. Another study regarding imaging-guided targeted drug delivery using HA-functionalized CDs was reported by Jia and coworkers, in which CDs were functionalized concomitantly with folate-terminated PEG-modified HA (FA-PEG-HA), yielding FA-PEG-HA-CDs (Fig. 11.7) [133] The nanohybrid exhibited perfect biocompatibility, dual receptor-mediated targeting function, ideal fluorescence property, and desirable tumor microenvironment-responsive controlled release performance, which are suitable for photoluminescent imaging and dual receptor-mediated targeting controlled drug delivery. In view of these results, HA-functionalized CDs provide a new strategy for developing imaging-guided targeted DDS for cancer treatment.

Despite encouraging results reported by several groups, the bench to bedside translation for CD nanomedicine remains to be a challenging task. First, CDs with high quantum yields remain rare. Therefore, reliable techniques to improve the photoluminescent quantum yield of CDs in the deep-red and NIR regions are needed [134]. In addition, tremendous research effort is still needed to enhance the sensitivity, selectivity, and robustness of CD-based sensing and bioimaging platforms [135]. Most notably, the issues of CD toxicity remain contentious to date. Thus, further research must be conducted before CQDs can be fully used in the biomedical field. In conclusion, considerable progress has been made in a wide variety of applications of CDs spanning bioimaging and targeted drug delivery and tracing. However, the imaging-guided drug delivery platforms based on HA-functionalized CDs remain relatively sparse. Therefore, tremendous effort is needed to develop nanoplatforms based on HA-functionalized CDs.

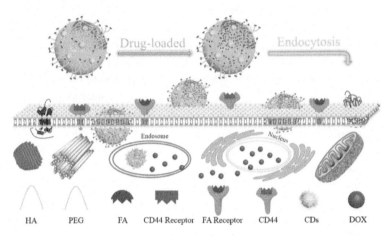

| HA | PEG | FA | CD44 Receptor | FA Receptor | CD44 | CDs | DOX |

Figure 11.7 Illustration of the proposed DOX-loaded FA-PEG-HA-CDs theranostic nanogels for the intracellular release [133].

11.6 Multifunctional Drug Delivery Platforms Based on HA-Functionalized Magnetic Nanoparticles (MNPs)

Over the years, MNPs have been used in an increasing number of biomedical applications. In particular, MNPs has have been extensively utilized as contrast agents for magnetic resonance imaging (MRI) and as carriers for drug delivery because of their intrinsic magnetic properties, excellent biocompatibility, chemical stability over physiological environments, and capability to function at the cellular and molecular levels of biological interactions [136–139]. MRI has emerged as a powerful noninvasive tool in cancer detection with the use of MNPs as strong T_2 imaging contrast agents [140, 141].

However, the utilizations of MNPs have been limited to their insolubility in the aqueous phase. Thus, one of the current challenges is to develop MNPs with high colloidal stability in harsh biological conditions [142]. In addition, MRI contrast agents using functionalized MNPs are still limited by their low tumor-targeting efficacy [143]. Therefore, modifying the surface of MNPs by using specific targeting moieties (such as an antibody or polysaccharide) is necessary. HA has been a widely used polymer for MNP functionalization. For example, Lee and coworkers [144] reported the modification of magnetite nanocrystal with dopamine-functionalized HA to form HA-DN/MCN complexes. This group demonstrated that these HA-functionalized MNPs were stable under physiological conditions, indicating their potential for in vivo applications. In another study, Lim and coworkers fabricated HA-functionalized magnetic nanoclusters (HA-MNCs) for the detection of CD44-overexpressing breast cancer cells using MRI [145]. In this study, the HA-functionalized MNPs exhibited superior targeting efficiency with MR sensitivity and high biocompatibility, which suggested that HA-MNPs can be potent cancer-specific molecular imaging agents. Similarly, iron oxide nanoparticles functionalized with PEG and HA can specifically label mesenchymal stem cells and show enhanced MRI contrast, which can be utilized as an imaging probe for biomedical diagnosis [146].

Aside from MRI contrast agents based on MNPs, DDS based on multifunctional MNPs have also been extensively investigated [138]. For example, nanohybrids composed of HA and iron oxide were synthesized using electrostatic interactions for the first time by Kumar and coworkers [147]. The HA-functionalized Fe_2O_3 can deliver drugs to the desired tissue by an external localized magnetic field gradient. HA-functionalized MNPs can be loaded with anticancer drugs while retaining their MRI property. Recently, El-Dakdouki and coworkers fabricated HA-conjugated superparamagnetic iron oxide nanoparticles (HA-SPION) (Fig. 11.8) as a nanoplatform to load DOX for targeted imaging and drug delivery [17]. In this work, DOX-loaded HA-SPION was more cytotoxic to drug-sensitive and drug-resistant cancer cells than free DOX. In a subsequent work from the same group, they further assessed the in vivo efficacy of DOX-loaded HA-SPION (Fig. 11.8) [148]. Their results confirmed that HA-SPION can act as an attractive theranostic platform to target CD44-expressing tumors. These findings are expected to introduce new promising strategies to design an efficient nanoplatform for targeted imaging and drug delivery.

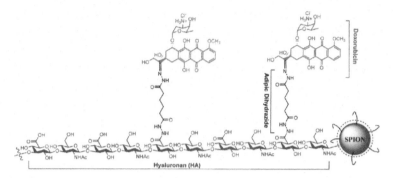

Figure 11.8 Schematic illustration of DOX-loaded HA-SPION [17, 148].

MNPs in the presence of an alternating magnetic field would induce hyperthermia. Such a phenomenon has been widely explored for magnetic hyperthermia cancer treatment [149–151]. In recent years, HA-functionalized MNPs have been explored to achieve MRI-guided hyperthermia treatment by taking advantage of their magnetic hyperthermia performance and MRI properties. For example, photosensitizer-conjugated HA-coated Fe_3O_4 was reported by Kim and coworkers and used as nanoplatform for tumor-targeted MRI imaging and photodynamic/hyperthermia combination therapy [152]. Similarly, SPION with HA functionalization for MRI imaging and hyperthermia therapy was developed by Thomas and coworkers [153]. They found that cancer tumors incubated with HA-SPION exhibited high MR T_2 contrast and low cell viability. Therefore, with HA-functionalized MNPs as nanoplatform for drug delivery, the HA-specific receptor CD44 was used to locally enhance their tumor accumulation and apply an alternating magnetic field to induce tumor heating to achieve further enhanced cancer diagnosis and therapy.

In this section, we have highlighted some exciting research progress on the utilization of HA-functionalized MNPs as tumor-targeted drug vehicles and MRI contrast agents for MRI imaging-guided drug delivery. These results are encouraging and show great potential for future biomedical applications. However, some important issues need to be resolved before we can transform HA-functionalized MSNs into clinical reality. For example, systematic studies on the long-term distribution behavior, degradation, and long-term toxicity of HA-functionalized MNPs are lacking. Thus, tremendous effort is needed to investigate their pharmacokinetics, long-term toxicity, and degradation properties before their further clinical application. Once the concern is addressed, HA-functionalized MNPs will hold great potential of providing new opportunities for early cancer detection and targeted therapies.

11.7 Conclusions and Future Outlook

In this chapter, we have highlighted the progress on HA-functionalized inorganic nanomaterials (such as MSNs, graphene, CNTs, QDs, CDs, and MNPs) that have been explored as nanocarriers for tumor-targeted drug delivery by using preclinical models. Many exciting research contributions are highlighted in this chapter to illustrate the important biomedical applications of HA-functionalized inorganic nanomaterials. Different aspects in the field of drug delivery have been discussed, which include stimuli-responsive (i.e., pH, redox, and enzyme) drug release, imaging-guided drug delivery, and photodynamic and photothermal cancer treatment.

Although significant progress has been made in the area of HA-functionalized inorganic nanoparticle-based drug delivery, many challenges have yet to be overcome before translation into clinical applications. First, many examples discussed in this chapter show that one of the most significant challenges is the lack of a series of general high-throughput methods for fabricating inorganic nanomaterials with controllable physicochemical properties (e.g., size, shapes, and architectures), thereby leaving considerable room for scientists to design sophisticated nanomaterials. Second, HA-functionalized inorganic nanomaterials inevitably face some challenges, such as premature cargo leakage, biodistribution, and long-term cytotoxicity. Thus, a comprehensive investigation should be considered, including pharmacokinetics, long-term toxicity, and degradation properties in vivo. Another challenging aspect is the availability of appropriate chemical modification methods because the extent of the chemical modification of HA influences the receptor-mediated uptake by cancer cells. Thus, tremendous effort is needed to investigate the influence of the chemical modification of HA on the CD44 receptor-mediated uptake behavior. Furthermore, more data from clinical trials are required to introduce HA-functionalized inorganic nanomaterials to clinical applications. We believe that the rapid development of modern technologies will help overcome these obstacles and accelerate the progress of HA-functionalized inorganic nanomaterials in the drug delivery field.

References

1. Xu, Z. P., Zeng, Q. H., Lu, G. Q., Yu, A. B., Inorganic nanoparticles as carriers for efficient cellular delivery. *Chem. Eng. Sci.* 2006, 61(3), 1027–1040.

2. Ganta, S., Devalapally, H., Shahiwala, A., Amiji, M., A review of stimuli-responsive nanocarriers for drug and gene delivery. *J. Control. Release* 2008, 126(3), 187–204.

3. Blum, A. P., Kammeyer, J. K., Rush, A. M., Callmann, C. E., Hahn, M. E., Gianneschi, N. C., Stimuli-responsive nanomaterials for biomedical applications. *J. Am. Chem. Soc.,* 2015, 137(6), 2140–2154.

4. Cai, Z., Zhang, H., Wei, Y., Cong, F., Hyaluronan-inorganic nanohybrid materials for biomedical applications. *Biomacromolecules* 2017, 18(6), 1677–1696.

5. Liu, J. Q., Cui, L., Losic, D., Graphene and graphene oxide as new nanocarriers for drug delivery applications. *Acta Biomater.* 2013, 9(12), 9243–9257.

6. Lanone, S., Boczkowski, J., Biomedical applications and potential health risks of nanomaterials: Molecular mechanisms. *Curr. Mol. Med.* 2006, 6(6), 651–663.

7. Sahoo, S. K., Labhasetwar, V., Nanotech approaches to delivery and imaging drug. *Drug Discov. Today* 2003, 8(24), 1112–1120.

8. Shi, J. J., Votruba, A. R., Farokhzad, O. C., Langer, R., Nanotechnology in drug delivery and tissue engineering: from discovery to applications. *Nano Lett.* 2010, 10(9), 3223–3230.

9. Morgan, T. T., Muddana, H. S., Altinoglu, E. I., Rouse, S. M., Tabakovic, A., Tabouillot, T., Russin, T. J., Shanmugavelandy, S. S., Butler, P. J., Eklund, P. C., Yun, J. K., Kester, M., Adair, J. H., Encapsulation of organic molecules in calcium phosphate nanocomposite particles for intracellular imaging and drug delivery. *Nano Lett.* 2008, 8(12), 4108–4115.

10. Ghosh, P., Han, G., De, M., Kim, C. K., Rotello, V. M., Gold nanoparticles in delivery applications. *Adv. Drug Deliv. Rev.* 2008, 60(11), 1307–1315.

11. Biju, V., Chemical modifications and bioconjugate reactions of nanomaterials for sensing, imaging, drug delivery and therapy. *Chem. Soc. Rev.* 2014, 43(3), 744–764.

12. Barbe, C., Bartlett, J., Kong, L. G., Finnie, K., Lin, H. Q., Larkin, M., Calleja, S., Bush, A., Calleja, G., Silica particles: A novel drug-delivery system. *Adv. Mater.* 2004, 16(21), 1959–1966.

13. Gupta, A. K., Gupta, M., Synthesis and surface engineering of iron oxide nanoparticles for biomedical applications. *Biomaterials* 2005, 26(18), 3995–4021.

14. Xu, Z. P., Lu, G. Q., Layered double hydroxide nanomaterials as potential cellular drug delivery agents. *Pure Appl. Chem.* 2006, 78(9), 1771–1779.

15. Probst, C. E., Zrazhevskiy, P., Bagalkot, V., Gao, X. H., Quantum dots as a platform for nanoparticle drug delivery vehicle design. *Adv. Drug Deliv. Rev.* 2013, 65(5), 703–718.

16. Son, S. J., Bai, X., Lee, S. B., Inorganic hollow nanoparticles and nanotubes in nanomedicine. Part 1. Drug/gene delivery applications. *Drug Discov. Today* 2007, 12(15–16), 650–656.

17. El-Dakdouki, M. H., Zhu, D. C., El-Boubbou, K., Kamat, M., Chen, J., Li, W., Huang, X., Development of multifunctional hyaluronan-coated nanoparticles for imaging and drug delivery to cancer cells. *Biomacromolecules* 2012, 13(4), 1144–1151.

18. Sapsford, K. E., Algar, W. R., Berti, L., Gemmill, K. B., Casey, B. J., Oh, E., Stewart, M. H., Medintz, I. L., Functionalizing nanoparticles with biological molecules: Developing chemistries that facilitate nanotechnology. *Chem. Rev.* 2013, 113(3), 1904–2074.

19. Cai, Z., Zhang, F., Wei, Y., Zhang, H., Freeze–Thaw-induced gelation of hyaluronan: Physical cryostructuration correlated with intermolecular associations and molecular conformation. *Macromolecules* 2017, 50(17), 6647–6658.

20. Hussain, Z., Thu, H. E., Katas, H., Bukhari, S. N. A., Hyaluronic acid-based biomaterials: a versatile and smart approach to tissue regeneration and treating traumatic, surgical, and chronic wounds. *Polym. Rev.* 2017, 57(4), 594–630.

21. Toole, B. P., Hyaluronan: From extracellular glue to pericellular cue. *Nat. Rev. Cancer* 2004, 4(7), 528–539.

22. Kim, H., Jeong, H., Han, S., Beack, S., Hwang, B. W., Shin, M., Oh, S. S., Hahn, S. K., Hyaluronate and its derivatives for customized biomedical applications. *Biomaterials* 2017, 123, 155–171.

23. Xu, W., Qian, J., Hou, G., Suo, A., Wang, Y., Wang, J., Sun, T., Yang, M., Wan, X., Yao, Y., Hyaluronic acid-functionalized gold nanorods with PH/NIR dual-responsive drug release for synergetic targeted photothermal chemotherapy of breast cancer. *ACS Appl. Mater. Interfaces* 2017, 9(42), 36533–36547.

24. Kramer, M. W., Escudero, D. O., Lokeshwar, S. D., Golshani, R., Ekwenna, O. O., Acosta, K., Merseburger, A. S., Soloway, M., Lokeshwar, V. B., Association of hyaluronic acid family members (has1, has2, and hyal-1) with bladder cancer diagnosis and prognosis. *Cancer Am. Cancer Soc.* 2011, 117(6), 1197–1209.

25. Knopf-Marques, H., Pravda, M., Wolfova, L., Velebny, V., Schaaf, P., Vrana, N. E., Lavalle, P., Hyaluronic acid and its derivatives in coating and delivery systems: Applications in tissue engineering, regenerative medicine and immunomodulation. *Adv. Healthc Mater.* 2016, 5(22), 2841–2855.

26. Li, J. C., Hu, Y., Yang, J., Wei, P., Sun, W. J., Shen, M. W., Zhang, G. X., Shi, X. Y., Hyaluronic acid-modified $Fe_3O_4@Au$ core/shell nanostars for multimodal imaging and photothermal therapy of tumors. *Biomaterials* 2015, 38, 10–21.

27. Wen, J., Yang, K., Liu, F. Y., Li, H. J., Xu, Y. Q., Sun, S. G., Diverse gatekeepers for mesoporous silica nanoparticle based drug delivery systems. *Chem. Soc. Rev.* 2017, 46(19), 6024–6045.

28. Slowing, I. I., Trewyn, B. G., Giri, S., Lin, V. S. Y., Mesoporous silica nanoparticles for drug delivery and biosensing applications. *Adv. Funct. Mater.* 2007, 17(8), 1225–1236.

29. Kageyama, K., Tamazawa, J., Aida, T., Extrusion polymerization: Catalyzed synthesis of crystalline linear polyethylene nanofibers within a mesoporous silica. *Science* 1999, 285(5436), 2113–2115.

30. Slowing, I. I., Vivero-Escoto, J. L., Wu, C. W., Lin, V. S. Y., Mesoporous silica nanoparticles as controlled release drug delivery and gene transfection carriers. *Adv. Drug Deliv. Rev.* 2008, 60(11), 1278–1288.

31. Argyo, C., Weiss, V., Brauchle, C., Bein, T., Multifunctional mesoporous silica nanoparticles as a universal platform for drug delivery. *Chem. Mater.* 2014, 26(1), 435–451.

32. Tang, F. Q., Li, L. L., Chen, D., Mesoporous silica nanoparticles: Synthesis, biocompatibility and drug delivery. *Adv. Mater.* 2012, 24(12), 1504–1534.

33. Meng, H., Xue, M., Xia, T., Ji, Z. X., Tarn, D. Y., Zink, J. I., Nel, A. E., Use of size and a copolymer design feature to improve the biodistribution

and the enhanced permeability and retention effect of doxorubicin-loaded mesoporous silica nanoparticles in a murine xenograft tumor model. *ACS Nano* 2011, 5(5), 4131–4144.

34. Perrault, S. D., Walkey, C., Jennings, T., Fischer, H. C., Chan, W. C. W., Mediating tumor targeting efficiency of nanoparticles through design. *Nano Lett.* 2009, 9(5), 1909–1915.

35. Zhang, Z. K., Berns, A. E., Willbold, S., Buitenhuis, J., Synthesis of poly(ethylene glycol) (PEG)-grafted colloidal silica particles with improved stability in aqueous solvents. *J. Colloid Interf. Sci.* 2007, 310(2), 446–455.

36. Cauda, V., Argyo, C., Bein, T., Impact of different PEGylation patterns on the long-term bio-stability of colloidal mesoporous silica nanoparticles. *J. Mater. Chem.* 2010, 20(39), 8693–8699.

37. Baeza, A., Colilla, M., Vallet-Regi, M., Advances in mesoporous silica nanoparticles for targeted stimuli-responsive drug delivery. *Exp. Opin. Drug Del.* 2015, 12(2), 319–337.

38. Luo, Z., Cai, K. Y., Hu, Y., Zhao, L., Liu, P., Duan, L., Yang, W. H., Mesoporous silica nanoparticles end-capped with collagen: Redox-responsive nanoreservoirs for targeted drug delivery. *Angew. Chem. Int. Edit.* 2011, 50(3), 640–643.

39. Ma, M., Chen, H. R., Chen, Y., Zhang, K., Wang, X., Cui, X. Z., Shi, J. L., Hyaluronic acid-conjugated mesoporous silica nanoparticles: Excellent colloidal dispersity in physiological fluids and targeting efficacy. *J. Mater. Chem.* 2012, 22(12), 5615–5621.

40. Yu, M. H., Jambhrunkar, S., Thorn, P., Chen, J. Z., Gu, W. Y., Yu, C. Z., Hyaluronic acid modified mesoporous silica nanoparticles for targeted drug delivery to CD44-overexpressing cancer cells. *Nanoscale* 2013, 5(1), 178–183.

41. Liu, K. W., Z., Wang, S., Liu, P., Qin, Y., Ma, Y., Li, X., Huo, Z., Hyaluronic acid-tagged silica nanoparticles in colon cancer therapy: therapeutic efficacy evaluation. *Int. J. Nanomed.* 2015, 10(1), 6445–6454.

42. Yin, T. J., Liu, J. Y., Zhao, Z. K., Zhao, Y. Y., Dong, L. H., Yang, M., Zhou, J. P., Huo, M. R., Redox sensitive hyaluronic acid-decorated graphene oxide for photothermally controlled tumor-cytoplasmselective rapid drug delivery. *Adv. Funct. Mater.* 2017, 27(14), 1604620.

43. Mura, S., Nicolas, J., Couvreur, P., Stimuli-responsive nanocarriers for drug delivery. *Nat. Mater.* 2013, 12(11), 991–1003.

44. Wu, X., Wang, Z. Y., Zhu, D., Zong, S. F., Yang, L. P., Zhong, Y., Cui, Y. P., PH and thermo dual-stimuli-responsive drug carrier based on

mesoporous silica nanoparticles encapsulated in a copolymer-lipid bilayer. *ACS Appl. Mater. Interfaces* 2013, 5(21), 10895–10903.

45. Cui, Y. N., Dong, H. Q., Cai, X. J., Wang, D. P., Li, Y. Y., Mesoporous silica nanoparticles capped with disulfide-linked PEG gatekeepers for glutathione-mediated controlled release. *ACS Appl. Mater. Interfaces* 2012, 4(6), 3177–3183.

46. Chen, Z. W., Li, Z. H., Lin, Y. H., Yin, M. L., Ren, J. S., Qu, X. G., Bioresponsive hyaluronic acid-capped mesoporous silica nanoparticles for targeted drug delivery. *Chem. Eur. J.* 2013, 19(5), 1778–1783.

47. Meng, F. H., Hennink, W. E., Zhong, Z., Reduction-sensitive polymers and bioconjugates for biomedical applications. *Biomaterials* 2009, 30(12), 2180–2198.

48. Zhao, Q. F., Geng, H. J., Wang, Y., Gao, Y. K., Huang, J. H., Wang, Y., Zhang, J. H., Wang, S. L., Hyaluronic acid oligosaccharide modified redox-responsive mesoporous silica nanoparticles for targeted drug delivery. *ACS Appl. Mater. Interfaces* 2014, 6(22), 20290–20299.

49. Zhao, Q. F., Liu, J., Zhu, W. Q., Sun, C. S., Di, D. H., Zhang, Y., Wang, P., Wang, Z. Y., Wang, S. L., Dual-stimuli responsive hyaluronic acid-conjugated mesoporous silica for targeted delivery to CD44-overexpressing cancer cells. *Acta Biomater.* 2015, 23, 147–156.

50. Rosenholm, J. M., Sahlgren, C., Linden, M., Towards multifunctional, targeted drug delivery systems using mesoporous silica nanoparticles - opportunities & challenges. *Nanoscale* 2010, 2(10), 1870–1883.

51. Gary-Bobo, M., Brevet, D., Benkirane-Jessel, N., Raehm, L., Maillard, P., Garcia, M., Durand, J. O., Hyaluronic acid-functionalized mesoporous silica nanoparticles for efficient photodynamic therapy of cancer cells. *Photodiagn Photodyn.* 2012, 9(3), 256–260.

52. Wang, D., Huang, J. B., Wang, X. X., Yu, Y., Zhang, H., Chen, Y., Liu, J. J., Sun, Z. G., Zou, H., Sun, D. X., Zhou, G. C., Zhang, G. Q., Lu, Y., Zhong, Y. Q., The eradication of breast cancer cells and stem cells by 8-hydroxyquinoline-loaded hyaluronan modified mesoporous silica nanoparticle-supported lipid bilayers containing docetaxel. *Biomaterials* 2013, 34(31), 7662–7673.

53. He, Q. J., Ma, M., Wei, C. Y., Shi, J. L., Mesoporous carbon@silicon-silica nanotheranostics for synchronous delivery of insoluble drugs and luminescence imaging. *Biomaterials* 2012, 33(17), 4392–4402.

54. Yang, K., Feng, L. Z., Liu, Z., Stimuli responsive drug delivery systems based on nano-graphene for cancer therapy. *Adv. Drug Deliv. Rev.* 2016, 105, 228–241.

55. Geim, A. K., Novoselov, K. S., The rise of graphene. *Nat. Mater.* 2007, 6(3), 183–191.

56. Chung, C., Kim, Y. K., Shin, D., Ryoo, S. R., Hong, B. H., Min, D. H., Biomedical applications of graphene and graphene Oxide. *Acc. Chem. Res.* 2013, 46(10), 2211–2224.

57. Shim, G., Kim, M. G., Park, J. Y., Oh, Y. K., Graphene-based nanosheets for delivery of chemotherapeutics and biological drugs. *Adv. Drug Deliv. Rev.* 2016, 105, 205–227.

58. Zhang, Y., Nayak, T. R., Hong, H., Cai, W. B., Graphene: A versatile nanoplatform for biomedical applications. *Nanoscale* 2012, 4(13), 3833–3842.

59. Liu, Z., Robinson, J. T., Sun, X. M., Dai, H. J., PEGylated nanographene oxide for delivery of water-insoluble cancer drugs. *J. Am. Chem. Soc.* 2008, 130(33), 10876–10877.

60. Goenka, S., Sant, V., Sant, S., Graphene-based nanomaterials for drug delivery and tissue engineering. *J. Control. Release* 2014, 173, 75–88.

61. Guo, Y. J., Guo, S. J., Ren, J. T., Zhai, Y. M., Dong, S. J., Wang, E. K., Cyclodextrin functionalized graphene nanosheets with high supramolecular recognition capability: Synthesis and host-guest inclusion for enhanced electrochemical performance. *ACS Nano* 2010, 4(7), 4001–4010.

62. Georgakilas, V., Otyepka, M., Bourlinos, A. B., Chandra, V., Kim, N., Kemp, K. C., Hobza, P., Zboril, R., Kim, K. S., Functionalization of graphene: Covalent and non-covalent approaches, derivatives and applications. *Chem. Rev.* 2012, 112(11), 6156–6214.

63. Zhang, L. M., Wang, Z. L., Lu, Z. X., Shen, H., Huang, J., Zhao, Q. H., Liu, M., He, N. Y., Zhang, Z. J., PEGylated reduced graphene oxide as a superior ssRNA delivery system. *J. Mater. Chem. B,* 2013, 1(6), 749–755.

64. Li, F., Park, S., Ling, D., Park, W., Han, J. Y., Na, K., Char, K., Hyaluronic acid-conjugated graphene oxide/photosensitizer nanohybrids for cancer targeted photodynamic therapy. *J. Mater. Chem. B* 2013, 1(12), 1678–1686.

65. Miao, W., Shim, G., Kang, C. M., Lee, S., Choe, Y. S., Choi, H. G., Oh, Y. K., Cholesteryl hyaluronic acid-coated, reduced graphene oxide nanosheets for anti-cancer drug delivery. *Biomaterials* 2013, 34(37), 9638–9647.

66. Wu, H. X., Shi, H. L., Wang, Y. P., Jia, X. Q., Tang, C. Z., Zhang, J. M., Yang, S. P., Hyaluronic acid conjugated graphene oxide for targeted drug delivery. *Carbon* 2014, 69, 379–389.

67. Song, E. Q., Han, W. Y., Li, C., Cheng, D., Li, L. R., Liu, L. C., Zhu, G. Z., Song, Y., Tan, W. H., Hyaluronic acid-decorated graphene oxide nanohybrids as nanocarriers for targeted and PH-responsive anticancer drug delivery. *ACS Appl. Mater. Interfaces* 2014, 6(15), 11882–11890.

68. Wang, Z. Q., Ciacchi, L. C., Wei, G., Recent advances in the synthesis of graphene-based nanomaterials for controlled drug delivery. *Appl. Sci. Basel* 2017, 7(11).

69. Karimi, M., Ghasemi, A., Zangabad, P. S., Rahighi, R., Basri, S. M. M., Mirshekari, H., Amiri, M., Pishabad, Z. S., Aslani, A., Bozorgomid, M., Ghosh, D., Beyzavi, A., Vaseghi, A., Aref, A. R., Haghani, L., Bahrami, S., Hamblin, M. R., Smart micro/nanoparticles in stimulus-responsive drug/gene delivery systems. *Chem. Soc. Rev.* 2016, 45(5), 1457–1501.

70. Chen, Y. W., Su, Y. L., Hu, S. H., Chen, S. Y., Functionalized graphene nanocomposites for enhancing photothermal therapy in tumor treatment. *Adv. Drug Deliv. Rev.* 2016, 105, 190–204.

71. Robinson, J. T., Tabakman, S. M., Liang, Y. Y., Wang, H. L., Casalongue, H. S., Vinh, D., Dai, H. J., Ultrasmall reduced graphene oxide with high near-infrared absorbance for photothermal therapy. *J. Am. Chem. Soc.* 2011, 133(17), 6825–6831.

72. Jiang, W., Mo, F., Jin, X., Chen, L., Xu, L. J., Guo, L., Fu, F., Tumor-targeting photothermal heating-responsive nanoplatform based on reduced graphene oxide/mesoporous silica/hyaluronic acid nanocomposite for enhanced photodynamic therapy. *Adv. Mater. Interfaces* 2017, 4(20), 1700425.

73. Fiorica, C., Mauro, N., Pitarresi, G., Scialabba, C., Palumbo, F. S., Giammona, G., Double-network-structured graphene oxide-containing nanogels as photothermal agents for the treatment of colorectal cancer. *Biomacromolecules* 2017, 18(3), 1010–1018.

74. Hwang, D. W., Kim, H. Y., Li, F., Park, J. Y., Kim, D., Park, J. H., Han, H. S., Byun, J. W., Lee, Y.-S., Jeong, J. M., Char, K., Lee, D. S., In vivo visualization of endogenous miR-21 using hyaluronic acid-coated graphene oxide for targeted cancer therapy. *Biomaterials* 2017, 121, 144–154.

75. Luo, Y. A., Cai, X. L., Li, H., Lin, Y. H., Du, D., Hyaluronic acid-modified multifunctional Q-graphene for targeted killing of drug-resistant lung cancer cells. *ACS Appl. Mater. Interfaces* 2016, 8(6), 4048–4055.

76. Liu, X., Cheng, X., Wang, F., Feng, L., Wang, Y., Zheng, Y., Guo, R., Targeted delivery of SNX-2112 by polysaccharide-modified graphene oxide nanocomposites for treatment of lung cancer. *Carbohyd. Polym.* 2018, 185, 85–95.

77. Guo, Y. F., Xu, H. X., Li, Y. P., Wu, F. Z., Li, Y. X., Bao, Y., Yan, X. M., Huang, Z. J., Xu, P. H., Hyaluronic acid and Arg-Gly-Asp peptide modified Graphene oxide with dual receptor-targeting function for cancer therapy. *J. Biomater. Appl.* 2017, 32(1), 54–65.

78. Ran, X., Du, Y., Wang, Z. Z., Wang, H., Pu, F., Ren, J. S., Qu, X. G., Hyaluronic acid-templated Ag nanoparticles/graphene oxide composites for synergistic therapy of bacteria infection. *ACS Appl. Mater. Interfaces* 2017, 9(23), 19717–19724.

79. Chen, Z. W., Li, Z. H., Wang, J. S., Ju, E. G., Zhou, L., Ren, J. S., Qu, X. G., A multi-synergistic platform for sequential irradiation-activated high-performance apoptotic cancer therapy. *Adv. Funct. Mater.* 2014, 24(4), 522–529.

80. Bottini, M., Rosato, N., Bottini, N., PEG-modified carbon nanotubes in biomedicine: Current status and challenges ahead. *Biomacromolecules* 2011, 12(10), 3381–3393.

81. Liu, Z., Chen, K., Davis, C., Sherlock, S., Cao, Q. Z., Chen, X. Y., Dai, H. J., Drug delivery with carbon nanotubes for in vivo cancer treatment. *Cancer Res.* 2008, 68(16), 6652–6660.

82. Lu, F. S., Gu, L. R., Meziani, M. J., Wang, X., Luo, P. G., Veca, L. M., Cao, L., Sun, Y. P., Advances in bioapplications of carbon nanotubes. *Adv. Mater.* 2009, 21(2), 139–152.

83. Prato, M., Kostarelos, K., Bianco, A., Functionalized carbon nanotubes in drug design and discovery. *Acc. Chem. Res.* 2008, 41(1), 60–68.

84. Zhang, Y., Bai, Y. H., Yan, B., Functionalized carbon nanotubes for potential medicinal applications. *Drug Discov. Today* 2010, 15(11–12), 428–435.

85. Liu, Z., Sun, X. M., Nakayama-Ratchford, N., Dai, H. J., Supramolecular chemistry on water-soluble carbon nanotubes for drug loading and delivery. *ACS Nano* 2007, 1(1), 50–56.

86. Cheng, J. P., Fernando, K. A. S., Veca, L. M., Sun, Y. P., Lamond, A. I., Lam, Y. W., Cheng, S. H., Reversible accumulation of PEGylated single-walled carbon nanotubes in the mammalian nucleus. *ACS Nano* 2008, 2(10), 2085–2094.

87. Chen, J. Y., Chen, S. Y., Zhao, X. R., Kuznetsova, L. V., Wong, S. S., Ojima, I., Functionalized single-walled carbon nanotubes as rationally designed vehicles for tumor-targeted drug delivery. *J. Am. Chem. Soc.* 2008, 130(49), 16778–16785.

88. Prakash, S., Malhotra, M., Shao, W., Tomaro-Duchesneau, C., Abbasi, S., Polymeric nanohybrids and functionalized carbon nanotubes as drug

delivery carriers for cancer therapy. *Adv. Drug Deliv. Rev.* 2011, 63(14–15), 1340–1351.

89. Meng, L. J., Zhang, X. K., Lu, Q. H., Fei, Z. F., Dyson, P. J., Single walled carbon nanotubes as drug delivery vehicles: Targeting doxorubicin to tumors. *Biomaterials* 2012, 33(6), 1689–1698.

90. Georgakilas, V., Kordatos, K., Prato, M., Guldi, D. M., Holzinger, M., Hirsch, A., Organic functionalization of carbon nanotubes. *J. Am. Chem. Soc.* 2002, 124(5), 760–761.

91. Liu, Z., Tabakman, S., Welsher, K., Dai, H. J., Carbon nanotubes in biology and medicine: In vitro and in vivo detection, imaging and drug delivery. *Nano Res.* 2009, 2(2), 85–120.

92. Zhao, Y. L., Stoddart, J. F., Noncovalent functionalization of single-walled carbon nanotubes. *Acc. Chem. Res.* 2009, 42(8), 1161–1171.

93. Hirsch, A., Functionalization of single-walled carbon nanotubes. *Angew. Chem. Int. Edit.* 2002, 41(11), 1853–1859.

94. Dvash, R., Khatchatouriants, A., Solmesky, L. J., Wibroe, P. P., Wei, M., Moghimi, S. M., Peer, D., Structural profiling and biological performance of phospholipid-hyaluronan functionalized single-walled carbon nanotubes. *J. Control. Release* 2013, 170(2), 295–305.

95. Yao, H. J., Sun, L., Liu, Y., Jiang, S., Pu, Y. Z., Li, J. C., Zhang, Y. G., Monodistearoylphosphatidylethanolamine-hyaluronic acid functionalization of single-walled carbon nanotubes for targeting intracellular drug delivery to overcome multidrug resistance of cancer cells. *Carbon* 2016, 96, 362–376.

96. Wong, B. S., Yoong, S. L., Jagusiak, A., Panczyk, T., Ho, H. K., Ang, W. H., Pastorin, G., Carbon nanotubes for delivery of small molecule drugs. *Adv. Drug Deliv. Rev.* 2013, 65(15), 1964–2015.

97. Zhang, X. K., Meng, L. J., Lu, Q. H., Fei, Z. F., Dyson, P. J., Targeted delivery and controlled release of doxorubicin to cancer cells using modified single wall carbon nanotubes. *Biomaterials* 2009, 30(30), 6041–6047.

98. Heister, E., Neves, V., Tilmaciu, C., Lipert, K., Beltran, V. S., Coley, H. M., Silva, S. R. P., McFadden, J., Triple functionalisation of single-walled carbon nanotubes with doxorubicin, a monoclonal antibody, and a fluorescent marker for targeted cancer therapy. *Carbon* 2009, 47(9), 2152–2160.

99. Yao, H. J., Zhang, Y. G., Sun, L., Liu, Y., The effect of hyaluronic acid functionalized carbon nanotubes loaded with salinomycin on gastric cancer stem cells. *Biomaterials* 2014, 35(33), 9208–9223.

100. Mo, Y. F., Wang, H. W., Liu, J. H., Lan, Y., Guo, R., Zhang, Y., Xue, W., Zhang, Y. M., Controlled release and targeted delivery to cancer cells of doxorubicin from polysaccharide-functionalised single-walled carbon nanotubes. *J. Mater. Chem. B* 2015, 3(9), 1846–1855.

101. Datir, S. R., Das, M., Singh, R. P., Jain, S., Hyaluronate tethered, "smart" multiwalled carbon nanotubes for tumor-targeted delivery of doxorubicin. *Bioconjug. Chem.* 2012, 23(11), 2201–2213.

102. Cao, X. Y., Tao, L., Wen, S. H., Hou, W. X., Shi, X. Y., Hyaluronic acid-modified multiwalled carbon nanotubes for targeted delivery of doxorubicin into cancer cells. *Carbohyd. Res.* 2015, 405, 70–77.

103. Wang, X., Xia, T., Ntim, S. A., Ji, Z. X., Lin, S. J., Meng, H., Chung, C. H., George, S., Zhang, H. Y., Wang, M. Y., Li, N., Yang, Y., Castranova, V., Mitra, S., Bonner, J. C., Nel, A. E., Dispersal state of multiwalled carbon nanotubes elicits profibrogenic cellular responses that correlate with fibrogenesis biomarkers and fibrosis in the murine lung. *ACS Nano* 2011, 5(12), 9772–9787.

104. Hussain, S., Ji, Z. X., Taylor, A. J., DeGraff, L. M., George, M., Tucker, C. J., Chang, C. H., Li, R. B., Bonner, J. C., Garantziotis, S., Multiwalled Carbon Nanotube Functionalization with High Molecular Weight Hyaluronan Significantly Reduces Pulmonary Injury. *ACS Nano* 2016, 10(8), 7675–7688.

105. Wang, G. H., Zhang, F., Tian, R., Zhang, L. W., Fu, G. F., Yang, L. L., Zhu, L., Nanotubes-embedded indocyanine green-hyaluronic acid nanoparticles for photoacoustic-imaging-guided phototherapy. *ACS Appl. Mater. Interfaces* 2016, 8(8), 5608–5617.

106. Bhirde, A. A., Chikkaveeraiah, B. V., Srivatsan, A., Niu, G., Jin, A. J., Kapoor, A., Wang, Z., Patel, S., Patel, V., Gorbach, A. M., Leapman, R. D., Gutkind, J. S., Walker, A. R. H., Chen, X. Y., Targeted therapeutic nanotubes influence the viscoelasticity of cancer cells to overcome drug resistance. *ACS Nano* 2014, 8(5), 4177–4189.

107. Liu, Z., Robinson, J. T., Tabakman, S. M., Yang, K., Dai, H. J., Carbon materials for drug delivery & cancer therapy. *Mater. Today* 2011, 14(7–8), 316–323.

108. Smith, A. M., Duan, H. W., Mohs, A. M., Nie, S. M., Bioconjugated quantum dots for in vivo molecular and cellular imaging. *Adv. Drug Deliv. Rev.* 2008, 60(11), 1226–1240.

109. Zrazhevskiy, P., Sena, M., Gao, X. H., Designing multifunctional quantum dots for bioimaging, detection, and drug delivery. *Chem. Soc. Rev.* 2010, 39(11), 4326–4354.

110. Medintz, I. L., Uyeda, H. T., Goldman, E. R., Mattoussi, H., Quantum dot bioconjugates for imaging, labelling and sensing. *Nat. Mater.* 2005, 4(6), 435–446.

111. Gao, X. H., Cui, Y. Y., Levenson, R. M., Chung, L. W. K., Nie, S. M., In vivo cancer targeting and imaging with semiconductor quantum dots. *Nat. Biotechnol.* 2004, 22(8), 969–976.

112. Goh, E. J., Kim, K. S., Kim, Y. R., Jung, H. S., Beack, S., Kong, W. H., Scarcelli, G., Yun, S. H., Hahn, S. K., Bioimaging of hyaluronic acid derivatives using nanosized carbon dots. *Biomacromolecules* 2012, 13(8), 2554–2561.

113. Derfus, A. M., Chan, W. C. W., Bhatia, S. N., Probing the cytotoxicity of semiconductor quantum dots. *Nano Lett.* 2004, 4(1), 11–18.

114. Mancini, M. C., Kairdolf, B. A., Smith, A. M., Nie, S. M., Oxidative quenching and degradation of polymer-encapsulated quantum dots: New insights into the long-term fate and toxicity of nanocrystals in vivo. *J. Am. Chem. Soc.* 2008, 130(33), 10836–10837.

115. Michalet, X., Pinaud, F. F., Bentolila, L. A., Tsay, J. M., Doose, S., Li, J. J., Sundaresan, G., Wu, A. M., Gambhir, S. S., Weiss, S., Quantum dots for live cells, in vivo imaging, and diagnostics. *Science* 2005, 307(5709), 538–544.

116. Li, N., Than, A., Wang, X. W., Xu, S. H., Sun, L., Duan, H. W., Xu, C. J., Chen, P., Ultrasensitive profiling of metabolites using tyramine-functionalized graphene quantum dots. *ACS Nano* 2016, 10(3), 3622–3629.

117. Hoshino, A., Fujioka, K., Oku, T., Suga, M., Sasaki, Y. F., Ohta, T., Yasuhara, M., Suzuki, K., Yamamoto, K., Physicochemical properties and cellular toxicity of nanocrystal quantum dots depend on their surface modification. *Nano Lett.* 2004, 4(11), 2163–2169.

118. Pellegrino, T., Manna, L., Kudera, S., Liedl, T., Koktysh, D., Rogach, A. L., Keller, S., Radler, J., Natile, G., Parak, W. J., Hydrophobic nanocrystals coated with an amphiphilic polymer shell: A general route to water soluble nanocrystals. *Nano Lett.* 2004, 4(4), 703–707.

119. Gerion, D., Pinaud, F., Williams, S. C., Parak, W. J., Zanchet, D., Weiss, S., Alivisatos, A. P., Synthesis and properties of biocompatible water-soluble silica-coated CdSe/ZnS semiconductor quantum dots. *J. Phys. Chem. B* 2001, 105(37), 8861–8871.

120. Bhang, S. H., Won, N., Lee, T. J., Jin, H., Nam, J., Park, J., Chung, H., Park, H. S., Sung, Y. E., Hahn, S. K., Kim, B. S., Kim, S., Hyaluronic acid-quantum dot conjugates for in vivo lymphatic vessel imaging. *ACS Nano* 2009, 3(6), 1389–1398.

121. Wang, H. N., Sun, H. F., Wei, H., Xi, P., Nie, S. M., Ren, Q. S., Biocompatible hyaluronic acid polymer-coated quantum dots for CD44(+) cancer cell-targeted imaging. *J. Nanopart Res.* 2014, 16(10), 2621.

122. Kim, K. S., Hur, W., Park, S. J., Hong, S. W., Choi, J. E., Goh, E. J., Yoon, S. K., Hahn, S. K., Bioimaging for targeted delivery of hyaluronic acid derivatives to the livers in cirrhotic mice using quantum dots. *ACS Nano* 2010, 4(6), 3005–3014.

123. Park, J., Lee, J., Kwag, J., Baek, Y., Kim, B., Yoon, C. J., Bok, S., Cho, S. H., Kim, K. H., Ahn, G. O., Kim, S., Quantum dots in an amphiphilic polyethyleneimine derivative platform for cellular labeling, targeting, gene delivery, and ratiometric oxygen sensing. *ACS Nano* 2015, 9(6), 6511–6521.

124. Cai, X., Luo, Y., Zhang, W., Du, D., Lin, Y., pH-sensitive ZnO quantum dots–doxorubicin nanoparticles for lung cancer targeted drug delivery. *ACS Appl. Mater. Interfaces* 2016, 8(34), 22442–22450.

125. Sun, Y. P., Zhou, B., Lin, Y., Wang, W., Fernando, K. A. S., Pathak, P., Meziani, M. J., Harruff, B. A., Wang, X., Wang, H. F., Luo, P. J. G., Yang, H., Kose, M. E., Chen, B. L., Veca, L. M., Xie, S. Y., Quantum-sized carbon dots for bright and colorful photoluminescence. *J. Am. Chem. Soc.* 2006, 128(24), 7756–7757.

126. Cao, L., Wang, X., Meziani, M. J., Lu, F. S., Wang, H. F., Luo, P. J. G., Lin, Y., Harruff, B. A., Veca, L. M., Murray, D., Xie, S. Y., Sun, Y. P., Carbon dots for multiphoton bioimaging. *J. Am. Chem. Soc.* 2007, 129(37), 11318–11319.

127. Yang, S. T., Cao, L., Luo, P. G. J., Lu, F. S., Wang, X., Wang, H. F., Meziani, M. J., Liu, Y. F., Qi, G., Sun, Y. P., Carbon dots for optical imaging in vivo. *J. Am. Chem. Soc.* 2009, 131(32), 11308–11309.

128. Zhu, S. J., Meng, Q. N., Wang, L., Zhang, J. H., Song, Y. B., Jin, H., Zhang, K., Sun, H. C., Wang, H. Y., Yang, B., Highly photoluminescent carbon dots for multicolor patterning, sensors, and bioimaging. *Angew. Chem. Int. Edit.* 2013, 52(14), 3953–3957.

129. Huang, P., Lin, J., Wang, X. S., Wang, Z., Zhang, C. L., He, M., Wang, K., Chen, F., Li, Z. M., Shen, G. X., Cui, D. X., Chen, X. Y., Light-triggered theranostics based on photosensitizer-conjugated carbon dots for simultaneous enhanced-fluorescence imaging and photodynamic therapy. *Adv. Mater.* 2012, 24(37), 5104–5110.

130. Tang, J., Kong, B., Wu, H., Xu, M., Wang, Y. C., Wang, Y. L., Zhao, D. Y., Zheng, G. F., Carbon nanodots featuring efficient FRET for real-time monitoring of drug delivery and two-photon imaging. *Adv. Mater.* 2013, 25(45), 6569–6574.

131. Feng, T., Ai, X., An, G., Yang, P., Zhao, Y., Charge-convertible carbon dots for imaging-guided drug delivery with enhanced in vivo cancer therapeutic efficiency. *ACS Nano* 2016, 10(4), 4410–4420.

132. Abdullah Al, N., Lee, J.-E., In, I., Lee, H., Lee, K. D., Jeong, J. H., Park, S. Y., Target delivery and cell imaging using hyaluronic acid-functionalized graphene quantum dots. *Mol. Pharm.* 2013, 10(10), 3736–3744.

133. Jia, X., Han, Y., Pei, M. L., Zhao, X. B., Tian, K., Zhou, T. T., Liu, P., Multi-functionalized hyaluronic acid nanogels crosslinked with carbon dots as dual receptor-mediated targeting tumor theranostics. *Carbohyd. Polym.* 2016, 152, 391–397.

134. Hola, K., Zhang, Y., Wang, Y., Giannelis, E. P., Zboril, R., Rogach, A. L., Carbon dots-Emerging light emitters for bioimaging, cancer therapy and optoelectronics. *Nano Today* 2014, 9(5), 590–603.

135. Lim, S. Y., Shen, W., Gao, Z. Q., Carbon quantum dots and their applications. *Chem. Soc. Rev.* 2015, 44(1), 362–381.

136. Sun, C., Lee, J. S. H., Zhang, M., Magnetic nanoparticles in MR imaging and drug delivery. *Adv. Drug Deliv. Rev.* 2008, 60(11), 1252–1265.

137. Chomoucka, J., Drbohlavova, J., Huska, D., Adam, V., Kizek, R., Hubalek, J., Magnetic nanoparticles and targeted drug delivering. *Pharmacol. Res.* 2010, 62(2), 144–149.

138. Arruebo, M., Fernández-Pacheco, R., Ibarra, M. R., Santamaría, J., Magnetic nanoparticles for drug delivery. *Nano Today* 2007, 2(3), 22–32.

139. Veiseh, O., Gunn, J. W., Zhang, M., Design and fabrication of magnetic nanoparticles for targeted drug delivery and imaging. *Adv. Drug Deliv. Rev.* 2010, 62(3), 284–304.

140. Lee, J.-H., Huh, Y.-M., Jun, Y.-W., Seo, J.-W., Jang, J.-T., Song, H.-T., Kim, S., Cho, E.-J., Yoon, H.-G., Suh, J.-S., Cheon, J., Artificially engineered magnetic nanoparticles for ultra-sensitive molecular imaging. *Nat. Med.* 2006, 13, 95.

141. Jun, Y.-W., Huh, Y.-M., Choi, J.-S., Lee, J.-H., Song, H.-T., KimKim, Yoon, S., Kim, K.-S., Shin, J.-S., Suh, J.-S., Cheon, J., Nanoscale size effect of magnetic nanocrystals and their utilization for cancer diagnosis via magnetic resonance imaging. *J. Am. Chem. Soc.* 2005, 127(16), 5732–5733.

142. Xiao, L., Li, J., Brougham, D. F., Fox, E. K., Feliu, N., Bushmelev, A., Schmidt, A., Mertens, N., Kiessling, F., Valldor, M., Fadeel, B., Mathur, S., Water-soluble superparamagnetic magnetite nanoparticles with

biocompatible coating for enhanced magnetic resonance imaging. *ACS Nano* 2011, 5(8), 6315–6324.

143. Lee, T., Son, H. Y., Choi, Y., Shin, Y., Oh, S., Kim, J., Huh, Y.-M., Haam, S., Minimum hyaluronic acid (HA) modified magnetic nanocrystals with less facilitated cancer migration and drug resistance for targeting CD44 abundant cancer cells by MR imaging. *J. Mater. Chem. B* 2017, 5(7), 1400–1407.

144. Lee, Y. H., Lee, H., Kim, Y. B., Kim, J. Y., Hyeon, T., Park, H., Messersmith, P. B., Park, T. G., Bioinspired surface immobilization of hyaluronic acid on monodisperse magnetite nanocrystals for targeted cancer imaging. *Adv. Mater.* 2008, 20(21), 4154–4157.

145. Lim, E. K., Kim, H. O., Jang, E., Park, J., Lee, K., Suh, J. S., Huh, Y. M., Haam, S., Hyaluronan-modified magnetic nanoclusters for detection of CD44-overexpressing breast cancer by MR imaging. *Biomaterials* 2011, 32(31), 7941–7950.

146. Chung, H. J., Lee, H., Bae, K. H., Lee, Y., Park, J., Cho, S. W., Hwang, J. Y., Park, H., Langer, R., Anderson, D., Park, T. G., Facile synthetic route for surface-functionalized magnetic nanoparticles: Cell labeling and magnetic resonance imaging studies. *ACS Nano* 2011, 5(6), 4329–4336.

147. Kumar, A., Sahoo, B., Montpetit, A., Behera, S., Lockey, R. F., Mohapatra, S. S., Development of hyaluronic acid–Fe_2O_3 hybrid magnetic nanoparticles for targeted delivery of peptides. *Nanomedicine: Nanotechnol. Biol. Med.* 2007, 3(2), 132–137.

148. El-Dakdouki, M. H., Xia, J. G., Zhu, D. C., Kavunja, H., Grieshaber, J., O'Reilly, S., McCormick, J. J., Huang, X. F., Assessing the in vivo efficacy of doxorubicin loaded hyaluronan nanoparticles. *ACS Appl. Mater. Interfaces* 2014, 6(1), 697–705.

149. Chen, X., Klingeler, R., Kath, M., El Gendy, A. A., Cendrowski, K., Kalenczuk, R. J., Borowiak-Palen, E., Magnetic silica nanotubes: Synthesis, drug release, and feasibility for magnetic hyperthermia. *ACS Appl. Mater. Interfaces* 2012, 4(4), 2303–2309.

150. Yang, Y., Liu, X., Lv, Y., Herng, T. S., Xu, X., Xia, W., Zhang, T., Fang, J., Xiao, W., Ding, J., Orientation mediated enhancement on magnetic hyperthermia of Fe_3O_4 nanodisc. *Adv. Funct. Mater.* 2015, 25(5), 812–820.

151. Kumar, C. S. S. R., Mohammad, F., Magnetic nanomaterials for hyperthermia-based therapy and controlled drug delivery. *Adv. Drug Deliv. Rev.* 2011, 63(9), 789–808.

152. Kim, K. S., Kim, J., Lee, J. Y., Matsuda, S., Hideshima, S., Mori, Y., Osaka, T., Na, K., Stimuli-responsive magnetic nanoparticles for tumor-targeted bimodal imaging and photodynamic/hyperthermia combination therapy. *Nanoscale* 2016, 8(22), 11625–11634.

153. Thomas, R. G., Moon, M. J., Lee, H., Sasikala, A. R. K., Kim, C. S., Park, I. K., Jeong, Y. Y., Hyaluronic acid conjugated superparamagnetic iron oxide nanoparticle for cancer diagnosis and hyperthermia therapy. *Carbohyd. Polym.* 2015, 131, 439–446.

Index